VOLUME FIVE HUNDRED AND FIFTY ONE

METHODS IN ENZYMOLOGY

Circadian Rhythms and Biological Clocks

Part A

METHODS IN ENZYMOLOGY

Editors-in-Chief

JOHN N. ABELSON and MELVIN I. SIMON
Division of Biology
California Institute of Technology
Pasadena, California

ANNA MARIE PYLE
Departments of Molecular, Cellular and Developmental Biology and Department of Chemistry Investigator
Howard Hughes Medical Institute
Yale University

Founding Editors

SIDNEY P. COLOWICK and NATHAN O. KAPLAN

VOLUME FIVE HUNDRED AND FIFTY ONE

METHODS IN ENZYMOLOGY

Circadian Rhythms and Biological Clocks

Part A

Edited by

AMITA SEHGAL
John Herr Musser Professor of Neuroscience; Director, Penn Program in Chronobiology, Perelman School of Medicine, University of Pennsylvania, Philadelphia, USA

AMSTERDAM • BOSTON • HEIDELBERG • LONDON
NEW YORK • OXFORD • PARIS • SAN DIEGO
SAN FRANCISCO • SINGAPORE • SYDNEY • TOKYO
Academic Press is an imprint of Elsevier

Academic Press is an imprint of Elsevier
225 Wyman Street, Waltham, MA 02451, USA
525 B Street, Suite 1800, San Diego, CA 92101-4495, USA
125 London Wall, London EC2Y 5AS, UK
The Boulevard, Langford Lane, Kidlington, Oxford OX5 1GB, UK

First edition 2015

Notices

Knowledge and best practice in this field are constantly changing. As new research and experience broaden our understanding, changes in research methods, professional practices, or medical treatment may become necessary.

Practitioners and researchers must always rely on their own experience and knowledge in evaluating and using any information, methods, compounds, or experiments described herein. In using such information or methods they should be mindful of their own safety and the safety of others, including parties for whom they have a professional responsibility.

To the fullest extent of the law, neither the Publisher nor the authors, contributors, or editors, assume any liability for any injury and/or damage to persons or property as a matter of products liability, negligence or otherwise, or from any use or operation of any methods, products, instructions, or ideas contained in the material herein.

ISBN: 978-0-12-801218-5
ISSN: 0076-6879

For information on all Academic Press publications visit our website at store.elsevier.com

CONTENTS

Part III
Circadian Regulation of Gene and Protein Expression

CONTRIBUTORS

Katharine Abruzzi
Department of Biology, Howard Hughes Medical Institute and National Center for Behavioral Genomics, Brandeis University, Waltham, Massachusetts, USA

J. Douglas Armstrong
Actual Analytics, Edinburgh, United Kingdom

Sofia Axelrod
Laboratory of Genetics, The Rockefeller University, New York, USA

Jasper Bosman
Department of Molecular Chronobiology, Groningen, The Netherlands

Joseph S. Boyd
Center for Circadian Biology, University of California, San Diego, La Jolla, California, USA

Joonseok Cha
Department of Physiology, University of Texas Southwestern Medical Center, Dallas, TX, USA

Xiao Chen
Department of Biology, Howard Hughes Medical Institute and National Center for Behavioral Genomics, Brandeis University, Waltham, Massachusetts, USA

Susan E. Cohen
Center for Circadian Biology, and Division of Biological Sciences, University of California, San Diego, La Jolla, California, USA

Rodolfo Costa
Department of Biology, University of Padova, Padova, Italy

Jay C. Dunlap
Department of Genetics, Geisel School of Medicine at Dartmouth, Hanover, New Hampshire, USA

Hao A. Duong
Department of Neurobiology, Harvard Medical School, Boston, Massachusetts, USA

Zheng Eelderink-Chen
Department of Molecular Chronobiology, Groningen, The Netherlands

Martin Egli
Department of Biochemistry, School of Medicine, Vanderbilt University, Nashville, Tennessee, USA

Marcella L. Erb
Division of Biological Sciences, University of California, San Diego, La Jolla, California, USA

Javier Espinosa
Division of Genetics, University of Alicante, Alicante, Spain

Jin-Yuan Fan
Division of Molecular Biology and Biochemistry, School of Biological Sciences, University of Missouri-Kansas City, Kansas City, Missouri, USA

Michael Gebert
Department of Neurobiology, Harvard Medical School, Boston, Massachusetts, USA

Susan S. Golden
Center for Circadian Biology, and Division of Biological Sciences, University of California, San Diego, La Jolla, California, USA

Gregory R. Grant
Department of Genetics, University of Pennsylvania, Philadelphia, Pennsylvania, USA, and Penn Center for Bioinformatics, University of Pennsylvania, Philadelphia, Pennsylvania, USA

Carla B. Green
Department of Neuroscience, University of Texas Southwestern Medical Center, Dallas, Texas, USA

Edward W. Green
Department of Genetics, University of Leicester, Leicester, United Kingdom

Ralph J. Greenspan
Center for Circadian Biology, and Kavli Institute for Brain and Mind, University of California, San Diego, California, USA

Paul E. Hardin
Department of Biology and Center for Biological Clocks Research, Texas A&M University, College Station, Texas, USA

Matthew M. Hindle
SynthSys and School of Biological Sciences, University of Edinburgh, Edinburgh, United Kingdom

Tsuyoshi Hirota
Molecular and Computational Biology Section, University of Southern California, Los Angeles, California, USA, and Institute of Transformative Bio-Molecules, Nagoya University, Nagoya, Japan

John B. Hogenesch
Department of Pharmacology, Institute for Translational Medicine and Therapeutics, University of Pennsylvania School of Medicine, Philadelphia, Pennsylvania, USA

Hung-Chung Huang
Department of Neuroscience, University of Texas Southwestern Medical Center, Dallas, Texas, USA

Michael E. Hughes
Department of Biology, University of Missouri-St. Louis, St. Louis, Missouri, USA

Jennifer Hurley
Department of Genetics, Geisel School of Medicine at Dartmouth, Hanover, New Hampshire, USA

Takeo Katsuki
Kavli Institute for Brain and Mind, University of California, San Diego, California, USA

Steve A. Kay
Molecular and Computational Biology Section, University of Southern California, Los Angeles, California, USA, and Institute of Transformative Bio-Molecules, Nagoya University, Nagoya, Japan

Andrew Keightley
Division of Molecular Biology and Biochemistry, School of Biological Sciences, University of Missouri-Kansas City, Kansas City, Missouri, USA

Jin Young Kim
Department of Neurobiology, Harvard Medical School, Boston, Massachusetts, USA

Tae-Kyung Kim
Department of Neuroscience, University of Texas Southwestern Medical Center, Dallas, Texas, USA

Yong-Ick Kim
Center for Circadian Biology, University of California, San Diego, La Jolla, California, USA

Nobuya Koike
Department of Neuroscience, University of Texas Southwestern Medical Center, Dallas, Texas, USA

Shihoko Kojima
Department of Biological Sciences, Virginia Tech, Blacksburg, VA, USA

Johanna Krahmer
SynthSys and School of Biological Sciences, University of Edinburgh, Edinburgh, United Kingdom

Vivek Kumar
Department of Neuroscience, and Howard Hughes Medical Institute, University of Texas Southwestern Medical Center, Dallas, Texas, USA

Pieter Bas Kwak
Department of Neurobiology, Harvard Medical School, Boston, Massachusetts, USA

Charalambos P. Kyriacou
Department of Genetics, University of Leicester, Leicester, United Kingdom

Thierry Le Bihan
SynthSys and School of Biological Sciences, University of Edinburgh, Edinburgh, United Kingdom

Tanya L. Leise
Department of Mathematics and Statistics, Amherst College, Amherst, Massachusetts, USA

Jiajia Li
Department of Biology, University of Missouri-St. Louis, St. Louis, Missouri, USA

Yi Liu
Department of Physiology, University of Texas Southwestern Medical Center, Dallas, TX, USA

Jennifer J. Loros
Department of Genetics, Geisel School of Medicine at Dartmouth, Hanover, New Hampshire, USA, and Department of Biochemistry, Geisel School of Medicine at Dartmouth, Hanover, New Hampshire, USA

Sarah F. Martin
SynthSys and School of Biological Sciences, University of Edinburgh, Edinburgh, United Kingdom

John C. Means
Division of Molecular Biology and Biochemistry, School of Biological Sciences, University of Missouri-Kansas City, Kansas City, Missouri, USA

Martha Merrow
Department of Molecular Chronobiology, Groningen, The Netherlands, and Institute of Medical Psychology, Munich, Germany

Andrew J. Millar
SynthSys and School of Biological Sciences, University of Edinburgh, Edinburgh, United Kingdom

Emi Nagoshi
Department of Genetics and Evolution, University of Geneva, Geneva, Switzerland

Prachi Nakashe
Department of Neuroscience, University of Texas Southwestern Medical Center, Dallas, Texas, USA

Emma K. O'Callaghan
Actual Analytics, Edinburgh, United Kingdom

Maria Olmedo
Institute of Medical Psychology, Munich, Germany

Mark L. Paddock
Center for Circadian Biology, University of California, San Diego, California, USA

Mirko Pegoraro
Department of Genetics, University of Leicester, Leicester, United Kingdom

Joe Pogliano
Division of Biological Sciences, University of California, San Diego, La Jolla, California, USA

Jeffrey L. Price
Division of Molecular Biology and Biochemistry, School of Biological Sciences, University of Missouri-Kansas City, Kansas City, Missouri, USA

Michael Rosbash
Department of Biology, Howard Hughes Medical Institute and National Center for Behavioral Genomics, Brandeis University, Waltham, Massachusetts, USA

Lino Saez
Laboratory of Genetics, The Rockefeller University, New York, USA

Ryan K. Shultzaberger
Center for Circadian Biology, and Kavli Institute for Brain and Mind, University of California, San Diego, California, USA

Joseph S. Takahashi
Department of Neuroscience, and Howard Hughes Medical Institute, University of Texas Southwestern Medical Center, Dallas, Texas, USA

Charles J. Weitz
Department of Neurobiology, Harvard Medical School, Boston, Massachusetts, USA

Michael W. Young
Laboratory of Genetics, The Rockefeller University, New York, USA

Wangjie Yu
Department of Biology and Center for Biological Clocks Research, Texas A&M University, College Station, Texas, USA

Abby Zadina
Department of Biology, Howard Hughes Medical Institute and National Center for Behavioral Genomics, Brandeis University, Waltham, Massachusetts, USA

Jian Zhou
Department of Biology and Center for Biological Clocks Research, Texas A&M University, College Station, Texas, USA

Mian Zhou
Department of Physiology, University of Texas Southwestern Medical Center, Dallas, TX, USA

PREFACE

In the 10 years since a previous circadian volume of *Methods in Enzymology* was published, the circadian field has evolved with the introduction of new concepts, new approaches, and, of course, many new investigators. As the previous volume was preceded by an explosion of discoveries and information about how clocks work in diverse organisms, from cyanobacteria to humans, one might have predicted a gradual plateauing of activity, but this was clearly not the case. Importantly, the new discoveries have continued to come from species across the phylogenetic tree, and so the field remains truly interdisciplinary, which is reflected in two volumes, Circadian Rhythms and Biological Clocks Part A and B.

We continue to see advances in the methods used to measure and analyze rhythms, to identify new circadian genes, and to characterize known clock components. In fact, clocks are now being modulated with a goal of therapeutic application. At the same time, there is considerable focus on the molecular mechanisms through which rhythms are transmitted by the clock. Major clock proteins are transcription factors that effectively drive rhythmic expression of many genes, which generally vary from tissue to tissue. The mechanism by which such rhythmic transcription occurs is a subject of intense investigation, and sophisticated techniques have been brought to bear on it. However, it is also evident that posttranscriptional or even posttranslational mechanisms alone can drive the cycling of RNA or protein. Indeed, proteins may not even need to cycle in terms of levels as their activity can be regulated through cyclic modifications, such as phosphorylation. Several clock proteins are kinases, and in cyanobacteria, the entire clock can be reconstituted as a rhythmic phosphorylation cycle. In addition, 24-h oscillations that are independent not only of transcription but also of known clock elements have been proposed for redox pathways in the cell. Thus, while genome-wide analyses are allowing identification of RNAs expressed rhythmically in different tissues, and even in small groups of brain neurons, proteomic studies have been initiated to pinpoint rhythms at this level. All of these have required development and use of the appropriate, anatomical, molecular, biochemical, and statistical tools described here.

Rhythms in molecules lead to rhythms in cellular and organismal physiology. As clocks are found throughout the body, they control many physiological processes and behaviors. In the brain, rhythms of electrical activity are synchronized across clock cells and transmitted across circadian circuits, likely comprised of neurons and glia, to generate rhythmic behavior. Cellular rhythms, such as in mitochondrial respiration, may occur in every cell in mammals. In addition, each organ/system, such as the cardiovascular system, has its own circadian physiology that can be measured through application of specific methods. Interestingly, while the light:dark cycle is the most powerful entraining stimulus for the clock in the brain, which then synchronizes other body clocks, rhythmic gene expression in peripheral tissues like the liver responds most strongly to the time of feeding. Thus, aberrant feeding schedules can have adverse effects on metabolic function, but the time of feeding can even be manipulated to produce beneficial consequences.

Overall, it is increasingly evident that circadian rhythms are critical for organismal fitness. Several approaches are now being used to assess circadian function in humans, its control by genetic factors, and its relevance to human health. Disrupted rhythms have been associated with neurological and psychiatric disorders and studies are now underway to address the significance of such association. These volumes provides methodological insight into circadian physiology and behavior in model organisms and in humans and touches upon the pathological implications of circadian dysfunction.

While I took on the job of putting these volumes together, the actual credit should go to all the contributing authors, who took time out of their extremely busy lives to make the volume representative of the best work in the circadian field. I am incredibly grateful for their efforts and their cooperation. Also, all of this was made possible by the constant help of Editorial Project Manager, Sarah Lay, who was a real pleasure to work with throughout the process.

Amita Sehgal

PART I

Organismal Rhythms as Read-Outs of Clock Function

CHAPTER ONE

Studying Circadian Rhythm and Sleep Using Genetic Screens in *Drosophila*

Sofia Axelrod, Lino Saez, Michael W. Young[1]
Laboratory of Genetics, The Rockefeller University, New York, USA
[1]Corresponding author: e-mail address: michael.young@rockefeller.edu

Contents

Abstract

The power of *Drosophila melanogaster* as a model organism lies in its ability to be used for large-scale genetic screens with the capacity to uncover the genetic basis of biological processes. In particular, genetic screens for circadian behavior, which have been performed since 1971, allowed researchers to make groundbreaking discoveries on multiple levels: they discovered that there is a genetic basis for circadian behavior, they identified the so-called core clock genes that govern this process, and they started to paint a detailed picture of the molecular functions of these clock genes and their encoded proteins. Since the discovery that fruit flies sleep in 2000, researchers have successfully been using genetic screening to elucidate the many questions surrounding this basic animal behavior. In this chapter, we briefly recall the history of circadian rhythm and sleep screens and then move on to describe techniques currently employed for mutagenesis and genetic screening in the field. The emphasis lies on comparing the newer approaches of transgenic RNA interference (RNAi) to classical forms of mutagenesis, in particular in their application to circadian behavior and sleep.

Methods in Enzymology, Volume 551
ISSN 0076-6879
http://dx.doi.org/10.1016/bs.mie.2014.10.026

We discuss the different screening approaches in light of the literature and published and unpublished sleep and rhythm screens utilizing ethyl methanesulfonate mutagenesis and transgenic RNAi from our lab.

1. INTRODUCTION: STUDYING CIRCADIAN BEHAVIOR IN THE FRUIT FLY, *DROSOPHILA MELANOGASTER*

Drosophila exhibits a multitude of innate and adaptive behaviors that allow researchers to study complex behaviors in a genetically tractable organism. Fruit flies, like all animals, need to correctly interpret and respond to their environment.

All life on earth is subject to the changes in light and temperature due to the earth's rotation. Many animals and plants exhibit diurnal or nocturnal behavior depending on their habitat and lifestyle. French scientist Jean-Jaques d'Ortous de Mairan discovered in 1729 that the daily opening and closing of plant leaves persisted in a dark room, indicating that this circadian behavior was not merely a reaction to light, but was effected by internal processes (de Mairan, 1729). It was not until over 200 years later that Konopka and Benzer analyzed the role of endogenous forces—genes—on the daily eclosion rhythm of the fruit fly *Drosophila melanogaster* (Konopka & Benzer, 1971). Since then, studies in *Drosophila* have played a prominent role in elucidating the genes and molecular mechanisms driving circadian behavior (Blau et al., 2007; Stanewsky, 2003). Analogous studies in mammals have revealed that these genes and mechanisms are largely conserved through evolution, indicating that these mechanisms are fundamental and underlie the conservation of animal behavior across evolution (Wager-Smith & Kay, 2000). Insights from *Drosophila* continue to have a broad impact on our understanding of circadian biology in vertebrates, including mechanisms of human circadian dysfunction that alter core clock components homologous to those characterized in *Drosophila* (Toh, Jones, He, Eide, & Hinz, 2001; Xu, Padiath, Shapiro, Jones, & Wu, 2005).

More recently, *Drosophila* has been used to study sleep, a behavior that is functionally linked to the circadian clock. Like other invertebrates that have been carefully examined (Campbell & Tobler, 1984), *Drosophila* displays the key behavioral attributes of sleep (Hendricks, Finn, Panckeri, & Chavkin, 2000; Shaw, Cirelli, Greenspan, & Tononi, 2000). These attributes include postural changes specific to sleep, immobility correlated with an increased arousal threshold, a homeostatic rebound in sleep duration and intensity after sleep deprivation, changes in brain electrical activity during sleep (Nitz, van Swinderen, Tononi, & Greenspan, 2002), and alterations in sleep by stimulants and hypnotics that parallel their effects in mammals (Hendricks et al., 2000; Shaw et al., 2000). Recently, it has been suggested that sleep in fruit flies, like that of humans, has different stages of depth during the sleep cycle (van Alphen, Yap, Kirszenblat, Kottler, & van Swinderen, 2013).

Although the adoption of *Drosophila* as a model organism to study sleep is relatively recent, considerable enthusiasm exists for its potential impact on our understanding of the molecular underpinnings of sleep regulation and function. Despite intensive studies over the past several decades, many aspects of sleep have remained elusive.

How sleep is regulated by circadian inputs and in a homeostatic manner (Borbély, 1982) is one focus of investigation. A second focus concerns the essential functions of sleep, as well as how sleep or lack thereof affects other physiological and behavioral processes. Theories for the functions of sleep invoke memory consolidation, synaptic downscaling, cell repair, metabolic and immune augmentation, and removal of toxins from the brain (Crocker & Sehgal, 2010; Xie et al., 2013). How sleep might function within the brain and somatic tissues to achieve these functions is still unclear, particularly at a molecular and cellular level, and these questions are the subject of several studies in *Drosophila*.

The impact of *Drosophila* in studies of circadian rhythms and sleep, as in other areas of biology, stems from the ability to perform large-scale and unbiased forward genetic screens and from powerful genetic tools that enable the fruits of these screens to be exploited (St Johnston, 2002). This chapter reviews recent genetic screens to gain further insight into the molecular basis circadian rhythm and sleep. We touch briefly on prior screens for rhythm and sleep mutants and proceed to the genetic screens for circadian rhythm and sleep that have been performed in recent years with an emphasis on transgenic mutagenesis in comparison with classical methods of genomic mutagenesis.

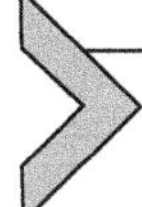

2. SCREENING FOR CIRCADIAN RHYTHM AND SLEEP MUTANTS

2.1. History of circadian rhythm screens

In their landmark 1971 study, Konopka and Benzer isolated the first mutants altering the rhythmicity of *Drosophila* circadian behavior (Konopka & Benzer, 1971). They conducted a screen with the goal of identifying genes for so-called free-running behavior in constant darkness (dark:dark, DD) and described mutants of a locus they named *period* (*per*), which shortened, lengthened, or abolished the rhythmicity of eclosion and locomotor activity in constant darkness. The cloning of the *per* gene in 1984 (Bargiello, Jackson, & Young, 1984; Zehring et al., 1984) marked the onset of a "clockwork explosion" in genetic screens identifying the genetic basis and molecular characteristics of the circadian clock. It has been over 15 years since most of these screens were completed and uncovered the majority of the circadian components. Extensive review of these earlier screens is not the subject of this review and can be found elsewhere (Blau et al., 2007, Price, 2005, Stanewsky, 2003).

While the first rhythm screens utilized measurement of eclosion behavior to identify mutants, later higher throughput screens monitored the rhythmicity of locomotor behavior in individual animals and its persistence in free-running conditions (Stanewsky, 2003). *Drosophila* means "dew-loving," and when put in a 12 h light–12 h dark cycle (12:12 LD), flies are indeed most active during dawn and dusk, and sleep most of the day and night (Fig. 1). In free-running conditions without any light or temperature cues, flies continue to wake at the beginning of the subjective day and sleep during the subjective night. Mutants deficient in clock components cannot maintain wild-type (~24 h) rhythmicity in DD and, depending on the type of mutation, display shortened or lengthened rhythms, or become completely arrhythmic.

A variety of mutagenesis methods have been used to identify clock mutants in *Drosophila*, reviewed in Stanewsky (2003). Chemical mutagenesis by feeding ethyl methanesulfonate (EMS) has been widely used to induce circadian mutants (Stanewsky, 2003). Other screens utilized gamma irradiation as a mutagen to screen for genes affecting the clock (Newby et al., 1991; van Swinderen & Hall, 1995). The advent of P-element screens, in which the integration of a transposable element disrupts a gene's expression, saw their adoption to identify circadian mutants (Kloss et al., 1998; Newby & Jackson, 1993; Price et al., 1998; Sehgal, Price, Man, & Young, 1994). Subsequent gain-of-function screens employed modified P-elements containing *UAS* elements (Rørth, 1996), enabling adjacent genes to be transcribed and overexpressed in the presence of the *Gal4* transcriptional activator (Brand & Perrimon, 1993). A neuroanatomically restricted screen in cells expressing the core clock gene *timeless* (*tim*) was performed using a *tim-Gal4* promoter (Martinek, Inonog, Manoukian, & Young, 2001), identifying a circadian role for the *glycogen synthase*

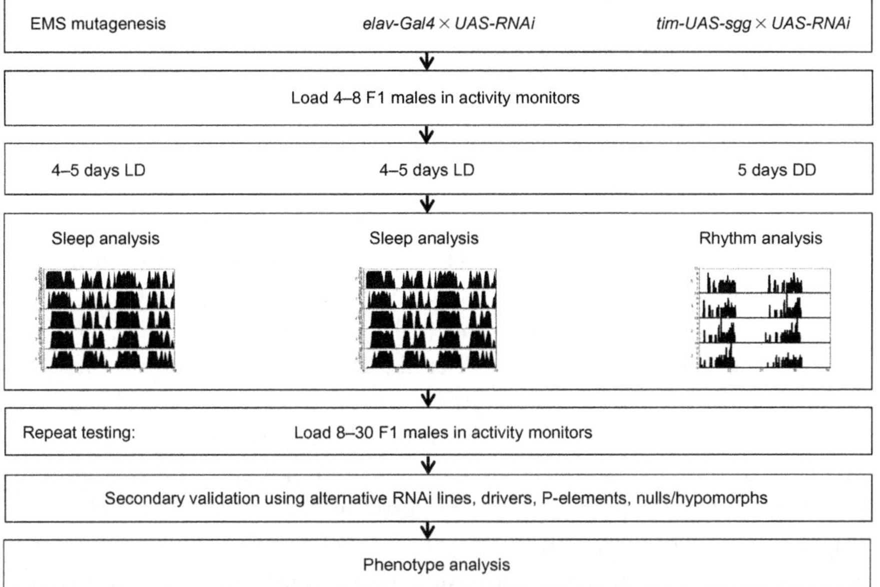

Figure 1 Workflow of three recent screens for circadian behavior. The workflow depicts differences and similarities in the screening process. We employed different strategies to obtain genetic nulls and hypomorphs, either in the whole fly or in specific cell types. Depending on the goal of the screen, behavioral assays were focused either on recording sleep, which is carried out in LD, or rhythmic behavior, which is conducted in DD. While circadian rhythms can be detected with data collections every 5 min or even every 30 min, measuring sleep requires a higher data resolution of at least 1 min bins. In all three screens, data were acquired in 1 min bins. Candidate genes were subjected to rounds of rescreening and secondary validation including available genetic tools. Phenotypic analysis of confirmed candidates includes identifying a gene's expression pattern, cells in which its function is required, effects on other behaviors, and molecular analyses of the protein function.

kinase-3 (*gsk-3*) /*shaggy* (*sgg*). As discussed below, recent screens for circadian rhythms and sleep continue to exploit the *Gal4/UAS* system in conjunction with current genetic tools for the manipulation of gene function.

2.2. History of sleep screens

Many years of research in mammals have begun to elucidate different aspects of sleep. These include the discovery of different sleep stages using electroencephalography (EEG), the role of various neurotransmitters and genetic pathways in promoting or inhibiting sleep, and the identification of brain loci important for sleep regulation (Crocker & Sehgal, 2010). The discovery that flies display key behavioral criteria for sleep has been followed by both reverse and forward genetic strategies to address the mechanisms underlying sleep regulation and function. Studies employing reverse genetics have focused on candidate genes and pathways, many of which are known to impact sleep in mammals (Bushey & Cirelli, 2011; Crocker & Sehgal, 2010). Genetic manipulation of neurotransmitter systems and neuropeptide signaling in *Drosophila* has shown that neurochemical modulation of sleep is similar in *Drosophila* and vertebrates. Neurotransmitters including dopamine (Kume, 2005), gamma-aminobutyric acid (GABA, Agosto et al., 2008), octopamine/norepinephrine (Crocker & Sehgal, 2008), serotonin (Yuan, Joiner, & Sehgal, 2006), histamine (Oh, Jang, Sonn, & Choe, 2013; Yi et al., 2013), and acetylcholine (Yi et al., 2013) impact sleep in *Drosophila*. Neuropeptide Y/F (He, Yang, Zhang, Price, & Zhao, 2013; Shang et al., 2013) and various signaling pathways including the CREB (Hendricks et al., 2001), extracellular signal-regulated kinase (ERK, Foltenyi, Greenspan, & Newport, 2007), and protein kinase A (Hendricks et al., 2001; Joiner, Crocker, White, & Sehgal, 2006) pathways impact sleep in *Drosophila* as in mammals. Alongside these candidate gene approaches, expression profiling (Cirelli, LaVaute, & Tononi, 2005; Williams, Sathyanarayanan, Hendricks, & Sehgal, 2007) and selective breeding (Seugnet, Suzuki, & Thimgan, 2009) have been used to identify genes that impact sleep, while neuroanatomically driven strategies, utilizing electrical and genetic manipulations, have implicated various neuronal populations in the regulation of sleep (Cavanaugh et al., 2014; Joiner et al., 2006; Pitman, McGill, Keegan, & Allada, 2006).

Several large-scale screens for sleep mutants have been reported (Cirelli, Bushey, et al., 2005; Koh et al., 2008; Liu et al., 2014; Pfeiffenberger & Allada, 2012; Rogulja & Young, 2012; Shi, Yue, Kuryatov, Lindstrom, & Sehgal, 2014; Stavropoulos & Young, 2011). These screens have largely focused on sleep duration and have led to the isolation of mutants that strongly reduce the length and consolidation of sleep. Two screens have yielded mutations in genes that regulate neuronal excitability, including the *Shaker* potassium channel (Cirelli, Bushey, et al., 2005) and *quiver/sleepless* (Koh et al., 2008), an extracellular membrane-linked peptide that alters *Shaker* expression and trafficking (Wu, Robinson, & Joiner, 2014). Mutations in *Hyperkinetic*, the beta-subunit of the *Shaker* channel, reduce sleep duration similarly (Bushey, Huber, Tononi, & Cirelli, 2007). The isolation of a short sleep mutant in the *redeye* allele of the α4 subunit of the nicotinic acetylcholine receptor (Shi et al., 2014) provides additional evidence for the modulation of arousal and sleep/wake states by neuronal excitability.

Other mutants implicate novel mechanisms by which sleep may be regulated. Cloning and characterization of *insomniac* have raised the possibility that neuronal protein degradation pathways may contribute to the control of sleep duration (Stavropoulos & Young, 2011), and the isolation of *cyclin A* (Rogulja & Young, 2012) has indicated a neuronal function of a broadly essential gene in regulating sleep.

Sleep timing is considered to be regulated by two inputs: the circadian clock determining a sensible daily sleep time and a yet elusive homeostat, which measures sleep pressure and could override circadian sleep timing (Borbély, 1982). This idea is supported by the observation that animals exhibit rebound sleep after sleep deprivation, which occurs independently of circadian timing and is accompanied with EEG changes in humans (Borberly & Achermann, 1999). Along with total daily sleep, researchers now can measure various sleep parameters including (a) number of sleep bouts and their length, which indicate sleep fragmentation; (b) sleep latency, pointing to difficulty to initiate sleep or circadian components; and (c) sleep homeostasis, which is measured as the amount of rebound sleep following sleep deprivation.

The first generation of screens identified mutants based on a reduction of total daily sleep, and recovered a handful of mutants that exhibit a reduction of daily sleep greater than 50% (Cirelli, Bushey, et al., 2005; Koh et al., 2008; Liu et al., 2014; Pfeiffenberger & Allada, 2012; Rogulja & Young, 2012; Shi et al., 2014; Stavropoulos & Young, 2011). The observed sleep reduction could be the result of either circadian or homeostatic disturbance. Recent studies are focusing on specific sleep parameters including sleep latency (Liu et al., 2014) and sleep homeostasis (Bushey & Cirelli, 2011) to tease apart each pathway's contribution to sleep. While the sleep patterns of published short-sleeping mutants appear fragmented compared to wild-type flies, they still show a normal circadian rhythm in constant darkness, suggesting that the circadian input into the timing of sleep is not affected (Cirelli, Bushey, et al., 2005; Koh et al., 2008; Liu et al., 2014; Rogulja & Young, 2012; Shi et al., 2014; Stavropoulos & Young, 2011). However, a detailed analysis of the short-sleeping mutant *wide awake* revealed an increased sleep latency (Liu et al., 2014). *wide awake* is strongly expressed in clock cells and its effect on latency is dependent on the core circadian clock gene *Clock*, despite exhibiting rhythmicity in constant darkness, pointing toward separate pathways for circadian activity and sleep mediated by *Clock*. Despite large-scale screening specifically looking for sleep homeostasis, so far no mutant has been found to specifically affect rebound sleep (Bushey & Cirelli, 2011). Of the published short-sleeping mutants, *sleepless* as well as *cyclinA* show reduced rebound sleep after sleep deprivation, suggesting a function in sleep homeostasis (Koh et al., 2008; Rogulja & Young, 2012).

For the most part, the precise mechanisms of the sleep genes' function remain unknown, although some mutants have been linked to mechanisms regulating synaptic transmission (Wu et al., 2014), to the GABA pathway (Chen et al., 2014), and to the circadian clock (Liu et al., 2014).

3. SCREENING TECHNIQUES

Various techniques have been used to induce genetic lesions in *Drosophila.* We are briefly reviewing EMS and transposon mutagenesis in the context of rhythm and sleep screens and then focus on transgenic techniques of gene inactivation. To illustrate some of the mutagenesis techniques, we discuss one EMS and two RNAi screens, one of which is unpublished, in more detail (Fig. 1).

3.1. EMS mutagenesis

EMS mutagenesis has been used for over 45 years to analyze gene function (Lewis & Bacher, 1968). Its power lies in the simplicity of administration, by feeding, and its ability to induce high mutation rates (Greenspan, 2004). Although there is increased lesion frequency in "hotspots" (Bentley, MacLennan, Calvo, & Dearolf, 2000), its ability to create various types of genomic lesions in an unbiased manner is paramount (Bökel, 2008). To illustrate the approach and workflow of EMS mutagenesis, we discuss a recent EMS screen for sleep mutants in the following section.

3.1.1 X-linked EMS screen for sleep mutants

We conducted a chemical mutagenesis screen to find novel genes affecting the flies' sleep. We selected a recently isogenized Canton S (CS) strain displaying well-consolidated nighttime sleep and screened the X-chromosome using a mating scheme in which we screened four F2 males from each mutagenized line (Fig. 2). Screening four animals for each mutant line permits increased screening throughput at the cost of additional rescreening. We screened more than 3500 lines and rescreened 471 lines (13.5%) assessed as exhibiting potentially different sleep patterns than wild-type animals. In most cases, a single round of rescreening is sufficient to discard lines. After multiple rounds of rescreening, approximately 50 stocks were assessed to

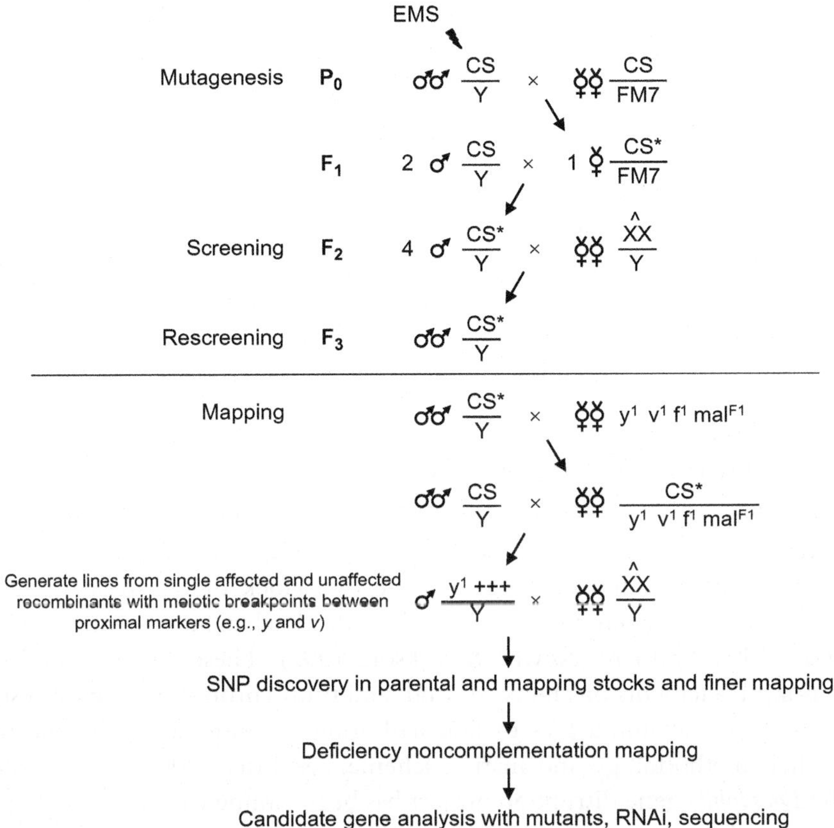

Figure 2 Crossing and mapping scheme employed in the EMS mutagenesis screen for sleep mutants. Wild-type Canton S (CS) males were mutagenized with 25–40 m*M* EMS and crossed en masse to FM7/CS virgins. F1 virgins were backcrossed to wild-type CS males. Four F2 males were assayed for sleep behavior and potential hits were backcrossed to isogenized attached-X virgins for rescreening. For genomic mapping of the *inc* mutation, balanced males were crossed to virgins from a mapping stock. Recombination analysis revealed that *inc* lies proximal to *y*. Further polymorphism mapping narrowed down the cytological location to a 250-kb to 1-mb window. Finally, deficiency noncomplementation analysis of the genomic region identified a 190-kb stretch that failed to complement the *inc* mutation and was used for further sequencing analysis. Asterisk (*) indicated mutated chromosome; iso, isogenized; *y*, *yellow*; *v*, *vermillion*; *f*, *forked*; *mal*, *maroon-like*.

potentially bear X-linked phenotypes of interest (1.4%). Sixteen candidates have extremely short sleep with less than 508 min of daily sleep, which is more than three standard deviations from the mean of 901 min/day, and five flies had extremely long sleep (>3 standard deviations above the mean) with more than 1295 min of daily sleep. Three mutant lines with a severe reduction of sleep were selected for further analysis, two of which are alleles of the already described *Shaker* mutation (Cirelli, Bushey, et al., 2005) as assessed by noncomplementation and shaking under ether anesthesia (data not shown). The third mutant was further characterized and mapped to an intergenic region between CG14795 and CG32810 (Fig. 2). Using a line carrying a transposable element disrupting CG32810 and replicating the phenotype, it was shown that this is the gene responsible for the sleep phenotype. We called the gene *insomniac* (*inc*) and published a detailed description of this novel gene, its phenotype, and cellular and molecular characteristics (Stavropoulos & Young, 2011).

3.2. Transposon mutagenesis

Mutagenesis via P-element transposition appeared in the 1990s and had the advantage that P-elements carry an identifiable sequence allowing for direct gene cloning. Such screens yielded the two clock mutants *tim0* (which, however, turned out to be unrelated to the P-element insertion, Sehgal et al., 1994) and *lark* (Newby & Jackson, 1993). These screens mobilized existing P-elements on the X, second, and third chromosomes by crossing P-element containing flies to flies harboring a transposase transgene (for explicit methodology and mating scheme, see Price, 2005). Since 1991, the *Drosophila* gene disruption project has been aiming to generate transposon insertions in all *Drosophila* genes and is now covering at least two-thirds of the *Drosophila* genome (Bellen et al., 2011). Ordering flies from this (http://flystocks.bio.indiana.edu/Browse/in/GDPtop.htm) and other (Kim et al., 2010; Ryder, 2004; Thibault et al., 2004) libraries has been allowing researchers to directly screen flies containing different P-element insertions for phenotypes without having to map or clone the mutation because P-elements' positions are annotated with single-nucleotide resolution (http://flybase.org/). Such screens identified one of the alleles of the clock gene *double-time* (Price et al., 1998) as well as the sleep genes *sleepless* and *wide awake* (Koh et al., 2008; Liu et al., 2014).

Some of the P-elements used to create libraries carry *UAS* gene activation sequences, which can be used to express neighboring DNA sequences if a *Gal4*-driver is also present. Martinek et al. used such a library of 2300 so-called EP lines (Rørth, 1996) to screen for circadian genes (Martinek et al., 2001). Crossing the EP lines to a *Gal4* driver line containing a *tim* promoter fused to *Gal4* (*tim*(*UAS*)-*Gal4*) led to F1 progeny where the DNA adjacent to the EP transgene was only expressed in *tim*-expressing clock neurons. This screen discovered a circadian function for the kinase *glycogen synthase kinase-3/shaggy* (*gsk-3/sgg*).

In addition to genomic mutants as produced by chemical and insertional mutagenesis, certain scientific questions benefit from a more targeted approach in which a gene of interest is removed from specific cells or at a specific time. To this end, different techniques have been developed. While EMS and P-element mutagenesis have been used for decades and extensively reviewed, we focus our discussion of the transgenic techniques of gene manipulation and inactivation.

3.3. Tools for conditional transgene expression

Many *Gal4* drivers have been created to drive transgene expression only in specific cells and tissues. For circadian research, a number of drivers exist that drive transgene expression in different subsets of neurons, including clock gene-expressing neurons (Yoshii, Rieger, & Helfrich-Frster, 2012). Moreover, researchers now have access to vast driver libraries containing drivers for randomly selected genes (8000 lines, Kvon et al., 2014) and many neuronal subgroups (7000 lines, Jenett et al., 2012). By combining *Gal4* expression with neuronal inactivation, researchers can screen for neuronal subpopulations required for a specific function (Cavanaugh et al., 2014; Joiner et al., 2006; Pitman et al., 2006). By using *Gal80ts*, a temperature-sensitive inhibitor of the gene expression driver *Gal4*, *Gal4* expression can be temporally and spatially limited (McGuire, 2003). Another method for conditional gene expression is the modified *Gal4* driver GeneSwitch, which is only activated after binding of the activator RU486, which is fed to flies (Osterwalder, Yoon, White, & Keshishian, 2001). In addition to *Gal4*, other systems exist for transgene expression, and all systems can be used interjectionally (Venken, Simpson, & Bellen, 2011) to express transgenes only in specific cells and/or at a specific time.

For genetic screens in specific tissues, transgenic RNA interference (RNAi) is widely used to study clock and sleep function on a genome-wide scale, both in *Drosophila* and other organisms (Chung, Kilman, Keath, Pitman, & Allada, 2009; Itoh & Matsumoto, 2012; Mandilaras & Missirlis, 2012; Rogulja & Young, 2012; Zhang et al., 2009).

It is believed that RNAi evolved to fight transposons and RNA viruses and comprises an intracellular machinery for the targeted destruction of specific mRNAs (Shabalina & Koonin, 2008). The discovery that this machinery can be exploited by researchers to specifically remove mRNAs of specific genes opened a new posttranscriptional method to analyze gene function without having to modify the genome itself (Dykxhoorn & Lieberman, 2005).

RNAi is based on the complementarity of a double-stranded RNA (dsRNA) molecule to an endogenous mRNA, which leads to destruction of complementary mRNA molecules. While in mammalian systems short interfering RNAs (siRNAs) have been used for RNAi, in *Drosophila* mostly long dsRNAs (300–600 bp long) have proven efficient. In addition to cell-based RNAi, researchers also have been making use of transgenic RNAi *in vivo* by fusing the complementary sequence to a *UAS* element thereby allowing knockdown of a given gene using tissue-specific *Gal4* drivers (Perrimon, Ni, & Perkins, 2010).

3.4. *Drosophila* RNAi libraries and screens

Now we have vast libraries with commercially available RNAi lines containing transgenic *UAS-RNAi* constructs, which are used for genome-wide RNAi screens. We want to provide an overview of the resources available for such screens as well as juxtapose RNAi with other types of mutagenesis.

To date, there are three main sources for RNAi lines: VDRC Vienna, Nig-Fly Japan, and TriP Harvard in Boston (Table 1). The VDRC stock center in Vienna consists of two libraries: the GD and the newer KK library together targeting ca. 12,000 genes or ca. 90% of the *Drosophila* protein-coding genome. The GD library was generated using random P-element-mediated transformation, regularly yielding multiple lines per construct, and covers 84% of the genome. The second-generation KK library covers 67% of genes and was generated using targeted insertion with the PhiC31 integrase into a defined attP landing site on the second chromosome to minimize position effects. These lines are considered in general more efficient than the GD lines (S. Axelrod, data not shown). The Japanese library was generated using random P-element-mediated transformation, with possible insertion sites on all three chromosomes. For many genes, multiple lines exist with variable RNAi efficiency depending on the genomic environment of the insertion site creating position effects suppressing or enhancing transgene expression. To test RNAi efficiency, all available RNAi lines have been crossed to an *actin-Gal4* driver thereby leading to ubiquitous RNAi expression. By assessing the survival of the next generation, NIG-Fly assessed the efficiency of a given RNAi line, lethality indicating strong expression of RNAi transgenes. *actin-Gal4*-induced lethality provides

Table 1 Resources for *in vivo* RNAi

Library	Number of lines	Type of insertion	Reference	Comments
Nig-Fly Japan	11,000	Random	http://www.shigen.nig.ac.jp/fly/nigfly	Multiple lines per gene, all lines crossed to *actin-Gal4* for lethality assessment
VDRC Vienna	25,259		http://stockcenter.vdrc.at, Dietzl et al. (2007)	
GD library	16,442	Random		Multiple lines per gene
KK library	9817	Site directed		One line per gene
TRiP Harvard	9128	Site directed	http://flyrnai.org/TRiP-HOME.html, Ni et al. (2009, 2011)	
TRiPSoma	2486			No germline expression
TRiPGermline	6638			1605 Expressed only in germline, 5033 in both

valuable information about the efficiency of a given RNAi line. The Harvard TriP library was generated using targeted insertion with the PhiC31 integrase in two characterized landing sites on the second and third chromosome. The landing sites had been chosen both for minimizing leaky expression of the *UAS-RNAi* transgenes in absence of *Gal4* and for maximizing RNAi expression in the presence of *Gal4*. There are two collections of lines, TriPSoma and TriPGermline, utilizing different vectors optimized for either somatic (TriPSoma) or germline/germline and somatic expression (TriPGermline) of RNAi (Table 1). While the TriPSoma lines utilize the regular long dsRNA method to induce RNAi, the second-generation TriPGermline lines utilize microRNA-mediated knockdown.

To illustrate the methodology and approach of RNAi, in the next section we outline two recent RNAi screens, one for rhythm and one for sleep mutants.

3.4.1 RNAi screen for suppressors and enhancers of shaggy

Many genes have been identified to be required for clock function, and progress has been made in delineating the molecular functions of the proteins involved. However, it is still largely unclear how different clock proteins interact with each other to achieve their effect. One of the main mysteries of the clock is how the exact timing of 24 h is established and how delays are built into the clock to prevent clock proteins from advancing the period. Posttranslational modifications such as phosphorylation and glycosylation are generally good candidates for such regulation of protein function (Petsko & Ringe, 2004). Indeed, three kinases have been identified to be required for correct clock timing: Double-time (Price et al., 1998), Casein kinase II (CKII, Akten et al., 2003), and Glycogen synthase kinase-3, Shaggy (GSK-3, SGG, Martinek et al., 2001). In addition, it has been shown that the core clock genes PER and Clock are glycosylated, which has been proposed as a mechanism for fine-tuning the clock (Kaasik et al., 2013; Kim et al., 2012; Li et al., 2013).

sgg was found in a screen that employed a library of transgenic lines called EP lines carrying P-element insertions (Martinek et al., 2001). DNA adjacent to the P-element insertion site was expressed in clock neurons using *tim(UAS)-Gal4*. One of the EP lines, when expressed in clock neurons, shortened the period by roughly 3–20.3 h. Inverse PCR analysis mapped the insertion to the *sgg* gene. *sgg* is an essential gene and null flies die during development, restricting analyses of *sgg*'s function in the circadian rhythm to conditional misexpression experiments. While overexpression of *sgg* shortens the period length, reducing *sgg* function lengthens the rhythms to 26 h. In addition, *sgg* overexpression advances nuclear entry and leads to increased phosphorylation of TIM, indicating that SGG advances the circadian clock through acting on TIM. However, in a 2007 paper, Stoleru et al. (2007) showed that when the circadian photoreceptor, *cryptochrome* (*cry*), is mutated, *sgg* overexpression only shortens the period by 1 h, instead of 3 h. How *sgg* function is affected by the presence of the light receptor *cry* is unclear. In 2010, Edery's group showed that SGG specifically phosphorylates a serine in PER, and that abolishing that site leads to longer behavioral rhythms (Ko et al., 2010).

The kinase encoded by *sgg* is intriguing because it is involved in several cellular and developmental processes including metabolism (Garofalo, 2002), growth (Woodgett, Plyte, Pulverer, Mitchell, & Hughes, 1993), cell fate determination (Siegfried, Chou, & Perrimon, 1992), and as the main target of lithium, a pharmaceutical drug used to treat bipolar disorder, potentially linking this mental illness to circadian dysfunction.

We are currently conducting a suppressor/enhancer screen to find genetic *sgg* interactors (Fig. 3). Such a modifier screen has been successfully

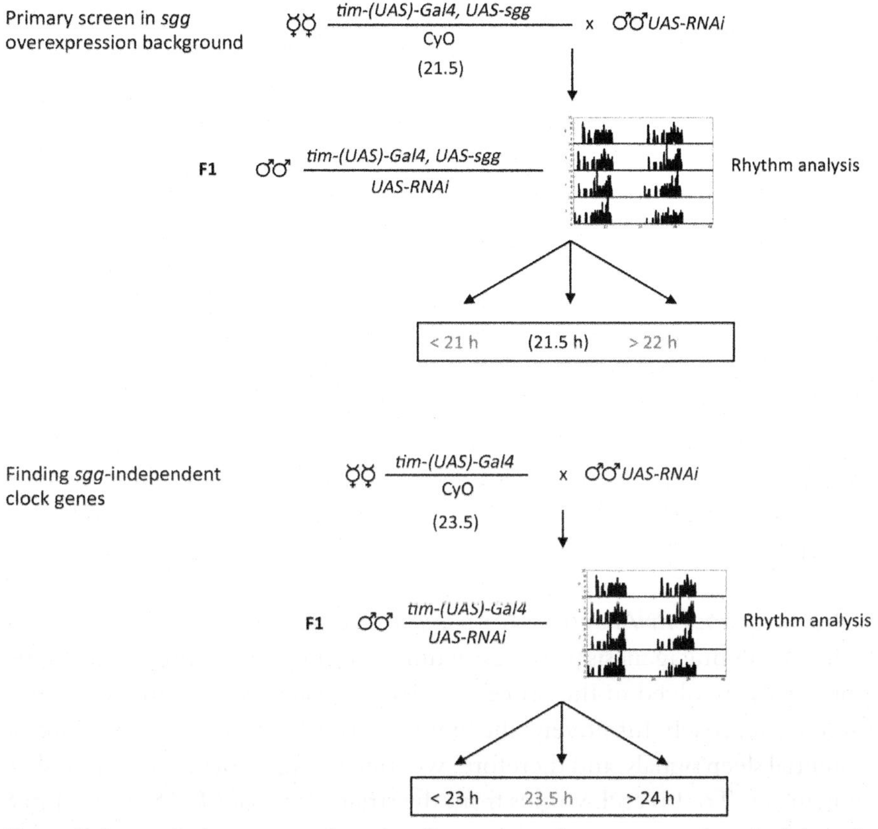

Figure 3 Screen for interactors of *sgg*. In a first step, random genes are knocked down in *tim*-expressing flies in a *sgg*-overexpression background. RNAi lines producing lengthening or shortening of this period are in a second stop recrossed to the *tim-Gal4* line alone to identify sgg-independent clock genes.

used in the past to find specific interactors of short and long period producing alleles of *per* (Rutila et al., 1996). We are using a *tim*(*UAS*)-*Gal4*, *UAS-sgg* line to overexpress *sgg* in *tim*-expressing neurons, which produces a period of 21.5 h. We then use this line to drive expression of RNAi in the same cells and monitor free-running activity rhythms in constant darkness. Gene knockdowns that change the period length in this assay (below 21 h or above 22 h) are candidates for *sgg*-specific suppressor or enhancers. To identify whether these candidates genes are general clock mutants, RNAi lines are retested without *sgg* overexpression. In some cases, the phenotype persists without *sgg* overexpression indicating potential direct involvement of such genes in the clock. To this day, we screened ca.

3000 lines representing ca. 2500 genes. In around 110 lines, we observed a change in rhythm (ca. 3% of tested lines). Notably, we recovered 10 known clock genes from the screen so far, which serves as proof of principle: *circadian trip*, *cycle*, *PAR-domain protein 1*, *per*, *ckIIα*, *ckIIβ*, *Clock*, *resistant to dieldrin*, *ras homolog enriched in brain ortholog*, and *cullin3*, a gene described to be involved in both sleep and circadian rhythms (Grima, Dognon, Lamouroux, Chélot, & Rouyer, 2012; Stavropoulos & Young, 2011). Among the other lines with changed rhythm are genes affecting several cellular processes including transcription, chromatin remodeling, ribosomal function, odorant binding, cytoskeleton, phosphorylation and dephosphorylation, Zn binding, splicing, and many genes whose function is unknown. This screen is proving to be a useful source to find new clock genes, as well as identify specific *sgg* interactors. To validate the results, we are using alternative RNAi lines both from Japan and VDRC. To narrow down the candidate genes' requirement in specific cells, we use different overlapping and nonoverlapping *Gal4* drivers.

3.4.2 Neuronal RNAi screen for sleep mutants

While EMS mutagenesis provides an unbiased approach to finding any genes potentially involved in the process of sleep, we also wanted to take a more targeted approach. Intuitively, the brain seems to be the most likely source of potential sleep signals, and therefore, we tried to find genes required for sleep in neurons. To this end, virgins from the strain *elav-Gal4;UAS-dicer2* (Lin & Goodman, 1994) were crossed to males from randomly selected RNAi lines (NIG-Fly, Japan).

If there were multiple RNAi lines available, we chose lines producing lethality when crossed to *actin-Gal4* because that indicates stronger RNAi efficiency than lines not producing lethal phenotypes. Approximately, 4000 lines covering ca. 3500 genes were tested. Males were collected after eclosion and aged for 1–5 days, and four males of each genotype were loaded into *Drosophila* Activity Monitoring Systems (Trikinetics). The screen was performed at a slightly elevated temperature of 26 °C/27 °C to reduce background levels of activity, and flies were assayed for at least 4 days in 12 h light, 12 h dark cycles (LD). To validate screen results, we retested initial hits three times with 8–16 flies. For interesting candidates, alternative RNAi lines were tested, which produced varying results. If available, null mutants were tested. Around 10% of screened neuronal genes produced lethality when knocked out in the nervous system. Of the nonlethal hits that survived the multiple rounds of retesting, 20 genes remain that show robust and reproducible phenotypes. Genes from this screen fall into various categories including membrane proteins, transcription factors, and RNA-binding proteins. Of the 20 genes, 14 show a reduction of sleep, 4 show an increase, and 2 show other activity changes, e.g., predominantly nocturnal activity. One of the genes whose knockdown in neurons severely reduced sleep is the *regulator of cyclin A1* as well as its target, *cyclin A*. We published an in-depth analysis of these phenotypes in 2012 (Rogulja & Young, 2012).

This screen demonstrates the usefulness of transgenic RNAi for *in vivo* studies of behavior and sleep in a spatially defined manner: while the genes discovered in this screen might have other functions or even be developmentally required, only removing their expression in specific neurons enables us to address their function in this specific context.

3.5. Advantages and drawbacks of screening with RNAi in comparison to chemical and transposon mutagenesis

The creation of *UAS-RNAi* libraries provides a quick and simple means to reduce gene function in conjunction with *Gal4* drivers. Large-scale RNAi screens offer several advantages over classical chemical and transposon mutagenesis screens. First, RNAi screens can be performed with a single cross of a *Gal4* driver of interest to *UAS-RNAi* lines, enabling F1 progeny to be screened. In contrast, chemical mutagenesis (Cirelli, Bushey, et al., 2005; Shi et al., 2014; Stavropoulos & Young, 2011) and transposon screens (Koh et al., 2008; Liu et al., 2014) require additional generations and longer breeding schemes to obtain progeny of interest. For "shelf" screens using existing transposon insertions, the sensitivity of sleep to genetic background and the possible accumulation of suppressor mutations require additional generations of backcrossing prior to homozygosis (Koh et al., 2008). A second, and important, advantage of RNAi screens is that they provide the immediate identity of targeted genes. Chemical mutagenesis, in contrast, generates unmarked lesions whose mapping and positional cloning is laborious and time consuming (Stavropoulos & Young, 2011). Mapping and positional cloning EMS-induced sleep mutants in the absence of other phenotypic hints (e.g., Cirelli, Bushey, et al., 2005) remain a challenge, even with the availability of polymorphism libraries (Berger et al., 2001) and whole-genome sequencing (Shi et al., 2014). A third advantage of RNAi screens is that they can be directed in an anatomically and temporally restricted manner as dictated by *Gal4* drivers. In particular, anatomically restricted RNAi screens can identify tissue-specific contributions of essential genes, as the lethality of null mutations in these genes would preclude their isolation from chemical mutagenesis screens (Rogulja & Young, 2012). A substantial fraction (25–30%) of all genes in *Drosophila* are essential (Miklos & Rubin, 1996).

Alongside these advantages, RNAi screens have several potential drawbacks that should be considered.

1. *False positives.* The long dsRNA is transcribed from the RNAi construct and processed by the cellular RNAi machinery to create the specific siRNAs targeting the desired mRNA. siRNAs can have off-target effects by binding not (only) to the desired sequence of the target gene but also to other homologous sequences in other genes. The observed phenotypic effect could then be the result of creating unwanted knockdowns in other genes and not of the gene in question (Kulkarni et al., 2006). A recent report from Green, Fedele, Giorgini, and Kyriacou (2014) shows that the host strain for the KK library from VDRC contains not one but two landing sites and that multiple stocks contain two transgenes creating nonspecific phenotypes.

   To mitigate against the possibility of false positives, different RNAi lines that target nonoverlapping portions of a transcript of interest should be used wherever possible (Yamamoto-Hino & Goto, 2013). Also, validation of RNAi at the level of reduced protein or mRNA levels as well as other means of secondary validation should be employed to rule out false positives (Echeverri et al., 2006; Perrimon et al., 2010). In particular, to ascertain that a phenotype stems from knockdown of a particular gene, RNAi can be combined with a recessive genomic loss-of-function allele of the same gene, with enhancement of the phenotype of RNAi pointing to further transcript loss of the same gene (Rogulja & Young, 2012). Also knockdown of genes in the same pathway yielding a similar phenotype is indicative of an on-target effect (Rogulja & Young, 2012). Another validation approach is rescuing the RNAi phenotype by expressing an RNAi-resistant version of the target gene (Yamamoto-Hino & Goto, 2013).
2. *False negatives.* The efficiency of RNAi is variable and may not reduce protein abundance sufficiently to induce phenotypes, even when genes have a role in the process under study.

   Only 60% of lines from the first GD library from VDRC reportedly produced a knockdown. Overexpression of the enzyme *dicer2*, which produces the siRNAs targeting specific mRNAs for degradation (Lee et al., 2004), enhances RNAi efficiency by ca. 50% and is routinely used by researchers to increase knockdown effects (e.g., Neely et al., 2010;

Neumüller et al., 2011; Rogulja & Young, 2012). Although newer KK and TRiP lines incorporate design elements for enhanced expression and RNAi transgenes are inserted in optimized genomic landing sites (Ni et al., 2009; Yamamoto-Hino & Goto, 2013), RNAi efficiency remains variable as it depends on many factors, only one of which is the actual siRNA production and their target affinity (Booker et al., 2011). Knockdown efficiency also depends on the transcript levels of a given gene, as well as the reduction of protein necessary to achieve a phenotype (Mohr & Perrimon, 2011). Protein turnover varies for different types of proteins, potentially hampering knockdown of very stable proteins (Scott et al., 2013). Using RNAi lines differing in construct sequence and/or insertion site (Yamamoto-Hino & Goto, 2013), enhancement of RNAi efficiency by adding *UAS-dicer2* (Dietzl et al., 2007) or using multiple copies of *Gal4* and/or *UAS* (S. Axelrod, data not shown) can be used to increase RNAi efficiency. To study highly redundant processes or paradigms where partial loss of function is unlikely to yield phenotypes, performing screens in a sensitized background could be useful. To validate candidate genes, comparing the phenotype to that of animals carrying null mutations or disruptive P-element insertions can be used to identify false-negative results.

3. *Target limitations*. All present RNAi libraries target protein-coding genes. For analysis of noncoding DNA regions, including regulatory regions and the various species of noncoding RNA, this approach cannot be employed and in this field the EMS and transposon mutagenesis methods are more useful (Sarin et al., 2010).

By combining unbiased and targeted approaches, investigators in the circadian rhythm and sleep fields are currently trying to expand our knowledge of behavior in different and complementary ways. Sleep screens help us to shed light on the molecular mechanisms required for proper sleep. Modifier screens like the *sgg* screen go back to the old question of how the clock works, and in particular how the 24 h rhythm is so precisely established. This systematic approach—utilizing the genetic power of the fruit fly to uncover the molecular and cellular basis of circadian rhythm and sleep—gives researchers the opportunity to gain mechanistic insight into two behaviors that are fundamental to all living organisms.

ACKNOWLEDGMENTS

We thank Nicholas Stavropoulos and Dragana Rogulja for helpful comments on the manuscript. This work was supported by NIH NS053087 and GM054339 (M. W. Y.).

REFERENCES

Agosto, J., Choi, J. C., Parisky, K. M., Stilwell, G., Rosbash, M., & Griffith, L. C. (2008). Modulation of GABAA receptor desensitization uncouples sleep onset and maintenance in Drosophila. *Nature Neuroscience*, *11*(3), 354–359.

Akten, B., Jauch, E., Genova, G. K., Kim, E. Y., Edery, I., Raabe, T., et al. (2003). A role for CK2 in the Drosophila circadian oscillator. *Nature Neuroscience*, *6*(3), 251–257.

Bargiello, T. A., Jackson, F. R., & Young, M. W. (1984). Restoration of circadian behavioural rhythms by gene transfer in Drosophila. *Nature*, *312*(5996), 752–754.

Bellen, H. J., Levis, R. W., He, Y., Carlson, J. W., Evans-Holm, M., Bae, E., et al. (2011). The Drosophila gene disruption project: Progress using transposons with distinctive site specificities. *Genetics*, *188*(3), 731–743.

Bentley, A., MacLennan, B., Calvo, J., & Dearolf, C. R. (2000). Targeted recovery of mutations in Drosophila. *Genetics*, *156*(3), 1169–1173.

Berger, J., Suzuki, T., Senti, K.-A., Stubbs, J., Schaffner, G., & Dickson, B. J. (2001). Genetic mapping with SNP markers in Drosophila. *Nature Genetics*, *29*(4), 475–481.

Blau, J., Blanchard, F., Collins, B., Dahdal, D., Knowles, A., Mizrak, D., et al. (2007). What is there left to learn about the Drosophila clock? *Cold Spring Harbor Symposia on Quantitative Biology*, *72*(1), 243–250.

Bökel, C. (2008). EMS screens: From mutagenesis to screening and mapping. *Methods in Molecular Biology (Clifton, N.J.)*, *420*, 119–138.

Booker, M., Samsonova, A. A., Kwon, Y., Flockhart, I., Mohr, S. E., & Perrimon, N. (2011). False negative rates in Drosophila cell-based RNAi screens: A case study. *BMC Genomics*, *12*(1), 50.

Borbély, A. A. (1982). A two process model of sleep regulation. *Human Neurobiology*, *1*(3), 195–204.

Borberly, A. A., & Achermann, P. (1999). Sleep homeostasis and models of sleep regulation. *Journal of Biological Rhythms*, *14*(6), 559–570.

Brand, A. H., & Perrimon, N. (1993). Targeted gene expression as a means of altering cell fates and generating dominant phenotypes. *Development (Cambridge, England)*, *118*(2), 401–415.

Bushey, D., & Cirelli, C. (2011). From genetics to structure to function: Exploring sleep in Drosophila. *International Review of Neurobiology*, *99*, 213–244.

Bushey, D., Huber, R., Tononi, G., & Cirelli, C. (2007). Drosophila hyperkinetic mutants have reduced sleep and impaired memory. *The Journal of Neuroscience: The Official Journal of the Society for Neuroscience*, *27*(20), 5384–5393.

Campbell, S. S., & Tobler, I. (1984). Animal sleep: A review of sleep duration across phylogeny. *Neuroscience and Biobehavioral Reviews*, *8*(3), 269–300.

Cavanaugh, D. J., Geratowski, J. D., Wooltorton, J. R. A., Spaethling, J. M., Hector, C. E., Zheng, X., et al. (2014). Identification of a circadian output circuit for rest: Activity rhythmsin Drosophila. *Cell*, *157*(3), 689–701.

Chen, W.-F., Maguire, S., Sowcik, M., Luo, W., Koh, K., & Sehgal, A. (2014). Glia interaction involving GABA transaminase contributes to sleep loss in sleepless mutants. *Molecular Psychiatry*, advance online publication, 18 March 2014, 1–12.

Chung, B. Y., Kilman, V. L., Keath, J. R., Pitman, J. L., & Allada, R. (2009). The GABAA receptor RDL acts in peptidergic PDF neurons to promote sleep in Drosophila. *Current Biology*, *19*(5), 386–390.

Cirelli, C., Bushey, D., Hill, S., Huber, R., Kreber, R., Ganetzky, B., et al. (2005). Reduced sleep in Drosophila Shaker mutants. *Nature*, *434*(7037), 1087–1092.

Cirelli, C., LaVaute, T. M., & Tononi, G. (2005). Sleep and wakefulness modulate gene expression in Drosophila. *Journal of Neurochemistry*, *94*(5), 1411–1419.

Crocker, A., & Sehgal, A. (2008). Octopamine regulates sleep in drosophila through protein kinase A-dependent mechanisms. *The Journal of Neuroscience: The Official Journal of the Society for Neuroscience, 28*(38), 9377–9385.

Crocker, A., & Sehgal, A. (2010). Genetic analysis of sleep. *Genes & Development, 24*(12), 1220–1235.

de Mairan, J. J. D. (1729). Observation botanique. *Histoire de Academie Royale Sciences*, pp. 35–36.

Dietzl, G., Chen, D., Schnorrer, F., Su, K.-C., Barinova, Y., Fellner, M., et al. (2007). A genome-wide transgenic RNAi library for conditional gene inactivation in Drosophila. *Nature, 448*(7150), 151–156.

Dykxhoorn, D. M., & Lieberman, J. (2005). The silent revolution: RNA interference as basic biology, research tool, and therapeutic. *Annual Review of Medicine, 56*(1), 401–423.

Echeverri, C. J., Beachy, P. A., Baum, B., Boutros, M., Buchholz, F., Chanda, S. K., et al. (2006). Minimizing the risk of reporting false positives in large-scale RNAi screens. *Nature Methods, 3*(10), 777–779.

Foltenyi, K., Greenspan, R. J., & Newport, J. W. (2007). Activation of EGFR and ERK by rhomboid signaling regulates the consolidation and maintenance of sleep in Drosophila. *Nature Neuroscience, 10*(9), 1160–1167.

Garofalo, R. S. (2002). Genetic analysis of insulin signaling in Drosophila. *Trends in Endocrinology and Metabolism, 13*(4), 156–162.

Green, E. W., Fedele, G., Giorgini, F., & Kyriacou, C. P. (2014). A Drosophila RNAi collection is subject to dominant phenotypic effects. *Nature Methods, 11*(3), 222–223.

Greenspan, R. J. (2004). *Fly pushing: The theory and practice of Drosophila genetics*. Cold Spring Harbor, NY: Cold Spring Harbor Laboratory Press.

Grima, B., Dognon, A., Lamouroux, A., Chélot, E., & Rouyer, F. (2012). CULLIN-3 controls TIMELESS oscillations in the Drosophila circadian clock. *PLoS Biology, 10*(8), e1001367.

He, C., Yang, Y., Zhang, M., Price, J. L., & Zhao, Z. (2013). Regulation of sleep by neuropeptide Y-like system in *Drosophila melanogaster*. *PLoS One, 8*(9), e74237.

Hendricks, J. C., Finn, S. M., Panckeri, K. A., & Chavkin, J. (2000). Rest in Drosophila is a sleep-like state. *Neuron, 25*(1), 129–138.

Hendricks, J. C., Williams, J. A., Panckeri, K., Kirk, D., Tello, M., Yin, J. C., et al. (2001). A non-circadian role for cAMP signaling and CREB activity in Drosophila rest homeostasis. *Nature Neuroscience, 4*(11), 1108–1115.

Itoh, T. Q., & Matsumoto, A. (2012). Genome-wide RNA interference screening for the clock-related gene of ATP-binding cassette transporters in *Drosophila melanogaster* (diptera: Drosophilidae). *Applied Entomology and Zoology, 47*(2), 79–86.

Jenett, A., Rubin, G. M., Ngo, T.-T. B., Shepherd, D., Murphy, C., Dionne, H., et al. (2012). A GAL4-driver line resource for Drosophila neurobiology. *Cell Reports, 2*(4), 991–1001.

Joiner, W. J., Crocker, A., White, B. H., & Sehgal, A. (2006). Sleep in Drosophila is regulated by adult mushroom bodies. *Nature, 441*(7094), 757–760.

Kaasik, K., Kivimäe, S., Allen, J. J., Chalkley, R. J., Huang, Y., Baer, K., et al. (2013). Glucose sensor O-GlcNAcylation coordinates with phosphorylation to regulate circadian clock. *Cell Metabolism, 17*(2), 291–302.

Kim, E. Y., Jeong, E. H., Park, S., Jeong, H. J., Edery, I., & Cho, J. W. (2012). A role for O-GlcNAcylation in setting circadian clock speed. *Genes & Development, 26*(5), 490–502.

Kim, Y.-I., Ryu, T., Lee, J., Heo, Y.-S., Ahnn, J., Lee, S.-J., et al. (2010). A genetic screen for modifiers of Drosophila caspase Dcp-1 reveals caspase involvement in autophagy and novel caspase-related genes. *BMC Cell Biology, 11*(1), 9.

Kloss, B., Price, J. L., Saez, L., Blau, J., Rothenfluh, A., Wesley, C. S., et al. (1998). The Drosophila clock gene double-time encodes a protein closely related to human casein kinase Iε. *Cell*, *94*(1), 97–107.

Ko, H. W., Kim, E. Y., Chiu, J., Vanselow, J. T., Kramer, A., & Edery, I. (2010). A hierarchical phosphorylation cascade that regulates the timing of PERIOD nuclear entry reveals novel roles for proline-directed kinases and GSK-3/SGG in circadian clocks. *The Journal of Neuroscience: The Official Journal of the Society for Neuroscience*, *30*(38), 12664–12675.

Koh, K., Joiner, W. J., Wu, M. N., Yue, Z., Smith, C. J., & Sehgal, A. (2008). Identification of SLEEPLESS, a sleep-promoting factor. *Science (New York, N.Y.)*, *321*(5887), 372–376.

Konopka, R. J., & Benzer, S. (1971). Clock mutants of *Drosophila melanogaster*. *Proceedings of the National Academy of Sciences of the United States of America*, *68*(9), 2112–2116.

Kulkarni, M. M., Booker, M., Silver, S. J., Friedman, A., Hong, P., Perrimon, N., et al. (2006). Evidence of off-target effects associated with long dsRNAs in *Drosophila melanogaster* cell-based assays. *Nature Methods*, *33*(1010), 833–838.

Kume, K. (2005). Dopamine is a regulator of arousal in the fruit fly. *The Journal of Neuroscience: The Official Journal of the Society for Neuroscience*, *25*(32), 7377–7384.

Kvon, E. Z., Kazmar, T., Stampfel, G., Yáñez-Cuna, J. O., Pagani, M., Schernhuber, K., et al. (2014). Genome-scale functional characterization of Drosophila developmental enhancers in vivo. *Nature*, *512*(7512), 91–95.

Lee, Y. S., Nakahara, K., Pham, J. W., Kim, K., He, Z., Sontheimer, E. J., et al. (2004). Distinct roles for Drosophila dicer-1 and dicer-2 in the siRNA/miRNA silencing pathways. *Cell*, *117*(1), 69–81.

Lewis, E. B., & Bacher, F. (1968). Method of feeding ethyl methane sulfonate (EMS) to Drosophila males. *Drosophila Information Service*, *43*, 193.

Li, M.-D., Ruan, H.-B., Hughes, M. E., Lee, J.-S., Singh, J. P., Jones, S. P., et al. (2013). GlcNAc signaling entrains the circadian clock by inhibiting BMAL1/CLOCK ubiquitination. *Cell Metabolism*, *17*(2), 303–310.

Lin, D. M., & Goodman, C. S. (1994). Ectopic and increased expression of Fasciclin II alters motoneuron growth cone guidance. *Neuron*, *13*(3), 507–523.

Liu, S., Lamaze, A., Liu, Q., Tabuchi, M., Yang, Y., Fowler, M., et al. (2014). WIDE AWAKE mediates the circadian timing of sleep onset. *Neuron*, *82*(1), 151–166.

Mandilaras, K., & Missirlis, F. (2012). Genes for iron metabolism influence circadian rhythms in *Drosophila melanogaster*. *Metallomics*, *4*(9), 928.

Martinek, S., Inonog, S., Manoukian, A. S., & Young, M. W. (2001). A role for the segment polarity gene shaggy/GSK-3 in the Drosophila circadian clock. *Cell*, *105*(6), 769–779.

McGuire, S. E. (2003). Spatiotemporal rescue of memory dysfunction in Drosophila. *Science (New York, N.Y.)*, *302*(5651), 1765–1768. http://dx.doi.org/10.1126/science.1089035.

Miklos, G., & Rubin, G. M. (1996). The role of the genome project in determining gene function: Insights from model organisms. *Cell*, *86*(4), 521–529.

Mohr, S. E., & Perrimon, N. (2011). RNAi screening: New approaches, understandings, and organisms. *Wiley Interdisciplinary Reviews. RNA*, *3*(2), 145–158.

Neely, G. G., Hess, A., Costigan, M., Keene, A. C., Goulas, S., Langeslag, M., et al. (2010). A genome-wide Drosophila screen for heat nociception identifies a2d3as an evolutionarily conserved pain gene. *Cell*, *143*(4), 628–638.

Neumüller, R. A., Richter, C., Fischer, A., Novatchkova, M., Neumüller, K. G., & Knoblich, J. A. (2011). Genome-wide analysis of self-renewalin Drosophila neural stem cells by transgenic RNAi. *Cell Stem Cell*, *8*(5), 580–593.

Newby, L. M., & Jackson, F. R. (1993). A new biological rhythm mutant of *Drosophila melanogaster* that identifies a gene with an essential embryonic function. *Genetics*, *135*(4), 1077–1090.

Newby, L. M., White, L., DiBartolomeis, S. M., Walker, B. J., Dowse, H. B., Ringo, J. M., et al. (1991). Mutational analysis of the Drosophila miniature-dusky (m-dy) locus: Effects on cell size and circadian rhythms. *Genetics*, *128*(3), 571–582.

Ni, J. Q., Liu, L. P., Binari, R., Hardy, R., Shim, H. S., Cavallaro, A., et al. (2009). A Drosophila resource of transgenic RNAi lines for neurogenetics. *Genetics*, *182*(4), 1089–1100.

Ni, J.-Q., Zhou, R., Czech, B., Liu, L.-P., Holderbaum, L., Yang-Zhou, D., et al. (2011). A genome-scale shRNA resource for transgenic RNAi in Drosophila. *Nature Methods*, *8*(5), 405–407.

Nitz, D. A., van Swinderen, B., Tononi, G., & Greenspan, R. J. (2002). Electrophysiological correlates of rest and activity in *Drosophila melanogaster*. *Current Biology*, *12*(22), 1934–1940.

Oh, Y., Jang, D., Sonn, J. Y., & Choe, J. (2013). Histamine-HisCl1 receptor axis regulates wake-promoting signals in *Drosophila melanogaster*. *PLoS One*, *8*(7), e68269.

Osterwalder, T., Yoon, K. S., White, B. H., & Keshishian, H. (2001). A conditional tissue-specific transgene expression system using inducible GAL4. *Proceedings of the National Academy of Sciences of the United States of America*, *98*(22), 12596–12601.

Perrimon, N., Ni, J.-Q., & Perkins, L. (2010). In vivo RNAi: Today and tomorrow. *Cold Spring Harbor Perspectives in Biology*, *2*(8), a003640.

Petsko, G. A., & Ringe, D. (2004). *Protein structure and function*. London: New Science Press.

Pfeiffenberger, C., & Allada, R. (2012). Cul3 and the BTB adaptor insomniac are key regulators of sleep homeostasis and a dopamine arousal pathway in Drosophila. *PLoS Genetics*, *8*(10), e1003003.

Pitman, J. L., McGill, J. J., Keegan, K. P., & Allada, R. (2006). A dynamic role for the mushroom bodies in promoting sleep in Drosophila. *Nature*, *441*(7094), 753–756.

Price, J. L. (2005). Genetic screens for clock mutants in Drosophila. *Methods in Enzymology*, *393*, 35–60.

Price, J. L., Blau, J., Rothenfluh, A., Abodeely, M., Kloss, B., & Young, M. W. (1998). double-time is a novel Drosophila clock gene that regulates PERIOD protein accumulation. *Cell*, *94*(1), 83–95.

Rogulja, D., & Young, M. W. (2012). Control of sleep by cyclin a and its regulator. *Science (New York, N.Y.)*, *335*(6076), 1617–1621.

Rørth, P. (1996). A modular misexpression screen in Drosophila detecting tissue-specific phenotypes. *Proceedings of the National Academy of Sciences of the United States of America*, *93*(22), 12418–12422.

Rutila, J. E., Zeng, H., Le, M., Curtin, K. D., Hall, J. C., & Rosbash, M. (1996). The timSL mutant of the Drosophila rhythm gene timeless manifests allele-specific interactions with period gene mutants. *Neuron*, *17*(5), 921–929.

Ryder, E. (2004). The DrosDel collection: A set of P-element insertions for generating custom chromosomal aberrations in *Drosophila melanogaster*. *Genetics*, *167*(2), 797–813.

Sarin, S., Bertrand, V., Bigelow, H., Boyanov, A., Doitsidou, M., Poole, R. J., et al. (2010). Analysis of multiple ethyl methanesulfonate-mutagenized caenorhabditis elegans strains by whole-genome sequencing. *Genetics*, *185*(2), 417–430.

Scott, J. G., Michel, K., Bartholomay, L. C., Siegfried, B. D., Hunter, W. B., Smagghe, G., et al. (2013). Towards the elements of successful insect RNAi. *Journal of Insect Physiology*, *59*(12), 1212–1221.

Sehgal, A., Price, J. L., Man, B., & Young, M. W. (1994). Loss of circadian behavioral rhythms and per RNA oscillations in the Drosophila mutant timeless. *Science*, *263*(5153), 1603–1606.

Seugnet, L., Suzuki, Y., & Thimgan, M. (2009). Identifying sleep regulatory genes using a Drosophila model of insomnia. *The Journal of Neuroscience*, *29*(22), 7148–7157.

Shabalina, S., & Koonin, E. (2008). Origins and evolution of eukaryotic RNA interference. *Trends in Ecology & Evolution*, *23*(10), 578–587. http://dx.doi.org/10.1016/j.tree.2008.06.005.

Shang, Y., Donelson, N. C., Vecsey, C. G., Guo, F., Rosbash, M., & Griffith, L. C. (2013). Short neuropeptide F is a sleep-promoting inhibitory modulator. *Neuron*, *80*(1), 171–183.

Shaw, P. J., Cirelli, C., Greenspan, R. J., & Tononi, G. (2000). Correlates of sleep and waking in *Drosophila melanogaster*. *Science*, *287*(5459), 1834–1837.

Shi, M., Yue, Z., Kuryatov, A., Lindstrom, J. M., & Sehgal, A. (2014). Identification of redeye, a new sleep-regulating protein whose expression is modulated by sleep amount. *Elife*, *3*, e01473.

Siegfried, E., Chou, T. B., & Perrimon, N. (1992). Wingless signaling acts through zeste-white 3, the Drosophila homolog of glycogen synthase kinase-3, to regulate engrailed and establish cell fate. *Cell*, *71*(7), 1167–1179.

Stanewsky, R. (2003). Genetic analysis of the circadian system in *Drosophila melanogaster* and mammals. *Journal of Neurobiology*, *54*(1), 111–147.

Stavropoulos, N., & Young, M. W. (2011). Insomniac and cullin-3 regulate sleep and wakefulness in Drosophila. *Neuron*, *72*(6), 964–976.

St Johnston, D. (2002). The art and design of genetic screens: *Drosophila melanogaster*. *Nature Reviews. Genetics*, *3*(3), 176–188.

Stoleru, D., Nawathean, P., Fernández, M. P., Menet, J. S., Ceriani, M. F., & Rosbash, M. (2007). The Drosophila circadian network is a seasonal timer. *Cell*, *129*(1), 207–219.

Thibault, S. T., Singer, M. A., Miyazaki, W. Y., Milash, B., Dompe, N. A., Singh, C. M., et al. (2004). A complementary transposon tool kit for *Drosophila melanogaster* using P and piggyBac. *Nature Genetics*, *36*(3), 283–287.

Toh, K. L., Jones, C. R., He, Y., Eide, E. J., & Hinz, W. A. (2001). An hPer2 phosphorylation site mutation in familial advanced sleep phase syndrome. *Science*, *291*(5506), 1040–1043.

van Alphen, B., Yap, M. H. W., Kirszenblat, L., Kottler, B., & van Swinderen, B. (2013). A dynamic deep sleep stage in Drosophila. *The Journal of Neuroscience: The Official Journal of the Society for Neuroscience*, *33*(16), 6917–6927.

van Swinderen, B., & Hall, J. C. (1995). Analysis of conditioned courtship in dusky-andante rhythm mutants of Drosophila. *Learning & Memory*, *2*(2), 49–61.

Venken, K. J. T., Simpson, J. H., & Bellen, H. J. (2011). Genetic manipulation of genes and cells in the nervous system of the fruit fly. *Neuron*, *72*(2), 202–230.

Wager-Smith, K., & Kay, S. A. (2000). Circadian rhythm genetics: From flies to mice to humans. *Nature Genetics*, *26*(1), 23–27.

Williams, J. A., Sathyanarayanan, S., Hendricks, J. C., & Sehgal, A. (2007). Interaction between sleep and the immune response in Drosophila: A role for the NFkappaB relish. *Sleep*, *30*(4), 389–400.

Woodgett, J. R., Plyte, S. E., Pulverer, B. J., Mitchell, J. A., & Hughes, K. (1993). Roles of glycogen synthase kinase-3 in signal transduction. *Biochemical Society Transactions*, *21*(4), 905–907.

Wu, M., Robinson, J. E., & Joiner, W. J. (2014). SLEEPLESS is a bifunctional regulator of excitability and cholinergic synaptic transmission. *Current Biology*, *24*(6), 621–629.

Xie, L., Kang, H., Xu, Q., Chen, M. J., Liao, Y., Thiyagarajan, M., et al. (2013). Sleep drives metabolite clearance from the adult brain. *Science (New York, N.Y.)*, *342*(6156), 373–377.

Xu, Y., Padiath, Q. S., Shapiro, R. E., Jones, C. R., & Wu, S. C. (2005). Functional consequences of a CKIδ mutation causing familial advanced sleep phase syndrome. *Nature*, *434*(7033), 640–644.

Yamamoto-Hino, M., & Goto, S. (2013). In vivo RNAi-based screens: Studies in model organisms. *Genes*, *4*(4), 646–665.

Yi, W., Zhang, Y., Tian, Y., Guo, J., Li, Y., & Guo, A. (2013). A subset of cholinergic mushroom body neurons requires go signaling to regulate sleep in Drosophila. *Sleep, 36*(12), 1809–1821.

Yoshii, T., Rieger, D., & Helfrich-Frster, C. (2012). *Two clocks in the brain: An update of the morning and evening oscillator model in Drosophila The neurobiology of circadian timing.* (1st ed., Vol. 199, pp. 59–82).

Yuan, Q., Joiner, W. J., & Sehgal, A. (2006). A sleep-promoting role for the Drosophila serotonin receptor 1A. *Current Biology, 16*(11), 1051–1062.

Zehring, W. A., Wheeler, D. A., Reddy, P., Konopka, R. J., Kyriacou, C. P., Rosbash, M., et al. (1984). P-element transformation with period locus DNA restores rhythmicity to mutant, arrhythmic *Drosophila melanogaster. Cell, 39*(2 Pt. 1), 369–376.

Zhang, E. E., Liu, A. C., Hirota, T., Miraglia, L. J., Welch, G., Pongsawakul, P. Y., et al. (2009). A genome-wide RNAi screen for modifiers of the circadian clock in human cells. *Cell, 139*(1), 199–210.

CHAPTER TWO

Dissecting the Mechanisms of the Clock in Neurospora

Jennifer Hurley*, Jennifer J. Loros*,†, Jay C. Dunlap*,[1]

*Department of Genetics, Geisel School of Medicine at Dartmouth, Hanover, New Hampshire, USA

†Department of Biochemistry, Geisel School of Medicine at Dartmouth, Hanover, New Hampshire, USA

[1]Corresponding author: e-mail address: Jay.C.Dunlap@Dartmouth.edu

Contents

Abstract

The circadian clock exists to synchronize inner physiology with the external world, allowing life to anticipate and adapt to the continual changes that occur in an organism's environment. The clock architecture is highly conserved, present in almost all major branches of life. Within eukaryotes, the filamentous fungus *Neurospora crassa* has consistently been used as an excellent model organism to uncover the basic circadian physiology and molecular biology. The Neurospora model has elucidated our fundamental understanding of the clock as nested positive and negative feedback loop, regulated by transcriptional and posttranscriptional processes. This review will examine the basics of circadian rhythms in the model filamentous fungus *N. crassa* as well as highlight the output of the clock in Neurospora and the reasons that *N. crassa* has continued to be a strong model for the study of circadian rhythms. It will also synopsize classical and emerging methods in the study of the circadian clock.

1. INTRODUCTION

Neurospora first emerged as a model organism for understanding circadian clocks and circadian systems in the late 1950s when Pittendrigh

Methods in Enzymology, Volume 551
ISSN 0076-6879
http://dx.doi.org/10.1016/bs.mie.2014.10.009

and coworkers described a rhythm in the events associated with overt asexual development in cultures grown on long hollow glass tubes (Pittendrigh, Bruce, Rosensweig, & Rubin, 1959). At the time, it was a natural choice; Neurospora was then and remains a premier genetic model, an organism whose fame stems from Beadle and Tatum's work on the one enzyme/one gene hypothesis and that is still the consensus model for filamentous fungi although first *Escherichia coli* and later yeast budding yeast eclipsed it for many studies (Davis & Perkins, 2002). Despite the existence of these other models, Neurospora has remained a useful model for the study of many problems, and especially for circadian rhythms because it is a wonderfully tractable genetic system, it is easy to use for biochemical follow ups, and enjoys a relatively large and heavily invested research community. As a result, nearly all of what we know about the molecular details of circadian rhythms in fungi stems from work on Neurospora. Beyond this, and because cells from fungi and animals share many aspects of regulation, much of what we know about circadian rhythms in animals can be traced to work on this system. Specifically, many of the proteins and most of the regulatory architecture of the core circadian oscillator in Neurospora and animals are quite similar and insights from cells in one system are generally applicable to the other. This review details the circadian system in Neurospora and the methods that are employed to study it.

1.1. Methods of analysis of circadian rhythms in *Neurospora crassa*

Neurospora has remained an excellent circadian model organism for a variety of reasons, not the least of which includes the ease and standardization of the methods used to examine the clock. Since the initial work on rhythms, Neurospora has become a highly tractable genetic organism, with a full genome sequence, transformation protocols yielding >98% efficiency in gene targeting, a variety of regulatable promoters and selectable markers, and a nearly complete knockout library (Collopy et al., 2010; Colot et al., 2006; Galagan et al., 2003; Ninomiya, Suzuki, Ishii, & Inoue, 2004). Neurospora is recognized by the NIH as an established model system (http://www.nih.gov/science/models/neurospora/). Genetic stocks including the full collection of gene knockouts are maintained by the Fungal Genetics Stock Center that distributes over 40,000 strains annually for nominal fees.

Neurospora has both sexual and asexual life cycles and rhythms are apparent in the regulation of both. Most studies employ the asexual cycle in which a developmental switch leads to the production of conidia in the subjective night. Although the developmental cycle can be seen on petri dish cultures inoculated at one point, this rhythmic development of conidia is generally assayed through use of a race tube, a long glass tube, bent at both ends, holding agar medium onto which mycelia or conidia are inoculated at one end (Fig. 1A). The individual clocks are synchronized by germinating the cultures in constant light for a day and then transferring the fungi to constant darkness. As the fungus grows down the tubes at about 2 cm per day, the growth front is marked each day. Because distance grown approximates time passed, the period length can be interpreted from the growth pattern. More specific instruction on growing a Neurospora strain on race tubes can be found at http://www.fgsc.net/Neurospora/neurospora.html. Once the race tube has been run, the period can be determined by hand or using the Chrono program (Roenneberg & Taylor, 2000).

Manifestation of the observed rhythm in asexual development and conidiation is not strong in wild-type cultures unless there is air exchange over the culture or they are chemically treated to result in mild oxidative stress (Belden, Larrondo, et al., 2007). However, identification of the *band* (*bd*) strain by Sargent made expression of the rhythm robust without the need for air flow (Sargent, Briggs, & Woodward, 1966), and now nearly all strains used for circadian studies incorporate this allele in the background. The causative mutation in the *band* strain was identified as a mutation in *ras-1* that increased the levels of reactive oxygen species (ROS) in Neurospora. This in turn, leads to the increased expression of a particular conidial regulation protein encoded by the gene *fluffy*, which accounts for the *bd* phenotype (Belden, Larrondo, et al., 2007). Because of the connection between asexual development, RAS signaling, and ROS levels in sporulation, genes, and pathways identified as under clock control in the *bd* strain may be RAS-responsive instead of clock-responsive and may not be clock regulated in a wild-type strain (Belden, Larrondo, et al., 2007).

Rhythms have also been followed in cultures grown in liquid medium, leading to the understanding of the clock on a molecular level in liquid cultures as opposed to the overt rhythms followed on solid media in the *bd* strain (Aronson, Johnson, Loros, & Dunlap, 1994; Loros, Denome, & Dunlap, 1989). Mycelial discs cut from syncytial mats retain their rhythmicity and phase and can be transferred to liquid media. These cultures are grown in the dark with gentle shaking and harvested at times throughout their circadian cycle to extract RNA or proteins enabling that assessment of overall marker levels. Care must be taken to normalize age and media of the culture as small changes are likely create large differences in macromolecular output (Loros et al., 1989).

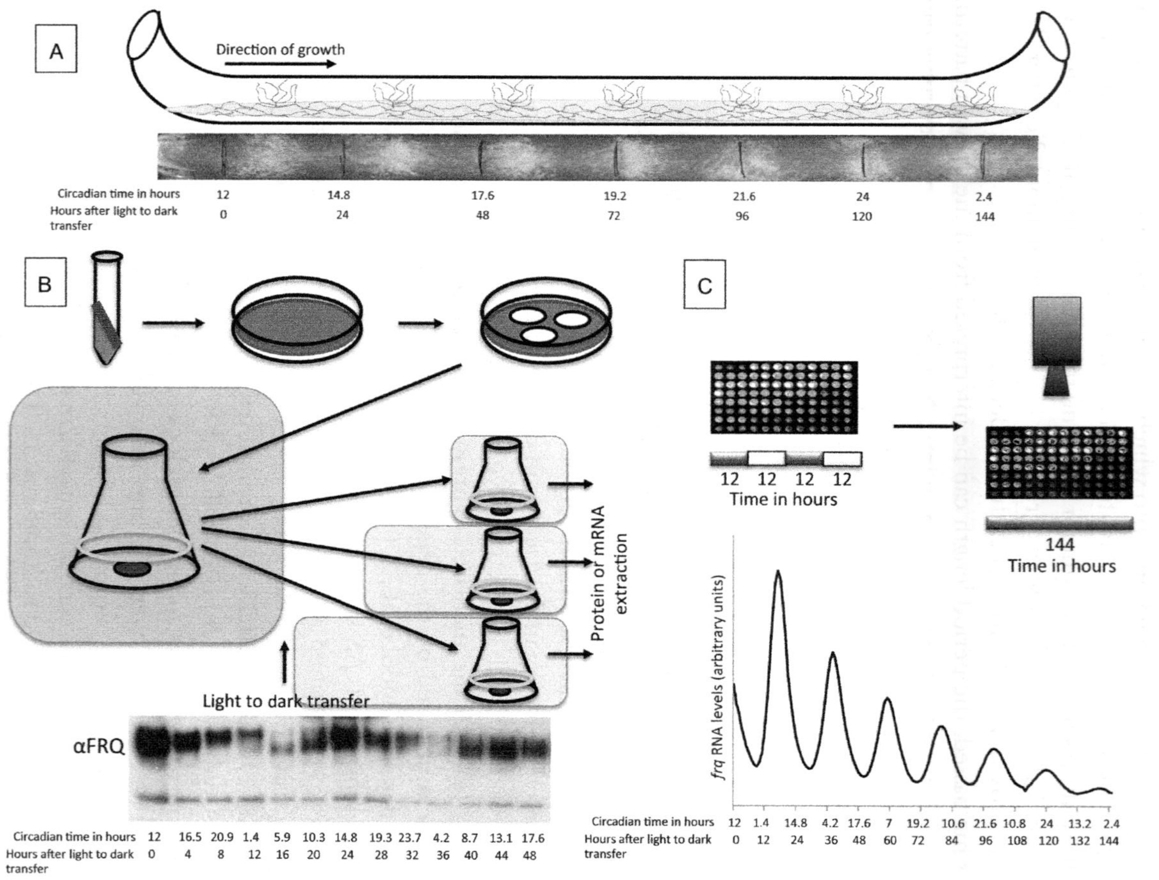

Figure 1 See legend on next page.

However, liquid media can quickly become depleted of nutrients and does not allow for long-term observation of the core circadian rhythm and due to this, liquid media may have a vastly different set of clock-controlled output genes (Perlman, Nakashima, & Feldman, 1981). To obviate this possible disparity, rhythms can be now monitored more directly using reporter expression of the luciferase gene from the firefly beetle *Photinus pyralis*, codon optimized for Neurospora, and driven by clock-regulated promoters such as *frequency* (*frq*) or *ccg-2* (Gooch et al., 2008; Morgan, Greene, & Bell-Pedersen, 2003). Tracking the clock with luciferase allows for direct tracking of the endogenous, core rhythms of the Neurospora clock on the transcriptional level, and this method has been used to understand temperature compensation as well as translational activity in the core Neurospora clock (Gooch et al., 2008; Larrondo, Loros, & Dunlap, 2012).

1.2. Circadian rhythms in other fungi

It should be noted that Neurospora is not the only fungus with a rhythm; there is documentation of a circadian circuit in other fungi. Similar rhythms in conidiospore formation have been reported in the Zygomycete Pilobolus (Bruce, Weight, & Pittendrigh, 1960), and less definite growth and developmental rhythms exist in a variety of Ascomycetes (reviewed in Dunlap & Loros, 2006; Greene, Keller, Haas, & Bell-Pedersen, 2003). Not surprisingly, the identification of conserved core clock components has been well investigated in fungi as well. A review of 42 sequenced fungal genomes noted that light sensing mechanisms, which commonly function as the positive arm of the clock, are widely conserved in the fungi as is the RNA helicase protein frequency interacting RNA helicase (FRH)

Figure 1 Methods of circadian analysis in Neurospora. (A) The basic outline of the use of a race tube in analyzing circadian rhythms in Neurospora along with the image of an actual race tube beneath it. Daily growth fronts are noted with vertical lines (sidereal time and in circadian time are noted below; J. Hurley, unpublished data). (B) The basics of analysis of the molecular rhythms for Neurospora liquid culture. The outline of the protocol to extract either mRNA or protein from Neurospora over circadian time. Western blot of FRQ protein tracked over 48 sidereal hours (time points taken every four sidereal hours) and labeled in circadian hours, highlighting the changes of phosphorylation state of FRQ protein over time (J. Emerson, unpublished data). (C) The outline of real-time analysis of molecular circadian rhythms. Ninety-six individual tubes of Neurospora are subjected to a 12:12 light/dark cycle and then allowed to free run in the dark. A luciferase trace of *frq* mRNA expression tracked over 144 sidereal hours using a CCD camera and the resulting traces are labeled in circadian hours, highlighting the changes of expression levels of *frq* mRNA over time (J. Emerson, unpublished data). (See the color plate.)

(see below). The clock-exclusive protein FRQ (the negative feedback element) is less conserved, however, and components for complete circadian feedback loops can be seen in the family Sordariacea (where Neurospora lies), suggesting that many plant and animal pathogens have a functional clock (e.g., Canessa, Schumacher, Hevia, Tudzynski, & Larrondo, 2013). In addition, recent data have revealed a FRQ ortholog in *Pyronema confluens* which shares a common ancestor with Neurospora on the order of 500 million years ago, thus extending the existence of FRQ, and probably of clocks, well before the divergence of Aspergillus and Neurospora and much farther back in time (Traeger et al., 2013). The unstructured nature of the negative arm protein FRQ, discussed later in this chapter, suggested that conservation of sequence might not be necessary to maintain the lack of structure needed to play the role of the negative arm protein (Dunlap & Loros, 2006; Hurley, Larrondo, Loros, & Dunlap, 2013; Salichos & Rokas, 2010); it is remarkable that sequence orthologs exist in such anciently diverged species, and their existence suggests conservation of interactions with FRQ.

2. MOLECULAR MECHANISM OF THE NEUROSPORA CIRCADIAN OSCILLATOR

At the core of circadian rhythms in Neurospora (and in all other known molecular oscillators) is a transcriptional/translational feedback loop run by a core clock complex that is strictly regulated by a series of ancillary interacting proteins. The core complex consists of two sets of protein pairs, the negative arm (in Neurospora this is comprised of the FRQ/FRH complex or FFC as well as CK1), and the positive arm (in Neurospora comprised of the White Collar Complex or WCC) which drives the expression of *frq*. The negative arm regulates its own expression on a time delay, the circadian period (Dunlap, 1999).

As a brief overview, the cycle begins late in the subjective night (Fig. 2), when the WCC binds to the *frq* promoter which leads to the induction of *frq* mRNA, a process that reaches its maximum around early subjective morning. FRQ protein takes around 4 h to translate after the start of *frq* expression, after which it binds rapidly to FRH and itself, enters the nucleus, and begins to form a complex with CK1 (Baker, Kettenbach, Loros, Gerber, & Dunlap, 2009; Dunlap & Loros, 2004; Hurley et al., 2013; Merrow, Garceau, & Dunlap, 1997). In the early circadian morning, new FRQ is also promptly phosphorylated in the PEST-1 and FFD domains (sites of FRQ/FRQ and FRQ/FRH interactions; see below), with further

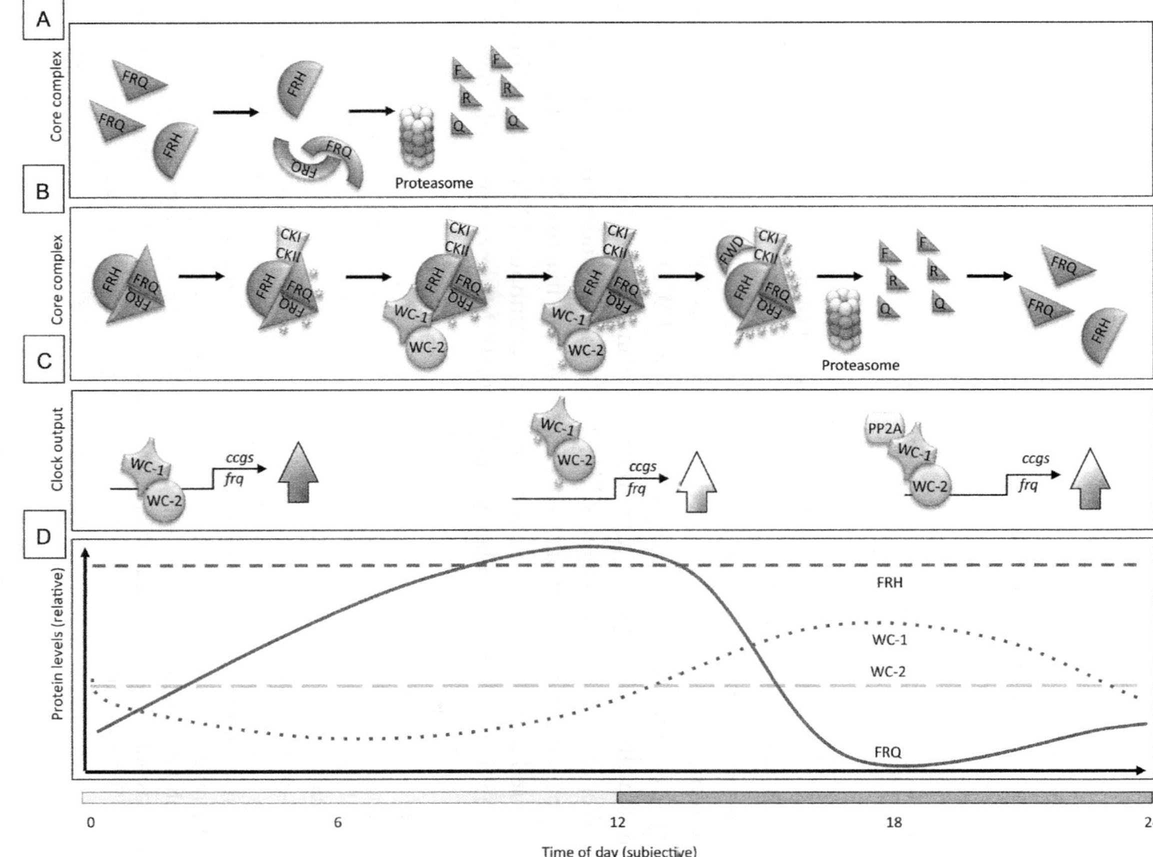

Figure 2 See legend on next page.

phosphorylation events occurring via interaction with several kinases in the C-terminal region shortly thereafter. These C-terminal phosphorylations have been shown to stabilize the protein (reviewed in Baker, Loros, & Dunlap, 2012; Heintzen & Liu, 2007).

Upon entry into the nucleus, the FFC autoregulates its own transcription by inhibiting the activity of the WCC while simultaneously increasing the levels of WC-1 (Dunlap & Loros, 2004). It is thought that direct interaction between the FFC and the WCC leads to the phosphorylation of the WCC, inactivating the WCC as well as clearing the WCC from the *frq* promoter (reviewed in Brunner & Kaldi, 2008; Liu & Bell-Pedersen, 2006). The WCC exits the nucleus at this point in the cycle, perhaps a reflection of the phosphorylation status of the WCC, further decreasing activation of *frq* expression (Hong, Ruoff, Loros, & Dunlap, 2008). Due to the lack of WCC activation, by late afternoon *frq* expression declines and so does FRQ synthesis (Merrow et al., 1997). FRQ in the FFC is increasingly phosphorylated throughout the circadian day at the PEST domain and the N-terminal domain leading to the recognition of FRQ by an SCF-ubiquitin ligase complex containing the F-box protein, FWD-1, FRQ ubiquitination, and finally targeting of FRQ to the proteasome for degradation (reviewed in Baker et al., 2012; Heintzen & Liu, 2007). In the standard model for fungal/animal clocks, the mass of WC-1 that was held inactive by the FFC is now released and the cycle restarts, with the now unbound WCC again binding to the *frq* promoter (Dunlap & Loros, 2004). More recent work, however,

Figure 2—Cont'd Neurospora circadian cycle at the molecular level. (A) If FRQ is not able to bind to its stabilizer, FRH, it is degraded by default due to the inherently disordered nature of FRQ and is unable to complete its function in the circadian clock. (B) During the late subjective night of the circadian cycle, the WCC induces expression of *frq* mRNA, leading to a rapid increase in FRQ translation. FRQ forms a homodimer and binds to its stabilizer FRH, allowing for the IDP FRQ to avoid degradation by default. As the circadian day progresses, FRQ is phosphorylated via interaction with several kinases. FRQ inhibits the activity of the WCC by promoting the phosphorylation of the WCC, turning off *frq* transcription. FRQ levels decrease as no new FRQ is made while old FRQ is increasingly phosphorylated, which leads to ubiquitination facilitated by FWD-1, leading to FRQ degradation. (C) Factors that drive the output of the circadian clock. Low FRQ levels cause WCC activity to increase which subsequently leads to the expression of *frq* mRNA as well as mRNAs from other *ccgs*. FRQ binds to the WCC, promoting phosphorylation of the WCC and causing the WCC to become inactive. Decreasing FRQ levels allow phosphatases to bind the WCC, dephosphorylating the WCC, and increasing WCC activation. (D) Protein levels of the core clock components. While FRH and WC-2 remain constant, FRQ and WC-1 oscillate in opposite phases to one another. Stars represent phosphorylation and lightning bolts represent ubiquitination. (See the color plate.)

has identified robust rhythms in FRQ-LUC expression in Δ*fwd*-1 strains, suggesting that complete turnover of FRQ is not necessary for reinitiation of synthesis but instead is correlated with the initiation of new synthesis. The real end of the cycle in this revised model occurs when FRQ is sufficiently posttranslationally modified that it becomes invisible to the circadian machinery (Larrondo et al., under revision). In any case, the tight regulation that leads to delays between *frq* expression and FRQ synthesis (3–6 h) and FRQ phosphorylation and eventual degradation (14–18 h) leads to the approximately 22.5 h rhythm in Neurospora and sets the specific circadian rhythm (Merrow et al., 1997).

3. CORE CLOCK COMPONENTS

3.1. The FRQ/FRH complex

The discovery of several mutants, each of which directly affects the period of banding in Neurospora, vaulted the organism to a key model for circadian rhythms at the molecular level. These mutants were all mapped to the *frq* locus and displayed long, short, or arrhythmic periods, including some interesting alleles which also altered or disrupted temperature compensation (Gardner & Feldman, 1980; Loros & Feldman, 1986). *frq* itself was cloned leading to the current in depth understanding of the molecular clock of Neurospora and a greater understanding of clocks in general (McClung, Fox, & Dunlap, 1989). The periodic change in FRQ levels and phosphorylation match the conidiation rhythm seen in the *bd* mutant (discussed Section 1.1). The demonstration that FRQ is the driver of the Neurospora period came with the observation that altering or inhibiting the FRQ rhythm had a direct and equivalent effect on the clock (Aronson et al., 1994; Belden, Larrondo, et al., 2007; Garceau, Liu, Loros, & Dunlap, 1997).

As noted above, FRQ constitutes one of two proteins that make up the negative arm of the clock. Full-length FRQ contains 989 amino acids and dimerizes via a coil–coil region near the N′-terminus (Aronson et al., 1994; Cheng, Yang, Heintzen, & Liu, 2001). The message, as well as the protein of *frq*, is rhythmically expressed in a 22.5 h cycle under constant conditions with a phase difference of approximately 4 h (Aronson et al., 1994; Garceau et al., 1997). FRQ is highly regulated at the transcriptional, post-transcriptional, translational, and posttranslational levels (see below; Baker et al., 2012). FRQ is found in both the nuclear and cytoplasmic fractions but most of the activity attributed to FRQ is nuclear, where it binds to and blocks the transcriptional activity of the WCC (Liu, He, & Cheng,

2003). FRQ is believed to increase WC-1 levels and this activity is probably the result of inhibiting the activity of WC-1, a protein believed to be unstable when it is active (Shi, Collett, Loros, & Dunlap, 2010, reviewed in Baker et al., 2012). FRQ has also been shown to increase the abundance of *wc-2* through an unknown mechanism (Liu et al., 2003).

In the regulation of FRQ transcription, it is the rhythmic binding of the relevant transacting factors that maintains a functional circadian clock. This regulation occurs at the *frq* promoter through the binding of the WCC proteins to two distinct *cis*-acting sequences termed the Clock box (C-box) and the proximal light-regulated element (PLRE) (Froehlich, Loros, & Dunlap, 2003). The role of the C-box is to regulate the rhythmic expression of *frq* and overall clock function in continual darkness, whereas the PLRE is essential to establish the proper phase when entrained by light (discussed below). While the combined function of these elements is responsible for high levels of light-induced *frq* expression via WCC influence on the *frq* promoter, each element acts differentially as chromatin is remodeled during the transcriptional activation and deactivation of *frq* (Belden, Loros, & Dunlap, 2007; Wang et al., 2014).

In order to properly regulate *frq* expression, the protein encoded by the gene *clockswitch* (*csw-1*) is required and also acts to negatively regulate WCC activity at *frq* by altering chromatin structure, creating a more compact chromatin structure at the C-box (Belden, Loros, et al., 2007). Chromodomain helicase DNA-binding (CHD-1) can also contribute to changes in chromatin structure at *frq* and is needed for normal *frq* expression. DNA methylation at *frq*, which is promoted by the loss of CHD-1, is transient, reversible, and catalyzed by the DNA methyltransferase DIM-2, which limits the onset of circadian regulated transcription via regulation of methylation at the *frq* promoter (Belden, Lewis, Selker, Loros, & Dunlap, 2011; Belden, Loros, et al., 2007). Recent results have added to the understanding of how *frq* expression is regulated in the light versus in the dark by the clock (Wang et al., 2014). When WCC binds to the C-box it recruits the SWI/SNF complex, a well-known chromatin modifying complex that is also involved in DNA bending. SWI/SNF in conjunction with other components removes a nucleosome from the C-box and also bends the DNA so that this region is brought into proximity with the transcription start site to initiate *frq* expression. Interestingly, and consistent with this model, loss of SWI/SNF abrogates circadian rhythms but has little to no effect on light-induced *frq* expression that is driven by the TSS PLRE.

There are a great many factors that affect *frq* mRNA regulation beyond transcriptional regulation. *frq* encompasses two translation initiation sites that result in the production of two distinct FRQ polypeptides: FRQ^{1-989} (L-FRQ) and $FRQ^{100-989}$ (S-FRQ; Garceau et al., 1997). Both forms of FRQ independently maintain rhythmicity at 25 °C, but the amplitude and robustness of the FRQ rhythms are affected across a range of physiological temperatures. At higher temperatures, a higher ratio of L-FRQ is produced while relatively even amounts of long- and short-FRQ proteins are maintained at lower temperatures (Liu, Garceau, Loros, & Dunlap, 1997). The phosphorylation of sites on the 100 amino acids on L-FRQ that are not present in S-FRQ decreases period length even in the presence of S-FRQ (discussed further below). It is believed that the ratio of FRQ polypeptides is an additional level of fine tuning of the clock which allows the period to respond to environmental cues while at the same time allowing it to remain a robust timekeeping mechanism in their absence (Baker et al., 2009; Diernfellner et al., 2007; Liu et al., 1997).

The determining factor in the selective transcription of either S-FRQ or L-FRQ is temperature, which triggers an alternative splicing event of a small intron encompassing the AUG of L-FRQ. At higher temperatures, this intron is retained resulting in the preferential use of the AUG from L-FRQ; lower temperatures trigger the removal of the intron, making the AUG from S-FRQ an equally likely start codon. Strains unable to splice this intron fail to produce S-FRQ (Colot, Loros, & Dunlap, 2005; Diernfellner, Schafmeier, Merrow, & Brunner, 2005). Alternative splicing events farther upstream in the 5′ UTR remove five upstream AUGs with four uORFs from all major *frq* transcripts to further regulate the expression of the *frq* transcript. Two AUGs remain and these uORFs may be differentially regulating S-FRQ and L-FRQ at the translational level by targeting transcripts for nonsense-mediated decay, either as a way to remove improperly spliced transcripts or as a mechanism for quantitative control of gene expression (Colot et al., 2005; Diernfellner et al., 2005).

A final level of regulation on the *frq* transcript is the *qrf* antisense transcript, which comprises the entire length of the FRQ open reading frame. Elimination of the *qrf* transcript leads to a slight period increases as well as rhythm loss at low physiological temperatures and earlier phase setting upon light to dark transfer. *qrf* antisense RNA may be an additional level of regulation on *frq* posttranscriptionally in order to further insulate the clock from environmental stresses (Kramer, Loros, Dunlap, & Crosthwaite, 2003).

While *frq* mRNA and protein levels change with a circadian periodicity, FRQ is also rhythmically phosphorylated, which, among other things, has a direct influence on FRQ turnover kinetics (Garceau et al., 1997;

Liu, Loros, & Dunlap, 2000). FRQ is phosphorylated rapidly upon translation and this phosphorylation continues in a highly regulated manner throughout the circadian day (Baker et al., 2009). When sites known to be phosphorylated are mutated to eliminate phosphorylation, FRQ stability is increased, which in turn leads to increased period lengths (Liu et al., 2000; Ruoff, Loros, & Dunlap, 2005). FRQ is also phosphorylated in constant light, though in a less specific and regulated manner (Baker et al., 2009; Tang et al., 2009).

The phosphorylations occur in clusters at specific times over the circadian day, with no specific phosphorylation event acting as the key determinant for any action in FRQ. At the start of the circadian day, FRQ is completely unphosphorylated. As time passes, FRQ is rapidly phosphorylated in the central regions, particularly between the PEST-1 and the FFD domain. The function of these central modifications has yet to be determined as mutations at these sites did not alter circadian rhythms. Next, the C-terminal regions are phosphorylated which increases FRQ protein stability. Mutations in this region result in a short-period rhythm. The PEST-1 domain shows a dramatic increase in phosphorylation midway through the circadian day. The phosphorylation of these residues is needed to promote turnover of FRQ as mutations of sites in this region showed an increase in period and more stable FRQ. Finally, phosphorylation of residues specific to L-FRQ, occurs late in the cycle. Mutations in the L-FRQ only region result in a longer period, suggesting a role in promoting turnover (Baker et al., 2009).

In total, FRQ has around 100 distinct modifications. To complete this extensive phosphorylation, there is a complex network of kinases and phosphatases. Many kinases are found to interact with the FFC, including casein kinases 1 and 2 (CK1a and CK2), a Neurospora homolog of checkpoint kinase-2 (PRD-4), as well as CAMK-1, and basophilic protein kinase A (Klengel et al., 2005); CK1a, CK2, and PRD-4 appear to directly interact with FRQ (Baker et al., 2012; Diernfellner & Schafmeier, 2011). The interaction of CK1a with FRQ is via two FRQ/CK1a interacting domains (FCDs) on FRQ. This interaction not only catalyzes the phosphorylation of FRQ (as many as 41 times) which is believed to lead to FRQ degradation, but may play a role in the clock-dependent phosphorylation of the WCC as well (He, Cha, Lee, Yang, & Liu, 2006; Querfurth et al., 2011). This lends credence to the hypothesis that FRQ acts as a scaffold for major components of the clock. CK2 also interacts with FRQ and these phosphorylations are involved in maintaining the temperature compensation function of the clock (Mehra et al., 2009). In addition to kinases, several phosphatases play a role in the clock, including protein phosphatase-1 (PP1), PP2a, and PP4. Phosphatases regulate FRQ stability, influencing *frq* transcription, dephosphorylate the WCC, and affect WCC subcellular localization (Baker et al., 2012).

In addition to effects on stability, it has been suggested that FRQ structure is directly affected by its phosphorylation (Querfurth et al., 2011). In the hypophosphorylated state, FRQ is in a closed conformation, which opens upon increasing phosphorylation presumably due to charge–charge repulsion, revealing a degradation signal in the middle portion of FRQ. New FRQ adopts preferentially the closed conformation as the positively charged N-terminal domain interacting with the negatively charged remainder portion of the protein due to the lack of phosphorylation. As the N-terminal domain of FRQ is progressively phosphorylated, it lowers the pI of the domain, increasing negative surface charge of the N-terminal domain and weakening the interaction with the negatively charged middle and C-terminal domains (Querfurth et al., 2011). However, this model fails to explain that N-terminal phosphorylations were previously shown to be among the last modifications during the circadian cycle rather than being among the first (Baker et al., 2009). Recently, FRQ was demonstrated to be an intrinsically disordered protein (IDP). Flexibility in FRQ structure allows for flexibility of binding, high levels of posttranslational modifications, ubiquitination, and a variety of protein–protein interactions to occur, all things that are necessary for proper FRQ function in the clock (Hurley et al., 2013).

In addition to phosphorylation, there are other posttranslational modifications that affect the degradation of FRQ, including ubiquitination (He & Liu, 2005). The F-box/WD40 repeat-containing protein FWD-1 has been shown to directly interact with the phosphorylated form of FRQ and is essential for FRQ's degradation. Phosphorylated FRQ appears to be a substrate for an FWD-1-containing SCF-type ubiquitin ligase complex that this SCF complex can recognize different phosphorylated motifs within FRQ. The more phosphorylated FRQ is the more potential FWD-1-binding sites are present on FRQ so this increases its affinity toward FWD-1 (He, Cheng, Yang, Yu, & Liu, 2003). This data lends credibility to the idea that progressive phosphorylation of FRQ may be a dynamic process that fine-tunes the stability of FRQ through its role in the ubiquitination of FRQ. Beyond the FWD-1 role in ubiquitination, it is believed that there may be other FRQ mechanisms of degradation including degradation by default (discussed later in this chapter; He et al., 2003; Hurley et al., 2013).

The second component of the FFC is FRH, a homolog of Mtr4p, which is a well-studied cofactor of the *Saccharomyces cerevisiae* exosome (Cheng, He, Wang, & Liu, 2005). Mtr4p is a member of the TRAMP complex and has been played a role in the exosome in yeast (LaCava et al., 2005). All FRQ is bound to FRH and when FRH levels are depleted via siRNA knockdown or use of the regulatable *qa-2* promoter (FRH is an essential gene in Neurospora), the clock loses rhythmicity completely and FRQ protein level decreases dramatically while mRNA increases (Cheng et al., 2005; Shi et al., 2010).

Due to its similarity to Mtr4p, initial studies were aimed at showing that FRH knockdown has an indirect effect on the clock because it regulates the levels of *frq* posttranscriptionally, as when FRH is knocked down, *frq* mRNA is stabilized (Guo, Cheng, Yuan, & Liu, 2009). A more direct role for FRH is in the complex interaction between FRQ and the WCC; FRH is essential to the interaction between the FFC and the WCC. FRH is also able to interact with the WCC in the absence of FRQ (Cheng et al., 2005; Guo, Cheng, & Liu, 2010; Shi et al., 2010). FRH has been implicated in the proper methylation of *frq* (Belden et al., 2011), as well as being an essential interactor of VVD (described later) in suppression of FRQ expression via interaction with the WCC (Hunt, Thompson, Elvin, & Heintzen, 2010). The association between FRQ and FRH is essential for the proper phosphorylation and stability of FRQ. In addition to this, it appears that FRH plays a role in the proper localization of FRQ protein (Cha, Yuan, & Liu, 2011; Guo et al., 2010) though it is important to note that FRH is cytoplasmic and the nuclear-cytoplasmic shuttling has been suggested to be dependent on the phosphorylation state of FRQ (Diernfellner, Querfurth, Salazar, Hofer, & Brunner, 2009). A point mutant of FRH was identified through a mutagenesis screen for negative feedback loop mutants. This mutation is outside the highly conserved helicase region of FRH and eliminates the interaction of FRH with the WCC but not with FRQ (Shi et al., 2010).

The mutation hints that the role of FRH that is specific to the clock may be different from its role in the TRAMP/exosome complex function that FRH plays for overall cell fitness. Recent work has been shown that the loss of the helicase function of FRH does not affect the running of the circadian cycle. IDPs, of which FRQ is one, follow two paths in stability: either they are degraded by default or they bind to a partner molecule to stabilize their structure. FRH has been shown to stabilize FRQ and recent work suggests that the role of FRH may be to act as a partner protein, or Nanny, to stabilize the IDP FRQ and allow it to perform its multitude of functions (Hurley et al., 2013). A competing theory posits that it is the ATPase function of FRH that regulates the function of CK1a, allowing for the proper phosphorylation of FRQ (Lauinger, Diernfellner, Falk, & Brunner, 2014); however, the retention of a clock in strains bearing FRH-lacking ATPase and helicase activity is hard to explain through such a model.

3.2. The White Collar Complex

WC-1 protein is GATA-like Zn-finger transcription factor that contains three PAS (Per-Ant-Sim) domains. The N-terminal-most PAS domain is of a special subclass, called the LOV domain. WC-1 interacts through its C-terminal-most PAS domain with the PAS domain of WC-2 to form a heterodimer (Cheng, Yang, Gardner, & Liu, 2002; Linden & Macino, 1997). WC-2 also contains a Zn-finger domain, but lacks a LOV domain for direct light sensing. The WCC binds to the C-box and PLRE where it functions as a transcriptional activator of FRQ. The WCC actually exists in two forms (Cheng, Yang, Wang, He, & Liu, 2003; Froehlich, Liu, Loros, & Dunlap, 2002). The first is the WC-1/2 heterodimer (small complex) that binds to promoter elements most strongly in the dark. *frq* expression driven by the small WCC is independent of the LOV domain of WC-1 and is thus independent of photosensing role of the WCC. Upon light exposure, the small complex is replaced on the DNA by a larger WCC, which consists of the WC-1/2 heterodimer with the addition of several more WC-1 proteins that interact via their LOV domains (Cheng et al., 2002; Collett, Garceau, Dunlap, & Loros, 2002; Linden, Ballario, & Macino, 1997).

When either of the WCC genes is knocked out, clock function is eliminated (Crosthwaite, Dunlap, & Loros, 1997). This is because the WCC complex binds the *frq* promoter and is responsible for the expression of *frq* mRNA (Froehlich et al., 2002). The WCC is then inhibited through direct interaction with the FFC. The interaction between the two main complexes of the circadian clock causes both WC-1 and WC-2 to be phosphorylated in a circadian manner and it is this phosphorylation that regulates the activity of the complex. The phosphorylation of WC-2 regulates the binding of the WCC to DNA. While currently only one site has been identified on WC-2 and it has been suggested that there are many more that could affect both the period and the stability of the protein (Sancar, Sancar, Brunner, & Schafmeier, 2009). WC-2 is necessary for interaction between the WCC and FRQ; although transcription of *wc-2* is weakly rhythmic (Hurley et al., in press) and it is both positively regulated by FRQ and negatively regulated by WC-1 (Liu et al., 2003), there is no rhythm to WC-2 content. As a complex, WCC stability is regulated by the CCR4-NOT complex (Huang, He, Guo, Cha, & Liu, 2013). The phosphorylation of WC-1 is also circadian; the phosphorylation at sites near to the Zn-finger DNA-binding domain is believed to regulate the ability of WC-1 to activate transcription (He et al., 2005). In addition to its activation role, WC-1 is also needed for the interaction between the WCC and FRQ. WC-1 level cycles, though this rhythm is not necessary for the clock and WC-1 is stabilized by WC-2.

3.3. The input and output of the clock

A circadian clock is beneficial because it allows the host organism to be sensitive to environmental input. The phase resetting and entrainment by light causes differential effects on *frq* expression and impacts the clock differently depending on the time in the ‘ that the light is seen. When *frq* expression is low, early in the circadian morning, exposure to light will increase *frq* levels and this will advance the clock to the time corresponding to the highest *frq* expression, mid-to-late circadian morning. The sharp increase of *frq* levels due to light exposure during the time of declining *frq* (late circadian afternoon) will lead to phase delays as *frq* levels are forced to return to their maximum mid-day levels after light exposure (Crosthwaite et al., 1997), in accordance with predictive phase response curves.

Light input to the clock is modified by the VIVID protein that is itself clock regulated and can gate the light response on the clock (Heintzen, Loros, & Dunlap, 2001). VVD is a small LOV-domain protein whose name stems from the phenotype of its loss-of-function mutants which display bright orange conidia when grown in constant light, attributed to the persistence activation of carotenoid pigments (Heintzen et al., 2001). The *vvd* promoter is a direct target of the WCC and transcript levels increase dramatically following illumination. VVD inhibits the WCC and this action sets the clock at the dusk transition as well as contributing to temperature compensation (Elvin, Loros, Dunlap, & Heintzen, 2005; Hunt, Elvin, Crosthwaite, & Heintzen, 2007). In addition, VVD levels in the dark inactivate any WCC induced by moonlight and keep the clock in phase during in the bright moonlight nights (Malzahn, Ciprianidis, Kaldi, Schafmeier, & Brunner, 2010).

Beyond light, temperature plays a role in both the entrainment as well as on period of the clock. When shifting to higher temperatures, levels of L-FRQ as well as levels of FRQ overall increase (Garceau et al., 1997; Liu et al., 1997). The FRQ levels at the shift are lower than the lowest

FRQ levels at the higher temperature, so the clock resets to subjective morning (Liu, Merrow, Loros, & Dunlap, 1998). Metabolism is known to play a role into clock input via a feedback loop on the positive arm by CSP-1 (Sancar, Sancar, & Brunner, 2012).

In order for the core molecular circuit to affect organismal behavior, there must be a method to provide rhythmic information to regulate the cell. To do this, the positive arm of the clock, the WCC, regulates a subset of rhythmically expressed genes termed the clock-controlled genes or *ccgs*. It is estimated that 5–15% of the genome is circadianly regulated in Neurospora (Dong et al., 2008; Dunlap & Loros, 2004) (Hurley et al in press). The expression of these genes is not synchronized but is actually staggered in their expression over circadian time, with late night to morning expression most common. While most WCC-driven/light-induced genes are also *ccgs*, there are distinct subsets of WCC-driven genes that are driven in the light and in the dark, showing that the dark expression is distinct from the role of the WCC in the light response (Dunlap & Loros, 2004).

Neurospora is the first system to establish a method of identification of *ccgs*, using subtractive hybridization to compare total nucleic acid levels between samples (Loros et al., 1989). This method and many of the methods used subsequently (differential hybridization, SAGE analysis, and microarrays) have many technical limitations, not the least of which was that the genes identified tended to be the most highly expressed genes (Bell-Pedersen, Shinohara, Loros, & Dunlap, 1996; Duffield et al., 2002; Zhu et al., 2001); these methods have been thoroughly reviewed (Duffield, Loros, & Dunlap, 2005). This is significant due to the low copy number of the most commonly tested *ccg*, *frq* (Merrow et al., 1997). Currently, the most common method of *ccg* discovery, and one that we have used successfully in our lab, is to follow mRNA levels using RNA deep sequencing. RNA extracted over circadian time is subjected to a standard RNA deep-sequencing analysis. The resulting data is normalized using RPKM values and then subjected to an analysis of cycling, i.e., JTK cycle (Hughes, Hogenesch, & Kornacker, 2010) (Hurley et al., in press).

The regulation of rhythmic gene expression occurs in part at transcription. The promoter of a well-studied *ccg*, *ccg-2*, is the perfect example of a circadianly regulated gene and contains several regulatory regions that individually confer light, developmental, and circadian regulation. The activating circadian element (ACE) is sufficient to confer clock regulation on this promoter and in other clock promoters (Bell-Pedersen, Dunlap, & Loros, 1996). The core ACE sequence is different from the core LRE sequence (discussed above) though both of these elements mediate clock control. Some known *ccgs* have neither element, suggesting hierarchical control in which the oscillator directly regulates oscillator proximal controllers that in turn regulate more downstream genes, or additional clock-control elements (Chen, Ringelberg, Gross, Dunlap, & Loros, 2009; Dunlap & Loros, 2004).

Another level of regulation can occur when the mitogen-activated protein kinase (MAPK) pathways are regulated by the clock at the transcriptional level by the WCC (Bennett, Beremand, Thomas, & Bell-Pedersen, 2013; Lamb, Finch, & Bell-Pedersen, 2012; Lamb, Goldsmith, Bennett, Finch, & Bell-Pedersen, 2011). In addition to this regulation, MAPK-1 has been shown to be phosphorylated in a circadian manner and its targets are *ccgs*, demonstrating the circadian clock signal can be propagated outside of the WCC regulation (Bennett et al., 2013). Recently, the Neurospora circadian cycle has been shown to play a role in cell cycle of the organism (Hong et al., 2014).

4. CONCLUSION

Neurospora remains a durable model organism for the study of circadian rhythms because it is so experimentally tractable and yet retains all the regulatory elements and regulatory architecture common to clocks in larger and more complicated organisms. Studies in Neurospora were the first to establish the essential nature of transcriptional negative feedback in the clock, to establish mechanisms for light resetting and for temperature resetting, and it was the first system in which a heterodimer of PAS-containing protein was proposed as the positive element in the feedback loop. The first systematic screens for genes regulated by the clock were performed in Neurospora, setting the stage for broadly envisioned analysis of output pathways. More recently work on Neurospora first showed the interconnection between cell cycle and circadian regulation and probed the involvement of phosphorylation in the mechanism of temperature compensation. Studies in Neurospora have highlighted the fact that many clock proteins may be IDPs and how this structure supports their role in maintaining a clock, and are also revealing the mechanisms through which antisense transcripts to clock genes play a role in rhythm persistence. Work in Neurospora has been shown that the circadian feedback loops can close through phosphorylations alone and do not need to close through phosphorylation-mediated clock protein turnover. These findings all presaged similar findings in animal circadian systems: it is the ability of Neurospora to predict how more complex systems work that makes it an excellent model.

REFERENCES

Aronson, B. D., Johnson, K. A., Loros, J. J., & Dunlap, J. C. (1994). Negative feedback defining a circadian clock: Autoregulation of the clock gene frequency. *Science, 263*, 1578–1584.

Baker, C. L., Kettenbach, A. N., Loros, J. J., Gerber, S. A., & Dunlap, J. C. (2009). Quantitative proteomics reveals a dynamic interactome and phase-specific phosphorylation in the Neurospora circadian clock. *Molecular Cell, 34*, 354–363.

Baker, C. L., Loros, J. J., & Dunlap, J. C. (2012). The circadian clock of Neurospora crassa. *FEMS Microbiology Review, 36*, 95–110.

Belden, W. J., Larrondo, L. F., Froehlich, A. C., Shi, M., Chen, C. H., Loros, J. J., et al. (2007). The band mutation in Neurospora crassa is a dominant allele of ras-1 implicating RAS signaling in circadian output. *Genes & Development, 21*, 1494–1505.

Belden, W. J., Lewis, Z. A., Selker, E. U., Loros, J. J., & Dunlap, J. C. (2011). CHD1 remodels chromatin and influences transient DNA methylation at the clock gene frequency. *PLoS Genetics, 7*, e1002166.

Belden, W. J., Loros, J. J., & Dunlap, J. C. (2007). Execution of the circadian negative feedback loop in Neurospora requires the ATP-dependent chromatin-remodeling enzyme CLOCKSWITCH. *Molecular Cell, 25*, 587–600.

Bell-Pedersen, D., Dunlap, J. C., & Loros, J. J. (1996). Distinct cis-acting elements mediate clock, light, and developmental regulation of the Neurospora crassa eas (*ccg*-2) gene. *Molecular and Cellular Biology, 16*, 513–521.

Bell-Pedersen, D., Shinohara, M. L., Loros, J. J., & Dunlap, J. C. (1996). Circadian clock-controlled genes isolated from Neurospora crassa are late night- to early morning-specific. *Proceedings of the National Academy of Sciences of the United States of America, 93*, 13096–13101.

Bennett, L. D., Beremand, P., Thomas, T. L., & Bell-Pedersen, D. (2013). Circadian activation of the mitogen-activated protein kinase MAK-1 facilitates rhythms in clock-controlled genes in Neurospora crassa. *Eukaryotic Cell, 12*, 59–69.

Bruce, V. G., Weight, F., & Pittendrigh, C. S. (1960). Resetting the sporulation rhythm in Pilobolus with short light flashes of high intensity. *Science, 131*, 728–730.

Brunner, M., & Kaldi, K. (2008). Interlocked feedback loops of the circadian clock of Neurospora crassa. *Molecular Microbiology, 68*, 255–262.

Canessa, P., Schumacher, J., Hevia, M. A., Tudzynski, P., & Larrondo, L. F. (2013). Assessing the effects of light on differentiation and virulence of the plant pathogen Botrytis cinerea: Characterization of the White Collar Complex. *PLoS One, 8*, e84223.

Cha, J., Yuan, H., & Liu, Y. (2011). Regulation of the activity and cellular localization of the circadian clock protein FRQ. *The Journal of Biological Chemistry, 286*, 11469–11478.

Chen, C. H., Ringelberg, C. S., Gross, R. H., Dunlap, J. C., & Loros, J. J. (2009). Genome-wide analysis of light-inducible responses reveals hierarchical light signalling in Neurospora. *EMBO Journal, 28*, 1029–1042.

Cheng, P., He, Q., Wang, L., & Liu, Y. (2005). Regulation of the Neurospora circadian clock by an RNA helicase. *Genes and Development, 19*, 234–241.

Cheng, P., Yang, Y., Gardner, K. H., & Liu, Y. (2002). PAS domain-mediated WC-1/WC-2 interaction is essential for maintaining the steady-state level of WC-1 and the function of both proteins in circadian clock and light responses of Neurospora. *Molecular and Cellular Biology, 22*, 517–524.

Cheng, P., Yang, Y., Heintzen, C., & Liu, Y. (2001). Coiled-coil domain-mediated FRQ-FRQ interaction is essential for its circadian clock function in Neurospora. *The EMBO Journal, 20*, 101–108.

Cheng, P., Yang, Y., Wang, L., He, Q., & Liu, Y. (2003). WHITE COLLAR-1, a multifunctional Neurospora protein involved in the circadian feedback loops, light sensing, and transcription repression of wc-2. *Journal of Biological Chemistry, 278*, 3801–3808.

Collett, M. A., Garceau, N., Dunlap, J. C., & Loros, J. J. (2002). Light and clock expression of the Neurospora clock gene frequency is differentially driven by but dependent on WHITE COLLAR-2. *Genetics, 160*, 149–158.

Collopy, P. D., Colot, H. V., Park, G., Ringelberg, C., Crew, C. M., Borkovich, K. A., et al. (2010). High-throughput construction of gene deletion cassettes for generation of Neurospora crassa knockout strains. *Methods in Molecular Biology, 638*, 33–40.

Colot, H. V., Loros, J. J., & Dunlap, J. C. (2005). Temperature-modulated alternative splicing and promoter use in the circadian clock gene frequency. *Molecular Biology of the Cell, 16*, 5563–5571.

Colot, H. V., Park, G., Turner, G. E., Ringelberg, C., Crew, C. M., Litvinkova, L., et al. (2006). A high-throughput gene knockout procedure for Neurospora reveals functions for multiple transcription factors. *Proceedings of the National Academy of Sciences of the United States of America, 103*, 10352–10357.

Crosthwaite, S. K., Dunlap, J. C., & Loros, J. J. (1997). Neurospora wc-1 and wc-2: Transcription, photoresponses, and the origins of circadian rhythmicity. *Science, 276*, 763–769.

Davis, R. H., & Perkins, D. D. (2002). Timeline: Neurospora: A model of model microbes. *Nature Reviews Genetics, 3*, 397–403.

Diernfellner, A., Colot, H. V., Dintsis, O., Loros, J. J., Dunlap, J. C., & Brunner, M. (2007). Long and short isoforms of Neurospora clock protein FRQ support temperature-compensated circadian rhythms. *FEBS Letters, 581*, 5759–5764.

Diernfellner, A. C., Querfurth, C., Salazar, C., Hofer, T., & Brunner, M. (2009). Phosphorylation modulates rapid nucleocytoplasmic shuttling and cytoplasmic accumulation of Neurospora clock protein FRQ on a circadian time scale. *Genes & Development, 23*, 2192–2200.

Diernfellner, A. C., & Schafmeier, T. (2011). Phosphorylations: Making the *Neurospora crassa* circadian clock tick. *FEBS Letters, 585*, 1461–1466.

Diernfellner, A. C., Schafmeier, T., Merrow, M. W., & Brunner, M. (2005). Molecular mechanism of temperature sensing by the circadian clock of Neurospora crassa. *Genes & Development, 19*, 1968–1973.

Dong, W., Tang, X., Yu, Y., Nilsen, R., Kim, R., Griffith, J., et al. (2008). Systems biology of the clock in Neurospora crassa. *PLoS One, 3*, e3105.

Duffield, G. E., Best, J. D., Meurers, B. H., Bittner, A., Loros, J. J., & Dunlap, J. C. (2002). Circadian programs of transcriptional activation, signaling, and protein turnover revealed by microarray analysis of mammalian cells. *Current Biology, 12*, 551–557.

Duffield, G., Loros, J. J., & Dunlap, J. C. (2005). Analysis of circadian output rhythms of gene expression in Neurospora and mammalian cells in culture. *Methods in Enzymology, 393*, 315–341.

Dunlap, J. C. (1999). Molecular bases for circadian clocks. *Cell, 96*, 271–290.

Dunlap, J. C., & Loros, J. J. (2004). The Neurospora circadian system. *Journal of Biological Rhythms, 19*, 414–424.

Dunlap, J. C., & Loros, J. J. (2006). How fungi keep time: Circadian system in Neurospora and other fungi. *Current Opinion in Microbiology, 9*, 579–587.

Elvin, M., Loros, J. J., Dunlap, J. C., & Heintzen, C. (2005). The PAS/LOV protein VIVID supports a rapidly dampened daytime oscillator that facilitates entrainment of the Neurospora circadian clock. *Genes and Development, 19*, 2593–2605.

Froehlich, A. C., Liu, Y., Loros, J. J., & Dunlap, J. C. (2002). White Collar-1, a circadian blue light photoreceptor, binding to the frequency promoter. *Science, 297*, 815–819.

Froehlich, A. C., Loros, J. J., & Dunlap, J. C. (2003). Rhythmic binding of a WHITE COLLAR-containing complex to the frequency promoter is inhibited by FREQUENCY. *Proceedings of the National Academy of Sciences of the United States of America, 100*, 5914–5919.

Galagan, J. E., Calvo, S. E., Borkovich, K. A., Selker, E. U., Read, N. D., Jaffe, D., et al. (2003). The genome sequence of the filamentous fungus Neurospora crassa. *Nature, 422*, 859–868.

Garceau, N. Y., Liu, Y., Loros, J. J., & Dunlap, J. C. (1997). Alternative initiation of translation and time-specific phosphorylation yield multiple forms of the essential clock protein FREQUENCY. *Cell, 89*, 469–476.

Gardner, G. F., & Feldman, J. F. (1980). The frq locus in Neurospora crassa: A key element in circadian clock organization. *Genetics, 96*, 877–886.

Gooch, V. D., Mehra, A., Larrondo, L. F., Fox, J., Touroutoutoudis, M., Loros, J. J., et al. (2008). Fully codon-optimized luciferase uncovers novel temperature characteristics of the Neurospora clock. *Eukaryotic Cell, 7*, 28–37.

Greene, A. V., Keller, N., Haas, H., & Bell-Pedersen, D. (2003). A circadian oscillator in Aspergillus spp. regulates daily development and gene expression. *Eukaryotic Cell, 2*, 231–237.

Guo, J., Cheng, P., & Liu, Y. (2010). Functional significance of FRH in regulating the phosphorylation and stability of Neurospora circadian clock protein FRQ. *The Journal of Biological Chemistry, 285*, 11508–11515.

Guo, J., Cheng, P., Yuan, H., & Liu, Y. (2009). The exosome regulates circadian gene expression in a posttranscriptional negative feedback loop. *Cell, 138*, 1236–1246.

He, Q., Cha, J., Lee, H. C., Yang, Y., & Liu, Y. (2006). CKI and CKII mediate the FREQUENCY-dependent phosphorylation of the WHITE COLLAR complex to close the Neurospora circadian negative feedback loop. *Genes & Development, 20*, 2552–2565.

He, Q., Cheng, P., Yang, Y., Yu, H., & Liu, Y. (2003). FWD1-mediated degradation of FREQUENCY in Neurospora establishes a conserved mechanism for circadian clock regulation. *The EMBO Journal, 22*, 4421–4430.

He, Q., & Liu, Y. (2005). Degradation of the Neurospora circadian clock protein FREQUENCY through the ubiquitin-proteasome pathway. *Biochemical Society Transactions, 33*, 953–956.

He, Q., Shu, H., Cheng, P., Chen, S., Wang, L., & Liu, Y. (2005). Light-independent phosphorylation of WHITE COLLAR-1 regulates its function in the Neurospora circadian negative feedback loop. *The Journal of Biological Chemistry, 280*, 17526–17532.

Heintzen, C., & Liu, Y. (2007). The *Neurospora crassa* circadian clock. *Advances in Genetics, 58*, 25–66.

Heintzen, C., Loros, J. J., & Dunlap, J. C. (2001). The PAS protein VIVID defines a clock-associated feedback loop that represses light input, modulates gating, and regulates clock resetting. *Cell, 104*, 453–464.

Hong, C. I., Ruoff, P., Loros, J. J., & Dunlap, J. C. (2008). Closing the circadian negative feedback loop: FRQ-dependent clearance of WC-1 from the nucleus. *Genes and Development, 22*, 3196–3204.

Hong, C. I., Zamborszky, J., Baek, M., Labiscsak, L., Ju, K., Lee, H., et al. (2014). Circadian rhythms synchronize mitosis in Neurospora crassa. *Proceedings of the National Academy of Sciences of the United States of America, 111*, 1397–1402.

Huang, G., He, Q., Guo, J., Cha, J., & Liu, Y. (2013). The Ccr4-not protein complex regulates the phase of the Neurospora circadian clock by controlling white collar protein stability and activity. *The Journal of Biological Chemistry, 288*, 31002–31009.

Hughes, M. E., Hogenesch, J. B., & Kornacker, K. (2010). JTK_CYCLE: An efficient nonparametric algorithm for detecting rhythmic components in genome-scale data sets. *Journal of Biological Rhythms, 25*, 372–380.

Hunt, S. M., Elvin, M., Crosthwaite, S. K., & Heintzen, C. (2007). The PAS/LOV protein VIVID controls temperature compensation of circadian clock phase and development in *Neurospora crassa. Genes and Development, 21*, 1964–1974.

Hunt, S. M., Thompson, S., Elvin, M., & Heintzen, C. (2010). VIVID interacts with the WHITE COLLAR complex and FREQUENCY-interacting RNA helicase to alter light and clock responses in Neurospora. *Proceedings of the National Academy of Sciences of the United States of America*, *107*, 16709–16714.

Hurley, J. M., Dasgupta, A., Emerson, J. M., Zhou, X., Ringelberg, C. S., Knabe, N., et al. (In press). Analysis of clock-regulated genes in Neurospora reveals widespread post-transcriptional control of metabolic potential. *Proceedings of the National Academy of Sciences of the United States of America.*

Hurley, J. M., Larrondo, L. F., Loros, J. J., & Dunlap, J. C. (2013). Conserved RNA helicase FRH acts nonenzymatically to support the intrinsically disordered Neurospora clock protein FRQ. *Molecular Cell*, *52*, 832–843.

Klengel, T., Liang, W.-J., Chaloupka, J., Ruoff, C., Schröppel, K., Naglik, J. R., et al. (2005). Fungal adenylyl cyclase integrates CO_2 sensing with cAMP signaling and virulence. *Current Biology*, *15*, 2021–2026.

Kramer, C., Loros, J. J., Dunlap, J. C., & Crosthwaite, S. K. (2003). Role for antisense RNA in regulating circadian clock function in Neurospora crassa. *Nature*, *421*, 948–952.

LaCava, J., Houseley, J., Saveanu, C., Petfalski, E., Thompson, E., Jacquier, A., et al. (2005). RNA degradation by the exosome is promoted by a nuclear polyadenylation complex. *Cell*, *121*, 713–724.

Lamb, T. M., Finch, K. E., & Bell-Pedersen, D. (2012). The Neurospora crassa OS MAPK pathway-activated transcription factor ASL-1 contributes to circadian rhythms in pathway responsive clock-controlled genes. *Fungal Genetics and Biology*, *49*, 180–188.

Lamb, T. M., Goldsmith, C. S., Bennett, L., Finch, K. E., & Bell-Pedersen, D. (2011). Direct transcriptional control of a p38 MAPK pathway by the circadian clock in Neurospora crassa. *PLoS One*, *6*, e27149.

Larrondo, L. F., Baker, C. L., Olivares-Yañez, C., Loros, J. J., & Dunlap, J.C. Decoupling circadian clock protein turnover from circadian period determination. (under revision).

Larrondo, L. F., Loros, J. J., & Dunlap, J. C. (2012). High-resolution spatiotemporal analysis of gene expression in real time: In vivo analysis of circadian rhythms in *Neurospora crassa* using a FREQUENCY-luciferase translational reporter. *Fungal Genetics and Biology*, *49*, 681–683.

Lauinger, L., Diernfellner, A., Falk, S., & Brunner, M. (2014). The RNA helicase FRH is an ATP-dependent regulator of CK1a in the circadian clock of Neurospora crassa. *Nature Communications*, *5*, 3598.

Linden, H., Ballario, P., & Macino, G. (1997). Blue light regulation in *Neurospora crassa*. *Fungal Genetics and Biology*, *22*, 141–150.

Linden, H., & Macino, G. (1997). White collar 2, a partner in blue-light signal transduction, controlling expression of light-regulated genes in Neurospora crassa. *EMBO Journal*, *16*, 98–109.

Liu, Y., & Bell-Pedersen, D. (2006). Circadian rhythms in *Neurospora crassa* and other filamentous fungi. *Eukaryotic Cell*, *5*, 1184–1193.

Liu, Y., Garceau, N. Y., Loros, J. J., & Dunlap, J. C. (1997). Thermally regulated translational control of FRQ mediates aspects of temperature responses in the Neurospora circadian clock. *Cell*, *89*, 477–486.

Liu, Y., He, Q., & Cheng, P. (2003). Photoreception in Neurospora: A tale of two White Collar proteins. *Cellular and Molecular Life Sciences*, *60*, 2131–2138.

Liu, Y., Loros, J., & Dunlap, J. C. (2000). Phosphorylation of the Neurospora clock protein FREQUENCY determines its degradation rate and strongly influences the period length of the circadian clock. *Proceedings of the National Academy of Sciences of the United States of America*, *97*, 234–239.

Liu, Y., Merrow, M., Loros, J. J., & Dunlap, J. C. (1998). How temperature changes reset a circadian oscillator. *Science*, *281*, 825–829.

Loros, J. J., Denome, S. A., & Dunlap, J. C. (1989). Molecular cloning of genes under control of the circadian clock in Neurospora. *Science*, *243*, 385–388.

Loros, J. J., & Feldman, J. F. (1986). Loss of temperature compensation of circadian period length in the frq-9 mutant of Neurospora crassa. *Journal of Biological Rhythms*, *1*, 187–198.

Malzahn, E., Ciprianidis, S., Kaldi, K., Schafmeier, T., & Brunner, M. (2010). Photoadaptation in Neurospora by competitive interaction of activating and inhibitory LOV domains. *Cell*, *142*, 762–772.

McClung, C. R., Fox, B. A., & Dunlap, J. C. (1989). The Neurospora clock gene frequency shares a sequence element with the Drosophila clock gene period. *Nature*, *339*, 558–562.

Mehra, A., Shi, M., Baker, C. L., Colot, H. V., Loros, J. J., & Dunlap, J. C. (2009). A role for casein kinase 2 in the mechanism underlying circadian temperature compensation. *Cell*, *137*, 749–760.

Merrow, M. W., Garceau, N. Y., & Dunlap, J. C. (1997). Dissection of a circadian oscillation into discrete domains. *Proceedings of the National Academy of Sciences of the United States of America*, *94*, 3877–3882.

Morgan, L. W., Greene, A. V., & Bell-Pedersen, D. (2003). Circadian and light-induced expression of luciferase in Neurospora crassa. *Fungal Genetics and Biology*, *38*, 327–332.

Ninomiya, Y., Suzuki, K., Ishii, C., & Inoue, H. (2004). Highly efficient gene replacements in Neurospora strains deficient for nonhomologous end-joining. *Proceedings of the National Academy of Sciences of the United States of America*, *101*, 12248–12253.

Perlman, J., Nakashima, H., & Feldman, J. F. (1981). Assay and characteristics of circadian rhythmicity in liquid cultures of Neurospora crassa. *Plant Physiology*, *67*, 404–407.

Pittendrigh, C. S., Bruce, V. G., Rosensweig, N. S., & Rubin, M. L. (1959). Growth patterns in Neurospora: A biological clock in Neurospora. *Nature*, *184*, 169–170.

Querfurth, C., Diernfellner, A. C., Gin, E., Malzahn, E., Hofer, T., & Brunner, M. (2011). Circadian conformational change of the Neurospora clock protein FREQUENCY triggered by clustered hyperphosphorylation of a basic domain. *Molecular Cell*, *43*, 713–722.

Roenneberg, T., & Taylor, W. (2000). Automated recordings of bioluminescence with special reference to the analysis of circadian rhythms. *Methods in Enzymology*, *305*, 104–119.

Ruoff, P., Loros, J. J., & Dunlap, J. C. (2005). The relationship between FRQ-protein stability and temperature compensation in the Neurospora circadian clock. *Proceedings of the National Academy of Sciences of the United States of America*, *102*, 17681–17686.

Salichos, L., & Rokas, A. (2010). The diversity and evolution of circadian clock proteins in fungi. *Mycologia*, *102*, 269–278.

Sancar, G., Sancar, C., & Brunner, M. (2012). Metabolic compensation of the Neurospora clock by a glucose-dependent feedback of the circadian repressor CSP1 on the core oscillator. *Genes & Development*, *26*, 2435–2442.

Sancar, G., Sancar, C., Brunner, M., & Schafmeier, T. (2009). Activity of the circadian transcription factor White Collar Complex is modulated by phosphorylation of SP-motifs. *FEBS Letters*, *583*, 1833–1840.

Sargent, M. L., Briggs, W. R., & Woodward, D. O. (1966). Circadian nature of a rhythm expressed by an invertaseless strain of Neurospora crassa. *Plant Physiology*, *41*, 1343–1349.

Shi, M., Collett, M., Loros, J. J., & Dunlap, J. C. (2010). FRQ-interacting RNA helicase mediates negative and positive feedback in the Neurospora circadian clock. *Genetics*, *184*, 351–361.

Tang, C. T., Li, S., Long, C., Cha, J., Huang, G., Li, L., et al. (2009). Setting the pace of the Neurospora circadian clock by multiple independent FRQ phosphorylation events. *Proceedings of the National Academy of Sciences of the United States of America*, *106*, 10722–10727.

Traeger, S., Altegoer, F., Freitag, M., Gabaldon, T., Kempken, F., Kumar, A., et al. (2013). The genome and development-dependent transcriptomes of Pyronema confluens: A window into fungal evolution. *PLoS Genetics*, *9*, e1003820.

Wang, B., Kettenbach, A. N., Gerber, S. A., Loros, J. J., & Dunlap, J. C. (2014). Neurospora WC-1 recruits SWI/SNF to remodel frequency and initiate a circadian cycle. *PLoS Genetics*, *10*, e1004599.

Zhu, H., Nowrousian, M., Kupfer, D., Colot, H. V., Berrocal-Tito, G., Lai, H., et al. (2001). Analysis of expressed sequence tags from two starvation, time-of-day-specific libraries of Neurospora crassa reveals novel clock-controlled genes. *Genetics*, *157*, 1057–1065.

CHAPTER THREE

High-Throughput and Quantitative Approaches for Measuring Circadian Rhythms in Cyanobacteria Using Bioluminescence

Ryan K. Shultzaberger*,†,1, Mark L. Paddock*,1, Takeo Katsuki†, Ralph J. Greenspan*,†, Susan S. Golden*,2

*Center for Circadian Biology, University of California, San Diego, La Jolla, California, USA
†Kavli Institute for Brain and Mind, University of California, San Diego, California, USA
[1]These authors contributed equally.
[2]Corresponding author: e-mail address: sgolden@ucsd.edu

Contents

Abstract

The temporal measurement of a bioluminescent reporter has proven to be one of the most powerful tools for characterizing circadian rhythms in the cyanobacterium *Synechococcus elongatus*. Primarily, two approaches have been used to automate this process: (1) detection of cell culture bioluminescence in 96-well plates by a photomultiplier

Methods in Enzymology, Volume 551
ISSN 0076-6879
http://dx.doi.org/10.1016/bs.mie.2014.10.010

tube-based plate-cycling luminometer (TopCount Microplate Scintillation and Luminescence Counter, Perkin Elmer) and (2) detection of individual colony bioluminescence by iteratively rotating a Petri dish under a cooled CCD camera using a computer-controlled turntable. Each approach has distinct advantages. The TopCount provides a more quantitative measurement of bioluminescence, enabling the direct comparison of clock output levels among strains. The computer-controlled turntable approach has a shorter set-up time and greater throughput, making it a more powerful phenotypic screening tool. While the latter approach is extremely useful, only a few labs have been able to build such an apparatus because of technical hurdles involved in coordinating and controlling both the camera and the turntable, and in processing the resulting images. This protocol provides instructions on how to construct, use, and process data from a computer-controlled turntable to measure the temporal changes in bioluminescence of individual cyanobacterial colonies. Furthermore, we describe how to prepare samples for use with the TopCount to minimize experimental noise and generate meaningful quantitative measurements of clock output levels for advanced analysis.

1. THEORY

In vivo bioluminescence measurements have been invaluable in the determination of circadian oscillations in many organisms, and especially in the cyanobacterium *Synechococcus elongatus*, for which no other visible circadian phenotype is evident (Mackey, Golden, & Ditty, 2011). One of the most successful approaches to identify the genetic components that determine a complex phenotype in any organism has been to systematically mutate the genome by targeted or random mutations and screen for phenotypic variants (Brenner, 1974; Mayer, Ruiz, Berleth, Miseéra, & Juürgens, 1991; Nolan, Kapfhamer, & Bućan, 1997). This method initially was used to identify many of the core genes involved in the *S. elongatus* clock (Kondo et al., 1994), and currently is being used to elucidate subtler features of the circadian gene network. Unfortunately, the only commercially available machine to monitor the temporal expression of bioluminescence in cyanobacteria is limited in the scale of mutants that it can assay, requires clones to be inoculated individually, and has a long set-up time (Mackey, Ditty, Clerico, & Golden, 2007). The lab of T. Kondo (Nagoya University) has shown that bioluminescence from individual *S. elongatus* colonies on Petri dishes can be reliably measured over time (Kondo & Ishiura, 1994). Kondo and colleagues built a computer-controlled turntable that iteratively rotates Petri dishes under a CCD camera for imaging, and significantly increased the throughput of cyanobacterial mutant screening (Kondo et al., 1994).

Construction of a similar apparatus by other labs has been limited because of technical hurdles involved in coordinating and controlling both the camera and the turntable, and in processing the resulting images. Applications developed for the software packages Matlab and R have simplified these challenges. Here, we describe how to construct and operate a computer-controlled turntable and how to analyze a temporal image series of bioluminescent cyanobacterial colonies. This system is reliable and can accurately characterize the periods of approximately 300 colonies per plate, or about 2700 colonies per run of 5 days, enabling the high-throughput screening for mutant colonies that have altered circadian phenotypes. This protocol can easily be adapted to other applications that require temporal image acquisition and processing.

While quantifiable differences in period, phase, and amplitude can be measured from individual colonies using the turntable described above, additional information is contained in the magnitude of bioluminescence, which can only be extracted from larger cell cultures that are carefully prepared. For example, the magnitude of bioluminescence can provide information directly related to whether the oscillator output is activating or repressing transcription (Paddock, Boyd, Adin, & Golden, 2013). Here, we also report on methods and internal tests to establish quantitatively informative bioluminescence measurements. Although this quantification can be done using either the computer-controlled turntable or a TopCount, we have found that using a TopCount is easier to control for culture density. Therefore, the protocol presented here is for use with a TopCount. General considerations for the measurement of bioluminescence have been described previously in detail (Mackey et al., 2007). Our focus will be on strain treatment and criteria for validation of quantitative bioluminescent measurements in cyanobacterial systems.

2. BUILD A COMPUTER-CONTROLLED TURNTABLE

The primary elements are: a sensitive cooled CCD camera that can detect bioluminescence from colonies, a precise stepper motor that can be externally controlled, a secure base to minimize movements during operation, a light source that uniformly illuminates the turntable's surface, and efficient light shielding of the Petri dish during image acquisition. Here, we present our design (Fig. 1), but as long as the above elements are satisfied, a machine that differs in some aspects should function correctly.

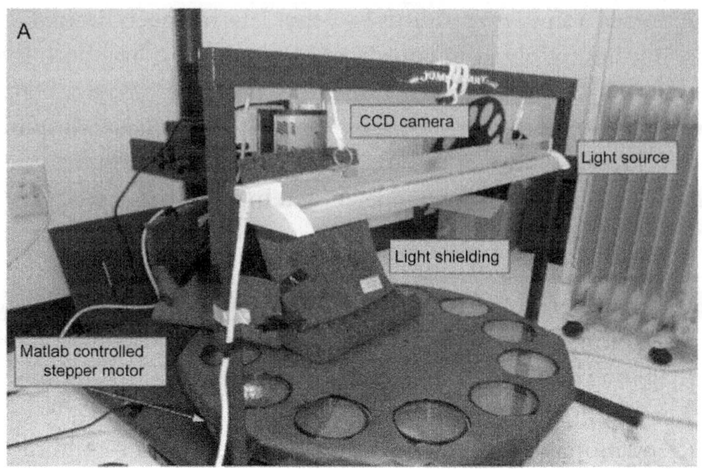

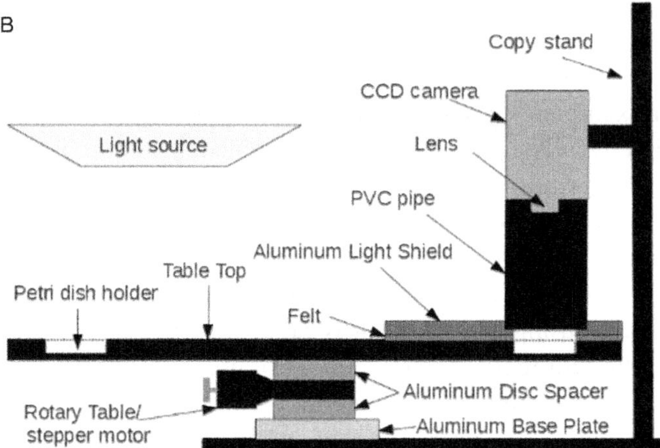

Figure 1 Fully constructed turntable. (A) Photograph of assembled computer-controlled turntable. (B) Cross-sectional schematic of assembled computer-controlled turntable. (See the color plate.)

2.1. Materials

1. Sherline P/N 8700 CNC Rotary Table (http://www.sherline.com/8700.htm)
2. Computer with parallel port that can run 32-bit Matlab
3. Large camera/Copy Stand
4. Pixis 1024B CCD Camera (Princeton Instruments)
5. 25 mm F0.95 Lens (Navitar)
6. 1-1/31″ thick by 23-3/4″ diameter edge-glued pine round (Home Depot)

7. 5″ Black PVC Female Adapter (screw type)
8. 5″ Black PVC Male Adapter (screw type)
9. Aluminum Base Plate (Fig. 2C)
10. Aluminum Light Shield (Fig. 2B)
11. Black butcher paper
12. Black paint
13. Two 4″ diameter Aluminum Disc Spacers (Fig. 2C)
14. Black electrical tape
15. Adhesive-backed black felt at least 8.5″ × 12″ × 1/4″
16. 4″–5″ rubber drain coupling
17. 8-pin Mini-Din Male MAC to DB25 Male Hayes-Compatible Model Cable (Cables to Go Part 02966)
18. Fluorescent light and stand at least 24″ wide

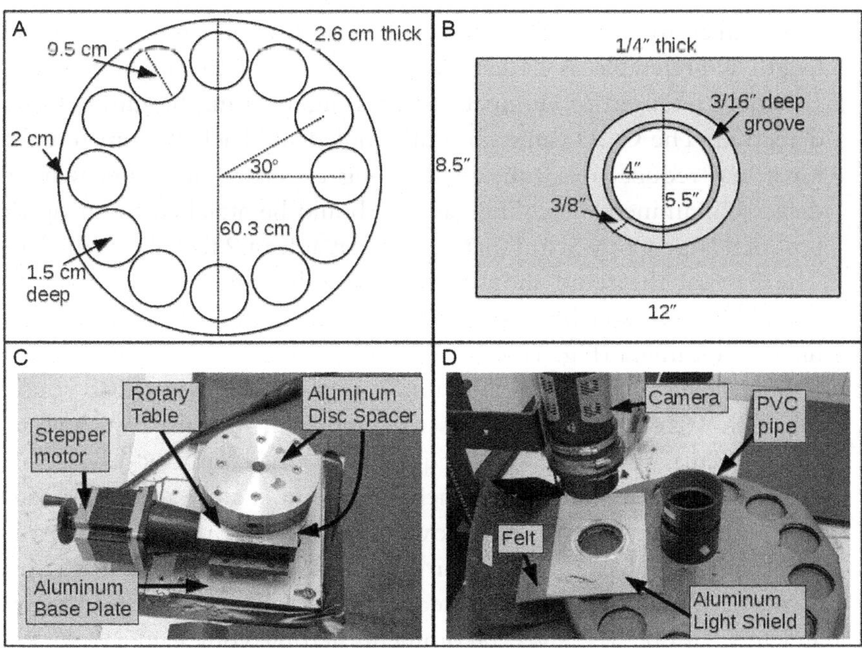

Figure 2 Turntable components. (A) Schematic of turntable surface (referred to in text as Table Top). (B) Schematic of Aluminum Light Shield. (C) Photograph of Rotary Table attached to Aluminum Disc Spacers, Aluminum Base Plate, and Copy Stand base. (D) Photograph of Aluminum Light Shield on turntable surface. The PVC pipe is moved from its final location to show the hole in the center of the Aluminum Light Shield. (See the color plate.)

2.2. Programs

1. Matlab (Mathworks)
2. `Data Acquisition Toolbox` for Matlab
3. PVCAM (Photometrics)
4. Micro Manager (Edelstein, Amodaj, Hoover, Vale, & Stuurman, 2010)

2.3. Protocol

1. *Download scripts and installation files*: We have written several scripts to control the rotary motor and cooled CCD camera with Matlab and to analyze the resulting images in R. These files along with additional useful software installation instructions can be downloaded from "http://golden.ucsd.edu/turn_table.html" and will be referred to throughout this protocol.
2. *Cooled CCD camera*: We use a Princeton Instruments Pixis 1024B CCD camera with a 25-mm F0.95 lens (Navitar Part DO-2595). This lens allows an entire Petri dish to be imaged sharply at 6 inches. A short focal length is preferable as it decreases the length of the Light Shielding Assembly required to eliminate external light during bioluminescence detection. The CCD camera is both cooled and back-illuminated, features that reduce noise during the long exposure times necessary to detect bioluminescence. The camera should be attached to an optical post or a large Copy Stand that positions it at least 2 feet above the base. The base of the stand should be sufficiently large that the turntable assembly can be attached to it while the edge of the turntable is directly under the camera (Fig. 1B).
3. *Controlling the camera*: Two additional programs are required to control the Pixis 1024B on a Windows machine with Matlab: PVCAM and Micro Manager. PVCAM is required to install camera drivers, and Micro Manager can set camera properties and acquire images (Edelstein et al., 2010). We provide specific instructions on how to install and use these programs in the file `Camera_setup.txt`, which is included in the files downloaded in the section "Download scripts and installation files." We recommend getting your camera working prior to final assembly of the turntable. This preparation will allow you to adjust camera focus during construction. Simple snapshots can be taken through the Micro Manager GUI interface.
4. *Turntable surface*: We could not find a prefabricated Table Top that could hold Petri dishes, so we had to have one machined. Twelve Petri

dish-sized holders were cut into a 1-1/31″ thick by 23-3/4″ diameter edge-glued pine round (Home Depot) as shown in Fig. 2A. Dimensions in this figure are given in cm rather than inches to be consistent with the units of a standard Petri dish. There are several considerations to keep in mind when making this part. (1) The holders should be tight around the plates to reduce translational movements during table rotation, which affect image analysis. (2) The centers of the holders need to be evenly spaced around the circumference of the platform. (3) There needs to be adequate space between plates, to ensure that only one plate is visualized at a time. (4) Holders should be far enough from the edge to prevent the incursion of external light during imaging. (5) Bolt holes need to be cut into the Table Top to connect the Table Top to the Rotary Table described below. After fabrication, cover the Table Top with black butcher paper to reduce light noise during imaging and to reduce friction between the Table Top and the Light Shielding Assembly. To further reduce light noise, you can paint the plate holders black.

5. *Rotary Table*: We use the Sherline P/N 8700 CNC Rotary Table and Motion Controller (Fig. 2C). This is a stepper-motor-based Rotary Table that can be externally controlled by Matlab as described below.
6. *Attaching the Table Top to Rotary Table and Copy Stand*: To attach the Rotary Table to the wooden Table Top, we had a 4″ diameter Aluminum Disc Spacer fabricated that had eight bolt holes in it: four of which were unthreaded and used to attach the Aluminum Disc Spacer to the Rotary Table and four threaded holes used to attach the Table Top to the Aluminum Disc Spacer. The stepper motor on the Rotary Table drops below the base of the assembly, preventing the Rotary Table from sitting flat on a uniform surface. We attached a second 4″ diameter Aluminum Disc Spacer to the base of the Rotary Table to raise the stepper motor. To this assembly we also attached a fabricated 5″ × 7″ Aluminum Base Plate with four oblong screw holes cut into each corner, increasing the overall stability of the turntable (Fig. 2C).
7. *Light Shielding Assembly*: One of the most difficult aspects of building this turntable is properly shielding the Petri dish from ambient light during imaging. Our approach was to attach a 5″ black PVC pipe to the camera. We used a threaded connector pipe, so that the height of the shielding could be adjusted. The PVC pipe was attached to the camera with a 4″–5″ rubber drain coupling. The seal between the PVC pipe and the table did not provide sufficient light shielding for imaging. To address this shortcoming, we fabricated an 8.5″ × 12″ × 1/4″ Aluminum Light

Shield with a Petri dish-sized hole cut into the middle of it (thanks to Carl H. Johnson at Vanderbilt University, for this suggestion). To the base of this Aluminum Light Shield, we attached a 1/4″ thick piece of felt, which both prevented light from entering the imaging chamber, and allowed the table to slide smoothly under the plate. The top of the Aluminum Light Shield had a 3/16″ deep circular groove that the PVC pipe could fit in (Fig. 2B and D). This part covers three Petri dish holders on the Table Top, and therefore reduces the number of plates that can be assayed to 9. The large footprint of the Aluminum Light Shield was necessary for us to get sufficient light shielding during imaging. The entire Light Shielding Assembly was wrapped in a heavy black curtain to further reduce light noise. If any aspects of the Light Shielding Assembly moves during table rotation, it can be stabilized using a ring stand.

8. *Lighting system*: We use a Jump Start 2 Foot Fluorescent Grow Light System (Hydrofarm) to illuminate the Table Top (Fig. 1), but any lighting system that provides strong uniform illumination across the table surface is acceptable. A ring-shaped light may be superior, but we have not tested one.
9. *Programming the Rotary Table*: To program the Motion Controller: (1) Plug it into the Rotary Table and turn it on. (2) Push the `Mode` button until the display says `Division Mode` and press `Enter`. (3) Enter the number of divisions that you want; it will be 12 if you use the Table Top described above. (4) Push the `Next` button and the table will rotate 30° clockwise. Each time you hit `Next` the table will rotate another 30°. Instead of hitting `Next`, you can also rotate the table by sending an electric TTL pulse into the `Interface` port that is located on the back of the controller. This pulse can be sent by Matlab as described below. To initially align your table, hit the `Stop/Jog` button to enter `Jog` mode, and then push either "`1`" or "`3`" on the number pad, to move the table left or right, respectively.

3. USE A COMPUTER-CONTROLLED TURNTABLE

3.1. Programs

1. Matlab (Mathworks)
2. `Data Acquisition Toolbox` for Matlab
3. Matlab script `RTinit.m`
4. Matlab script `RTturn.m`
5. Matlab script `RTfull.m`

6. Matlab script `RTexp.m`
7. PVCAM (Photometrics)
8. Micro Manager (Edelstein et al., 2010)

3.2. Protocol

1. *Cyanobacterial strains*: Two different luciferase reporters have been used in *S. elongatus*: the bacterial *luxAB* operon and the firefly *luc* gene (Andersson et al., 2000; Kondo et al., 1993). Although both work well, the substrate for the bacterial reporter can be synthesized within *S. elongatus* by expressing the *luxCDE* operon, whereas bioluminescence from Luc is dependent upon the addition of D-luciferin. Moreover, the absolute signal strengths are higher with Lux. To ensure continuous bioluminescence over the course of the experiment without substrate reapplication, and to achieve the highest sensitivity of detection, we suggest using a strain that contains the bacterial Lux reporter.
2. *Preparing plates*: *S. elongatus* strains containing the reporter and the genes necessary for substrate synthesis are plated on Petri dishes containing the BG11 solid medium previously described (Mackey et al., 2007), and grown until colonies are 1 mm in diameter. Plates are then entrained for two 12:12 light/dark cycles before testing. The raised upper edge on the top of many Petri dish lids can scatter external light across the plate, obscuring the bioluminescent signal. To mitigate these effects, we wrap the edge of the Petri dish in black electrical tape. To allow for air exchange on the plate, it is necessary to cut ventilation slits into the tape with a razor blade between the Petri dish base and lid. Place the plates in the holders on the turntable. The plates should be snug in the holders, so they do not rotate during the course of the experiment. If the plates can easily rotate, they can be stabilized by sliding a small piece of hard plastic, or part of a metal twist tie, between the plate edge and the holder.
3. *Initializing the Rotary Table for use with Matlab*: As previously mentioned, the Motion Controller can be used as a programmable interface between Matlab and the Rotary Table. For this to work, you will need to install the `Data Acquisition Toolbox` for Matlab, which can control the parallel port on the computer. Using an 8-pin Mini-Din Male MAC to DB25 Male Hayes-Compatible Model Cable (Cables to Go Part 02966), connect the 25-pin parallel port on the back of the computer to the 8-pin interface port on the back of the Motion Controller while it is

turned OFF. Initialize the parallel port for use with the `RTinit.m` script in Matlab. After initialization, turn the Motion Controller on, choose `Division` mode, and pick 12 divisions as described in Programming the Rotary Table above. The `RTturn.m` script sends a single TTL pulse to the interface and triggers the Motion Controller to turn one division. `RTfull.m` turns the table 12 divisions, resulting in one full rotation.

4. *Controlling the CCD camera with Matlab*: The CCD camera can be controlled using Matlab and the Micro Manager Matlab library. We provide the Matlab script `micro2.m`, which takes a single picture with a 3-min exposure. Exposure length within this script can be modified by changing the value in the function `core.setExposure`.
5. *Running a time-course experiment*: To run a full experiment, use the Matlab script `RTexp.m`. This script uses `RTturn.m` and `micro2.m` to turn the table

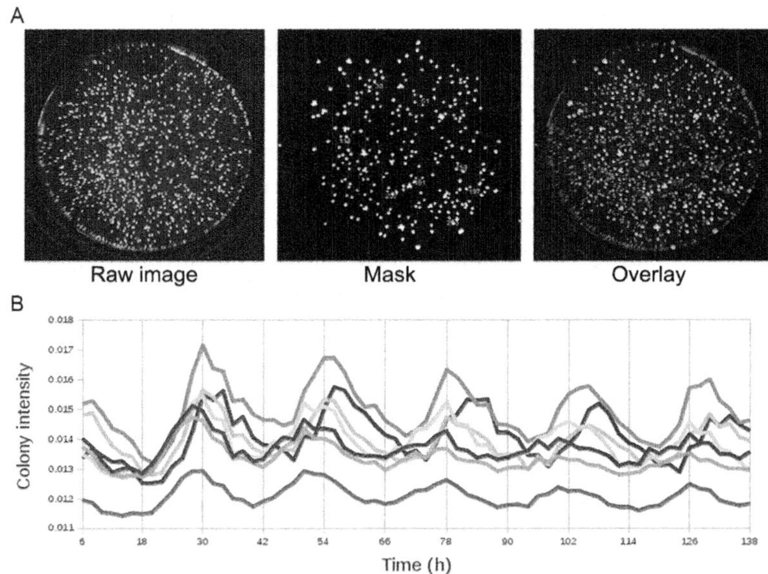

Figure 3 Example data from time course experiment. (A) The image on the left is a raw image of a plate with *luxAB-luxCDE* expressing cyanobacteria. The image in the center is the mask generated by the `RCFinder.R` script. Each white spot represents an identified colony. Those spots that are numbered and circled in red were identified as rhythmic. The number is displaced down and to the right of the spot. The right image is an overlay of the first two images to show which colonies on the plate are rhythmic. (B) Bioluminescence data for five rhythmic colonies found in (A). Colony intensity is a measure of the average pixel intensity for a colony object and varies between 0 and 1 (arbitrary units). (See the color plate.)

and take pictures respectively. Parameters within this script can be modified to adjust the number of time points taken and the duration between time points. The script is currently set to take pictures once every 2 h for 10 days. Five days worth of data is a sufficient sample to get reliable period predictions using the programs described below. To adjust the interval at which pictures are taken, adjust the value in the `pause` function at the end of the script, which is currently set to 5242 s. Images are saved as tiff files and named according to plate number and time point (i.e., `plate_1_001.tiff`). An image of a plate is shown in Fig. 3A.

4. ANALYZING DATA FROM TURNTABLE

4.1. Programs

1. R (CRAN)
2. R package `EBImage`
3. R package `biOps`
4. R package `Rwave`
5. R package `waveclock`
6. R script `RCFinder.R`
7. R script `WC.R`
8. R script `EBI2biOps.R`
9. ImageJ (Rasband, 1997)

4.2. Protocol

1. *Install R libraries*: Our R scripts for image analysis and period quantification are dependent upon several R packages: `EBImage`, `biOps`, `Rwave`, and `waveclock`. Instructions on how to install these packages and links to packages are given in the `R_Package_Install.txt` file included with those downloaded in "Download scripts and installation files." These scripts were tested and work with R version 3.0.1. We used the following versions of each of the other packages: `EBImage version 4.2` (Pau, Fuchs, Sklyar, Boutros, & Huber, 2010), `biOps version 0.2.2`, `Rwave version 2.2`, and `waveclock version 1.04` (Price, Baggs, Curtis, FitzGerald, & Hogenesch, 2008).
2. *Process plates*: Move all plate images and the `RCFinder.R`, `wc.R`, and `EBI2biOps.R` files into a new directory for processing. `RCFinder.R` is the main plate processing script that identifies individual colonies on a plate, calculates the intensity of each colony for each time point,

and determines the period of its bioluminescence. To do this, it sums all images for a given plate and generates a mask of colony objects. An example mask is shown in Fig. 3A. This mask is applied to all images, and the pixel intensity within each colony object is calculated. The period of each colony is then determined using `waveclock` (Price et al., 2008). Bioluminescence data from individual colonies are shown in Fig. 3B. Three image files are generated for each plate in the `results` subdirectory. The first is `plateX_numbered.tiff` which shows the mask and object ID numbers for all identified colony objects. The second is `plateX_rc.tiff` which is the same as the first image, except it shows only the colony ID numbers for those colonies that have a rhythmic circadian phenotype. These colonies are also circled in red (center image in Fig. 3A). The third is `plateX_per.tiff` which is the same as the second image, except it reports the period of the colony instead of the object ID number. To overlay any of these masks with a raw image of a plate, like in Fig. 3A, we use the `Overlay` function in ImageJ (Rasband, 1997). Finally, the period of each colony and all bioluminescence data is reported in the file `all_rc.xls`, also in the `results` subdirectory.

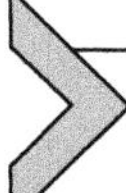

5. STEPS TO EXTRACT RELIABLE QUANTITATIVE INFORMATION FROM BIOLUMINESCENCE LEVELS

The method described above is extremely useful for identifying mutants that have altered circadian properties, but the signal from individual colonies is low, and more quantitative data can be attained from a greater number of cells. A stronger signal can be acquired using the computer-controlled turntable by streaking colonies into larger patches, but to get comparable measurements of bioluminescence levels between strains, you need to start with liquid cultures that have the same cell density. Here, we present a strategy to achieve meaningful bioluminescent measurements with the TopCount Microplate Reader, which is better suited to handle a large number of liquid cultures, enabling the direct comparison of clock output levels between strains (Fig. 4). Corrections for sample size are necessary and can be made easily in bacterial cultures. Additional concerns about interpreting bioluminescence levels have arisen because bioluminescence is a function of not only the level of luciferase, but also its substrate and metabolites such as ATP or $FMNH_2$, depending on the species source of the enzyme. Therefore, comparison of quantitative values for bioluminescence in cyanobacterial studies has rarely been emphasized. These concerns can be addressed by carefully controlling cell counts, growth rates, and luciferase

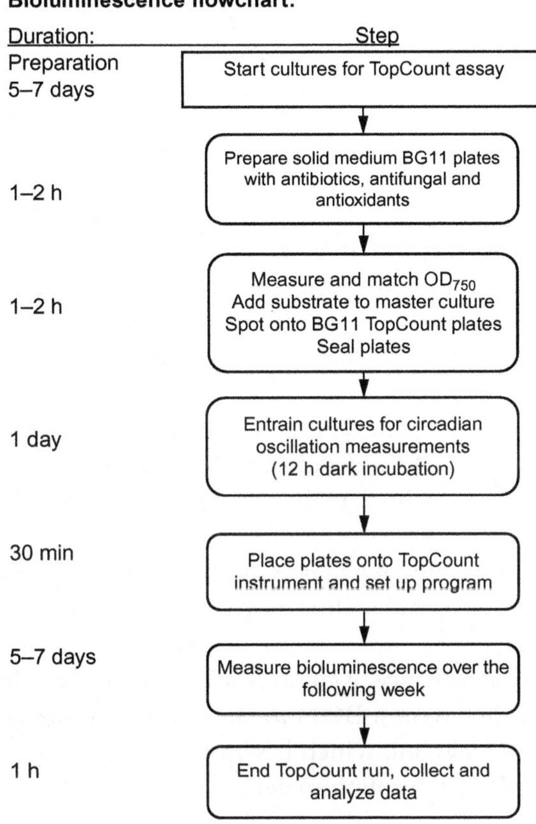

Figure 4 Flowchart for quantitative bioluminescence sample preparation.

substrate levels. Errors in any of these variables will result in changes to the measured level of bioluminescence that do not necessarily reflect the genotype of the mutant strains. This protocol describes how to minimize these errors through careful sample preparation.

5.1. Equipment

1. Laminar flow hood with ultraviolet light
2. Packard TopCount Microplate Scintillation and Luminescence Counter (Perkin Elmer Life Sciences, Boston, MA)
3. Black 96-well microtiter plates and clear plastic lids (ThermoLabsystems, Franklin, MA)
4. Packard Topseal (Perkin Elmer Life Sciences)
5. Clear 96-well plates (ThermoLabsystems)

5.2. Programs

1. Excel (Microsoft)
2. BRASS: Biological Rhythms Analysis Software System (Millar Lab)

5.3. Protocol

1. *Strain growth (5–7 days)*: Start cultures from colonies on an agar plate following transformation or recovery. Pick five colonies using a sterile toothpick from each transformation and patch (spread inoculum with toothpick) onto a new plate that includes the appropriate antibiotics. Two to three of these that pass growth criteria described below will be used as biological replicates to ensure that the data are best representative of the genotype, and are not subject to spontaneous secondary mutations. More details for media preparation: how to setup TopCount runs and how to handle strains are described in Mackey et al. (2007).
   - **1A.** After several days, when colony color develops, pick cells and start 5 ml BG11 liquid cultures. Grow for 2–3 days until color is developed.
   - **1B.** Measure absorbance of cultures at 750 nm (optical density, OD_{750}). Dilute cells into new BG11 medium with appropriate antibiotics to about 10^8 cells/ml, which has an $OD_{750} \sim 0.2$ (Beckman Coulter DU 640B Spectrophotometer).

     Tip: Note that OD measurements for cultures can be different for different instruments because the OD is measuring a scattering from the culture. Thus, the measured value will depend on the details of the detection system, in particular, the cross-section of the scattering that the detector captures. Calibration of the OD_{750} with cell count may be necessary for accurate density measurements. However, comparisons can be made between cultures at standardized readings even if absolute cell counts are not known. As a general rule, work with samples in the range of OD_{750} 0.1–0.5.
   - **1C.** Grow for 2 days, monitoring OD_{750} to make sure cells are healthy and that the growth rates, which will vary depending on light penetration, are the same for all of the cultures. Discard cultures that do not meet this criterion. If comparing mutants that have distinctly different growth rates, it may not be possible to make quantitative comparisons among strains. If a clone of a strain that is usually

impaired in growth relative to wild type (WT) unexpectedly grows robustly, this improved growth is evidence of a suppressor mutation and should be considered with caution.

1D. Differences in growth rates will be evident as changes in the slope of the bioluminescence data with time. If there are differences, then the best window for quantitative comparison should be at the early time points (days 1, 2, and 3) when the cell densities are still similar.

1E. Aim for final density of 2×10^8 cells/ml ($OD_{750} = 0.3$ with our instrument).

Tip: It is best to use cultures that begin the experiment with similar growth histories. We grow precultures that have been recently diluted and grown for 2–3 days under standard conditions to use as the inocula for the samples that will go onto the TopCount. Avoid comparing samples from cultures that are significantly different in culture density, as those cultures would have been growing under different conditions such as lower overall light intensity, possible limitation of some nutrients, and differences in entry into stationary phase.

Tip: Avoid a high cell culture density as it can result in significant shading and an effective lower light intensity and slower growth. The reduction in light penetration into a cyanobacterial culture as density increases is dramatic; remember that these cells actively absorb photons for a living! The center of a stationary phase 100 ml culture in a 250-ml flask is essentially dark, regardless of the intensity of the lights in your chamber. All cultures should be in a similar state of growth at the start of the bioluminescence experiment.

2. *TopCount Microplate Preparation (1–2 days)*: The TopCount measures bioluminescence from the top of up to 8×96-well black Microplates that carry samples. Clear plates that allow light penetration through the plate stacker are also present. The general considerations for using a TopCount to measure bioluminescence from cyanobacteria are presented elsewhere (Mackey et al., 2007). To set up sample plates for measurement, liquid cyanobacterial cultures are pipetted onto pads of BG11 solid medium in each well as described below. Using solid medium in the well places the entire sample near the top of the well, where its bioluminescence can be readily detected; liquid cultures that distribute and scatter the emission are not suitable.

2A. Prepare BG11 solid medium as described in Mackey et al. (2007). Equilibrate to ~60 °C before making 96-well plate for the TopCount. Add antibiotics (if appropriate for the strains), the antifungal benomyl to 10 μg/ml, and filter-sterilized antioxidant Na_2SO_3 to 1 m*M*.

Tip: Because the cells will divide only once or twice per experiment, it is not necessary to maintain all antibiotics during the course of the experiment. Previous measurements of growth rates during the culture preparation would reveal any significant differences between WT and mutants' growth rates. Consider that even if there were 0.1% of the culture contaminated with a revertant that had a growth rate that is twice that of the parental strain, the level of background would increase to between 0.2% and 0.4% by the end of the experiment, a level that is insignificant compared to other uncertainties. Maintain antibiotic selection for the most critical genetic elements in the strains; experience with your strains and thoughtful planning are key.

2B. Use a multichannel pipette to fill each well of a black 96-well microtiter plate with 280 μl melted BG11 solid media. Let dry for 30–60 min.

Tip: White opaque plates produce background signals and are not suitable.

2C. Measure the OD_{750} of each sample immediately prior to use, and dilute the more dense cultures with sterile BG11 liquid medium so that all cultures have the same cell density. Try to maintain cell densities $\geq 2 \times 10^8$ cells/ml ($OD_{750} = 0.3$ with our instrument) for sufficient signal intensity.

2D. Test aliquots of each culture for possible contamination during Microplate set-up. Add 10 μl on a rich medium plate such as an OMNI plate (Mackey et al., 2007) and incubate in the dark overnight. Most contaminating bacteria can utilize the carbon sources whereas *S. elongatus* cannot.

2E. *Inoculation*: Ideally, 30 μl of each culture is tested in at least 10 wells on the plate, requiring a 300 μl sample. If you use the Luc reporter instead of Lux, add 10 μl of 100 m*M* firefly D-luciferin to each 300 μl aliquot prior to plating.

Tip: It is best to prepare a slightly larger sample of inoculum than needed in case of accidental sample loss; scale D-luciferin appropriately.

2F. Distribute the samples into wells in a pattern across the plate to span the row from positions 2 to 11 (see tip below), as there is a gradient of light intensity across the plate if an external light source is used. Averaging the wells (provided they remain alive and healthy) provides a first-order assessment of the circadian phenotypes of the strains.

Tip: Arrange strains such that you have a positive and negative control on the same plate with the strains to be tested.

Tip: Avoid using the wells at the perimeter of the plates for samples as they are most subject to drying out during a 5- to 7-day run. Cell death can be detected by a loss in the bioluminescence over time; any wells that exhibit cell death should be discarded from the analysis, else they will lead to a systematic decrease in the bioluminescence levels which is not accounted for in statistical error treatments.

Tip: When using the TopCount to directly compare the circadian periods of different strains, the pattern of sample distribution must be considered so that only samples equidistant from the edges of the plate nearest the light source are compared. Cyanobacteria follow "Aschoffs Rule" and exhibit slight variations in period as a function of incident light intensity. This consideration is covered more fully elsewhere (Mackey et al., 2007).

2G. Seal the plate with a TopSeal cover. Using a 16-gauge sterile needle, poke a hole in the plastic seal above each well, being careful not to touch the samples. The hole allows gas exchange throughout the TopCount run.

3. *Run TopCount and analyze data*

3A. For circadian measurements, entrain the TopCount sample plates with a 12-h dark incubation. For experiments that include mutants that are sensitive to light/dark cycles, such as *rpa*A- strains, keep the light intensity during light periods relatively low (50 $\mu E/m^2/s$) and limit to one dark pulse.

Tip: Administer dark incubations during local solar night. You should assume that unentrained cultures already have their clocks generally set to local time, and a major change in day/night cycle would add an unintentional phase shift to the experiment.

3B. Load plates onto the TopCount instrument as described in Holtman et al. (2005).

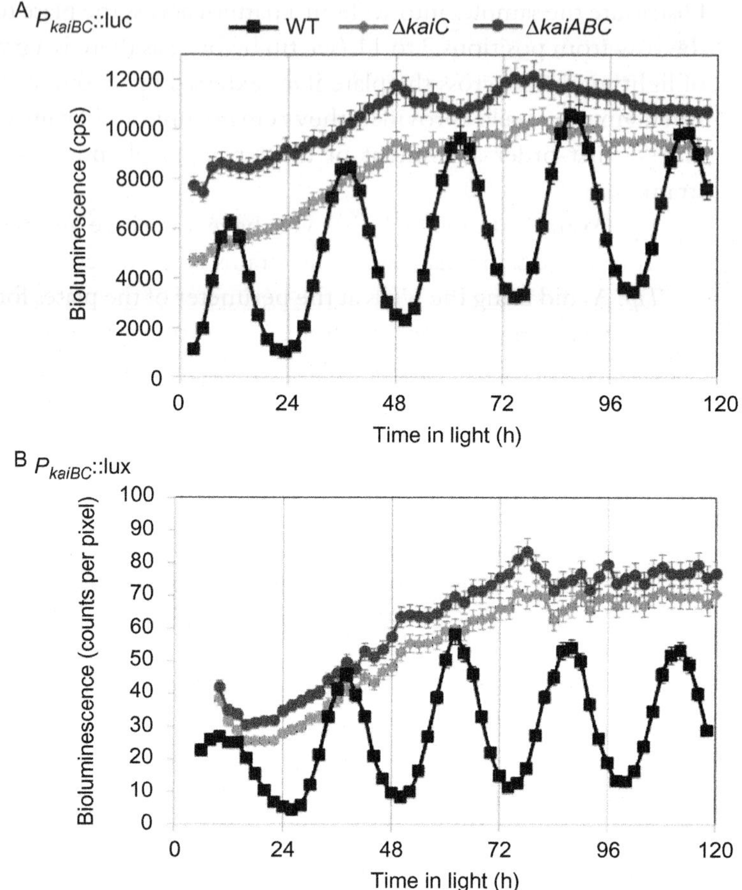

Figure 5 Example data from time course experiment. Default high bioluminescence level for class 1 promoter P_{kaiBC} measured from both the *luc* (A) and *lux* constructs (B). Time dependence of bioluminescence from the WT (black squares) and mutants that carry disruptions of KaiC (blue diamonds (light gray in the print version)) or KaiABC (purple circles (dark gray in the print version)) following a 1- and 2-day entrainment period for (A) and (B), respectively. The bioluminescence using *luc* was measured with the TopCount (Mackey et al., 2007), whereas the bioluminescence using *lux* was measured on the turntable as described above. Averages of the replicates and the standard error of the means for the bioluminescence values are indicated. Bioluminescence was converted to counts per pixel in (B) for easier comparison with the TopCount (Paddock et al., 2013). Each genetic background showed the same general behavior with both reporter systems, even though the luciferase systems and the detection systems are distinct. Thus, the constitutively high levels of bioluminescence observed in the knockouts strains are attributed to the genetic lesions and are not a consequence of detection methods or luciferase reporter constructions. *This figure was adapted from Paddock et al. (2013); Copyright 2013, with permission from the National Academy of Sciences.*

3C. After the run is complete analyze data using BRASS (Biological Rhythms Analysis Software System, http://millar.bio.ed.ac.uk/PEBrown/BRASS/BrassPage.htm; A.J. Millar Laboratory, University of Edinburgh, Scotland, United Kingdom). Check that each well remained healthy over the course of the experiment and exclude samples that showed a dramatic decline in bioluminescence signal during the run.

3D. Average the data from usable wells of the same strain (ideally all 10 wells). Normalize data to the peak signal amplitude of the WT strain at some time early in the run, such as at or near the 36-h time point. We found this time frame to yield the lowest variance among experiments, avoiding variations at the light/dark transition very early in the run and changes due to growth or drying after several cycles. Plot the averages of the time-dependent bioluminescence with standard deviations of the mean (SEM). We present data for WT and arrhythmic mutants to show reproducibility in measurements (Fig. 5; Paddock et al., 2013).

Tip: Bioluminescence magnitudes vary from run to run. Thus, comparison of absolute bioluminescence values among different runs or even different plates in a single run is subject to systematic errors. However, relative values among plates and runs, normalized in the manner presented here, are very reproducible when cell numbers and growth conditions are standardized.

ACKNOWLEDGMENTS

The computer-controlled turntable described here was modeled on the one originally designed by Takao Kondo, which was instrumental in the revolutionary discovery of the *kai* genes. We take this opportunity to acknowledge Dr. Kondo's unmatched contribution to the molecular understanding of cyanobacterial circadian rhythms through bold methods and insightful findings. We owe special thanks to Carl H. Johnson for advice in constructing our version of the "Kondotron." This work was supported by grants from the W. M. Keck Foundation and AFOSR 13RSL031 (R. J. G.), an NRSA fellowship F32GM097977-01 (R. K. S.), and NIGMS—NIH Award R01GM062419 (S. S. G.).

REFERENCES

Andersson, C., Tsinoremas, N., Shelton, J., Lebedeva, N., Yarrow, J., Min, H., et al. (2000). Application of bioluminescence to the study of circadian rhythms in cyanobacteria. *Methods in Enzymology*, *305*, 527–542.

Brenner, S. (1974). The genetics of *Caenorhabditis elegans*. *Genetics*, 77(1), 71–94.

Edelstein, A., Amodaj, N., Hoover, K., Vale, R., & Stuurman, N. (2010). Computer control of microscopes using μManager. *Current Protocols in Molecular Biology*, 14–20.

Holtman, C. K., Chen, Y., Sandoval, P., Gonzales, A., Nalty, M. S., Thomas, T. L., et al. (2005). High-throughput functional analysis of the *Synechococcus elongatus pcc 7942* genome. *DNA Research, 12*(2), 103–115.

Kondo, T., & Ishiura, M. (1994). Circadian rhythms of cyanobacteria: Monitoring the biological clocks of individual colonies by bioluminescence. *Journal of Bacteriology, 176*(7), 1881–1885.

Kondo, T., Strayer, C., Kulkarni, R., Taylor, W., Ishiura, M., Golden, S., et al. (1993). Circadian rhythms in prokaryotes: Luciferase as a reporter of circadian gene expression in cyanobacteria. *Proceedings of the National Academy of Sciences of the United States of America, 90*(12), 5672–5676.

Kondo, T., Tsinoremas, N., Golden, S., Johnson, C., Kutsuna, S., & Ishiura, M. (1994). Circadian clock mutants of cyanobacteria. *Science, 266*(5188), 1233–1236.

Mackey, S. R., Ditty, J. L., Clerico, E. M., & Golden, S. S. (2007). Detection of rhythmic bioluminescence from luciferase reporters in cyanobacteria. *Circadian rhythms* (pp. 115–129). Totowa: Humana Press.

Mackey, S. R., Golden, S. S., & Ditty, J. L. (2011). The itty-bitty time machine: Genetics of the cyanobacterial circadian clock. *Advances in Genetics, 74,* 13–53.

Mayer, U., Ruiz, R. A. T., Berleth, T., Miseéra, S., & Juürgens, G. (1991). Mutations affecting body organization in the *Arabidopsis* embryo. *Nature, 353,* 402–407.

Nolan, P., Kapfhamer, D., & Bućan, M. (1997). Random mutagenesis screen for dominant behavioral mutations in mice. *Methods, 13*(4), 379–395.

Paddock, M. L., Boyd, J. S., Adin, D. M., & Golden, S. S. (2013). Active output state of the *Synechococcus* Kai circadian oscillator. *Proceedings of the National Academy of Sciences of the United States of America, 110*(40), E3849–E3857.

Pau, G., Fuchs, F., Sklyar, O., Boutros, M., & Huber, W. (2010). EBImage an R package for image processing with applications to cellular phenotypes. *Bioinformatics, 26*(7), 979–981.

Price, T., Baggs, J., Curtis, A., FitzGerald, G., & Hogenesch, J. (2008). WAVECLOCK: Wavelet analysis of circadian oscillation. *Bioinformatics, 24*(23), 2794–2795.

Rasband, W. S. (1997). *ImageJ.* Bethesda, Maryland, USA: US National Institutes of Health.

CHAPTER FOUR

Using Circadian Entrainment to Find Cryptic Clocks

Zheng Eelderink-Chen*[,2], Maria Olmedo†[,3], Jasper Bosman*[,4], Martha Merrow*[,†,1]

*Department of Molecular Chronobiology, Groningen, The Netherlands
†Institute of Medical Psychology, Munich, Germany
[1]Corresponding author: e-mail address: merrow@lmu.de

Contents

Abstract

Three properties are most often attributed to the circadian clock: a ca. 24-h free-running rhythm, temperature compensation of the circadian rhythm, and its entrainment to zeitgeber cycles. Relatively few experiments, however, are performed under entrainment conditions. Rather, most chronobiology protocols concern constant conditions. We have turned this paradigm around and used entrainment to study the circadian clock in organisms where a free-running rhythm is weak or lacking. We describe two examples therein: *Caenorhabditis elegans* and *Saccharomyces cerevisiae*. By probing the system with zeitgeber cycles that have various structures and amplitudes, we can demonstrate the establishment of systematic entrained phase angles in these organisms. We conclude that entrainment can be utilized to discover hitherto unknown circadian clocks and we discuss the implications of using entrainment more broadly, even in model systems that show robust free-running rhythms.

[2] Present address: Research Group of Chromatin Biochemistry, MPI for Biophysical Chemistry, Göttingen, Germany
[3] Present address: Andalusian Center for Developmental Biology, Carretera de Utrera Km 1, 41013 Seville, Spain
[4] Present address: Groningen Research Institute of Pharmacy, Antonius Deusinglaan 1, 9713 AV, Groningen, The Netherlands

Methods in Enzymology, Volume 551
ISSN 0076-6879
http://dx.doi.org/10.1016/bs.mie.2014.10.028

1. INTRODUCTION

Colin Pittendrigh characterized circadian clocks according to a list of 16 "generalizations" (Pittendrigh, 1960). We commonly use a distilled version of this list, describing them as having a ca. 24-h period in constant conditions, showing temperature compensation of that period and possessing a mechanism by which the free-running circadian rhythm is entrained to exactly 24 h by zeitgebers. A survey of the chronobiology literature quickly indicates that most experimental protocols in our research field rely on constant conditions, or at least an attempt to achieve them. The main reason to do this is that it simplifies experiments substantially. Scientific experimentation is all about standardizing and controlling conditions, usually keeping them unchanging. By adding entraining conditions (zeitgeber cycles) into the mix, we potentially introduce noise into our phenotype that stems from the physical stimulus (i.e., masking), not from the biological clock. A near universal response to zeitgebers is masking (Mrosovsky, 1999; Mrosovsky, Lucas, & Foster, 2001), which is sometimes difficult to discern from unmasked entrainment. Activity in mice is an example of negative masking because it is suppressed by light (Mrosovsky, 1999). *frq* RNA expression in *Neurospora* shows positive masking because it is induced with light exposure at any time of day (Crosthwaite, Loros, & Dunlap, 1995), whereas the timing of FRQ protein expression and of clock-regulated conidiation (asexual spore formation) is an integration of photoperiod, cycle length (T), and free-running period (Remi, Merrow, & Roenneberg, 2010). FRQ accumulation depends on the structure of the zeitgeber (i.e., photoperiod and/or scotoperiod; Tan, Dragovic, Roenneberg, & Merrow, 2004).

Despite many reasons to use free-running period as a clock output, we note that the clock exists in nature—and thus it evolved—in the entrained state. One might therefore expect that entrainment of circadian systems is a more robust property of circadian clocks. We also propose that a fuller understanding of the principles of entrainment—from cells to societies—will give important insights into other clock properties. We start by reviewing some entrainment protocols and what they can reveal about the circadian clock. We then describe two examples of how we used entrainment to show clock properties in the absence of a robust, free-running rhythm. Finally, we discuss additional implications of entrainment and constant condition protocols that are not actually constant.

1.1. Entrainment protocols

What are some of the entrainment methods that can be used to characterize circadian clocks? They all use the trick of changing zeitgeber structure to show a highly systematic set of responses. A very early conceptual breakthrough in circadian biology was the observation that the biological clock behaved like a physical oscillator in that it synchronizes or entrains differently if characteristics of the entraining oscillator change (Bruce, 1960). This is clearly an evolved response to, for example, seasons and other predictable features of the light/dark (LD) cycle.

Just as entraining cycles vary from day to day, free-running, circadian clocks in different individuals have different inherent properties (e.g., amplitude or period). This creates differences in entrainment between individuals in a given zeitgeber cycle, yielding a distribution of chronotypes (Dominoni, Helm, Lehmann, Dowse, & Partecke, 2013; Roenneberg, Wirz-Justice, & Merrow, 2003). Chronotype or phase of entrainment is systematic and depends on many aspects of the zeitgeber, as described below.

1.1.1 Photoperiod—Longer or shorter days, shorter or longer nights

One of the most basic and natural ways to characterize entrainment is to mimic the changing of the seasons by simply changing the photoperiod within a 24-h cycle. The relevance of this protocol lies in the profound, seasonal reproduction observed in some animals, in plants, and also in some fungi (Bünning, 1960; Elliott & Goldman, 1981; Roenneberg & Merrow, 2001). A masked response would follow the zeitgeber transition (either lights-on or lights-off), whereas a clock-regulated one will often change phase relationships, sometimes entraining relative to midnight or midday, regardless of photo- and scotoperiod. An example of these phenomena is found in *Neurospora*, where *frequency* (*frq*) mRNA transcription is induced every time the lights are turned on (masking), whereas FRQ protein is produced rapidly following lights-on in a short photoperiod (after a long night) or after as much as 8 h after lights-on in a long photoperiod entraining cycle (after a short night; Tan et al., 2004). When spore formation in *Neurospora* was investigated, the entrained phase was highly systematic and could follow either a LD transition or midnight/midday, depending on for instance endogenous versus exogenous period lengths (Remi et al., 2010). When mice are given light pulses of different length and are then analyzed for their phase response curves (suggesting entrainment characteristics), the midpoint of the light "pulse"—not the onset and not the offset—corresponds best to the resulting phase shift (Comas, Beersma, Spoelstra, & Daan, 2006). Interestingly, a large-scale study of the timing of human sleep behavior shows that people sleep at different times in summer and winter, indicating that we also show alternative entrainment as photoperiod/scotoperiod changes (Roenneberg, Kumar, & Merrow, 2007).

1.1.2 Zeitgeber strength

It is clear that photoperiod and scotoperiod are both important for entrainment. One of the most dramatic experiments to show this compared activity in *Drosophila* held in LD cycles compared to those entrained in the more realistic light/moonlight cycles (Bachleitner, Kempinger, Wulbeck, Rieger, & Helfrich-Forster, 2007). Flies exposed to light at night showed more overall activity with the two peaks of activity shifting earlier and later (they are generally biphasic, almost crepuscular) and the siesta becoming longer. The presence of light at night thus leads to different timing of and amount of activity/behavior. The implications for modern life, where we choose between sleeping with open curtains (with exposure to artificial lights) and closed curtains (shielding us from moonlight), are numerous. Interestingly, exposure to light at night has been suggested to be associated with various pathologies based on studies of shift workers (Haus & Smolensky, 2013; Schernhammer & Thompson, 2011). Obviously, this form of light at night may be entirely different than what we are exposed to on a night with a full moon, presumably a condition that we have experienced over evolutionary time. Light at night may alter entrained phase by changing zeitgeber strength. The numerical difference between no light and a number of nanoeinsteins of light as occurs with a full moon is minimal but if this is expressed as a ratio or relative amount, the difference can be substantial. Part of the mechanism by which light at night may change entrainment could be by interfering with adaptation characteristics of the light input system. Exposure to light at night should change the responsivity of the system to light during the photoperiod. A quantitative demonstration of this principle can be seen in the type II phase response curves of Aschoff (Aschoff & North Atlantic Treaty Organization. Scientific Affairs Division, 1965). In these experiments, zeitgeber pulses were delivered in the first 24 h of constant conditions, expressly because the magnitude of the phase shift was much less in a light-adapted circadian system than if it was held in constant conditions for longer.

Zeitgeber strength can also be investigated experimentally by changing characteristics of the photoperiod—both amount and duration of light—on a dark background. The relevance to human behavior is evident, as exposure to high-amplitude zeitgeber cycles (more daylight hours spent outside) results in an earlier chronotype (Roenneberg & Merrow, 2007).

1.1.3 Dawn and dusk transitions

The sudden nature of the lights/on and lights/off of a typical laboratory LD cycle ignores the potential complexity of signals in sunrise and sunset. In addition to a gradual change in the amount of total light, which would lead to different adaptation characteristics, the spectral quality of light changes drastically throughout dusk and dawn (Lythgoe, 1979). Experiments comparing dusk and dawn transitions (quantitatively not qualitatively defined) in the lab relative to square-wave lighting conditions suggest that the former condition amounts to a stronger zeitgeber (Boulos, Macchi, & Terman, 2002; Comas & Hut, 2009).

1.1.4 T *cycles and phase angles*

An important protocol in the circadian repertoire has been *T* cycles. "*T*" represents the exogenous cycle length. A given individual entrains later in a shorter cycle and earlier in a longer one (Aschoff, 1978; Hoffmann, 1965; Merrow, Brunner, & Roenneberg, 1999). This observation has led to a number of important extensions. First, a noncircadian, driven rhythm—one that is simply masking or responding directly to the zeitgeber transitions—will show no such phase angles and will synchronize to the same time relative to the zeitgeber, independent of cycle length. Second, the protocol can be turned around to understand what happens with entrainment of individuals with different free-running periods. In theory, individuals with shorter free-running periods should entrain earlier and those with longer ones should entrain later. This is generally correct (Duffy, Dijk, Hall, & Czeisler, 1999; Merrow et al., 1999) although there are other possibilities besides period (e.g., amplitude) that could explain an alternative entrained phase (Granada, Bordyugov, Kramer, & Herzel, 2013). Finally, we also suggest that entrainment behavior in *T* cycles reveals the inherent clock property of robustness. A weak system would show drivenness, whereas only a robust one would change phase angles in concert with small changes in endogenous period. Eventually, lack of entrainment results as the range of entrainment is exceeded (Aschoff, 1978).

2. METHODS

What follows are two examples of using entrainment to deduce a circadian clock in organisms where a free-running rhythm has been difficult to discern. In both cases, protocols showing circadian entrainment were first optimized. Experiments to show free-running rhythms followed.

2.1. *Saccharomyces cerevisiae*

A haploid yeast strain was grown in 1 L chemostat cultures in YPD (Eelderink-Chen et al., 2010). This is in contrast to most laboratory experiments where diploid *Saccharomyces cerevisiae* strains are grown continuously in Satroutdinov medium. In these other protocols, pH is strictly controlled (Keulers, Suzuki, Satroutdinov, & Kuriyama, 1996; Lloyd, Lemar, Salgado, Gould, & Murray, 2003; Murray, Engelen, Keulers, Kuriyama, & Lloyd, 1998; Satroutdinov, Kuriyama, & Kobayashi, 1992; Tu, Kudlicki, Rowicka, & McKnight, 2005) and dissolved oxygen (dO_2) in the media is the measured "output." Under these conditions, ultradian rhythms were observed. Rather, we primarily used extracellular pH to monitor the state of the yeast cultures, as *Gonyaulax* cultures showed a circadian oscillation in pH (Eisensamer & Roenneberg, 2004). We determined optimal conditions in an iterative procedure. Chemostat cultures were subjected to two different temperature cycle conditions (18/25 °C and 21/28 °C) and to a variety of dilution rates. The changing phase angle between the pH rhythm and zeitgeber cycles was taken as an indication of a yeast circadian clock.

2.1.1 *Identification of the optimal dilution rate*

The temperature cycles consisted of an 11-h cold phase (18 °C or 21 °C) followed by a 60-min temperature transition to an 11-h warm phase (25 °C or 28 °C), which was followed by a 60-min temperature transition to the cold phase. The entire zeitgeber cycle was thus 24 h. This zeitgeber structure was implemented to avoid a "square wave" temperature cycle. Dilution rates describe the rate at which fresh media is pumped into the culture, while simultaneously an equivalent volume of conditioned or depleted media with cells is pumped out of the culture. Dilution rates of between 0.025 and 0.125 h^{-1} were tested (dilution rates refer to liters of media per hour.) In the cycles at the lower temperature, a great variety of waveforms were observed (Fig. 1). At 0.078 h^{-1}, for instance, two peaks in the pH oscillation occurred. At higher dilution rates, the pH oscillation shifted

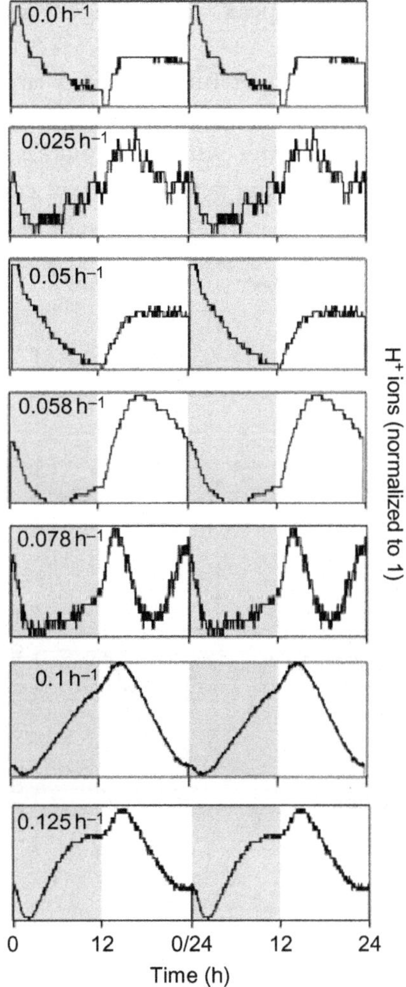

Figure 1 Yeast chemostat cultures in temperature cycles (18–25 °C). Oscillations in H^+ concentrations at different dilution rates. Gray and white areas indicate cool and warm phase (18 or 25 °C). The cycle was structured with 11-h at the cool temperature, a 60 min transition to the warm temperature, 11 h at the warm temperature and then a 60 min transition to the cool temperature. Here, the pH is converted to proton concentration. The H^+ oscillations were calculated without smoothing and trend correction. The dilution rate represents L/h (e.g., 0.078 h^{-1} means 0.078 L/h were circulated through the 1 L culture each hour). The data are double plotted.

slightly earlier although the peak in hydrogen ion concentration still occurred in the warm phase.

Experiments at higher temperatures showed larger changes in phase angles (Fig. 2). Starting with the dilution rate of 0.054 h^{-1}, the phase of the pH oscillation became earlier with each increase in dilution rate. At the two highest dilution rates, the peak of the ion concentration switched

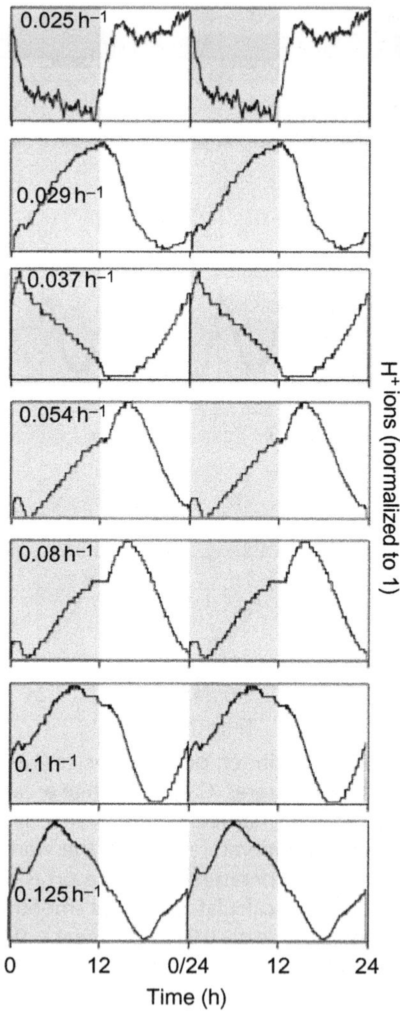

Figure 2 Yeast chemostat cultures in temperature cycles (21–28 °C). Oscillations in H^+ concentrations at different dilution rates in cycles at high temperature. See Fig. 1 for description of graphs except that here the temperatures used for the zeitgeber cycle are 21 and 28 °C. The data are double plotted.

Table 1 H^+ oscillations in a 12/12-h temperature cycle (18–25 °C) with various dilution rates

Dilution rate (h^{-1})	Amplitude (H^+ ions, $M \times 10^{-6}$)	r
0.025	9.71	0.95
0.05	18.16	0.92
0.058	31.15	0.99
0.078	5.02	0.78
0.1	34.63	1.0
0.125	9.08	0.97

Table 2 H^+ oscillations in a 12/12-h temperature cycle (21–28 °C) with various dilution rates

Dilution rate (h^{-1})	Amplitude (H^+ ions, $M \times 10^{-6}$)	r
0.025	7.23	0.89
0.029	18.08	0.99
0.037	114.26	0.99
0.054	26.24	0.99
0.08	26.13	0.99
0.1	27.32	0.99
0.125	26.4	0.99

to the cool phase. The experiments were additionally analyzed for robustness by comparing amplitude of the H^+ oscillation (Tables 1 and 2). The dilution rates that showed the most robust pH rhythm as defined by the r-value and the amplitude were 0.058 and 0.1 h^{-1} in the 18/25 °C cycles and 0.1 h^{-1} in the 21/28 °C cycle.

2.1.2 T cycles

We investigated whether yeast synchronizes to symmetrical T cycles with systematically changing phase angles. In temperature cycles of 18/25 °C with a dilution rate of 0.058 h^{-1}, fermenter cultures were subjected to T cycles with a period of 26, 24, 16, and 14 h (Fig. 3). The oscillation in ion concentration shifted later as the T cycles transitioned from long to short. The peak of the H^+ oscillation moves from the beginning of the warm phase in the 26-h cycle to the warm to cold transition in cycles of 14 h.

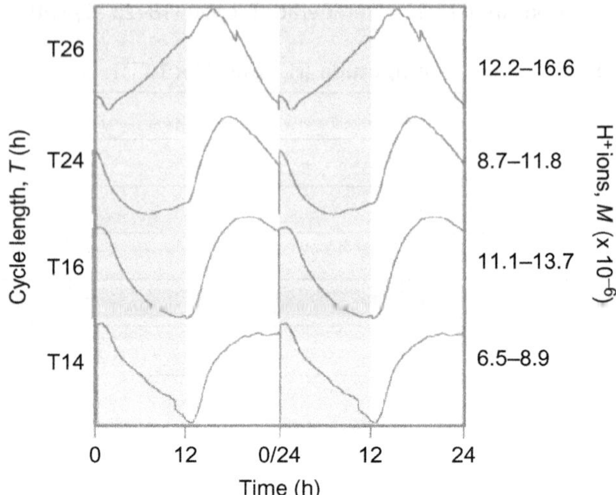

Figure 3 *T* cycles. Yeast chemostat cultures were held in temperature cycles of 18–25 °C with a dilution rate of 0.058 h^{-1}. Gray and white areas indicate cold and warm phase. The cycle lengths are indicated on the left and the concentrations of H^+ in each cycle are indicated on the right. The units on the *x*-axis represent each cycle, independent of its actual length, divided into 24 h of equal length. The phase of the H^+ oscillations changes systematically with cycle length. The waveform of the proton concentrations obviously changes in different cycle lengths. The data are double plotted.

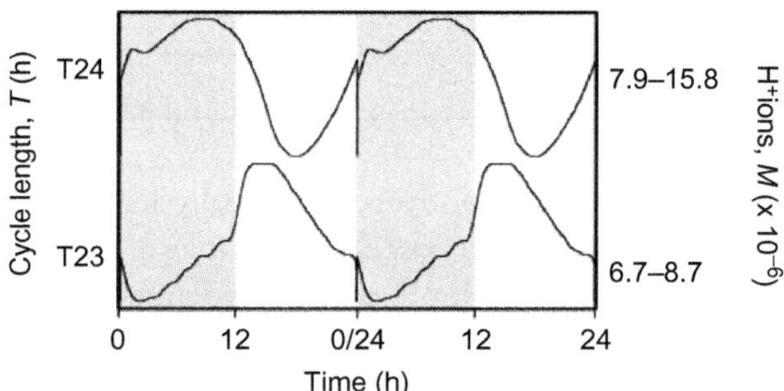

Figure 4 *T* cycles. Yeast chemostat cultures were held in temperature cycles of 21–28 °C with a dilution rate of 0.1 h^{-1}. Gray and white areas indicate cold and warm phase. The cycle lengths are indicated on the left and the concentrations of H^+ in each cycle are indicated on the right. The phase of the proton concentration oscillations varied with cycle length: the shorter cycle ($T=23$ h) leads to a delayed phase. The data are double plotted.

We compared a 24-h cycling yeast culture with one in a 23-h cycle at the higher temperature conditions (21/28 °C, dilution rate of 0.1 h^{-1}; Fig. 4). In this case, the phase of the oscillation shifted by much more than the 1-h change in *T*. The peak of H^+ ions occurred in either the cold or the warm phase in the longer or shorter cycles, respectively. The troughs also switched position. In summary, in both conditions (18/25 °C with a dilution rate of 0.058 h^{-1} and 21/28 °C with a dilution rate of 0.1 h^{-1}), the phase of the pH rhythm changed with shortening of the *T* cycle by 1 h. Such an observation is assumed to indicate that there is an underlying oscillator—in yeast—that has a period that resonates with the entraining cycle. This oscillator must also possess sufficient robustness to deliver these very different phase angles in different *T* cycles. A very weak oscillator would give very small ones. In addition, for the entraining cycle to be "read" by the yeast, we assume that yeast possesses a zeitgeber sensory system, in this case, obviously for temperature.

2.1.3 Zeitgeber strength and entrainment of yeast

Zeitgeber strength was altered in two ways: (1) as referred to above, simply shifting the temperature cycle to lower mean level (18/25 °C vs. 21/28 °C) and (2) changing amplitude of temperature cycles (16/27, 18/25, 19/23, and 20/22 °C; Fig. 5). The 20/22 °C cycle failed to yield a rhythmic pH oscillation, suggesting that this zeitgeber condition is too weak to entrain a yeast circadian clock (data not shown). In the entraining temperature cycles with increasing amplitude, the phase of the pH rhythm becomes later. At lower temperatures, the phase of the H^+ oscillation also moves later than the phase in higher temperatures. This suggests that, at least within the conditions tested, lower temperature (18/25 °C) is perceived as a stronger zeitgeber than the higher temperature condition (21/28 °C). That is, entrainment at the lower temperature is more like the higher amplitude temperature cycles.

2.1.4 Constant conditions: Free-running rhythm?

We interpreted the systematic entrainment response of the H^+ oscillations in the chemostat cultures as evidence of an underlying circadian clock. We further hypothesized that by entraining the yeast in the chemostat in temperature cycles, we might be synchronizing the system—within cells, organizing their metabolism and gene expression relative to the zeitgeber cycle, and between cells, bringing them to the same phase throughout the culture. In this case, one might expect to see free-running rhythms on release to constant conditions. The cultures were entrained for at least 1 week in 24-h temperature cycles until characteristic phase angles had been established. Once again, the pH was monitored. When cultures were released to the lower constant temperature (18 °C), the pH rhythm damped after a single peak in H^+ (Fig. 6A). When the cultures were released to constant high temperature (25 °C), they showed one ca. 24-h oscillation before damping

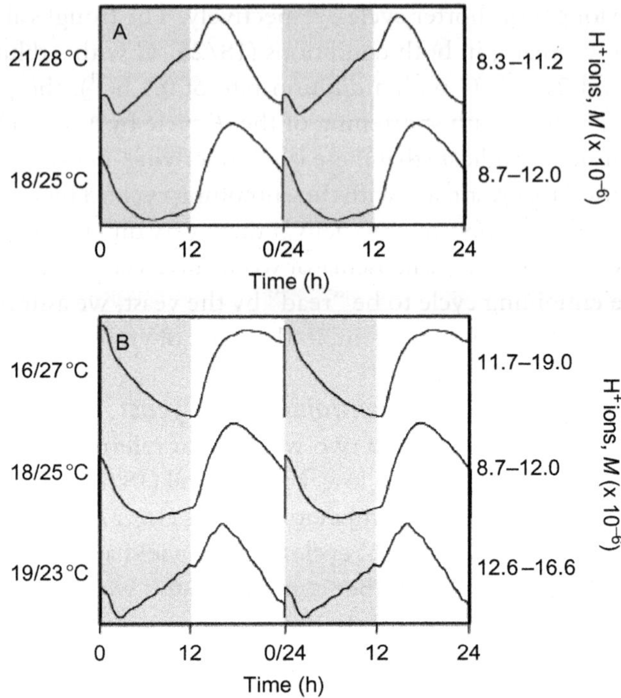

Figure 5 Phase relationships change with zeitgeber strength. Gray panels indicate cool phase; white panels indicate warm phase. (A) The phase of the H^+ oscillation in temperature cycles with the same amplitude but different absolute values. In lower temperature cycles (18–25 °C; lower panel), the peak of the H^+ oscillation shifted to the later position than the peak in warmer cycles (21–28 °C; upper panel). (B) The phase of the H^+ oscillation in temperature cycles with different amplitudes. The peak of the H^+ oscillation shifted to a later phase in the high-amplitude cycles. The data are double plotted.

(Fig. 6B). This may mean that yeast has no machinery that supports a free-running rhythm, that the potential of the system to free-run under these conditions is limited, or that individual cells remain rhythmic but a free-running rhythm in the bulk culture rapidly damps. The latter condition would lead to an apparent arrhythmicity in constant conditions, as has been clearly documented in tissue culture cells (Leise, Wang, Gitis, & Welsh, 2012).

2.2. *Caenorhabditis elegans*

Although many groups have published studies suggestive of a circadian clock in this nematode (Kippert, Saunders, & Blaxter, 2002; Migliori et al., 2012; Migliori, Simonetta, Romanowski, & Golombek, 2011; Saigusa et al., 2002;

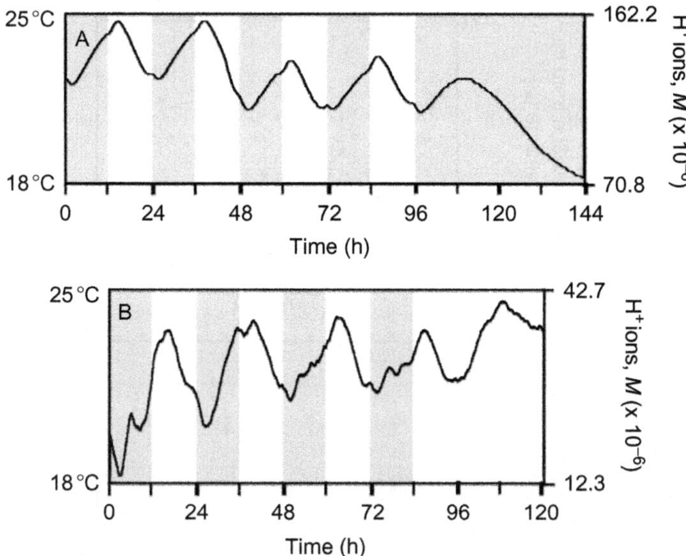

Figure 6 Entrained yeast cultures released to constant temperature. Chemostat cultures were entrained in 24-h temperature cycles of 18–25 °C with dilution rate of 0.1 h^{-1}. Gray and white areas indicate cold and warm phase. (A) A culture released to constant low temperature (18 °C). (B) A culture as released to constant high temperature (25 °C).

Simonetta, Romanowski, Minniti, Inestrosa, & Golombek, 2008; van der Linden et al., 2010), quantitative protocols—which would be necessary to fully exploit this model system with genetic methods—are still lacking. As an approach to the study of circadian rhythms in *C. elegans*, we measured the daily regulation of olfaction. The circadian clock regulates sensory input pathways at many levels. Olfaction is under circadian control in insects and mammals (Granados-Fuentes, Tseng, & Herzog, 2006; Krishnan, Dryer, & Hardin, 1999). The nematode *C. elegans* shows attractive and repulsive responses to a wide range of chemicals, and these are typically robust and lend themselves to quantitative assessment (Bargmann, Hartwieg, & Horvitz, 1993).

The olfactory response to a repellant (1-octanol) was measured in temperature cycles (12/12 h, 13/16 °C) at hourly intervals on the sixth day of entrainment. At this time, the nematodes are young adults and have a completely developed nervous system. Under these conditions, there is a more pronounced response to 1-octanol during the warm phase of the cycle (Fig. 7A). The temperature transition from cold to warm coincided with increased olfactory responses. The stronger response in warm conditions

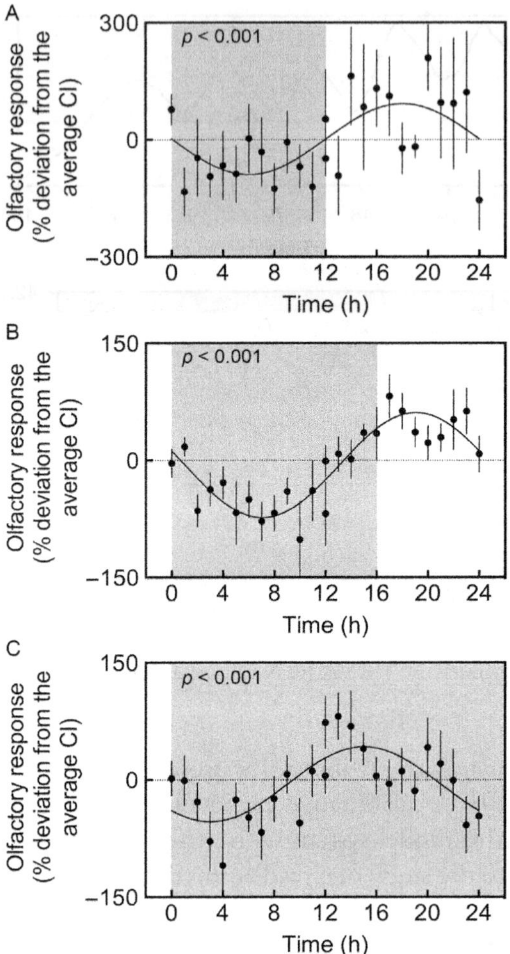

Figure 7 Olfaction in *C. elegans* in response to 1-octanol. Gray and white panels represent cold and warm phases of a temperature cycle (13/16 °C), respectively. The response to 1-octanol is measured as percentage deviation from the average chemotaxis index of the experiment. The timing of harvest (*x*-axis) is expressed as time from the end of the warm to cold transition at the beginning of the 6th cycle. A sinusoidal curve was fitted to the data ($P < 0.001$) using Circwave. (A) Response to 1-octanol over 24 h in a 12/12 temperature cycle. (B) Response to 1-octanol over 24 h in a 16/8 cool/warm temperature cycle. (C) Response to 1-octanol over 24 h in constant conditions (14.5 °C).

might be a direct reaction to the temperature increase, be it due to a higher volatility of the chemical, a higher locomotor activity of the nematodes or other to factors that are not directly related to the olfactory sensitivity of the animal.

We thus changed the structure of the zeitgeber cycle to 16/8 h (still at 13/16 °C). A higher response at high temperature was again observed; however, importantly, the increase in the olfactory response starts during the cold period rather than with the transition to warm (Fig. 7B). This response pattern is typical for the quality called "anticipation" that is often referred to in circadian biology. Using the entrainment protocols to synchronize the system, we looked for sustained rhythms upon release to a free run. The measurement of the response during the first day in constant conditions after five cycles of entrainment (16/8 h at 13/16 °C) reveals a free-running rhythm with an earlier peak, suggesting a short free-running period and also the contribution of temperature-induced masking to the entrained phenotype (Fig. 7C).

3. DISCUSSION

In this chapter, we discuss how and why entrainment can be used as a quantitative measure for circadian behavior, even in the absence of other overt clock properties. The clock evolved under highly predictable zeitgeber conditions and—as a phenotype—it may be more robust than free-running rhythms. The circadian system is tuned to accommodate and interpret changing seasons and environmental conditions using an active entrainment mechanism. Entrainment is furthermore the property that leads to chronotypes (Roenneberg et al., 2007), or the differences in the timing of human behavior, and also jetlag conditions. Without entrainment, adjustment to shift work would be trivial and presumably less dangerous for health!

Competition experiments clearly demonstrate the power of zeitgeber conditions to drive selection for traits that concern temporal organization (Ouyang, Andersson, Kondo, Golden, & Johnson, 1998). It is thus reasonable to hypothesize that all organisms continually exposed to 24-h zeitgebers will possess a system—a biological clock—for coping with daily exogenous structures. A battery of entrainment protocols exist that are useful for demonstrating systematic circadian entrainment—different states or phases for different conditions. Our key assumption in applying entrainment protocols to discover circadian clocks is the expectation that the entrained phase changes with each alteration in zeitgeber structure or amount. In the absence of the systematic changes, we would assume that we are not using conditions that support circadian entrainment. Conversely, when we do see phase angles, we hypothesize that we are synchronizing a circadian clock.

Here, we describe applying entrainment protocols to valuable genetic model systems (*C. elegans* and *S. cerevisiae*) that are not yet in the circadian toolkit. Anecdotally, over the years many attempts have been made to show free-running, circadian rhythms in these organisms, only yielding limited success. In the case of the nematodes, subtle rhythms were apparent (Kippert et al., 2002; Migliori et al., 2012, 2011; Saigusa et al., 2002; Simonetta et al., 2008; van der Linden et al., 2010) but these were generally not robust and robustness of phenotype is a requirement for full exploitation of genetics methods. A rhythm in growth rate was observed in yeast when it was cultured at very low temperatures (12 °C) in bulk cultures in an LD cycle (Edmunds, Apter, Rosenthal, Shen, & Woodward, 1979). The rhythm persisted when constant conditions were initiated. Unfortunately, these experiments were never repeated. Over the years, we have experimented with many methods for culturing *S. cerevisiae* that would allow visualization of circadian or clock-regulated properties. Just to list a few of these, we tried to see morphological differences in yeast cultures that were induced to grow as pseudohyphae by stressing nutritional state (Gimeno, Ljungdahl, Styles, & Fink, 1992). Changing nutrition had revealed a new set of clock-regulated behaviors in *Gonyaulax* (Rehman, Lindgren, & Roenneberg, 1996). We monitored pH of bulk cultures of unicellular yeast in entraining cycles and in constant conditions, most successfully growing *Neurospora* instead of our yeast. We tried a molecular approach, reasoning (in the 1990s) that the heat and cold shock system would be involved in normal circadian entrainment with temperature but failed to see systematic regulation of these genes in our cultures (all unpublished data). Without having any good clock gene candidates in yeast, the development of good protocols also lacked a genetic approach.

When two publications showed ultradian rhythms in yeast (with periods of ~40 min and ~4–5 h; Murray et al., 1998; Murray, Klevecz, & Lloyd, 2003; Tu et al., 2005), we hypothesized that these might be related to circadian oscillations. Using the same culturing conditions as Murray et al., we were able to show ultradian rhythms in budding yeast. The conditions used by Tu et al. require air pressure levels that are incompatible with our equipment. Interestingly, by increasing pressure within equipment specifications, we successfully extended the ~40-min cycle of Murray et al. into the hours range. Additional adjustment of the culturing conditions failed to extend the ultradian rhythms to circadian ones. Hence, we turned to ca. 24-h entrainment protocols.

In these protocols, yeast showed circadian entrainment but failed to show robust, self-sustained circadian rhythms. Why is this? Formally speaking, even systems that show systematic entrainment need not have a self-sustained free-running rhythm. Anticipation of a zeitgeber could be programmed based on the previous zeitgeber exposure. There are other explanations that may be more likely. For instance, the growth conditions for most model organisms in the lab are drastically different than those in nature. We rarely consider the ecology of the organism when designing our artificial growth media in the lab. Rather, considerations include short generation time and cost control. It is worth noting that in well-studied circadian model systems, the free-running rhythm is dependent on the particular set of constant conditions. In constant darkness (DD) the period is different than in dim LL, and in bright LL almost all nonphotosynthetic organisms become arrhythmic (Roenneberg & Hastings, 1991; Somers, Devlin, & Kay, 1998). Changing nutritional conditions can lead to changes in the period, despite metabolic compensation (Dragovic, Tan, Gorl, Roenneberg, & Merrow, 2002). Similarly, changes in temperature lead to changes in period despite temperature compensation (Barrett & Takahashi, 1995; Dibner et al., 2009; Gardner & Feldman, 1981).

It may be time to reconsider how or why we use a free-running rhythm. If there is no single free-running rhythm (as elaborated above), then what is the value therein? For modeling purposes, why is the period in DD used rather than in dim LL? After all, in the entrained state, there is some darkness and some light. Therefore, which is relevant? Furthermore, many protocols that claim constant conditions cannot deliver this. Consider the case of mice. We know much about the genetics of free-running rhythms due to the robustness of the behavioral clock phenotype in this tiny workhorse of a model organism. The catalog of free-running rhythms includes that of body temperature. Thus, even in animals in DD, all tissues will receive ca. 24-h temperature cycles. Thus the liver, heart, lung, etc., are harvested at "circadian" timepoints from these animals will be experiencing a physiologically relevant entraining temperature cycle, not constant conditions in any sense. Furthermore, the internal zeitgeber strength will be altered on the days when the system is released from an LD cycle to DD, leading to new internal phase relationships. It may be that most of these experiments are harvested at a time when phases are volatile—in transition—and much less stable than if they had been maintained in a standardized LD cycle. We predict that this state would lead to a loss of information through underrepresentation of rhythmic genes as they shift their entrained state from one set of zeitgeber cycles and amplitudes to another. Genes that might be robustly rhythmic may be judged nonrhythmic simply due to lack of statistical power in a system that has more noise due to a less stable, transition state.

Given the importance of entrainment of the circadian system for the health and wellbeing of living organisms, it is important to look critically at our experimental paradigms and understand what they are delivering to us. We suggest that standardized entrainment protocols should replace free-running conditions, a practice that will deliver at once more relevant information in protocols that actually are what they are supposed to be.

ACKNOWLEDGMENTS

We thank Till Roenneberg, Serge Daan, and Gabrielle Mazzotta for many helpful discussions. Our work is supported by the Dutch Science Foundation (NWO), the Rosalind Franklin Fellowship Program of the University of Groningen, and the Ludwig-Maximilians-Univeristät in Munich.

REFERENCES

Aschoff, J. (1978). Circadian rhythms within and outside their ranges of entrainment. In I. Asenmacher, & D. S. Farner (Eds.), *Environmental endocrinology* (pp. 172–181). Berlin/Heidelberg/New York: Springer Verlag.

Aschoff, J., & North Atlantic Treaty Organization. Scientific Affairs Division (1965). *Circadian clocks: Proceedings*. Amsterdam: North-Holland Pub. Co.

Bachleitner, W., Kempinger, L., Wulbeck, C., Rieger, D., & Helfrich-Forster, C. (2007). Moonlight shifts the endogenous clock of Drosophila melanogaster. *Proceedings of the National Academy of Sciences of the United States of America*, *104*(9), 3538–3543. http://dx.doi.org/10.1073/pnas.0606870104.

Bargmann, C. I., Hartwieg, E., & Horvitz, H. R. (1993). Odorant-selective genes and neurons mediate olfaction in C. elegans. *Cell*, *74*(3), 515–527.

Barrett, R. K., & Takahashi, J. S. (1995). Temperature compensation and temperature entrainment of the chick pineal cell circadian clock. *The Journal of Neuroscience*, *15*(8), 5681–5692.

Boulos, Z., Macchi, M. M., & Terman, M. (2002). Twilights widen the range of photic entrainment in hamsters. *Journal of Biological Rhythms*, *17*(4), 353–363.

Bruce, V. (1960). Environmental entrainment of circadian rhythms. *Cold Spring Harbor Symposia on Quantitative Biology*, *25*, 29–48.

Bünning, E. (1960). Circadian rhythms and the time measurement in photoperiodism. *Cold Spring Harbor Symposia on Quantitative Biology*, *25*, 249–256.

Comas, M., Beersma, D. G., Spoelstra, K., & Daan, S. (2006). Phase and period responses of the circadian system of mice (Mus musculus) to light stimuli of different duration. *Journal of Biological Rhythms*, *21*(5), 362–372. http://dx.doi.org/10.1177/0748730406292446.

Comas, M., & Hut, R. A. (2009). Twilight and photoperiod affect behavioral entrainment in the house mouse (Mus musculus). *Journal of Biological Rhythms*, *24*(5), 403–412. http://dx.doi.org/10.1177/0748730409343873.

Crosthwaite, S. K., Loros, J. J., & Dunlap, J. C. (1995). Light-induced resetting of a circadian clock is mediated by a rapid increase in frequency transcript. *Cell*, *81*(7), 1003–1012.

Dibner, C., Sage, D., Unser, M., Bauer, C., d'Eysmond, T., Naef, F., et al. (2009). Circadian gene expression is resilient to large fluctuations in overall transcription rates. *The EMBO Journal*, *28*(2), 123–134. http://dx.doi.org/10.1038/emboj.2008.262.

Dominoni, D. M., Helm, B., Lehmann, M., Dowse, H. B., & Partecke, J. (2013). Clocks for the city: Circadian differences between forest and city songbirds. *Proceedings of the Biological Sciences*, *280*(1763), 20130593. http://dx.doi.org/10.1098/rspb.2013.0593.

Dragovic, Z., Tan, Y., Gorl, M., Roenneberg, T., & Merrow, M. (2002). Light reception and circadian behavior in 'blind' and 'clock-less' mutants of Neurospora crassa. *The EMBO Journal, 21*(14), 3643–3651. http://dx.doi.org/10.1093/emboj/cdf377.

Duffy, J. F., Dijk, D. J., Hall, E. F., & Czeisler, C. A. (1999). Relationship of endogenous circadian melatonin and temperature rhythms to self-reported preference for morning or evening activity in young and older people. *Journal of Investigative Medicine, 47*(3), 141–150.

Edmunds, L. N., Jr., Apter, R. I., Rosenthal, P. J., Shen, W.-K., & Woodward, J. R. (1979). Light effects in yeast: Persisting oscillations in cell division activity and amino acid transport in cultures of *Saccharomyces cerevisiae* entrained by light-dark cycles. *Photochemistry and Photobiology, 30*, 595–601.

Eelderink-Chen, Z., Mazzotta, G., Sturre, M., Bosman, J., Roenneberg, T., & Merrow, M. (2010). A circadian clock in Saccharomyces cerevisiae. *Proceedings of the National Academy of Sciences of the United States of America, 107*(5), 2043–2047. http://dx.doi.org/10.1073/pnas.0907902107.

Eisensamer, B., & Roenneberg, T. (2004). Extracellular pH is under circadian control in *Gonyaulax polyedra* and forms a metabolic feedback loop. *Chronobiology International, 21*(1), 27–41.

Elliott, J. A., & Goldman, B. D. (Eds.), (1981). *Seasonal reproduction: Photoperiodism and biological clocks*. New York: Plenum Press.

Gardner, G. F., & Feldman, J. F. (1981). Temperature compensation of circadian period length in clock mutants of Neurospora crassa. *Plant Physiology, 68*(6), 1244–1248.

Gimeno, C. J., Ljungdahl, P. O., Styles, C. A., & Fink, G. R. (1992). Unipolar cell divisions in the yeast S. cerevisiae lead to filamentous growth: Regulation by starvation and RAS. *Cell, 68*(6), 1077–1090.

Granada, A. E., Bordyugov, G., Kramer, A., & Herzel, H. (2013). Human chronotypes from a theoretical perspective. *PLoS One, 8*(3), e59464. http://dx.doi.org/10.1371/journal.pone.0059464.

Granados-Fuentes, D., Tseng, A., & Herzog, E. D. (2006). A circadian clock in the olfactory bulb controls olfactory responsivity. *The Journal of Neuroscience, 26*(47), 12219–12225. http://dx.doi.org/10.1523/JNEUROSCI.3445-06.2006.

Haus, E. L., & Smolensky, M. H. (2013). Shift work and cancer risk: Potential mechanistic roles of circadian disruption, light at night, and sleep deprivation. *Sleep Medicine Reviews, 17*(4), 273–284. http://dx.doi.org/10.1016/j.smrv.2012.08.003.

Hoffmann, K. (Ed.), (1965). *Overt circadian frequencies and circadian rule*. Amsterdam: North-Holland Publ. Co.

Keulers, M., Suzuki, T., Satroutdinov, A. D., & Kuriyama, H. (1996). Autonomous metabolic oscillation in continuous culture of Saccharomyces cerevisiae grown on ethanol. *FEMS Microbiology Letters, 142*(2–3), 253–258.

Kippert, F., Saunders, D. S., & Blaxter, M. L. (2002). Caenorhabditis elegans has a circadian clock. *Current Biology, 12*(2), R47–R49.

Krishnan, B., Dryer, S. E., & Hardin, P. E. (1999). Circadian rhythms in olfactory responses of Drosophila melanogaster. *Nature, 400*(6742), 375–378. http://dx.doi.org/10.1038/22566.

Leise, T. L., Wang, C. W., Gitis, P. J., & Welsh, D. K. (2012). Persistent cell-autonomous circadian oscillations in fibroblasts revealed by six-week single-cell imaging of PER2::LUC bioluminescence. *PLoS One, 7*(3), e33334. http://dx.doi.org/10.1371/journal.pone.0033334.

Lloyd, D., Lemar, K. M., Salgado, L. E., Gould, T. M., & Murray, D. B. (2003). Respiratory oscillations in yeast: Mitochondrial reactive oxygen species, apoptosis and time: A hypothesis. *FEMS Yeast Research, 3*(4), 333–339.

Lythgoe, J. N. (1979). *The ecology of vision*. Oxford: Clarendon Press.

Merrow, M., Brunner, M., & Roenneberg, T. (1999). Assignment of circadian function for the Neurospora clock gene frequency. *Nature*, *399*(6736), 584–586. http://dx.doi.org/10.1038/21190.

Migliori, M. L., Romanowski, A., Simonetta, S. H., Valdez, D., Guido, M., & Golombek, D. A. (2012). Daily variation in melatonin synthesis and arylalkylamine N-acetyltransferase activity in the nematode Caenorhabditis elegans. *Journal of Pineal Research*, *53*(1), 38–46. http://dx.doi.org/10.1111/j.1600-079X.2011.00969.x.

Migliori, M. L., Simonetta, S. H., Romanowski, A., & Golombek, D. A. (2011). Circadian rhythms in metabolic variables in Caenorhabditis elegans. *Physiology & Behavior*, *103*(3–4), 315–320. http://dx.doi.org/10.1016/j.physbeh.2011.01.026.

Mrosovsky, N. (1999). Masking: History, definitions, and measurement. *Chronobiology International*, *16*(4), 415–429.

Mrosovsky, N., Lucas, R., & Foster, R. (2001). Persistence of masking responses to light in mice lacking rods and cones. *Journal of Biological Rhythms*, *16*, 585–587.

Murray, D. B., Engelen, F. A., Keulers, M., Kuriyama, H., & Lloyd, D. (1998). NO +, but not NO., inhibits respiratory oscillations in ethanol-grown chemostat cultures of Saccharomyces cerevisiae. *Biochemical Society Transactions*, *26*(4), S339.

Murray, D. B., Klevecz, R. R., & Lloyd, D. (2003). Generation and maintenance of synchrony in Saccharomyces cerevisiae continuous culture. *Experimental Cell Research*, *287*(1), 10–15.

Ouyang, Y., Andersson, C. R., Kondo, T., Golden, S. S., & Johnson, C. H. (1998). Resonating circadian clocks enhance fitness in cyanobacteria. *Proceedings of the National Academy of Sciences of the United States of America*, *95*(15), 8660–8664.

Pittendrigh, C. S. (1960). Circadian rhythms and the circadian organization of living systems. *Cold Spring Harbor Symposia on Quantitative Biology*, *25*, 159–184.

Rehman, J., Lindgren, K., & Roenneberg, T. (1996). Light and nitrate: Interactions of two zeitgebers in the marine unicellular alga Gonyaulax. In *Paper presented at the meeting of the society for research in biological rhythms, Amelia Island, FL, USA*.

Remi, J., Merrow, M., & Roenneberg, T. (2010). A circadian surface of entrainment: Varying T, tau, and photoperiod in Neurospora crassa. *Journal of Biological Rhythms*, *25*(5), 318–328. http://dx.doi.org/10.1177/0748730410379081.

Roenneberg, T., & Hastings, J. W. (1991). Are the effects of light on phase and period of the Gonyaulax clock mediated by different pathways? *Photochemistry and Photobiology*, *53*(4), 525–533.

Roenneberg, T., Kumar, C. J., & Merrow, M. (2007). The human circadian clock entrains to sun time. *Current Biology*, *17*(2), R44–R45. http://dx.doi.org/10.1016/j.cub.2006.12.011.

Roenneberg, T., & Merrow, M. (2001). Seasonality and photoperiodism in fungi. *Journal of Biological Rhythms*, *16*, 403–414.

Roenneberg, T., & Merrow, M. (2007). Entrainment of the human circadian clock. *Cold Spring Harbor Symposia on Quantitative Biology*, *72*, 293–299. http://dx.doi.org/10.1101/sqb.2007.72.043.

Roenneberg, T., Wirz-Justice, A., & Merrow, M. (2003). Life between clocks—Daily temporal patterns of human chronotypes. *Journal of Biological Rhythms*, *18*(1), 80–90.

Saigusa, T., Ishizaki, S., Watabiki, S., Ishii, N., Tanakadate, A., Tamai, Y., et al. (2002). Circadian behavioural rhythm in Caenorhabditis elegans. *Current Biology*, *12*(2), R46–R47.

Satroutdinov, A. D., Kuriyama, H., & Kobayashi, H. (1992). Oscillatory metabolism of Saccharomyces cerevisiae in continuous culture. *FEMS Microbiology Letters*, *77*(1–3), 261–267.

Schernhammer, E. S., & Thompson, C. A. (2011). Light at night and health: The perils of rotating shift work. *Occupational and Environmental Medicine*, *68*(5), 310–311. http://dx.doi.org/10.1136/oem.2010.058222.

Simonetta, S. H., Romanowski, A., Minniti, A. N., Inestrosa, N. C., & Golombek, D. A. (2008). Circadian stress tolerance in adult Caenorhabditis elegans. *Journal of Comparative Physiology. A, Neuroethology, Sensory, Neural, and Behavioral Physiology*, *194*(9), 821–828. http://dx.doi.org/10.1007/s00359-008-0353-z.

Somers, D. E., Devlin, P. F., & Kay, S. A. (1998). Phytochromes and cryptochromes in the entrainment of the Arabidopsis circadian clock. *Science*, *282*(5393), 1488–1490.

Tan, Y., Dragovic, Z., Roenneberg, T., & Merrow, M. (2004). Entrainment dissociates transcription and translation of a circadian clock gene in Neurospora. *Current Biology*, *14*(5), 433–438. http://dx.doi.org/10.1016/j.cub.2004.02.035.

Tu, B. P., Kudlicki, A., Rowicka, M., & McKnight, S. L. (2005). Logic of the yeast metabolic cycle: Temporal compartmentalization of cellular processes. *Science*, *310*(5751), 1152–1158. http://dx.doi.org/10.1126/science.1120499.

van der Linden, A. M., Beverly, M., Kadener, S., Rodriguez, J., Wasserman, S., Rosbash, M., et al. (2010). Genome-wide analysis of light- and temperature-entrained circadian transcripts in Caenorhabditis elegans. *PLoS Biology*, *8*(10), e1000503. http://dx.doi.org/10.1371/journal.pbio.1000503.

CHAPTER FIVE

Wavelet-Based Analysis of Circadian Behavioral Rhythms

Tanya L. Leise[1]
Department of Mathematics and Statistics, Amherst College, Amherst, Massachusetts, USA
[1]Corresponding author: e-mail address: tleise@amherst.edu

Contents

Abstract

The challenging problems presented by noisy biological oscillators have led to the development of a great variety of methods for accurately estimating rhythmic parameters such as period and amplitude. This chapter focuses on wavelet-based methods, which can be quite effective for assessing how rhythms change over time, particularly if time series are at least a week in length. These methods can offer alternative views to complement more traditional methods of evaluating behavioral records. The analytic wavelet transform can estimate the instantaneous period and amplitude, as well as the phase of the rhythm at each time point, while the discrete wavelet transform can extract the circadian component of activity and measure the relative strength of that circadian component compared to those in other frequency bands. Wavelet transforms do not require the removal of noise or trend, and can, in fact, be effective at removing noise and trend from oscillatory time series. The Fourier periodogram and spectrogram are reviewed, followed by descriptions of the analytic and discrete wavelet transforms. Examples illustrate application of each method and their prior use in chronobiology is surveyed. Issues such as edge effects, frequency leakage, and implications of the uncertainty principle are also addressed.

Methods in Enzymology, Volume 551
ISSN 0076-6879
http://dx.doi.org/10.1016/bs.mie.2014.10.011

1. INTRODUCTION

Measuring rhythmic parameters such as period, amplitude, and phase of time course data from biological oscillators can be a challenging endeavor due to the noisy and often nonstationary nature of biological rhythms. Two essential tasks must be carried out: determining whether a significant rhythm is present and, if so, accurately assessing that rhythm's period and other rhythmic parameters. For behavioral data, researchers may want to determine, in addition to these rhythmic parameters, quantities like duration of daily activity, variability in period or time of onset, ultradian patterns within a circadian rhythm, or multiday patterns.

A wide variety of methods have been developed for these tasks, including autocorrelation, sine-fitting, Fourier-related methods, the Hilbert transform, and wavelet-based methods. Different methods can be best-suited to particular types of time series data. Refinetti, Cornélissen, and Halberg (2007) summarize procedures for detection of circadian rhythmicity. Dowse (2009) provides a general overview of methods to assess rhythmicity, with a focus on autocorrelation and maximum entropy spectral analysis (MESA) as applied to biological data, as well as a discussion of conditioning of time series via filtering and an example of discrete wavelet analysis. Application of filtering and digital signal analysis to behavioral and molecular rhythms is described in Levine, Funes, Dowse, and Hall (2002). Nelson, Tong, Lee, and Halberg (1979) and Cornélissen (2014) discuss cosinor analysis for chronobiological data, including 95% confidence regions for amplitude and acrophase. Zielinski, Moore, Troup, Halliday, and Millar (2014) compare the accuracy of six methods commonly applied to circadian data, including the Lomb–Scargle periodogram and MESA, and offer general advice on how to select an appropriate period estimation method. Reviews of wavelet-based methods for analysis of circadian rhythms can be found in Leise and Harrington (2011) and Leise (2013). Other less frequently used methods that can offer distinct insights into circadian data include Bayesian spectral analysis (Cohen, Leise, & Welsh, 2012), serial analysis (Díez-Noguera, 2013), and detrended fluctuation analysis (Hu, Scheer, Ivanov, Buijs, & Shea, 2007), as well as newly developed methods like the phasegram, which can be interpreted as a "bifurcation diagram in time" to reveal qualitative changes in periodicity over time (Herbst, Herzel, Svec, Wyman, & Fitch, 2013).

Certain general principles tend to apply across methods. In particular, the number of cycles is typically more important than the sampling rate for accuracy of period estimation. For instance, the uncertainty in the optimal frequency estimate under the assumption of a single sinusoid plus white noise is reduced 65% by doubling the number of cycles recorded, but only reduced 29% by doubling the sampling rate (Bretthorst, 1988); also see the discussion in Cohen et al. (2012). The nonstationarity typical of biological time series (Refinetti, 2004) implies that any period estimate is inherently associated with relatively large uncertainty unless a large number of cycles can be recorded Cohen et al. (2012). Fortunately, this is often possible with behavioral records. Note that the problem of period estimation from a time series composed of multiple cycles of a rhythm is quite distinct from the problem of detecting the presence of a rhythm in very short (1–2 cycles) data from microarrays or other non-bioluminescence molecular methods. In contexts where it is not possible to sample more than 1 or 2 cycles, it is usually not possible to estimate period, but increasing sampling frequency can assist in detection of rhythmicity through methods like JTK_Cycle (Hughes, Hogenesch, & Kornacker, 2010) and RAIN (Thaben & Westermark, 2014) that look for a rising and falling pattern in the data.

Behavioral records pose some specific problems distinct from those that arise in analysis of molecular or genetic data. In particular, activity tends to be highly discontinuous in nature, in contrast to the more continuous sinusoidal waveforms typical in some other types of circadian data. Activity can sometimes be sparse or erratic, for example, in a condition like constant light under which some nocturnal animals may be relatively inactive. Masking under conditions such as light–dark (LD) cycles can also make it difficult to assess the animal's internal clock through external measures like wheel-running, which itself may affect the clock. Some period estimation methods that work well on molecular data may not work as well on certain behavioral records. For instance, autocorrelation typically yields poor results on activity data composed of short isolated bouts. Detecting ultradian rhythms present in circadian behavioral data has posed a particularly interesting challenge. Stephenson, Lim, Famina, Caron, and Dowse (2007) applied autocorrelation and MESA to assess ultradian rhythms in sleep–wake behavior of rats; a similar analysis was used to compare ultradian periodicities in different mouse strains (Dowse, Umemori, & Koide, 2010). The Lomb–Scargle periodogram and cosinor analysis have been employed to measure changes in ultradian and circadian rhythms in locomotor activity across the female hamster reproductive cycle and under different photoperiods for male hamsters (Prendergast, Beery, Paul, & Zucker, 2012; Prendergast, Cisse, Cable, & Zucker, 2012). Application of wavelet transforms to detect ultradian patterns in behavioral records has been explored in Leise (2013).

In this chapter, we focus on wavelet-based methods useful for analysis of behavioral data. Applications can include isolating a frequency band of interest such as the circadian component of activity, detecting onset of activity, tracking peak of activity, decomposing the variance in different frequency bands to assess strength of circadian rhythmicity, and estimating fluctuations in period, phase, or amplitude over time in order to detect multiday patterns in the period or changes in ultradian rhythms. The underlying mathematical formulations will be briefly described and illustrated, with an emphasis on applications to analysis of behavioral records but also to chronobiology more generally.

2. FOURIER AND WAVELET METHODS FOR TIME SERIES ANALYSIS

2.1. Discrete Fourier transform

2.1.1 Background and theory

A fundamental tool for assessing the spectrum of a signal is the discrete Fourier transform (DFT), which is also a building block for time–frequency methods like wavelet transforms. The term *signal* here refers to an observed time series, for example, the wheel-running record of a caged rodent. The DFT can be an efficient means to determine the dominant frequencies occurring in the signal, for instance, whether circadian rhythmicity is present. The DFT of a signal $x=(x_0, x_1, x_2, \ldots, x_{N-1})$ with N points sampled every Δt time units has coefficients X_k corresponding to frequency $\omega_k=2\pi k/(N\Delta t)$ radians per time unit defined by:

$$X_k=\sum_{n=0}^{N-1} x_n e^{-2\pi i k n/N}, \quad \text{where } e^{i\theta}=\cos\theta+i\sin\theta.$$

Note that frequency ω corresponds to period $2\pi/\omega$ in this context. The Fourier periodogram is a plot of the relative energy or power:

$$E_k=|X_k|^2/\sum_{n=0}^{N-1}|X_n|^2,$$

corresponding to the period $2\pi/\omega_k=N\Delta t/k$.

While the DFT provides an efficient means of evaluating frequency content of a signal and is a powerful theoretical tool, it has several drawbacks for practical use in chronobiology. Unless the number of cycles is quite large, the frequency resolution will be poor (although variations like the Lomb–Scargle and Schuster periodograms partially address this issue) and harmonics may be prominent. The Fourier periodogram also cannot indicate how the period might be changing over time: it reveals *what* frequencies are present, but cannot tell us *when* they occurred.

To illustrate the DFT and the other methods discussed in this chapter, consider a simulated time series oscillating with a 5-day pattern in the period and amplitude (average period of 24 h) and a 4.5 h ultradian rhythm during the first half of each day, sampled every $\Delta t = 15$ min, as shown in Fig. 1A. The Fourier periodogram of this time series is shown in Fig. 1B. The average period of 24.0 h is clearly prominent, and there is also evidence of an approximately 4.5 h rhythm. However, the DFT cannot pinpoint at what

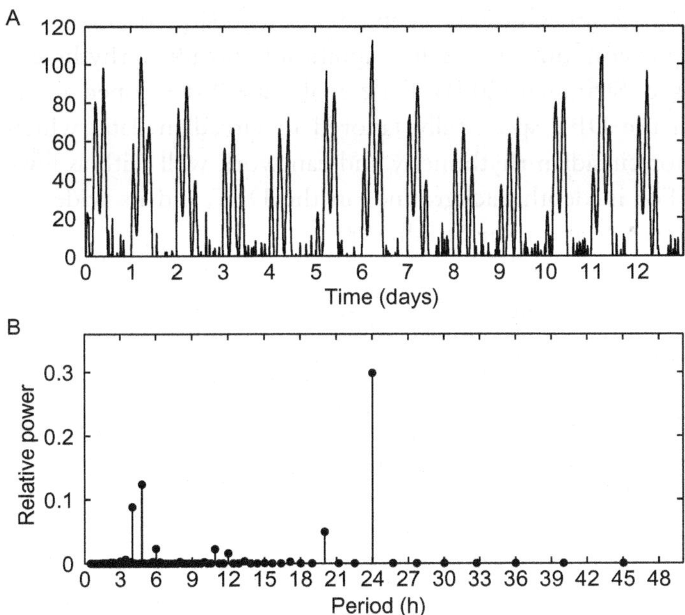

Figure 1 Simulated time series. (A) A simulated signal was generated for illustrative purposes with a 5-day rhythm in the period and amplitude of the underlying circadian rhythm, as well as a 4.5 h ultradian rhythm during the first half of each day and 5% noise. (B) The Fourier periodogram (power spectral density with respect to period) of the simulated signal shows a strong spike at 24 h and several smaller spikes between 4 and 6 h.

time of day the 4.5 h rhythm occurred (in this time series, it is only present during the first half of each day) or detect the multiday changes in the period and amplitude.

2.1.2 Applications to chronobiology

Despite these limitations, the DFT can be useful in a variety of contexts. Granada, Cambras, Díez-Noguera, and Herzel (2011) applied spectral analysis to detect multiple periods in the locomotor activity of rats exhibiting desynchronization under a 22 h LD cycle. The periodogram (power spectral density) clearly revealed a ~25 h-period carrier component plus a 22 h-period fast component and a 28 h-period slow component, as predicted by their theoretical analysis. The power spectral density can be also used as a rhythmicity criterion, for example, to differentiate wild-type rhythms (with strong DFT power at circadian periods) from those of arrhythmic mutants (which exhibit only background levels) (Ko et al., 2010). Various statistical tests of significance level for the Fourier periodogram have been derived; for instance, Fisher's procedure is described in Refinetti et al. (2007). This approach can be adapted to screen large data sets of very short gene expression time series for significant circadian rhythms (Wichert, Fokianos, & Strimmer, 2004). Leise et al. (2012) developed a statistical test based on the DFT specifically tailored to circadian data, which yields a *p*-value for circadian rhythmicity and can work well with as few as 3 days of data. For in-depth background on the DFT and its wide applications to time series analysis, see Smith (2007).

2.2. Short-time Fourier transform

2.2.1 Background and theory

To address the DFT's lack of time localization, we can take the DFT of different time intervals to attempt to localize frequency information with respect to time. Given a window function w, for instance, a Gaussian window with:

$$w_n = exp\left(-\frac{1}{2}(n/\sigma)^2\right),$$

the windowed DFT (or discrete short-time Fourier transform) is defined by:

$$X_{k,m} = \sum_{n=0}^{N-1} x_n w_{n-m(\bmod N)} e^{-2\pi i k n/N}.$$

The essential idea is that the window function isolates a particular portion of the time series, typically tapering at the edges to minimize numerical artifacts. The coefficient $X_{k,m}$ corresponds to the kth coefficient in the DFT of a segment of the time series centered at the mth time point. A heat map of the energy $|X_{k,m}|^2$ is called the spectrogram, a way of visualizing the energy distribution of a signal with respect to time and frequency. While the Fourier periodogram shows how energy (or power) is distributed among various frequencies, the spectrogram improves on this by generating a periodogram for each windowed segment. Think of the spectrogram as a compilation of periodograms displayed as color-coded vertical strips, aligned along the time axis according to the center of each segment. See Fig. 2, for spectrograms of the simulated time series shown in Fig. 1A.

The role of a window parameter like σ is to control the width of the window. Figure 2 illustrates how the width of the window affects the spectrogram. A shorter window yields better time localization, e.g., we can more clearly detect the 5-day pattern in Fig. 2A, while a wider window yields tighter frequency estimates (horizontal bands are narrower) but smears out the time information, as seen in Fig. 2B. A variety of window functions with different properties exist, including Blackman, B-splines, Hanning, and Hamming windows. Such window functions can also be applied to the time series as a whole to improve performance of the DFT, by essentially tapering the edges of the signal to avoid boundary effects. See Mallat (2009) for further details on the spectrogram and other time–frequency methods.

While an improvement over the periodogram if time localization is desired, the spectrogram has some disadvantages. Because the DFT is taken over shortened portions of the time series rather than over the entire duration, the spectrogram will have worsened frequency resolution. It also tends to exhibit strong harmonics (note the horizontal bands at each integer frequency in the sample spectrograms, particularly in Fig. 2B).

2.2.2 Applications to chronobiology

Relatively little use has been made of the spectrogram in the study of circadian rhythms. It has been applied to visualize rhythmicity, e.g., in *Bmal*$^{-/-}$ SCN explants and cells (Ko et al., 2010). A method involving the short-time Fourier transform has been developed as an electrocardiogram-based measure of sleep (Thomas, 2013). For chronobiological applications requiring higher resolution in frequency and time, a better option may be the wavelet transform described in Section 2.3.

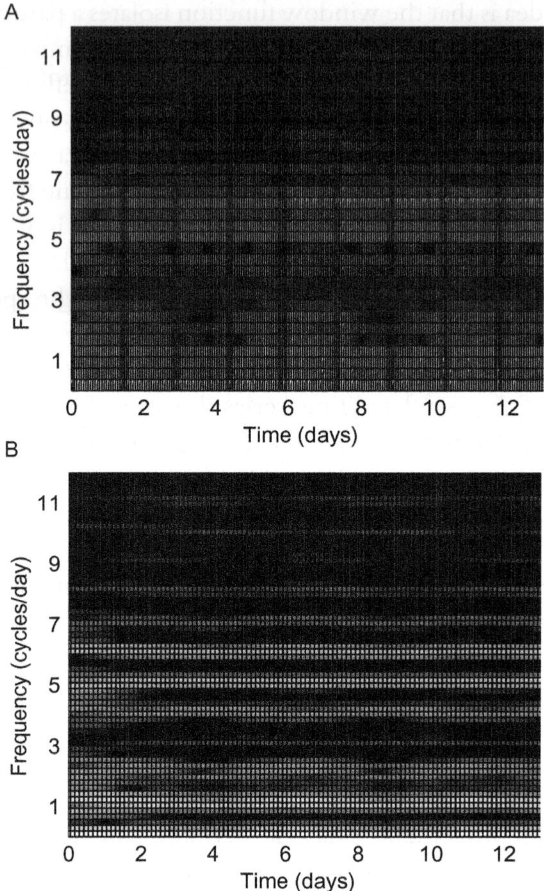

Figure 2 Spectrograms of simulated time series shown in Fig. 1A. Window lengths of (A) 64 h and (B) 128 h were employed to generate the spectrograms, both with a Hamming window, in MATLAB. The shorter window more clearly reveals the 5-day pattern in the circadian rhythm, while the longer window has tighter frequency resolution. (See the color plate.)

2.3. Analytic wavelet transform

2.3.1 Background and theory

To improve the time localization, we need to scale appropriately with respect to frequency: a cycle at a low frequency spans a wider time interval than a cycle at a high frequency. That is, measuring low frequencies requires wide time windows, while a narrow window suffices for observing high frequencies. Wavelet transforms operate on this principle, resulting in excellent resolution (but note that no method can yield perfect resolution with respect to both time and frequency; see discussion of the uncertainty principle below). Wavelet transforms localize in time, like the windowed DFT, but scale the "windows" appropriately according to the frequency (wider for low frequencies, narrower for high frequencies), thereby improving the time–frequency estimates.

We focus here on a particular type of continuous wavelet transform called the analytic wavelet transform (AWT), which involves a complex-valued analytic wavelet function. The formal definition of the AWT coefficient $W(t,s)$ at time t and scale s is:

$$W(t,s) = \int_{-\infty}^{\infty} \frac{1}{s}\psi^*\left(\frac{u-t}{s}\right)x(u)\mathrm{d}u,$$

where the asterisk denotes the complex conjugate. As will be demonstrated below, the scale s essentially codes for the period, so the AWT wavelet coefficient $W(t,s)$ indicates how well the period associated with scale s matches the time series near the timepoint t. A common choice for the function $\psi(t)$ is the Morlet (or Gabor) wavelet, which is essentially a Gaussian window multiplied by a complex exponential $e^{i\nu t}$. An excellent alternative is the Morse wavelet, defined in Lilly and Olhede (2010), where choice of parameters $\beta = 10$ and $\gamma = 3$ produces a function similar to the Morlet; see Fig. 3A. The Morlet wavelet is more widely used, but is only approximately analytic, while the Morse wavelet has the advantage of being exactly analytic (has no negative frequencies). The consequence of the Morlet wavelet not being analytic is leakage to negative frequencies in the AWT, leading to spurious fluctuations in its estimate of the instantaneous frequency. The Morse wavelets do not experience this problem (Lilly & Olhede, 2009).

The AWT is more efficiently calculated in the frequency domain, by integrating the Fourier transform $X(\omega)$ of the signal $x(t)$ against the Fourier transform of the scaled and shifted wavelet function:

$$W(t,s) = \frac{1}{2\pi}\int_{-\infty}^{\infty} \Psi^*(s\omega)X(\omega)e^{i\omega t}\mathrm{d}\omega.$$

From this point of view, the AWT is also windowing with respect to the frequency: the Fourier transform $\Psi(\omega)$ of the wavelet function looks very much like a window function, operating in the frequency domain rather than the time domain (see Fig. 3B). We can also use this frequency domain formula to understand how the AWT works. Given a sinusoidal signal $x_0(t) = a_0 \cos(\omega_0 t + \phi_0)$ with amplitude a_0 and period $2\pi/\omega_0$, the Fourier transform is $X_0(\omega) = \pi a_0\left(e^{i\phi_0}\delta(\omega - \omega_0) - e^{-i\phi_0}\delta(\omega + \omega_0)\right)$, where δ is the

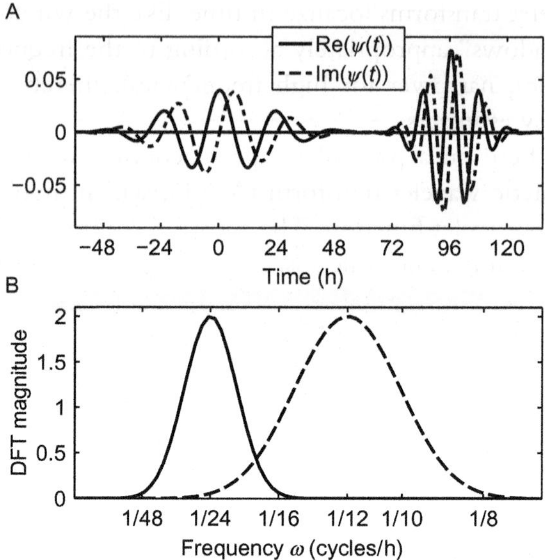

Figure 3 The Morse complex-valued wavelet function $\psi(t)$ and its Fourier transform at two different times and scales, using parameter values $\beta=9$ and $\gamma=3$. (A) On the left is the Morse wavelet function scaled to yield 1 cycle per 24 h, while that on the right is scaled to 1 cycle per 12 h and shifted by 96 h. (B) The corresponding Fourier transforms, with the left curve (solid) having peak frequency at 1 cycle per 24 h and the right curve (dashed) having peak frequency at 1 cycle per 12 h. The wavelet on the left is broader in time but narrower in frequency, while that on the right is narrower in time but broader in frequency, illustrating the essential consequence of the uncertainty principle.

Dirac delta function. Hence, the AWT of $x_0(t)$ according to the frequency domain expression is:

$$W_0(t,s) = \frac{1}{2} a_0 e^{i(\omega_0 t + \phi_0)} \Psi^*(s\omega_0).$$

The wavelet ridge follows the maximum value with respect to scale s of the AWT magnitude at each time t, which for the simple sinusoidal signal is:

$$|W_0(t,s)| = \frac{1}{2} a_0 |\Psi(s\omega_0)|.$$

The function $|\Psi(s\omega_0)|$ takes its maximum at $s\omega_0 = \omega_\psi$, so $\omega_0 = \omega_\psi / s_{\max}$, where ω_ψ is the peak frequency of $\Psi(\omega)$ and $s_{\max}$ is the scale that maximizes $|W_0(t,s)|$. As a result, we interpret the ridge occurring for a general oscillatory signal x at scale $s = s_{\max}$ (that maximizes the magnitude of the AWT) as corresponding to frequency $\omega = \omega_\psi / s_{\max}$. The formula for $W_0(t,s)$ also demonstrates that the angle associated with the complex-valued AWT coefficient yields the phase angle $\omega_0 t + \phi_0$ at time t. The function $\Psi(\omega)$ is normalized to have magnitude 2 at its peak frequency ω_ψ, so the magnitude at the ridge will equal a_0, the amplitude. In this manner, the wavelet ridge yields instantaneous period, phase, and amplitude estimates at each time point. For further details on the AWT and wavelet ridge analysis, see Lilly and Olhede (2009, 2010). Other resources to consult include Mallat (2009), Quotb, Bornat, and Renaud (2011), and Torrence and Compo (1998).

Because data are not continuous, a discretized version of the AWT is applied to time series data, but the same principles hold. To illustrate its use, consider Fig. 4A, which shows the scalogram (heat map of the AWT magnitude) with wavelet ridges for the simulated signal. The wavelet ridges follow the 5-day rhythm in amplitude and period for the ~24 h portion of the rhythm, as well as the 4.5 h ultradian rhythm present during the first half of each day.

A drawback of the AWT, as with the other methods discussed in this chapter, is the potential for distortion near each boundary, i.e., edge effects, which typically requires removal of 1–2 cycles from each edge. Because of this loss of up to 4 cycles worth of information, wavelet-based methods may not work well for records with fewer than 5 cycles of the rhythm. The issue is that the signal must be extended past each edge in some manner in order to

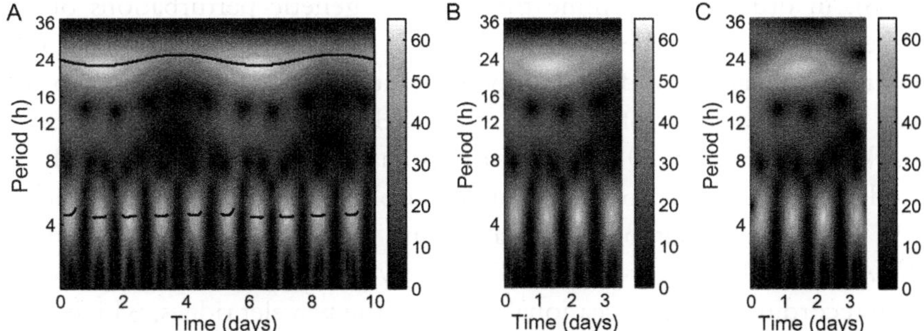

Figure 4 Analytic wavelet transform of the simulated time series. (A) Heat map (scalogram) of the AWT coefficient magnitudes for the simulated signal shown in Fig. 1A. The black curves mark the wavelet ridges, indicating what periods are present in the signal at each time point. The amplitude of each component is indicated by the color of the heat map along the corresponding ridge curve. (B) Accurate scalogram of first 3.5 days of simulated data. (C) Scalogram of the same segment of the time series, exhibiting significant edge effects due to poor choice of boundary conditions. Scalograms were generated in MATLAB using JLAB (Lilly, 2012). (See the color plate.)

compute the transform near the boundary. Common choices are to pad with zeros or the mean value of the time series, to reflect the time series at each end, or to periodically extend it. See Fig. 4B and C, for an illustration of edge effects. Careful handling of boundaries is important to reduce the distortion near the edges and so minimize loss of information at the beginning and end of the time series. For typical activity data, beginning and ending behavioral records at midpoints of rest intervals and then reflecting at each edge may help minimize edge effects.

Another potential difficulty in applying the AWT to behavioral records is that disruptions or large changes in activity can lead to gaps in the wavelet ridge curve, with large artifacts near the breaks. These issues can limit the usefulness of the AWT in calculating the mean period or variability in period unless the rhythm is sufficiently steady to yield a continuous ridge curve over the entire interval.

2.3.2 Applications to chronobiology

The use of continuous wavelet transforms in chronobiology is steadily growing, as demonstrated by the following examples, chosen to illustrate different ways in which wavelets have proven useful. The AWT works very naturally with molecular data, which tend to be fairly sinusoidal in waveform. Baggs et al. (2009) applied the AWT with the Morlet wavelet to measure period and amplitude in cell luminescence data, using the WAVECLOCK package (Price, Baggs, Curtis, Fitzgerald, & Hogenesch, 2008), in order to determine the effect of genetic perturbations on the molecular clock. Etchegaray, Yu, Indic, Dallman, and Weaver (2010) applied a similar method to SCN explant bioluminescence rhythms to analyze the relative roles of casein kinase 1 delta and epsilon in the circadian clock. Meeker et al. (2011) used wavelet analysis combined with stochastic modeling to study period instabilities in isolated SCN neurons. While the AWT offers a means of deeply analyzing molecular data, application of the AWT in this context can be limited by the typically short duration of such recordings and by occasional breaks in the wavelet ridges, which complicate assessment of the period and tend to be associated with numerical artifacts leading to errors in the period estimate.

The AWT has also been applied quite effectively to behavioral records, despite their noisy and discontinuous nature. An earlier study used the AWT to examine the effect of cage size on ultradian rhythms (Poon et al., 1997). More recently, Nakamura, Takumi, Takano, Hatanaka, and Yamamoto (2013) computed a spectral density for circadian and ultradian rhythms of mouse locomotor activity using the AWT with the Morlet wavelet to reveal significant amplification of ultradian rhythms in BMAL1-deficient mice and instability in *Per2* mutants. Examples demonstrating detection of ultradian and multiday patterns of activity using the Morse wavelet are given in Leise (2013). Paul, Indic, and Schwartz (2014) used real-valued continuous wavelet transforms to determine amplitude over time (with a real-valued Morlet wavelet function) and phases of rhythm onset and offset (with a Mexican hat wavelet function) to assess impact of cohabitation on temperature rhythms of hamsters. A real-valued Morlet wavelet transform has also been used to study multiscale characteristics of human motility data, to aid in diagnosing certain psychiatric disorders (Indic et al., 2012, 2011). Further applications of continuous wavelet transforms to study behavioral rhythms are likely to be developed as the field progresses. The availability of accessible wavelet software for analysis of circadian rhythms is also growing, e.g., WAVECLOCK (Price et al., 2008) and WAVOS (Harang, Bonnet, & Petzold, 2012).

2.4. Discrete wavelet transform

2.4.1 Background and theory

Let us return to the Fourier transform, considered from an alternative perspective. Rather than viewing the power spectrum of a signal (its periodogram), we can interpret the DFT as decomposing a signal into a sum of sinusoids with amplitudes given by the coefficients X_k, through the inverse Fourier transform:

$$x_n = \frac{1}{N}\sum_{k=0}^{N-1} X_k e^{2\pi ikn/N}.$$

The drawback to sinusoids is that their frequency is fixed for all time (see the example DFT-derived sinusoidal decomposition in Fig. 5A), while we may want time-localized information about period and amplitude. An alternative method to decompose a signal is to apply a discrete wavelet transform (DWT). The translation-invariant (also called stationary or maximal overlap) DWT works particularly well in the context of circadian data, so we focus on that type of DWT.

The underlying idea is to decompose a signal into components associated with dyadic scales by repeatedly applying a pair of wavelet (high pass) and scaling (low pass) filters ψ and φ of length L, which replace the window function w of the spectrogram (for time localization) and the sinusoidal

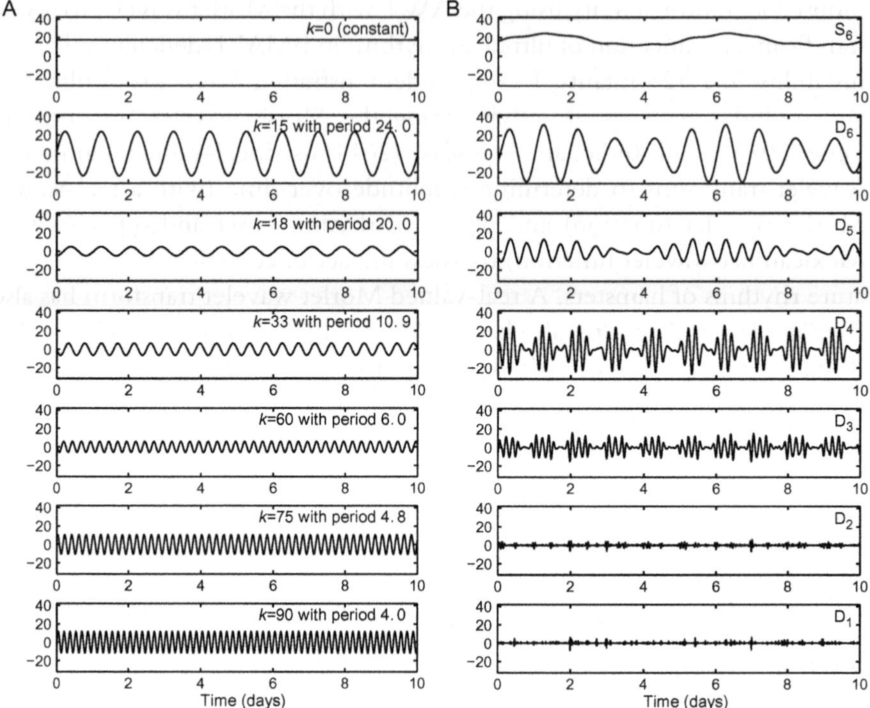

Figure 5 Comparison of two additive decompositions of the simulated signal. (A) Decomposition into sinusoids using frequencies with at least 2% of the DFT energy. (B) Multiscale resolution analysis (MRA) decomposition using the translation-invariant DWT with the Daubechies least asymmetric scaling and wavelet filters of length 12 for 6 scale levels, yielding wavelet details D_1–D_6 and wavelet smooth S_6, calculated in MATLAB using the WMTSA package (Cornish, 2006). The circadian component corresponds to $k=15$ for the DFT and wavelet detail D_6 for the DWT. The 4.5 h rhythm during the first half of each day is reflected in wavelet detail D_4, but is dispersed among several sinusoidal components in the DFT decomposition. In both cases, the sum of the components reconstitutes the original signal, but only the MRA decomposition reveals the changes over time occurring in the period and amplitude at each scale.

expression $e^{2\pi ikn/N}$ in the DFT (for frequency localization). At the jth level, the new wavelet coefficients $W_{j,m}$ and scaling coefficients $V_{j,m}$ are obtained by applying the pair of filters to the preceding level's scaling coefficients $V_{j-1,m}$:

$$W_{j,m} = \sum_{\ell=0}^{L-1} \psi_\ell V_{j-1,m-2^{j-1}\ell \bmod N},$$

$$V_{j,m} = \sum_{\ell=0}^{L-1} \varphi_\ell V_{j-1,m-2^{j-1}\ell \bmod N}.$$

Note that the original signal is treated as the 0th level set of scaling coefficients. The main difference from the standard DWT is that downsampling is not done, with the resulting advantage that the translation-invariant DWT of a signal remains the same when the indexing is shifted to the left or right (hence the name). The idea is analogous to that underlying the AWT: by rescaling and shifting the filters, we can determine which frequencies are present in the signal near each point in time. The major difference between the AWT and the DWT is that the AWT identifies instantaneous frequency over time, while the DWT decomposes the signal into components associated with different frequency bands. After iterating the process through J levels, applying the inverse DWT to the wavelet coefficients $W_{j,m}$ yields the wavelet details D_j for $j = 1, \ldots, J$, while the wavelet smooth S_J is obtained from the top level scaling coefficients $V_{J,m}$. This algorithm leads to a multiresolution analysis of the signal. That is, it decomposes the signal x into a sum of components associated with different scales, where the jth level scale roughly corresponds to periods $2^j \Delta t$ to $2^{j+1} \Delta t$:

$$x = S_J + \sum_{j=1}^{J} D_j.$$

This decomposition is illustrated in Fig. 5B for the simulated signal. Observe that the wavelet detail D_6 clearly shows the 5-day variation in amplitude and period of the circadian component of the signal, and D_4 shows the 4.5 h ultradian activity pattern during the first half of each day. We can similarly decompose the energy in a signal with respect to scale, where *energy* here refers to the sum of squares, $||x||^2 = \sum_{n=0}^{N-1} x_n^2$ (equivalent to the variance if the signal has mean zero). The energy decomposition involves the scaling and wavelet coefficients,

$$||x||^2 = ||V_J||^2 + \sum_{j=1}^{J} ||W_j||^2,$$

and can be useful for measuring the strength of the circadian component and for characterizing behavioral rhythms.

As with other filtering methods, the DWT suffers from edge effects, which can be reduced with a good choice of boundary condition. Options include padding with zeros or the mean value, reflecting at each end, or periodically extending the signal. For instance, if the signal begins and ends at a peak or trough of the cycle, reflection can be a good choice, as shown in Fig. 6. Shorter length scaling and wavelet filters typically decrease edge effects, but do not separate low and high frequencies as well as longer length filters (Fig. 7) and so tend to experience frequency leakage between scales (Fig. 8). Selecting the length of filter to use requires a trade-off between minimizing edge effects and reducing frequency leakage; a longer filter increases edge effects, while a shorter filter increases frequency leakage between scales. For circadian data, a filter of length 12 often works well, balancing these two objectives.

For a comprehensive source on discrete wavelet analysis, see Percival and Walden (2000); for a discussion of wavelets and development of an alternative method, see Selesnick, Baraniuk, and Kingsbury (2005). Also note that a wide variety of wavelet filters have been developed which can be explored to test which works best for a given data set.

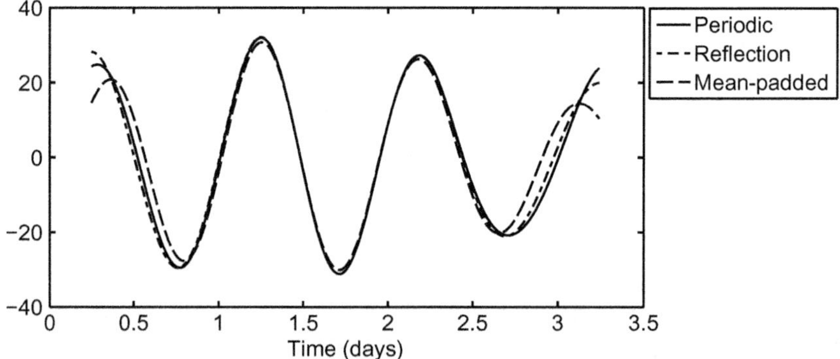

Figure 6 Comparison of the influence of boundary conditions on DWT edge effects. The wavelet detail D_6 for a segment of the simulated signal is shown for three common boundary conditions. Reflection yields the best results in this case, and in general reflection can minimize edge effects when starting and ending at peaks or troughs (for instance, the middle of a rest interval in an activity record). Periodic extensions and padding typically cause discontinuities at the ends, as can reflection at non-optimal points, thereby worsening edge effects.

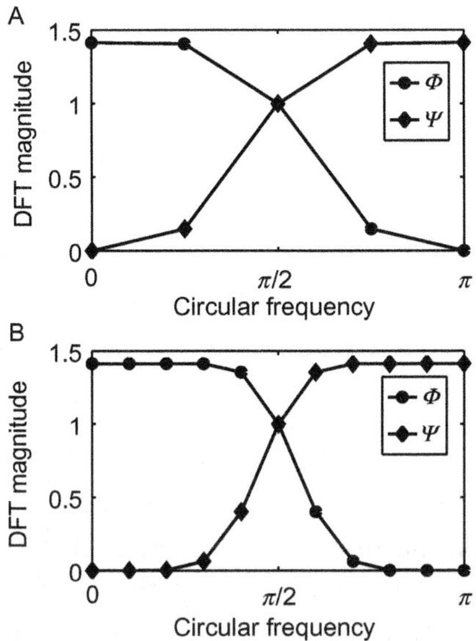

Figure 7 Comparison of the DFTs of (A) length 8 and (B) length 20 Daubechies least asymmetric wavelet filters (symmlets). The scaling filter's DFT Φ covers the lower half, while the wavelet filter's DFT Ψ covers the upper half of the frequency range. Longer filters separate low from high frequencies with less overlap.

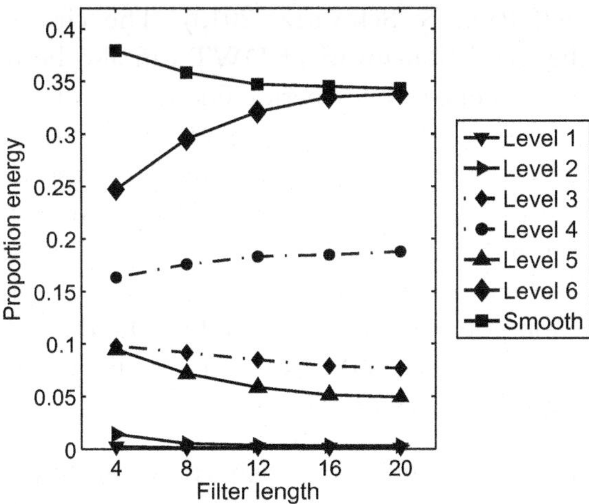

Figure 8 Comparison of frequency leakage by Daubechies filters of different lengths. As filter length increases, the decomposition of energy among the wavelet details at the six levels (D_1–D_6) and the wavelet smooth (S_6) stabilizes, demonstrating that longer length filters reduce leakage between levels and more cleanly separate frequency bands representing different scales.

2.4.2 Applications to chronobiology

Discrete wavelet analysis serves a different purpose than a continuous transform like the AWT. The translation-invariant DWT is an excellent tool for extracting the circadian component of a signal, for removing noise or trend, and for decomposing into different scales. For instance, this method was used to preprocess time series in Evans, Leise, Castanon-Cervantes, and Davidson (2013); a two-dimensional DWT was also applied to remove local background from bioluminescence images to aid in identification of regions of interest. Chan, Wu, Lam, Poon, and Poon (2000) developed a tree-based multiscale characterization of locomotor activity of mice under different lighting conditions. Leise and Harrington (2011) applied an alternative approach using the DWT energy decomposition to characterize activity patterns at different scales. In that study, the proportion of energy at each DWT level provided a quantification of the strength of the circadian component to contrast with the fragmentation of activity that tended to occur under constant light (LL). A similar approach was taken in Leise, Harrington, et al. (2013) to quantify how voluntary exercise strengthened the circadian rhythms of aged mice. As a quite distinct application, the DWT can be adapted to determine activity onsets of locomotor activity by using a Daubechies wavelet filter of length 4, which excels at detecting discontinuities in the first derivative of a signal, as well as the center of each day's activity (Leise, Indic, Paul, & Schwartz, 2013). The circadian component extracted by the translation-invariant DWT can also be used to estimate cycle lengths, as was done to study period fluctuations of cellular oscillations in (Leise et al., 2012). This approach provides a reliable alternative to the AWT ridge curves, which are not guaranteed to exist at all time points and can exhibit distortions near breaks in the ridge. An excellent overview of wavelet theory and how to apply the translation-invariant DWT in ways likely to prove useful beyond the specific application addressed (efficient detection of action potentials) can be found in Quotb et al. (2011). These examples demonstrate that the DWT offers a flexible tool that can assist with a wide variety of signal processing tasks.

2.5. Example with wavelet analysis of a behavioral record

To illustrate the wavelet transforms, we apply both wavelet methods to the wheel-running record of a female mouse entrained to a 12:12 LD cycle for 25 days, followed by weekly 8 h advances of the LD cycle, shown in Fig. 9A, from an experiment described in Leise and Harrington (2011). The circadian component extracted by the translation-invariant DWT is shown in Fig. 9B,

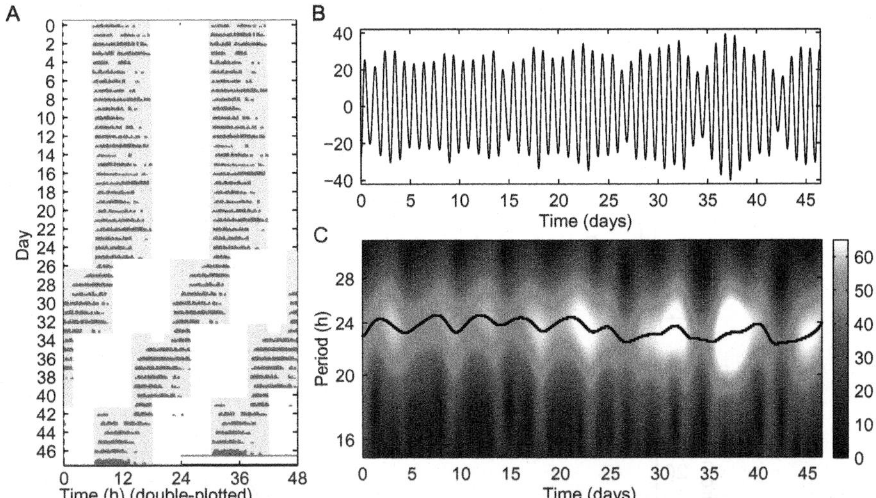

Figure 9 Wheel-running record of female mouse entrained to a 12:12 LD cycle, followed by repeated 8 h advances of the light–dark cycle, record from study described in Leise and Harrington (2011). (A) Actogram of wheel-running activity. (B) Circadian component of activity derived by the DWT. (C) AWT scalogram with ridge. (See the color plate.)

and the scalogram with wavelet ridge is shown in Fig. 9C. Both approaches clearly show the changes in amplitude due to the scalloping of activity and during re-entrainment after each LD shift. The wavelet ridges redrawn in Fig. 10A track how the changes in period and amplitude are coordinated over time. The AWT also provides phase information, allowing calculation of how the mouse's actual phase differs from what would be expected if it had a constant 24 h period. Figure 10B shows this phase difference over time, with roughly 5-day oscillations during the first 25 days due to scalloping of activity, followed by weekly jumps in response to the 8 h advances of the LD cycle.

2.6. Implications of the uncertainty principle for time–frequency analysis

While wavelet transforms can provide excellent period estimates with respect to both time and frequency, no time–frequency method can provide perfect or truly instantaneous estimates, due to the limitations imposed by the Heisenberg uncertainty principle (as applied to signal processing, which is analogous mathematically to the quantum physics version, but with a different interpretation). The Heisenberg uncertainty principle says that the product of the dispersion (a measure of spread) of a signal x about a point

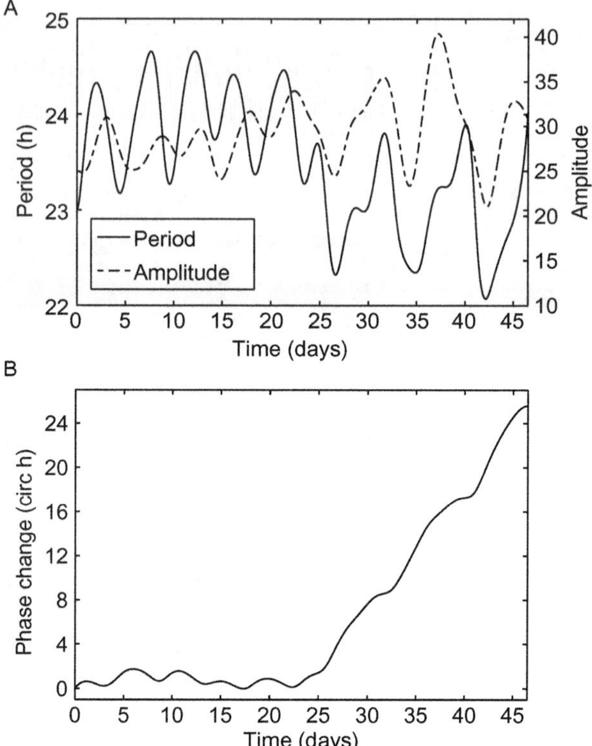

Figure 10 Wavelet ridge curves from the AWT of the wheel-running record shown in Fig. 9. (A) Wavelet ridges showing period and amplitude over time. During the first 25 days under the 12:12 LD cycle, scalloping of activity leads to a regular 5-day oscillation in both period and amplitude. In the second half of the record, weekly 8-h advances of the LD cycle lead to shortened periods while the mouse re-entrains. (B) The difference over time between the wavelet ridge-estimated phase and the predicted phase if the mouse ran with exactly a 24 h period. Scalloping of activity leads to roughly 5-day oscillations in this phase difference during the first 25 days, followed by weekly 8 h increases in response to the 8 h advances of the LD cycle.

in time and the dispersion of its Fourier transform about any given frequency is bounded below by a fixed constant. Think of this product as measuring the area shadowed by a signal in the time–frequency plane, that is, its width in the time domain times its width in the frequency domain. As a consequence of the uncertainty principle, short windows provide good time localization but poor frequency localization (reducing uncertainty about timing increases the uncertainty about frequency), while wide windows provide better frequency localization but reduced time localization. The uncertainty principle forces us to make trade-offs between time and frequency localization, but we can try to do so optimally.

Therefore, let us consider how best to subdivide the time–frequency plane, given the constraint of the uncertainty principle, which essentially says that the areas of the subdivided regions cannot fall below a fixed bound. For instance, examine the spectrograms in Fig. 2. The spectrogram in Fig. 2A is broken into boxes narrower in time but wider in frequency compared to those in Fig. 2B, but the area of both types of boxes is the same. The DFT has good frequency localization with no time localization (divides up plane as shown in Fig. 11A), and the spectrogram uses the same window size

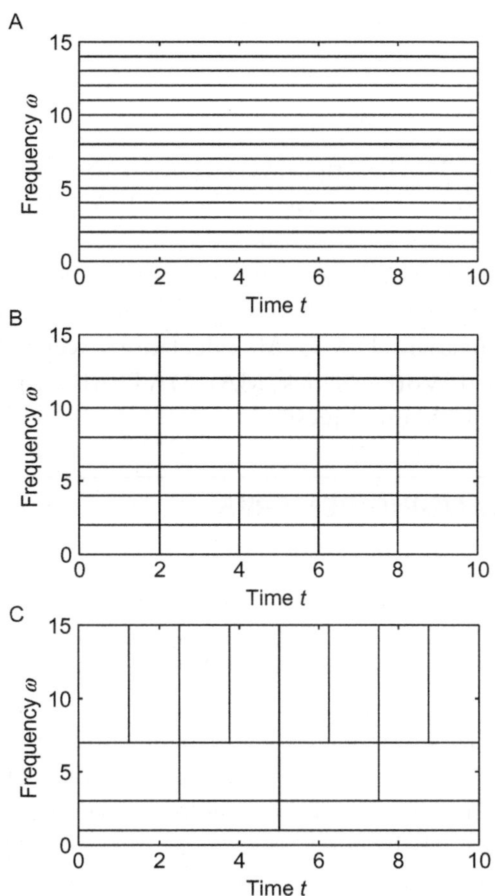

Figure 11 Schematics for division of the time–frequency plane for (A) the discrete Fourier transform, (B) the windowed or short-time Fourier transform (STFT), and (C) the discrete wavelet transform (DWT). The Fourier transform has no time localization, while the STFT applies a window function to obtain frequency estimates local to each time point. The DWT improves on this by adjusting the window length to the frequency: high frequencies can be estimated using a narrow window, while lower frequencies require a wider window.

for all frequencies (Fig. 11B), providing a compromise between frequency and time localization. More optimal, however, is the approach taken by the DWT, which adapts window size through scaling to implement the layout in Fig. 11C, so that low frequencies are associated with wide windows and high frequencies with narrow windows. The fine-grained resolution of the AWT scalogram may make it appear to beat the limitations of the uncertainty principle, but this is an illusion fostered by the beautiful images; in general, the AWT will smooth out fluctuations in period and amplitude and may underestimate changes. See Mallat (2009) for further discussion on the uncertainty principle and its implications.

3. COMPUTATIONS

Custom MATLAB scripts were written to run the computations, making use of two freely available toolboxes: JLAB (Lilly, 2012) for computing the AWT and ridges, using the Morse wavelet function with $\beta=12$ and $\gamma=3$ for analysis of activity records, and WMTSA (Cornish, 2006), companion software to Percival and Walden (2000), for computing the translation-invariant DWT. All calculations were run in MATLAB R2014a (The MathWorks, Natick, MA). MATLAB scripts to run the computations are available on request.

4. CONCLUDING REMARKS

Wavelet-based methods are not necessarily the best choice for analysis of all types of chronobiological data, but they do offer a powerful set of tools. While many period estimation methods require removal of noise and trend to be most effective, wavelet-based methods work well without such preprocessing. In fact, the DWT can be used to extract components of interest, such as the circadian component, thereby efficiently removing any noise or trend that may be present in the time series. Applications of discrete and continuous wavelet transforms in chronobiology have been steadily growing, from characterizing behavioral rhythms at different scales to measuring changes in circadian and ultradian activity patterns. Future directions could include advances such as wavelet packets, which offer further sophistication in discriminating features of a signal and have been employed, for instance, to assist in automating classification of sleep stages in EEG recordings (Ebrahimi, Mikaeili, Estrada, & Nazeran, 2008). Other well-established methods like multitaper techniques (Percival & Walden, 1993) could provide an alternative to continuous wavelet transforms, for instance, to measure ultradian rhythms.

The uncertainty principle tells us that we cannot achieve perfect localization in both time and frequency, with increased time localization leading to worse frequency localization and vice versa. Shannon information theory also tells us that we cannot detect frequencies higher than the Nyquist frequency (half the sampling rate) and that uniformly sampled signals suffer from aliasing (different frequencies can yield the same sampled signal). However, moving beyond the classic theory might be possible by considering alternatives to traditional sampling, e.g., compressed sensing, if the signal is known to be "sparse" with respect to some basis, for instance, only involving a few frequencies (Bryan & Leise, 2013; Tropp, Laska, Duarte, Romberg, & Baraniuk, 2010). Adaptive representations computed through methods like basis pursuit with an "overcomplete waveform dictionary" (Chen, Donoho, & Saunders, 2001) could open up exciting new directions in the analysis of biological oscillations. The field of chronobiology can benefit from the ongoing advances in time–frequency analysis, as we have witnessed with the recent strides in applying wavelet-based methods.

REFERENCES

Baggs, J., Price, T., DiTacchio, L., Panda, S., Fitzgerald, G., & Hogenesch, J. (2009). Network features of the mammalian circadian clock. *PLoS Biology*, *7*, e52.

Bretthorst, G. L. (1988). *Bayesian spectrum analysis and parameter estimation (lecture notes in statistics 48)*. Berlin: Springer.

Bryan, K., & Leise, T. (2013). Making do with less: An introduction to compressed sensing. *SIAM Review*, *55*(3), 547–566.

Chan, F., Wu, B., Lam, F., Poon, P., & Poon, A. (2000). Multiscale characterization of chronobiological signals based on the discrete wavelet transform. *IEEE Transactions on Biomedical Engineering*, *47*, 88–95.

Chen, S., Donoho, D., & Saunders, M. (2001). Atomic decomposition by basis pursuit. *SIAM Review*, *43*(1), 129–159.

Cohen, A., Leise, T., & Welsh, D. (2012). Bayesian statistical analysis of circadian oscillations in fibroblasts. *Journal of Theoretical Biology*, *314*, 182–191.

Cornélissen, G. (2014). Cosinor-based rhythmometry. *Theoretical Biology and Medical Modelling*, *11*, 16.

Cornish, C. (2006). *WMTSA wavelet toolkit for MATLAB 0.2.6*. http://www.atmos.washington.edu/wmtsa/.

Díez-Noguera, A. (2013). Methods for serial analysis of long time series in the study of biological rhythms. *Journal of Circadian Rhythms*, *11*, 7.

Dowse, H. (2009). Analyses for physiological and behavioral rhythmicity. In M. L. Johnson, & L. Brand (Eds.), *Methods in enzymology: Vol. 454*. (pp. 141–174). Burlington: Academic Press.

Dowse, H., Umemori, J., & Koide, T. (2010). Ultradian components in the locomotor activity rhythms of the genetically normal mouse, Mus musculus. *The Journal of Experimental Biology*, *213*, 1788–1795.

Ebrahimi, F., Mikaeili, M., Estrada, E., & Nazeran, H. (2008). Automatic sleep stage classification based on EEG signals by using neural networks and wavelet packet coefficients. In *Paper presented at the Engineering in Medicine and Biology Society, 30th Annual Conference of the IEEE.*

Etchegaray, J.-P., Yu, E., Indic, P., Dallman, R., & Weaver, D. (2010). Casein kinase 1 delta (CK1δ) regulates period length of the mouse suprachiasmatic circadian clock in vitro. *PLoS One, 5*(4), e10303.

Evans, J., Leise, T., Castanon-Cervantes, O., & Davidson, A. (2013). Dynamic interactions mediated by nonredundant signaling mechanisms couple circadian clock neurons. *Neuron, 80*, 973–983.

Granada, A., Cambras, T., Díez-Noguera, A., & Herzel, H. (2011). Circadian desynchronization. *Interface Focus, 1*(1), 153–166.

Harang, R., Bonnet, G., & Petzold, L. (2012). WAVOS: A MATLAB toolkit for wavelet analysis and visualization of oscillatory systems. *BMC Research Notes, 5*, 163.

Herbst, C., Herzel, H., Svec, J., Wyman, M., & Fitch, W. (2013). Visualization of system dynamics using phasegrams. *Journal of the Royal Society Interface, 10*, 20130288.

Hu, K., Scheer, F., Ivanov, P., Buijs, R., & Shea, S. (2007). The suprachiasmatic nucleus functions beyond circadian rhythm generation. *Neuroscience, 149*, 508–517.

Hughes, M., Hogenesch, J., & Kornacker, K. (2010). JTK_CYCLE: An efficient nonparametric algorithm for detecting rhythmic components in genome-scale data sets. *Journal of Biological Rhythms, 25*(5), 372–380.

Indic, P., Murray, G., Maggini, C., Amore, M., Meschi, T., Borghi, L., et al. (2012). Multiscale motility amplitude associated with suicidal thoughts in major depression. *PLoS One, 7*(6), e38761. http://dx.doi.org/10.1371/journal.pone.0038761.

Indic, P., Salvatore, P., Maggini, C., Ghidini, S., Ferraro, G., Baldessarini, R. J., et al. (2011). Scaling behavior of human locomotor activity amplitude: Association with bipolar disorder. *PLoS One, 6*(5), e20650. http://dx.doi.org/10.1371/journal.pone.0020650.

Ko, C. H., Yamada, Y. R., Welsh, D. K., Buhr, E., Liu, A., Zhang, E., et al. (2010). Emergence of noise-induced oscillations in the central circadian pacemaker. *PLoS Biology, 8*, e1000513.

Leise, T. (2013). Wavelet analysis of circadian and ultradian behavioral rhythms. *Journal of Circadian Rhythms, 11*, 5.

Leise, T., & Harrington, M. (2011). Wavelet-based time series analysis of circadian rhythms. *Journal of Biological Rhythms, 26*(5), 454–463.

Leise, T., Harrington, M. E., Molyneux, P. C., Song, I., Queenan, H., Zimmerman, E., et al. (2013). Voluntary exercise can strengthen the circadian system in aged mice. *Age (Dordrecht, Netherlands), 35*(6), 2137–2152. http://dx.doi.org/10.1007/s11357-012-9502-y.

Leise, T., Indic, P., Paul, M., & Schwartz, W. (2013). Wavelet meets actogram. *Journal of Biological Rhythms, 28*, 62–68.

Leise, T., Wang, C., Gitis, P., & Welsh, D. (2012). Persistent cell-autonomous circadian oscillations in fibroblasts revealed by six-week single-cell imaging of PER2::LUC bioluminescence. *PLoS One, 73*(3), e33334. http://dx.doi.org/10.1371/journal.pone.0033334.

Levine, J. D., Funes, P., Dowse, H., & Hall, J. C. (2002). Signal analysis of behavioral and molecular cycles. *BMC Neuroscience, 3*, 1.

Lilly, J. (2012). *JLAB: Matlab freeware for data analysis, Version 0.94.* http://www.jmlilly.net/jmlsoft.html.

Lilly, J., & Olhede, S. (2009). Higher-order properties of analytic wavelets. *IEEE Transactions on Signal Processing, 57*(1), 146–160.

Lilly, J., & Olhede, S. (2010). On the analytic wavelet transform. *IEEE Transactions on Information Theory, 56*, 4135–4156.

Mallat, S. (2009). *A wavelet tour of signal processing: The sparse way* (3rd ed.). Burlington, MA: Academic Press.

Meeker, K., Harang, R., Webb, A., Welsh, D., Doyle, F., Bonnet, G., et al. (2011). Wavelet measurement suggests cause of period instability in mammalian circadian neurons. *Journal of Biological Rhythms*, *26*, 353–362.

Nakamura, T., Takumi, T., Takano, A., Hatanaka, F., & Yamamoto, Y. (2013). Characterization and modeling of intermittent locomotor dynamics in clock gene-deficient mice. *PLoS One*, *8*(3), e58884.

Nelson, W., Tong, Y., Lee, J., & Halberg, F. (1979). Methods for cosinor-rhythmometry. *Chronobiologia*, *6*, 305–323.

Paul, M., Indic, P., & Schwartz, W. (2014). Social forces can impact the circadian clocks of cohabiting hamster. *Proceedings of the Royal Society B*, *281*, 20132535.

Percival, D., & Walden, A. (1993). *Multitaper and conventional univariate techniques*. Cambridge, England: Cambridge University Press.

Percival, D., & Walden, A. (2000). *Wavelet methods for time series analysis*. New York: Cambridge University Press.

Poon, A., Wu, B., Poon, P., Cheung, E., Chan, F., & Lam, F. (1997). Effect of cage size on ultradian locomotor rhythms of laboratory mice. *Physiology & Behavior*, *62*, 1253–1258.

Prendergast, B., Beery, A., Paul, M., & Zucker, I. (2012). Enhancement and suppression of ultradian and circadian rhythms across the female hamster reproductive cycle. *Journal of Biological Rhythms*, *27*, 246–256.

Prendergast, B., Cisse, Y., Cable, E., & Zucker, I. (2012). Dissociation of ultradian and circadian phenotypes in female and male Siberian hamsters. *Journal of Biological Rhythms*, *27*, 287–298.

Price, T. S., Baggs, J. E., Curtis, A. M., Fitzgerald, G. A., & Hogenesch, J. B. (2008). WAVECLOCK: Wavelet analysis of circadian oscillation. *Bioinformatics*, *24*(23), 2794–2795.

Quotb, A., Bornat, Y., & Renaud, S. (2011). Wavelet transform for real-time detection of action potentials in neural signals. *Frontiers in Neuroengineering*, *4*(7), 1–10, http://dx.doi.org/10.3389/fneng.2011.00007.

Refinetti, R. (2004). Non-stationary time series and the robustness of circadian rhythms. *Journal of Theoretical Biology*, *227*, 571–581.

Refinetti, R., Cornélissen, G., & Halberg, F. (2007). Procedures for numerical analysis of circadian rhythms. *Biological Rhythm Research*, *38*, 275–325.

Selesnick, I., Baraniuk, R., & Kingsbury, N. (2005). The dual-tree complex wavelet transform. *IEEE Signal Processing Magazine*, *22*, 123–151.

Smith, J. (2007). *Mathematics of the Discrete Fourier Transform (DFT)* (2nd ed.). USA: W3K Publishing.

Stephenson, R., Lim, J., Famina, S., Caron, A., & Dowse, H. (2007). Sleep-wake behavior in the rat: Ultradian rhythms in a light–dark cycle and continuous bright light. *Biological Rhythm Research*, *38*, 275–325.

Thaben, P., & Westermark, P. (2014). *RAIN (Rhythmicity analysis incorporating non-parametric methods)*. http://rain.biologie.hu-berlin.de/rain/.

Thomas, R. (2013). The electrocardiogram-spectrogram. In S. Chokroverty, & R. Thomas (Eds.), *Atlas of sleep medicine* (2nd ed.). Philadelphia: Elsevier Saunders.

Torrence, C., & Compo, G. (1998). A practical guide to wavelet analysis. *Bulletin of the American Meteorological Society*, *69*, 61–78.

Tropp, J., Laska, J., Duarte, M., Romberg, J., & Baraniuk, R. (2010). Beyond Nyquist: Efficient sampling of sparse bandlimited signals. *IEEE Transactions on Information Theory*, *56*(1), 520–544.

Wichert, S., Fokianos, K., & Strimmer, K. (2004). Identifying periodically expressed transcripts in microarray time series data. *Bioinformatics*, *20*, 5–20.

Zielinski, T., Moore, A., Troup, E., Halliday, K., & Millar, A. (2014). Strengths and limitations of period estimation methods for circadian data. *PLoS One*, *9*, e96462.

CHAPTER SIX

Genetic Analysis of *Drosophila* Circadian Behavior in Seminatural Conditions

Edward W. Green*, Emma K. O'Callaghan†, Mirko Pegoraro*, J. Douglas Armstrong†, Rodolfo Costa‡, Charalambos P. Kyriacou*,[1]

*Department of Genetics, University of Leicester, Leicester, United Kingdom
†Actual Analytics, Edinburgh, United Kingdom
‡Department of Biology, University of Padova, Padova, Italy
[1]Corresponding author: e-mail address: cpk@leicester.ac.uk

Contents

Abstract

The study of circadian behavior in model organisms is almost exclusively confined to the laboratory, where rhythmic phenotypes are studied under highly simplified conditions such as constant darkness or rectangular light–dark cycles. Environmental cycles in nature are far more complex, and recent work in rodents and flies has revealed that when placed in natural/seminatural situations, circadian behavior shows unexpected features that are not consistent with laboratory observations. In addition, the recent observations of clockless mutants, both in terms of their circadian behavior and their Darwinian fitness, challenge some of the traditional beliefs derived from laboratory studies about what constitutes an adaptive circadian phenotype. Here, we briefly summarize the results of these newer studies and then describe how *Drosophila* behavior can be studied in the wild, pointing out solutions to some of the technical problems associated with extending locomotor monitoring to this unpredictable environment. We also briefly describe how to generate sophisticated simulations of natural light and temperature cycles that can be used to successfully mimic the fly's natural circadian behavior. We further clarify some misconceptions that have been raised in recent studies of natural fly behavior and show how these can be overcome with appropriate methodology. Finally, we describe some recent technical developments that will enhance the naturalistic study of fly circadian behavior.

Methods in Enzymology, Volume 551
ISSN 0076-6879
http://dx.doi.org/10.1016/bs.mie.2014.10.001

1. INTRODUCTION

The study of circadian behavior of model organisms and the genetic and molecular analysis of biological clocks provide one of the flagship stories of gene regulation of recent decades (Ozkaya & Rosato, 2012; Partch, Green, & Takahashi, 2014). Animals—be they rodents, fish, or insects—are usually studied under either constant darkness or under entraining light–dark conditions in constant temperature, with lights coming on and off suddenly every 12 h, and with complete darkness representing nighttime. These simplified entraining conditions are artificial compared to the experiences of animals in nature, yet they allow the study of behavioral and molecular responses to a single entraining variable (or "zeitgeber") while all others are held constant. In nature however, temperature, light, humidity, moonlight, social factors, predators, food resources, and even tides are all changing in a dynamic daily and seasonal fashion that will impact on the organism's circadian phenotype. With this in mind, in the past few years, chronobiologists have begun to study circadian behavior of model organisms in their natural habitats, or at least in seminatural environments, and with flies and mice at least, they have also investigated the behavior of clockless mutants under natural conditions (Daan et al., 2011; De, Varma, Saha, Sheeba, & Sharma, 2013; Menegazzi, Yoshii, & Helfrich-Forster, 2012; Vanin et al., 2012). The results of these (admittedly few) studies have nevertheless been very surprising and suggest that some of our ideas about what is adaptive circadian behavior may need to be revised.

While it is not the intention to review these studies in any detail, some mention of their results will place the rest of this chapter into the broader context. Historically, we should not forget that before circadian genetics, there was circadian neuroanatomy, in which ablation of the mammalian suprachiasmatic nuclei (SCN) was used to study the adaptive implications of rhythmicity. The investigations of Pat DeCoursey and collaborators (DeCoursey & Krulas, 1998; DeCoursey, Krulas, Mele, & Holley, 1997; DeCoursey, Walker, & Smith, 2000) involved lesioning the SCN of various rodent species, before studying them and their sham and intact controls for extended periods. DeCoursey observed that predation was significantly higher for the SCN-lesioned animals compared to the controls, and in particular, in the studies of chipmunks, she suggested that the inappropriate nocturnal activity of the lesioned animals served as a stimulus for predation, probably by weasels (DeCoursey et al., 2000). Consequently, the survival value of being rhythmic as opposed to being arrhythmic was reasonably clear, with obvious implications for adaptation and Darwinian fitness.

More recent experiments with clock mutant mice have questioned whether circadian rhythmicity has survival value (Daan et al., 2011). In this study, mutant *mPer2* laboratory mice and their wild-type and heterozygous littermates were placed outside in seminatural environments composed of 400-m^2 pens and followed for 2 years. Two major findings were reported, both of which were surprising. First, the wild-type mice, which are nocturnal in the laboratory, appeared to be predominantly diurnal in nature. This result resonates with an earlier study of golden hamsters, also nocturnal in the laboratory, but almost exclusively diurnal in natural conditions (Gattermann et al., 2008). Second, in terms of survival, there were no differences between mutant and nonmutant mice, and although the clock mutant allele decreased in frequency during the first year, it recovered to its initial value by the end of the second year of the study. The authors concluded that "the results caution against inferences from laboratory experiments on fitness consequences in the natural environment" (Daan et al., 2011).

These results are as exciting as they are unexpected. The most obvious solution to this conundrum (to us at least) is that selection pressures in the wild—such as predation or food availability—are such that they push the animals' "endogenous" behavior toward the daytime. Perhaps, in the relatively luxurious laboratory environment where food is plentiful and *ad libitum* and the stresses of predation are absent, locomotor behavior simply reverts to an "endogenous" nocturnal pattern. Yet, there is something circular and inherently unsatisfactory about this "solution" because animals did not evolve in such stress-free environments. In addition, the survival value of major clock genes such as *mPer2* is also challenged by Daan's study, particularly as mutations in clock genes affect so many, if not all, circadian phenotypes. However, it is possible that the intact paralogous *mPer1/mPer3* genes in these animals compensate for *mPer2* defects (Clayton, Kyriacou, & Reppert, 2001).

In contrast to mice, *Drosophila* generally do not carry more than one copy of each clock gene, and where they do have paralogous copies, as for *timeless* and *timeless2* (or *timeout*), the two have different and nonoverlapping functions (Benna et al., 2010). Given the results of the Daan et al. study (Daan et al., 2011), it was of considerable interest when the circadian locomotor behavior of per^{01} and tim^{01} mutants was reported in seminatural conditions to be practically identical to that of a number of wild-type strains (Vanin et al., 2012). In this study, flies were placed in the same TriKinetics activity monitors used in the laboratory, but these were sited in sheltered positions outdoors between April and November in two locations: Leicester, UK and Treviso, Italy. Thus, the flies were exposed to natural daily, lunar, and seasonal changes in light, temperature, and humidity.

Among a plethora of unexpected results, it was observed that per^{01} and tim^{01} mutants showed quasinormal locomotor profiles compared to the wild type, suggesting that clock gene activity from these cardinal negative regulators was not required for entrainment. Other deeply held laboratory-generated assumptions concerning fly rhythms challenged by the results of this study included such fundamental behavioral features as "morning anticipation"; the rise in locomotor activity that precedes "lights-on" in the laboratory. In nature, this sudden "lights-on" signal is replaced by the slow emergence of civil from nautical twilight, and under these conditions, the morning rise in locomotor behavior was shown to be highly temperature-dependent, effectively disappearing under colder temperatures when morning activity was delayed. Also, the "siesta," a period during which flies rested during the hottest parts of the summer day, was replaced by a large afternoon (A) burst of activity; this "A" component had never been observed in the laboratory paradigm, even under high constant temperatures (Fig. 1). Furthermore, the flies behavior in the wild was not "crepuscular" as often cited (e.g., Rieger et al., 2007), but diurnal, and the long-held belief that light–dark cycles provide the most important

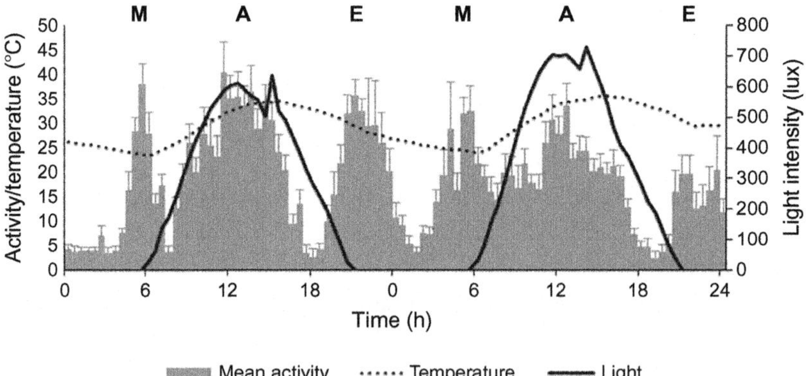

Figure 1 TriKinetics-generated natural circadian locomotor profile collected from Treviso (Italy) between June 24 and 25, 2009. Mean and SEMs of locomotor activity from 15-min time bins are shown. The M, A, and E components are shown in addition to the light intensity (yellow, right-hand *Y*-axis) and temperature (red, left-hand *Y*-axis). The photoperiod is approximately LD16:8 and the temperature maximum is 35.6 °C, minimum 24.1 °C. (See the color plate.)

environmental zeitgeber was not supported, in that temperature appeared to be more important for the phasing of the major locomotor components, "M," "A," and "E" (Vanin et al., 2012).

Further, seminatural studies have largely confirmed and extended these results in interesting ways. Menegazzi et al. (2012) suggested that under natural conditions, the amplitude of the main locomotor components was repressed by PER so that in the *per* null mutants, the "M" and "A" peaks were higher, revealing a putative adaptive role for these clock genes in repressing inappropriately high levels of entrained activity. We have confirmed this effect by mining our natural dataset, but with the caveat that the clock mutants were from different genetic backgrounds compared to the three wild-type lines we used (Green et al., in preparation). When we repeatedly backcrossed per^{01} into a common natural wild type background, we failed to reveal any significant differences in amplitude of the main locomotor components between wild-type and the mutant (Green et al., in preparation). Consequently, we believe that the locomotor amplitude repression by PER is probably background dependent, which is nevertheless an interesting observation that might allow mapping of the *trans*-acting factors involved. However, the message that emerges from these studies is that comparison of clock mutants and wild-types should be done on congenic backgrounds, particularly when assessing the more subtle characteristics of circadian behavior.

Another study, carried out in tropical latitudes, suggested that the "A" component might be an artifact of the recording system (De et al., 2013), but further analysis (Green et al., in preparation) has revealed that this interpretation is incorrect (see below). Nevertheless, both Vanin et al. (2012) and Menegazzi et al. (2012) suggested that the "A" component may be a clock-regulated escape response to high temperature, with the latter paper having some experimental support for this view. Similar conclusions have been drawn about the "A" response in four Drosophilid species, including *Drosophila melanogaster* (Prabhakaran & Sheeba, 2014). Indeed, recent work from our laboratory has revealed that the "A" component is mediated by the TrpA1 channel (Green et al., in preparation), which responds to noxious stimuli such as high temperatures (Neely et al., 2011).

Menegazzi et al. (2013) have also examined PER and TIM cycling in clock neurons under natural conditions. Unexpectedly, they observed that during the Italian midsummer, PER and TIM cycling were decoupled in clock neurons, such that while TIM still peaked at night, PER cycling could be up to 8 h out of phase with TIM, both in the wild and in simulated natural environments in the laboratory. This paints a very different picture from laboratory studies in which PER and TIM levels generally rise at approximately the same time (e.g., Saez & Young, 1996). Menegazzi et al. also confirmed a preliminary observation of Vanin et al. (2012), where PER and TIM cycle in the Dorsal Neurons (DNs) in advance of the other neurons, suggesting a faster oscillation in these cells. This has been experimentally supported in the laboratory using a functional assay of the DNs in which these cells' influence on the clock cellular network is amplified, resulting in a shorter behavioral period (Dissel et al., 2014). Finally, while locomotor behavior in the wild appears to be dominated by the temperature cycle, the natural light cycle appears to be the more dominant zeitgeber for the cellular expression of PER and TIM (Menegazzi et al., 2013; Vanin et al., 2012).

These studies of locomotor cycles have also been buttressed by work with pupal–adult eclosion rhythms. For example, different species of *Drosophila* showed species-specific emergence patterns in the laboratory, yet under more natural conditions, these species differences were not evident, and the emergence cycles were dictated by the prevailing environmental conditions (Prabhakaran, De, & Sheeba, 2013).

Given the current interest in natural and seminatural studies of *Drosophila* circadian behavior, it is timely to provide an overview of the methods that are commonly employed for these observations. We shall focus initially on how our groups in Leicester/Padova performed these studies, embellishing them with further useful methods from other groups.

2. CONSIDERATIONS FOR STUDIES OUTSIDE

The work by Vanin et al. (2012) was carried out at two European latitudes, one in Leicester, UK (latitude 52°38′N) and the other in Treviso, Italy (45°65′N). In both locations, a number of factors had to be taken into account when setting up the apparatus to best recapitulate the shaded natural habitat flies occupy in the wild. The first was that light pollution from the two city centers at night had to be avoided; the suburban locations of both recording centers meant that this could be largely accomplished. The second was that the TriKinetics locomotor monitors had to be kept out of direct sunlight, which not only dessicates the flies but can also lead to spurious triggering of the infrared sensors of DAM2 activity monitors as a result of sudden changes in illumination, for example, when clouds move across the sun. We were able to address these points in Leicester by placing the activity monitors in a child's playhouse (colored green) which was itself covered by a large

Wisteria plant, while in Treviso we placed monitors on a large window shelf placed directly under the roof of the house. This had the additional benefit of protecting the electrical recording apparatus from rain—especially a problem in Leicester. Furthermore, slight changes were made to the sensitivity of the infrared sensors to ensure that sudden changes in illumination were not problematic, and as a safeguard each 32-channel activity monitor contained a number of empty channels to check for any such spurious activation.

We assessed the wavelengths transmitted through a number of activity tubes made of different materials. TriKinetics provides polycarbonate or Pyrex tubes, the former material absorbing wavelengths up to 410 nm (which is in the blue), whereas Pyrex transmits wavelengths >310 nm (Fig. 2). As *Drosophila* cryptochrome, the dedicated circadian photoreceptor in the fly is maximally photoactivated in the 350–400 nm range (VanVickle-Chavez & Van Gelder, 2007), polycarbonate glass is not suitable for studies of light entrainment so we used Pyrex. One could envisage a scenario in which any effects of a *cry-null* mutant on entrainment might be diminished in polycarbonate but not Pyrex tubes. We also continuously monitored environmental illumination levels and temperature but found that locomotor tubes were buffered from the considerable daily environmental

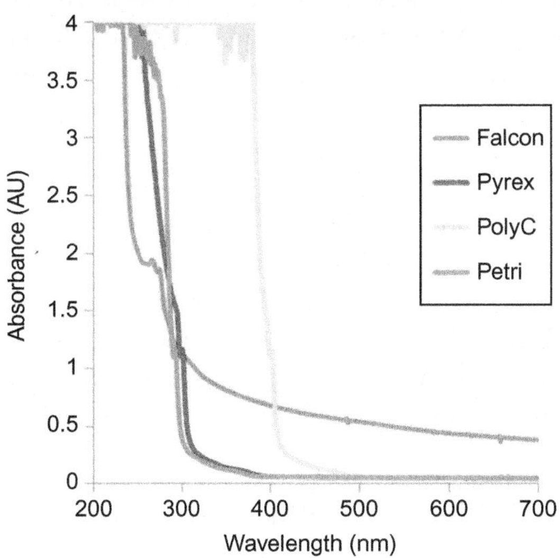

Figure 2 Absorbance (in arbitrary units, AU) of different types of transparent materials. Note how PolyC (polycarbonate) absorbs UVA and blue wavelengths. *Data from M. Pegoraro.* (See the color plate.)

fluctuation in humidity by being sealed at one end with rubber bungs and with a cotton wool ball at the other.

For experiments, male flies aged between 2 and 3 days were placed in the tubes, and allowed to acclimatize for 24 h before the recording apparatus was switched on, and, depending on the particular experiments, recordings were made for 3–7 days. The genotypes studied were mostly the classic clock mutants, but a few Gal4/UAS genotypes were also examined. There is an inherent problem however with using this binary expression system in that Gal4 is a yeast transcription factor, so is most active at warmer temperatures (Brand, Manoukian, & Perrimon, 1994). This is not a problem under midsummer conditions in Italy, for example, when the daily temperature range is quite often between 25 and 35 °C, but at the beginning and end of the summer when temperatures are much lower, the activity of GAL4 is reduced and consequently so will be the expression levels of the *UAS* construct. This is not a concern if one is interested in making a developmental manipulation, such as an ablation, as for example with *Pdf-Gal4*>*hid, reaper*, because these flies are raised in the laboratory at 25 °C, the Pdf-expressing neurons are eliminated, and the resulting flies can then be placed in the wild (Vanin et al., 2012). However, for a gene such as *per*, in which developmental expression is not required for later adult rhythmicity (Ewer, Hamblen-Coyle, Rosbash, & Hall, 1990), any GAL4-mediated manipulation such as expression of *UAS-per-RNAi* will inevitably be less efficient under colder conditions in nature. As the promoters used to drive GAL4 may also be quite temperature sensitive in their expression, considerable caution (and appropriate controls) should be exercised when interpreting such data.

Finally, the data from natural experiments can be analyzed by the same types of programs and algorithms used for laboratory experiments. In our study, the data were analyzed in Excel using BeFly!, a custom-made suite of programs designed and developed by E. W. G. This is a flexible set of routines that allows rapid analysis of locomotor records, including automated phase shift analyses, estimates of periodicity using cosinor curve-fitting, and the plotting of circadian actograms and histograms together with corresponding temperature and light intensity data (Allebrandt et al., 2013).

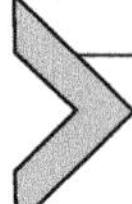

3. SIMULATING NATURAL CONDITIONS IN THE LABORATORY

The extensive natural data of Vanin et al. (2012) was collected over three seasons and provides a database of fly circadian behavior collected in northern and southern Europe that we are still mining (e.g., Green et al., in preparation). However, when one wishes to study further a particular phenotype at specific conditions, or extend the study with other genotypes in certain environmental conditions, the uncertainties of the weather mean that this is extremely difficult. To circumvent this problem, we have simulated natural conditions in the laboratory. This is not as easy as it sounds. Many laboratory incubators, for example, can cycle temperature or lights in a basic manner that means the two variables can be increased or decreased under timer control. However, these manipulations require step-ups or step-downs, and for both light and temperature, this generates startle responses in a fly's activity which can mask the underlying behavior. To solve these problems, we used the Memmert IPP 500 peltier incubators in which the temperature can be cycled very gradually at a resolution of 0.1 °C, while also producing much less noise and vibration than a comparable compressor incubator, thereby avoiding startle effects.

The light cycle was generated by a sophisticated custom-made piece of hard/software designed by Stefano Bastianello from Euritmi, a spin-out company from the Venetian Institute of Molecular Medicine. The light simulator has six batteries of LEDs of different wavelengths that can be combined to approximate the natural spectral composition of daylight as well as moonlight. Four replicate batteries of these LEDs are mounted behind masks that block various proportions of the emitted light; by combining different masks and varying the current supplied to the LEDs, the light intensity within the incubator can be sufficiently finely controlled to emulate the very low-amplitude light changes that distinguish astronomical, nautical, and civil twilight at dawn, and vice versa at dusk. This fine control of low intensity light can also be used to simulate moonlight, while the maximum light intensity of ~1500 lux—although much lower than the peak intensities experienced in direct sunlight—is more than sufficient to simulate the shaded conditions flies prefer in the wild (Rieger et al., 2007). We should emphasize here that our natural recordings of daily European light and temperature changes reveal that light intensity peaks 2–4 h before temperature peaks, so while light intensity comes down in the afternoon, temperature is still rising (Fig. 1). This needs to be programmed into the simulation as this phase difference between the two variables may have important implications for locomotor behavior (Vanin et al., 2012).

One recent criticism of activity records generated using TriKinetics DAM2 activity monitors is that the infrared beams monitor a part of the apparatus that can be shaded, particularly if under direct sunlight (De et al., 2013). In this work, flies were monitored using DAM2 TriKinetics monitors, which were simultaneously visually inspected between 07:00 and 19:00 hours to determine the flies' positions in the glass tubes. They claimed that their visually inspected data did not show an "A" component but only the "M" and "E" activity peaks, and presumed that the "A" component from the automated recording was generated by the fly hiding and moving in the shaded area, interrupting the beam. However, even a cursory examination of these results reveals that the visually inspected "M" component occurred several hours after the true M component that had been registered in the automated recordings. Indeed, this visually determined "M" was actually in phase with the DAM2 recorded "A" component (Green et al., in preparation), so in other words, De et al. had misinterpreted their own data and had visually confirmed the "A" component! Consequently, natural studies using TriKinetics DAM2 monitors need not concern themselves about any such shade effects, particularly if, like Vanin et al. (2012) or Menegazzi et al. (2012), the TriKinetics monitors are always kept in the shade.

However, by monitoring only one portion of the activity arena, the DAM2 beam-based activity monitor may underestimate true activity levels, especially in the context of sleep (Zimmerman, Raizen, Maycock, Maislin, & Pack, 2008). Our older custom made Drosophix system had three infrared sensors distributed along the glass tube (e.g., Collins, Dissel, Gaten, Rosato, & Kyriacou, 2005; Collins, Rosato, & Kyriacou, 2004), and TriKinetics currently produce a number of other monitors, including the MB5, which has 17 infrared beams spaced along a 51-mm tube. Needless to say, this particular monitor is more than three times as expensive as the standard DAM2s that have two beams at right angles to each other in the same location. Perhaps, rather than use monitors with many sensors, an alternative approach might be to use tracking software. For example, an interesting recent development has been to place individual flies in 96-well plates and use the Noldus Ethovision tracking system to monitor activity in a high-throughput manner (A. Reddy, personal communication). Since the flies are continuously observed, small movements (down to 0.2 mm) can be recorded and thus high-resolution and accurate monitoring of both activity and sleep (which is defined by a lack of movement) can be achieved. Moreover, video tracking can also delineate behavioral changes that cannot be discerned by beam-crossing methods (e.g., rotation of the fly in the *xy*-plane, wing movements, etc.).

We have also attempted to more closely study the natural behavior of flies by video tracking their activity in open fields. This both removes some of the constraints of single recording chambers such as with TriKinetics or with 96-well plates, and allows for the monitoring of groups of flies in which social cues have been shown to play an important role in the synchronization of circadian locomotor behavior (Bloch et al., 2013; Levine et al. 2002). A number of relevant software packages have been developed to analyze activity videos, including the open source programs pySolo-video (Gilestro, Tononi, & Cirelli, 2009) and Ctrax (Branson, Robie, Bender, Perona, & Dickinson, 2009), and commercial solutions EthovisionXT (Noldus Information Technology). We have used ActualTrack from Actual Analytics, a spin-out company from the University of Edinburgh, whose tracking software we have validated in a circadian context using both single fly activity tubes and open-field observations using mixed sex groups (Green et al., in preparation).

Circadian open field observations in flies have been reported previously, again by De et al. (2013), in which they placed flies in Petri dishes and claimed that the "A" component was not present under warm summer conditions. A problem with using such dishes for behavioral observations noted by both Simon and Dickinson (2010) and Soibam et al. (2012) is that the adult flies prefer to explore the edge of the dish. This complicates automated or visual tracking because the flies frequently occlude each other and are difficult to distinguish from the dish edge itself. Our recent studies (Green et al., in preparation) have adopted the gently sloping 12-cm dish developed by Simon and Dickinson, in which flies move freely throughout the whole arena. Under such conditions, in simulated summer days, open field behavior reveals a dominant "A" component, just as in TriKinetics DAM2 monitors maintained in the wild or in a laboratory summer simulation (Green et al., in preparation; Vanin et al., 2012). In addition, video tracking flies in simulated summer conditions using the same glass tubes as used in TriKinetics DAM2 monitors (but without using the monitors so no potential shaded areas), again revealed the "A" component as the major part of the daily activity profile (Green et al., in preparation). Consequently, the "A" component is not an artifact of any recording method, but *absence of "A"* is an artifact of either using standard Petri dishes or of misinterpreting the "A" component as the "M" component as mentioned above (De et al., 2013; Green et al., in preparation).

In conclusion, we have described here how *Drosophila* circadian behavior can be studied under natural/seminatural conditions, but a number of precautions, some obvious, some rather not, need to be taken into consideration in order for the adaptive phenotype to be studied in a valid and sensible manner. We anticipate that in future, more such studies, not only in *Drosophila*, but also in other model and non-model organisms, will enrich the chronobiological literature. These studies may use available clock mutants or other useful genetic variants, for example, those that may interfere with entrainment pathways. We suspect that there will be many surprises along the way, as there have been already, and we predict that these types of studies will provide a more realistic ecological framework in which to understand the adaptive significance of circadian behavior.

ACKNOWLEDGMENTS

E. W. G. was supported by a BBSRC grant to C. P. K. and the award of a Marie Curie Initial Training Network "INsecTIME" to C. P. K., R. C. and J. D. A. supported E. O. C., M.P. was supported by EC grant EUCLOCK (018741) to R.C. and C.P.K.

REFERENCES

Allebrandt, K. V., Amin, N., Muller-Myhsok, B., Esko, T., Teder-Laving, M., Azevedo, R. V., et al. (2013). A K(ATP) channel gene effect on sleep duration: From genome-wide association studies to function in Drosophila. *Molecular Psychiatry, 18*, 122–132.

Benna, C., Bonaccorsi, S., Wulbeck, C., Helfrich-Forster, C., Gatti, M., Kyriacou, C. P., et al. (2010). Drosophila timeless2 is required for chromosome stability and circadian photoreception. *Current Biology, 20*, 346–352.

Bloch, G., Herzog, E. D., Levine, J. D., & Schwartz, W. J. (2013). *Proceedings Biological sciences/The Royal Society, 280*, 20130035. http://dx.doi.org/10.1098/rspb.2013.0035.

Brand, A. H., Manoukian, A. S., & Perrimon, N. (1994). Ectopic expression in Drosophila. *Methods in Cell Biology, 44*, 635–654.

Branson, K., Robie, A. A., Bender, J., Perona, P., & Dickinson, M. H. (2009). High-throughput ethomics in large groups of Drosophila. *Nature Methods, 6*, 451–457.

Clayton, J. D., Kyriacou, C. P., & Reppert, S. M. (2001). Keeping time with the human genome. *Nature, 409*, 829–831.

Collins, B. H., Dissel, S., Gaten, E., Rosato, E., & Kyriacou, C. P. (2005). Disruption of cryptochrome partially restores circadian rhythmicity to the arrhythmic *period* mutant of Drosophila. *Proceedings of the National Academy of Sciences of the United States of America, 102*, 19021–19026.

Collins, B. H., Rosato, E., & Kyriacou, C. P. (2004). Seasonal behavior in *Drosophila melanogaster* requires the photoreceptors, the circadian clock, and phospholipase C. *Proceedings of the National Academy of Sciences of the United States of America, 101*, 1945–1950.

Daan, S., Spoelstra, K., Albrecht, U., Schmutz, I., Daan, M., Daan, B., et al. (2011). Lab mice in the field: Unorthodox daily activity and effects of a dysfunctional circadian clock allele. *Journal of Biological Rhythms, 26*, 118–129.

De, J., Varma, V., Saha, S., Sheeba, V., & Sharma, V. K. (2013). Significance of activity peaks in fruit flies, *Drosophila melanogaster*, under seminatural conditions. *Proceedings of the National Academy of Sciences of the United States of America, 110*, 8984–8989.

DeCoursey, P. J., & Krulas, J. R. (1998). Behavior of SCN-lesioned chipmunks in natural habitat: A pilot study. *Journal of Biological Rhythms, 13*, 229–244.

DeCoursey, P. J., Krulas, J. R., Mele, G., & Holley, D. C. (1997). Circadian performance of suprachiasmatic nuclei (SCN)-lesioned antelope ground squirrels in a desert enclosure. *Physiology & Behavior, 62*, 1099–1108.

DeCoursey, P. J., Walker, J. K., & Smith, S. A. (2000). A circadian pacemaker in free-living chipmunks: Essential for survival? *Journal of Comparative Physiology. A, Sensory, Neural, And Behavioral Physiology, 186*, 169–180.

Dissel, S., Hansen, C. N., Ozkaya, O., Hemsley, M., Kyriacou, C. P., & Rosato, E. (2014). The logic of circadian organisation in Drosophila. *Current biology: CB, 24*, 2257–2266.

Ewer, J., Hamblen-Coyle, M., Rosbash, M., & Hall, J. C. (1990). Requirement for *period* gene expression in the adult and not during development for locomotor activity rhythms of imaginal *Drosophila melanogaster*. *Journal of Neurogenetics*, 7, 31–73.

Gattermann, R., Johnston, R. E., Yigit, N., Fritzsche, P., Larimer, S., Ozkurt, S., et al. (2008). Golden hamsters are nocturnal in captivity but diurnal in nature. *Biology Letters, 4*, 253–255.

Gilestro, G. F., Tononi, G., & Cirelli, C. (2009). Widespread changes in synaptic markers as a function of sleep and wakefulness in Drosophila. *Science, 324*, 109–112.

Levine, J. D., Funes, P., Dowse, H. B., & Hall, J. C. (2002). Resetting the circadian clock by social experience in Drosophila melanogaster. *Science, 298*, 2010–2012.

Menegazzi, P., Vanin, S., Yoshii, T., Rieger, D., Hermann, C., Dusik, V., et al. (2013). Drosophila clock neurons under natural conditions. *Journal of Biological Rhythms, 28*, 3–14.

Menegazzi, P., Yoshii, T., & Helfrich-Forster, C. (2012). Laboratory versus nature: The two sides of the Drosophila circadian clock. *Journal of Biological Rhythms, 27*, 433–442.

Neely, G. G., Keene, A. C., Duchek, P., Chang, E. C., Wang, Q. P., Aksoy, Y. A., et al. (2011). TrpA1 regulates thermal nociception in Drosophila. *PLoS One, 6*, e24343.

Ozkaya, O., & Rosato, E. (2012). The circadian clock of the fly: A neurogenetics journey through time. *Advances in Genetics*, 77, 79–123.

Partch, C. L., Green, C. B., & Takahashi, J. S. (2014). Molecular architecture of the mammalian circadian clock. *Trends in Cell Biology, 24*, 90–99.

Prabhakaran, P. M., De, J., & Sheeba, V. (2013). Natural conditions override differences in emergence rhythm among closely related Drosophilids. *PLoS One, 8*, e83048.

Prabhakaran, P. M., & Sheeba, V. (2014). Simulating natural light and temperature cycles in the laboratory reveals differential effects on activity/rest rhythm of four Drosophilids. *Journal of Comparative Physiology. A, Neuroethology, Sensory, Neural, and Behavioral Physiology, 200*, 849–862.

Rieger, D., Fraunholz, C., Popp, J., Bichler, D., Dittmann, R., & Helfrich-Forster, C. (2007). The fruit fly *Drosophila melanogaster* favors dim light and times its activity peaks to early dawn and late dusk. *Journal of Biological Rhythms, 22*, 387–399.

Saez, L., & Young, M. W. (1996). Regulation of nuclear entry of the Drosophila clock proteins period and timeless. *Neuron, 17*, 911–920.

Simon, J. C., & Dickinson, M. H. (2010). A new chamber for studying the behavior of Drosophila. *PLoS One, 5*, e8793.

Soibam, B., Mann, M., Liu, L., Tran, J., Lobaina, M., Kang, Y. Y., et al. (2012). Open-field arena boundary is a primary object of exploration for Drosophila. *Brain and Behavior, 2*, 97–108.

Vanin, S., Bhutani, S., Montelli, S., Menegazzi, P., Green, E. W., Pegoraro, M., et al. (2012). Unexpected features of Drosophila circadian behavioural rhythms under natural conditions. *Nature, 484*, 371–375.

VanVickle-Chavez, S. J., & Van Gelder, R. N. (2007). Action spectrum of Drosophila cryptochrome. *The Journal of Biological Chemistry, 282*, 10561–10566.

Zimmerman, J. E., Raizen, D. M., Maycock, M. H., Maislin, G., & Pack, A. I. (2008). A video method to study Drosophila sleep. *Sleep, 31*, 1587–1598.

DeCoursey, P. J., Krulas, J. R., Mele, G., & Holley, D. C. (1997). Circadian performance of suprachiasmatic nuclei (SCN)-lesioned antelope ground squirrels in a desert enclosure. *Physiology & Behavior*, *62*, 1099–1108.
DeCoursey, P. J., Walker, J. K., & Smith, S. A. (2000). A circadian pacemaker in free-living chipmunks: essential for survival? [illegible]
[illegible] The logic of circadian organization in *Drosophila*. [illegible]
[illegible]
Gattermann, R., Johnston, R. E., Yigit, N., Fritzsche, P., Larimer, S., Ozkurt, S., et al.

PART II

Characterization of Molecular Clock Components

CHAPTER SEVEN

Methods to Study Molecular Mechanisms of the *Neurospora* Circadian Clock

Joonseok Cha, Mian Zhou, Yi Liu[1]
Department of Physiology, University of Texas Southwestern Medical Center, Dallas, TX, USA
[1]Corresponding author: e-mail address: yi.liu@utsouthwestern.edu

Contents

Abstract

Eukaryotic circadian clocks are comprised of interlocked autoregulatory feedback loops that control gene expression at the levels of transcription and translation. The filamentous fungus *Neurospora crassa* is an excellent model for the complex molecular network of regulatory mechanisms that are common to all eukaryotes. At the heart of the network, posttranslational regulation and functions of the core clock elements are of major interest. This chapter discusses the methods used currently to study the regulation of clock molecules in *Neurospora*. The methods range from assays of gene expression to phosphorylation, nuclear localization, and DNA binding of clock proteins.

1. INTRODUCTION

Circadian clocks are self-sustaining timekeepers found in almost all organisms on earth (Dunlap, 1999; Young & Kay, 2001). Eukaryotic

Methods in Enzymology, Volume 551
ISSN 0076-6879
http://dx.doi.org/10.1016/bs.mie.2014.10.002

circadian oscillators employ complex networks of molecules to form interlocked feedback loops. Despite the evolutionary distance, the mechanism of the circadian oscillator of the filamentous fungus *Neurospora crassa* is very similar to that of higher eukaryotes, and *Neurospora* has served as an outstanding model organism for the field (Heintzen & Liu, 2007; Liu & Bell-Pedersen, 2006). Working with fungi is relatively simple, and analysis of this organism allowed identification of clock components and mechanisms through genetic, biochemical, and molecular biological studies. Furthermore, the availability of a whole genome knockout collection and a biolu minescence reporter provided technical versatility that enabled dissection of the *Neurospora* circadian clock (Colot et al., 2006; Gooch et al., 2008).

In the *Neurospora* circadian clock, FREQUENCY (FRQ) forms an FFC (FRQ–FRH complex) with its partner, FRQ-interacting RNA helicase (FRH), to function as the negative limb in the core negative feedback loop (Aronson, Johnson, Loros, & Dunlap, 1994; Cheng, He, He, Wang, & Liu, 2005). The transcription of *frq* gene is activated by the positive element, WHITE COLLAR complex (WCC), which consists of two PER-ARNT-SIM domain-containing transcription factors WC-1 and WC-2 (Cheng, Yang, Gardner, & Liu, 2002; Cheng, Yang, Wang, He, & Liu, 2003; Crosthwaite, Dunlap, & Loros, 1997; He & Liu, 2005b). Circadian expression of the *frq* gene is achieved by rhythmic binding of WCC to its promoter and requires timely modulation of chromatin structure by multiple factors (Belden, Lewis, Selker, Loros, & Dunlap, 2011; Belden, Loros, & Dunlap, 2007; Cha, Zhou, & Liu, 2013; Froehlich, Loros, & Dunlap, 2003). It was also recently shown that RCO-1-mediated suppression of WC-independent transcription of *frq* is essential for clock function (Zhou, Liu, et al., 2013).

FFC interacts with WCC to promote phosphorylation of WCs; this phosphorylation is primed by protein kinase A (PKA) in an FRQ-independent manner (He, Cha, Lee, Yang, & Liu, 2006; Huang et al., 2007; Schafmeier et al., 2005). The inhibition of WCC by FFC is the critical step in circadian negative feedback, and the mechanism involves not only physical interactions but also enzymatic reactions to modify WC proteins posttranslationally. Phosphorylation events stabilize WCs, decrease their DNA-binding ability, and result in export of these proteins to the cytoplasm (Cha, Chang, Huang, Cheng, & Liu, 2008; Diernfellner, Querfurth, Salazar, Hofer, & Brunner, 2009; He, Shu, et al., 2005; Hong, Ruoff, Loros, & Dunlap, 2008). In addition to the well-known circadian oscillations of *frq* mRNA and FRQ protein, the WCC phosphorylation status, its occupancy of the *frq* promoter, and nuclear localization all display circadian rhythms in constant darkness. Protein phosphatases PP2A and PP4 are known to counterbalance the relevant kinases to regulate the timely reactivation and relocation of WCC for a new cycle (Cha et al., 2008; Diernfellner et al., 2009; Yang et al., 2004).

After its synthesis, FRQ forms a homodimer and interacts with FRH, which is important for the stability and proper structure (Cha, Yuan, & Liu, 2011; Cheng et al., 2005; Guo, Cheng, & Liu, 2010; Guo, Cheng, Yuan, & Liu, 2009; Shi, Collett, Loros, & Dunlap, 2010). Like its animal homolog PERIOD (Per), FRQ is progressively targeted by kinases through the subjective day and evening, leading to its extensive phosphorylation and eventual degradation by the ubiquitin/proteasome pathway (Gorl et al., 2001; He & Liu, 2005a; He, Cheng, He, & Liu, 2005; He et al., 2006, 2003; Liu, Loros, & Dunlap, 2000; Pregueiro, Liu, Baker, Dunlap, & Loros, 2006; Yang, Cheng, He, Wang, & Liu, 2003; Yang, Cheng, Zhi, & Liu, 2001). FWD-1, the E3 ubiquitin ligase for FRQ, recognizes phosphorylated FRQ and facilitates its degradation, whereas the COP9 signalosome regulates the activity and stability of the SCF^{FWD-1} (SKP-1/CUL-1/FWD-1) complex to indirectly affect FRQ stability. PKA also phosphorylates FRQ, but in contrast to the other kinases, it stabilizes FRQ; dephosphorylation of FRQ by PP1, PP2A, and PP4 is also important for its stability (Cha et al., 2008; Huang et al., 2007; Yang et al., 2004). We and others have identified more than 100 phosphorylation sites on FRQ by analyzing *in vitro* phosphorylation by casein kinases and by purification of phosphorylated FRQ from *Neurospora* (Baker, Kettenbach, Loros, Gerber, & Dunlap, 2009; Tang et al., 2009). Extensive phosphorylation of FRQ may change the protein conformation allowing better access by SCF^{FWD-1}, and the presence of multiple motifs that interact with CKI suggests that conformational changes facilitate the degradation of FRQ (Querfurth et al., 2011). The phosphorylation of FRQ also modulates its interactions with other proteins to affect its function in the negative feedback loop (Cha et al., 2011). Thus, the phosphorylation of FRQ is crucial for regulating the circadian negative feedback loop, and these modifications are fine-tuned by a series of regulators to determine FRQ stability and the period length of the clock. The roles of phosphorylation events in the core circadian negative feedback loop are described in Fig. 1.

Combinations of biochemical, genetic, and molecular approaches were used effectively to study the mechanisms of circadian clock (Liu, 2005). Here, we will describe some of newly developed methods used in the studies

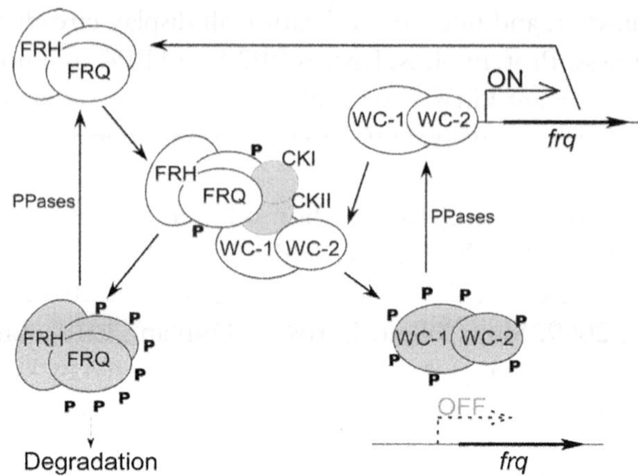

Figure 1 Phosphorylation-mediated regulation in the core *Neurospora* circadian negative feedback loop. In the subjective morning, hypophosphorylated WCC binds to *frq* promoter and activates the transcription of *frq*. Newly synthesized FRQ homodimer forms FFC with FRH. The formation of FFC is essential for FRQ stability. FFC then recruits casein kinases and interacts with WCC to promote WC phosphorylation. Phosphorylation of WCC inhibits its DNA binding activity and sequesters WCC in the cytoplasm. Phosphorylations on FRQ by the casein kinases inhibit the interactions between FRQ and CKs/WCC. Hyperphosphorylated FRQ triggers ubiquitination mediated by $SCF^{FWD\text{-}1}$ complex and is degraded by ubiquitin/proteasome pathway.

of the *Neurospora* circadian clock. These techniques may be useful for future studies and in other model organisms.

2. DESCRIPTION OF METHODS

2.1. Purification of epitope-tagged proteins and interacting partners from *Neurospora* extracts

Purification of a protein of interest enables investigation of the molecular context of its cellular function by allowing identification of putative interaction partners. Epitope-tagged proteins can avoid the difficulty of purifying endogenous proteins. We have successfully used a tandem affinity tag made of c-Myc and 6-His to demonstrate interactions of clock proteins and to perform biochemical assays of the purified enzymes. The construction of tandem repeats was previously described (He, Cheng, et al., 2005), and the strain of interest can be generated using homologous recombination to integrate the transgene expressing the epitope-tagged protein at the *his-3* locus.

To set up an affinity column for the 6 × His-tag, wash an empty column with MilliQ water and add desired amount of Ni-NTA beads (Qiagen) or modified metal affinity resin such as TALON (Clontech). Let the column stand for 20–25 min, so that all beads sediment to the bottom. Wash the column with 10 column volumes (CVs) of MilliQ water and then with 10 CVs of extraction buffer (50 m*M* HEPES, pH 7.4, 137 m*M* NaCl, 10% glycerol (v/v)). Protease inhibitors (1 mg/ml pepstatin A, 1 mg/ml leupeptin, and 100 m*M* phenylmethylsulphonyl fluoride) should be added to this buffer just before protein extraction. After harvesting the *Neurospora* mycelia cultures, they are ground into fine powder in liquid nitrogen. Keeping the mortar and pestle cold is absolutely necessary to prevent denaturation of sensitive proteins during purification. Add three volumes of extraction buffer to the powder, mix well by vortexing with frequent chilling on ice, and centrifuge at 12,000 rpm at 4 °C for 15 min. Transfer the supernatant to a fresh tube and centrifuge again (45,000 rpm, 4 °C, 30 min). Transfer the supernatant to a clean tube, measure the protein concentration, dilute to 2–4 mg/ml, and add imidazole to 0–20 m*M* depending on the target proteins. Load the extract onto the column and elute at a flow rate of around 1 ml/min. Collect the flow-through. Wash the column with 10 CVs of extraction buffer (salt concentration may need to be adjusted to improve the purity) and elute the attached proteins with 1 CV of elution buffer (50 m*M* HEPES, pH 7.4, 137 m*M* NaCl, 200 m*M* imidazole, and 20% glycerol (v/v)); this may require five to seven washes with elution buffer. Each fraction should be analyzed by Western blot using an anti-c-Myc antibody, and the most enriched fractions should be combined for the immunoprecipitation (IP) step. For IP, mix 1 ml of eluted fraction with 40 μl anti-c-Myc agarose beads (10 μl beads; Santa Cruz Biotechnology), and incubate on a rotator at 4 °C for 4 h. Collect beads by centrifugation (4000 rpm, 4 °C, 1 min) and wash the beads with 1 ml of high salt buffer (20 m*M* Tris–HCl, pH 7.5, 500 m*M* NaCl) and then with low salt buffer (20 m*M* Tris–HCl, pH 7.5, 50 m*M* NaCl).

The beads should be resuspended in the desired buffer: appropriate assay buffer for enzyme assays or 1 × SDS-PAGE loading buffer to run a gel. To identify interacting proteins, the eluted proteins are separated by a 4–15% gradient gel (Bio-Rad) and the gel is stained with silver (GE Healthcare). Bands of interest are excised and proteins are eluted/digested from the gel for mass spectrometry (MS) analysis using protocols suggested by the MS operator.

2.2. Identification of phosphorylated residues of clock proteins

As mentioned earlier, FRQ is progressively and extensively phosphorylated over circadian time. To identify phosphorylated residues, we analyzed both *in vitro* and *in vivo* phosphorylated FRQ samples by MS and concluded that CKI and CKII are the major kinases that phosphorylate FRQ (Tang et al., 2009). Furthermore, quantitative methods were used to reveal preferentially phosphorylated residues in the hyperphosphorylated FRQ species, which is the species targeted for degradation.

2.2.1 *Mapping* in vitro *phosphorylation sites*

His-tagged full-length FRQ was expressed and purified from Sf9 cells. CK-1a and CKA were also His-tagged at their N-termini, expressed in *Escherichia coli*, and purified by Ni-NTA column. To perform *in vitro* phosphorylation, 6–8 μg of FRQ protein were incubated with 1–2 μg of kinase(s) in phosphorylation buffer (25 m*M* HEPES, pH 7.9, 10 m*M* $MgCl_2$, 2 m*M* $MnCl_2$, 25 μ*M* ATP, and 10 μCi/ml [γ-^{32}P]ATP) at 37 °C for 2–4 h and then subjected to SDS-PAGE. After electrophoresis, the bands were visualized by colloidal blue staining. FRQ bands were excised and analyzed by MS.

2.2.2 *Mapping* in vivo *phosphorylation sites*

To purify FRQ from *Neurospora* for mapping of the *in vivo* phosphorylation sites, we inserted 5 × c-Myc and 9 × His tags at the C-terminus of FRQ. The construct was transformed into an *frq*-null strain and an *fwd-1*^{RIP} strain. The latter displays elevated levels of FRQ protein in hyperphosphorylated forms. Epitope-tagged FRQ was purified to near homogeneity, and the bands corresponding to FRQ were excised from the gel for trypsin digestion and MS analyses.

2.2.3 *MS analyses*

Protein was digested in gel with 10 ng/l sequencing-grade trypsin in 50 m*M* NH_4HCO_3 (pH 7.8) at 37 °C overnight. The resulting peptides were extracted sequentially with 5% formic acid/50% acetonitrile and 0.1% formic acid/75% acetonitrile and concentrated to about 10 μl for the following steps. The trypsin-digested peptides were loaded onto a precolumn packed with 5–15 μm spherical C18, reversed-phase particles (YMC). The precolumn was connected by a piece of Teflon tubing to a homemade analytical column packed with YMC, 5-μm spherical, C18 reversed-phase particles. The eluted peptides were sprayed directly into a QSTAR XL mass spectrometer (MDS SCIEX) equipped with a nano-ESI ion source. The spectra were collected in Information Dependent Acquisition mode. Spray voltage was 2.1 kV, MS scan range was 400–2000 Da, resolution was low for precursor ion isolation, and the top three most abundant precursor ions were selected for MS/MS scans with enhance-all mode. Database searches were performed using an in-house Mascot server. After the database searches, all of the recognized phosphopeptides were manually checked to exclude false positives.

2.2.4 Quantitative MS

To understand how the extensive phosphorylation on FRQ is regulated in a circadian cycle or to distinguish FRQ-dependent from FRQ-independent phosphorylation on WCs, we developed a quantitative method of MS employing whole-cell ^{15}N metabolic labeling in *Neurospora* (Huang et al., 2007; Tang et al., 2009). For the purification of Myc-His-FRQ for the quantitative MS experiments, NH_4Cl or $^{15}NH_4Cl$ (Cambridge Isotope Labs) was used to replace NH_4NO_3 in Vogel's medium. Unlabeled or labeled cultures of 2–3 l were used to prepare extracts, and both extracts were mixed in equal protein amounts before the purification step. Epitope-tagged FRQ was analyzed by MS, and the resulting files from the Mascot search were imported into the open-source software MSQuant (http://msquant.sourceforge.net). The ^{15}N-labeled and -unlabeled peptide pairs were recognized automatically by MSQuant based on the information from Mascot search results and their differences in mass-to-charge ratios. Peptide ratios were obtained by calculating the extracted ion chromatograms of the peptide pairs for quantification, and the results were also manually verified. The average ^{15}N to ^{14}N ratios of unphosphorylated peptides were used as the correction factor to determine the ratios of the phosphorylated peptides. If a certain phosphopeptide of WC-1 in the wild-type strain is more abundant than the *frq*-null strain over the ratio of unphophorylated peptide, such a phosphorylation is regarded as FRQ-dependent.

2.3. Isolation of *Neurospora* nuclei to analyze localization of clock proteins

Posttranslational regulation of WCs and FRQ is involved in the mechanism controlling the nucleocytoplasmic trafficking of these proteins (Cha et al., 2008, 2011; Diernfellner et al., 2009; Hong et al., 2008; Luo, Loros, & Dunlap, 1998; Schafmeier et al., 2008). Given that the transcriptional activation of *frq* by WCC and the FRQ-mediated inhibition of DNA binding of WCC are essential for the circadian negative feedback loop, analyses of nuclear localization of clock proteins are critical to our understanding of the molecular mechanisms of the clock.

To prepare nuclear extract, grind frozen *Neurospora* cells with acid-washed glass beads (Sigma-Aldrich) in liquid nitrogen. We use equal weights of glass beads and dehydrated tissues. Slowly pour the cell powder into 10 ml of buffer A (1 *M* sorbitol, 7% Ficoll, 20% glycerol, 5 m*M* magnesium acetate, 3 m*M* $CaCl_2$, 50 m*M* Tris–HCl, pH 7.5) on ice while stirring. Filter the resuspended sample through cheesecloth (prewet with buffer A) into a fresh flask on ice. Add 2 volumes of cold buffer B (10% glycerol, 5 m*M* magnesium acetate, 25 m*M* Tris–HCl, pH 7.5) slowly with gentle shaking on ice. Layer the mixed solution onto the bed of 10 ml of cold buffer A/B (4:6.6) in the centrifuge tube, and centrifuge (3000 × *g*, 4 °C, 7 min). Layer the supernatant (total extract) onto a bed of 5 ml of buffer D (1 *M* sucrose, 10% glycerol, 5 m*M* magnesium acetate, 25 m*M* Tris–HCl, pH 7.5) and centrifuge (9400 × *g*, 4 °C, 15 min). Discard the supernatant (cytosolic fraction) and resuspend the pellet (nuclear fraction) in half volume of buffer D. Add the same volume of the extraction buffer and sonicate briefly to disrupt the nuclear membrane. Centrifuge (12,000 rpm, 4 °C, 15 min), resolve the supernatant by SDS-PAGE, and analyze by Western blot.

2.4. Chromatin immunoprecipitation

The association of transcription factors, histones, and RNA polymerase II with target sites can provide important information on how they regulate gene expression. WCC rhythmically binds to the *frq* promoter to drive its circadian transcription (Froehlich et al., 2003; He et al., 2006). CSW-1 and CATP are required to generate a circadian rhythm of chromatin state at the *frq* locus to ensure proper WCC-driven transcription (Belden, Loros, et al., 2007; Cha et al., 2013). The chromatin immunoprecipitation (ChIP) assay has been widely used to determine whether a protein of interest is associated with a specific genomic region in the cell. We analyzed formaldehyde-fixed chromatin for occupancies by WCC and modified histones at the *frq* gene and at other clock-controlled gene loci.

To cross-link the proteins with the chromatin, add 1% formaldehyde directly into the liquid culture and incubate for 15 min. To stop the cross-linking, add 125 m*M* glycine (pH 7.5) and incubate 5 min. Then wash the mycelia by transferring to wash buffer (50 m*M* HEPES, pH 7.5, 137 m*M* NaCl). Harvest the culture and grind cells in liquid nitrogen. Add 1 ml of lysis buffer (50 m*M* HEPES, pH 7.5, 137 m*M* NaCl, 1 m*M* EDTA, 1% Triton X-100, 0.1% deoxycholate, 0.1% SDS) with protease inhibitors to around 200 μl of cell powder and mix thoroughly. Sonicate the chromatin using Bioruptor (Diagenode) for 15 min with a 30 s:30 s cycle. The sonication conditions should be adjusted based on the size of the sheared chromatin. We obtained <200-bp fragments with the conditions above. Centrifuge (10,000 × *g*, 4 °C, 15 min) and transfer the supernatant to the fresh tubes. Measure the protein concentration, and dilute 1 mg protein into 0.5 ml lysis buffer, add preblocked Gammabind G Sepharose (GE Healthcare; incubated with 100 mg/ml BSA and 100 mg/ml salmon sperm DNA at 4 °C with rotation for 4 h), and incubate at 4 °C for 1 h to remove nonspecific binding. Spin down the beads (4000 rpm, 2 min), and transfer the supernatant to fresh tubes. Remove 50 μl to assay "input," and add appropriate antibody to the remaining chromatin. Incubate at 4 °C for 4 h (or overnight), add 40 μl (10 μl beads only) Gammabind G Sepharose, and incubate for 1 h. Collect the beads (4000 rpm, 2 min), and wash the beads sequentially with low-salt washing buffer (0.1% SDS, 1% Triton X-100, 2 m*M* EDTA, 150 m*M* NaCl, 20 m*M* Tris–HCl, pH 8), high-salt washing buffer (0.1% SDS, 1% Triton X-100, 2 m*M* EDTA, 500 m*M* NaCl, 20 m*M* Tris–HCl, pH 8), LNDET buffer (0.25 *M* LiCl, 1% NP40, 0.1% deoxycholate, 1 m*M* EDTA,10 m*M* Tris–HCl, pH 8), and TE buffer (1 m*M* EDTA, 10 m*M* Tris–HCl, pH 8).

To elute DNA from the ChIP samples, we use Chelex 100 resin (Bio-Rad) and incubate at 94 °C for 15 min. Chill the samples on ice and then spin down (1000 × *g*, 1 min); the supernatant is the ChIP DNA. "Input" samples are incubated at 65 °C for more than 4 hours to reverse the cross-link, and DNAs are extracted using phenol/chloroform. Quantitative real-time PCR is used to quantify the enrichment of the target site, and the efficiency of primers is verified by analysis of a dilution series of input DNAs.

2.5. Monitoring bioluminescence reporter expression during the circadian cycle

Codon-optimized luciferase has been successfully used as a bioluminescence reporter in *Neurospora* (Gooch et al., 2008). Luciferase reporter activity can be greatly improved by codon optimization of the firefly luciferase gene based on the codon usage of *Neurospora*. With this optimization, the light produced from luciferase driven by a highly expressed promoter can be visible to naked eyes. The *frq*-promoter-driven luciferase activity rhythm can be monitored in the conventional race tube assay setting. A liquid nitrogen-cooled or electronically cooled camera can be used to capture the image of the race tube every 10–30 min over the time course, and the time-lapse video shows an oscillating intensity of luminescence. For the quantitative analysis, the region of interest—which can be the whole tube or a small part of the growth front—can be analyzed from the time-lapse images, and the period of different strains can be compared with or without the presence of conidial banding of the strains. Thus, the circadian rhythms of the strains defective in conidiation can be monitored by this relatively simple platform. This method avoids the generation of mutant strains in the back ground of *ras-1*^{bd} by a sexual cross (Belden, Larrondo, et al., 2007), which takes more than 4 weeks and in some cases, sexual reproduction of some strains may be difficult or impossible.

We also constructed a pfrq-luc-I (the precursor was a generous gift from Dr. Jay Dunlap) reporter in a plasmid that contains the *basta/ignite resistance* (*bar*) gene, which can be used to transform *Neurospora* strains independent of *his*$^{+}$ marker. The transformants from either *his-3* or *bar* transformation are inoculated on autoclaved FGS-Vogel's (AFV) media (0.05% fructose, 0.05% glucose, 2% sorbose, 1 × Vogel's, 50 μg/l biotin, and 1.8% agar) containing 50 μ*M* firefly luciferin (BioSynt). The culture is incubated in constant light overnight and then transferred to a LumiCycle (Actimetrics), which has four independent photomultiplier tubes to record data from 32 samples simultaneously. The data are then analyzed using LumiCycle Analysis software supplied by the manufacturer. The activity curve is normalized by subtracting the baseline luciferase signal, which changes as cells grow.

The *frq*-driven reporter construct has been modified to perform additional assays. The translational fusion of luciferase to the FRQ ORF displays a rhythm as robust as that of the transcriptional reporter and expands the application of the reporter to the analysis of posttranslational regulation on FRQ protein (Larrondo, Loros, & Dunlap, 2012). More recently, it was shown that introduction of the PEST sequence of FRQ, which contains the major phosphorylation sites to induce FRQ degradation, significantly destabilized the luciferase protein (Cesbron, Brunner, & Diernfellner, 2013). The destabilized luciferase reporter reveals rapid luciferase activity changes that otherwise would be masked by the long-lasting reporter activity. Furthermore, the development of a 96-well plate-based high-throughput platform can accelerate the high-throughput analysis of circadian promoters or even replace RNA analysis, which demands more time and suffers from lack of reproducibility.

2.6. Analysis of protein conformation changes by limited digestion and freeze–thaw cycles

It has been previously suggested that the conformation adopted by FRQ affects its stability (Guo et al., 2010; Querfurth et al., 2011). A growing number of proteases are available with varied amino acid recognition sequences that can be used for digestion analyses to probe protein structure changes. For example, trypsin cuts after the amino acids arginine and lysine, and the environment in which these sites reside (loose or tight, exposed or buried structures) impacts the trypsin sensitivity of the protein. Other commercially available proteases have different preferential amino acid sequences. Repeated freeze–thaw cycles are also known to destroy proteins, probably through irreversible denaturation and loss of secondary and tertiary structures (Strambini & Gabellieri, 1996; Zhang, Qi, Singh, & Fernandez, 2011). We and others have used these methods to demonstrate that posttranslational regulation or changes to codons in the *frq* ORF affect conformation and thereby change the protein stability (Querfurth et al., 2011; Zhou, Guo, et al., 2013).

2.6.1 Limited protease digestion

Prepare *Neurospora* protein extracts in extraction buffer (50 m*M* HEPES, pH 7.4, 137 m*M* NaCl, and 10% glycerol, no protease inhibitors). Dilute the protein extract with extraction buffer to a total protein concentration of 2.5 μg/μl. Treat 100 μl extract with trypsin (final concentration depends on protein, ~0.3 μg/ml for wild-type FRQ is recommended) at 25 °C in a water bath with gentle shaking. Take a 20 μl sample from the reaction at each time point (0, 5, 15, and 30 min) after addition of trypsin and mix each 20-μl sample with protein loading buffer. Protein samples are resolved on a 12.5% SDS-PAGE gel. Western blot is performed to examine the target protein levels at each time point. ImageJ is used to quantify the results and plot the degradation curve.

A similar method can be also applied with C-terminal TAP-tagged FRQ protein purified from *Neurospora* culture in different phosphorylation states (Querfurth et al., 2011). After purification using a tandem affinity column, treat the same amount of hypo- and hyperphosphorylated FRQ with 30 ng of Factor Xa protease (New England Biolabs) at 20 °C for 10 min. Stop the reaction by adding SDS-PAGE sample buffer and boiling at 95 °C for 3 min, and subject the samples to SDS-PAGE. The pattern of digested fragments can be visualized by Western blot using a polyclonal anti-FRQ antibody.

2.6.2 Freeze–thaw assay

Prepare *Neurospora* protein extracts and dilute to a total protein concentration of 2.5 μg/μl (the same extraction buffer described in Section 2.6.1 is used). For each culture, 200 μl of protein extract is used in this assay. In each freeze–thaw cycle, the extract is snap frozen by immersion in liquid nitrogen and thawed at 37 °C for 15 min. A 20-μl sample is taken from the extract before the first freeze–thaw, and after the second and fourth freeze–thaw cycles, respectively. If the target protein is quite stable, additional freeze–thaw cycles should be used. Each 20-μl sample is mixed with protein loading buffer, and subjected to SDS-PAGE and Western blot analysis (Zhou, Guo, et al., 2013).

3. CONCLUDING REMARKS

Neurospora has been an outstanding model organism for circadian clock research and enables use of a wide range of methodologies to study the molecular mechanisms of eukaryotic circadian clocks. Various techniques have been used to identify the mechanisms that are critical for the core negative feedback loop. These mechanisms regulate *frq* expression and FRQ activity and stability at transcriptional, posttranscriptional, cotranslational, and posttranslational levels. The knowledge provided from the *Neurospora* studies has helped us understand how circadian period length and clock entrainment are determined at the molecular level. The remarkable similarities between the *Neurospora* clock and clocks of higher eukaryotes can help us address more complex problems in animal and plant clocks.

ACKNOWLEDGMENT

This work was supported by grants from the National Institutes of Health and the Welch Foundation (I-1560) to Yi Liu.

REFERENCES

Aronson, B. D., Johnson, K. A., Loros, J. J., & Dunlap, J. C. (1994). Negative feedback defining a circadian clock: Autoregulation of the clock gene frequency. *Science*, *263*, 1578–1584.

Baker, C. L., Kettenbach, A. N., Loros, J. J., Gerber, S. A., & Dunlap, J. C. (2009). Quantitative proteomics reveals a dynamic interactome and phase-specific phosphorylation in the *Neurospora* circadian clock. *Molecular Cell*, *34*, 354–363.

Belden, W. J., Larrondo, L. F., Froehlich, A. C., Shi, M., Chen, C. H., Loros, J. J., et al. (2007). The band mutation in Neurospora crassa is a dominant allele of ras-1 implicating RAS signaling in circadian output. *Genes and Development*, *21*, 1494–1505.

Belden, W. J., Lewis, Z. A., Selker, E. U., Loros, J. J., & Dunlap, J. C. (2011). CHD1 remodels chromatin and influences transient DNA methylation at the clock gene frequency. *PLoS Genetics*, 7, e1002166.

Belden, W. J., Loros, J. J., & Dunlap, J. C. (2007). Execution of the circadian negative feedback loop in Neurospora requires the ATP-dependent chromatin-remodeling enzyme CLOCKSWITCH. *Molecular Cell*, *25*, 587–600.

Cesbron, F., Brunner, M., & Diernfellner, A. C. (2013). Light-dependent and circadian transcription dynamics *in vivo* recorded with a destabilized luciferase reporter in Neurospora. *PLoS One*, *8*, e83660.

Cha, J., Chang, S. S., Huang, G., Cheng, P., & Liu, Y. (2008). Control of WHITE COLLAR localization by phosphorylation is a critical step in the circadian negative feedback process. *The EMBO Journal*, *27*, 3246–3255.

Cha, J., Yuan, H., & Liu, Y. (2011). Regulation of the activity and cellular localization of the circadian clock protein FRQ. *The Journal of Biological Chemistry*, *286*, 11469–11478.

Cha, J., Zhou, M., & Liu, Y. (2013). CATP is a critical component of the Neurospora circadian clock by regulating the nucleosome occupancy rhythm at the frequency locus. *EMBO Reports*, *14*, 923–930.

Cheng, P., He, Q., He, Q., Wang, L., & Liu, Y. (2005). Regulation of the Neurospora circadian clock by an RNA helicase. *Genes and Development*, *19*, 234–241.

Cheng, P., Yang, Y., Gardner, K. H., & Liu, Y. (2002). PAS domain-mediated WC-1/WC-2 interaction is essential for maintaining the steady-state level of WC-1 and the function of both proteins in circadian clock and light responses of Neurospora. *Molecular and Cellular Biology*, *22*, 517–524.

Cheng, P., Yang, Y., Wang, L., He, Q., & Liu, Y. (2003). WHITE COLLAR-1, a multifunctional neurospora protein involved in the circadian feedback loops, light sensing, and transcription repression of wc-2. *The Journal of Biological Chemistry*, *278*, 3801–3808.

Colot, H. V., Park, G., Turner, G. E., Ringelberg, C., Crew, C. M., Litvinkova, L., et al. (2006). A high-throughput gene knockout procedure for Neurospora reveals functions for multiple transcription factors. *Proceedings of the National Academy of Sciences of the United States of America*, *103*, 10352–10357.

Crosthwaite, S. K., Dunlap, J. C., & Loros, J. J. (1997). Neurospora wc-1 and wc-2: Transcription, photoresponses, and the origins of circadian rhythmicity. *Science*, *276*, 763–769.

Diernfellner, A. C., Querfurth, C., Salazar, C., Hofer, T., & Brunner, M. (2009). Phosphorylation modulates rapid nucleocytoplasmic shuttling and cytoplasmic accumulation of Neurospora clock protein FRQ on a circadian time scale. *Genes and Development*, *23*, 2192–2200.

Dunlap, J. C. (1999). Molecular bases for circadian clocks. *Cell*, *96*, 271–290.

Froehlich, A. C., Loros, J. J., & Dunlap, J. C. (2003). Rhythmic binding of a WHITE COLLAR-containing complex to the frequency promoter is inhibited by FREQUENCY. *Proceedings of the National Academy of Sciences of the United States of America*, *100*, 5914–5919.

Gooch, V. D., Mehra, A., Larrondo, L. F., Fox, J., Touroutoutoudis, M., Loros, J. J., et al. (2008). Fully codon-optimized luciferase uncovers novel temperature characteristics of the Neurospora clock. *Eukaryotic Cell*, 7, 28–37.

Gorl, M., Merrow, M., Huttner, B., Johnson, J., Roenneberg, T., & Brunner, M. (2001). A PEST-like element in FREQUENCY determines the length of the circadian period in Neurospora crassa. *The EMBO Journal*, *20*, 7074–7084.

Guo, J., Cheng, P., & Liu, Y. (2010). Functional significance of FRH in regulating the phosphorylation and stability of Neurospora circadian clock protein FRQ. *The Journal of Biological Chemistry*, *285*, 11508–11515.

Guo, J., Cheng, P., Yuan, H., & Liu, Y. (2009). The exosome regulates circadian gene expression in a posttranscriptional negative feedback loop. *Cell, 138*, 1236–1246.

He, Q., Cha, J., Lee, H. C., Yang, Y., & Liu, Y. (2006). CKI and CKII mediate the FREQUENCY-dependent phosphorylation of the WHITE COLLAR complex to close the Neurospora circadian negative feedback loop. *Genes and Development, 20*, 2552–2565.

He, Q., Cheng, P., He, Q., & Liu, Y. (2005). The COP9 signalosome regulates the Neurospora circadian clock by controlling the stability of the SCFFWD-1 complex. *Genes and Development, 19*, 1518–1531.

He, Q., Cheng, P., Yang, Y., He, Q., Yu, H., & Liu, Y. (2003). FWD1-mediated degradation of FREQUENCY in Neurospora establishes a conserved mechanism for circadian clock regulation. *The EMBO Journal, 22*, 4421–4430.

He, Q., & Liu, Y. (2005a). Degradation of the Neurospora circadian clock protein FREQUENCY through the ubiquitin-proteasome pathway. *Biochemical Society Transactions, 33*, 953–956.

He, Q., & Liu, Y. (2005b). Molecular mechanism of light responses in Neurospora: From light-induced transcription to photoadaptation. *Genes and Development, 19*, 2888–2899.

He, Q., Shu, H., Cheng, P., Chen, S., Wang, L., & Liu, Y. (2005). Light-independent phosphorylation of WHITE COLLAR-1 regulates its function in the Neurospora circadian negative feedback loop. *The Journal of Biological Chemistry, 280*, 17526–17532.

Heintzen, C., & Liu, Y. (2007). The Neurospora crassa circadian clock. *Advances in Genetics, 58*, 25–66.

Hong, C. I., Ruoff, P., Loros, J. J., & Dunlap, J. C. (2008). Closing the circadian negative feedback loop: FRQ-dependent clearance of WC-1 from the nucleus. *Genes and Development, 22*, 3196–3204.

Huang, G., Chen, S., Li, S., Cha, J., Long, C., Li, L., et al. (2007). Protein kinase A and casein kinases mediate sequential phosphorylation events in the circadian negative feedback loop. *Genes and Development, 21*, 3283–3295.

Larrondo, L. F., Loros, J. J., & Dunlap, J. C. (2012). High-resolution spatiotemporal analysis of gene expression in real time: *In vivo* analysis of circadian rhythms in Neurospora crassa using a FREQUENCY-luciferase translational reporter. *Fungal Genetics and Biology, 49*, 681–683.

Liu, Y. (2005). Analysis of posttranslational regulations in the Neurospora circadian clock. *Methods in Enzymology, 393*, 379–393.

Liu, Y., & Bell-Pedersen, D. (2006). Circadian rhythms in Neurospora crassa and other filamentous fungi. *Eukaryotic Cell, 5*, 1184–1193.

Liu, Y., Loros, J., & Dunlap, J. C. (2000). Phosphorylation of the Neurospora clock protein FREQUENCY determines its degradation rate and strongly influences the period length of the circadian clock. *Proceedings of the National Academy of Sciences of the United States of America, 97*, 234–239.

Luo, C., Loros, J. J., & Dunlap, J. C. (1998). Nuclear localization is required for function of the essential clock protein FRQ. *The EMBO Journal, 17*, 1228–1235.

Pregueiro, A. M., Liu, Q., Baker, C. L., Dunlap, J. C., & Loros, J. J. (2006). The Neurospora checkpoint kinase 2: A regulatory link between the circadian and cell cycles. *Science, 313*, 644–649.

Querfurth, C., Diernfellner, A. C., Gin, E., Malzahn, E., Höfer, T., & Brunner, M. (2011). Circadian conformational change of the Neurospora clock protein FREQUENCY triggered by clustered hyperphosphorylation of a basic domain. *Molecular Cell, 43*, 713–722.

Schafmeier, T., Diernfellner, A., Schafer, A., Dintsis, O., Neiss, A., & Brunner, M. (2008). Circadian activity and abundance rhythms of the Neurospora clock transcription factor WCC associated with rapid nucleo-cytoplasmic shuttling. *Genes and Development, 22*, 3397–3402.

Schafmeier, T., Haase, A., Kaldi, K., Scholz, J., Fuchs, M., & Brunner, M. (2005). Transcriptional feedback of Neurospora circadian clock gene by phosphorylation-dependent inactivation of its transcription factor. *Cell, 122*, 235–246.

Shi, M., Collett, M., Loros, J. J., & Dunlap, J. C. (2010). FRQ-interacting RNA helicase mediates negative and positive feedback in the Neurospora circadian clock. *Genetics, 184*, 351–361.

Strambini, G. B., & Gabellieri, E. (1996). Proteins in frozen solutions: Evidence of ice-induced partial unfolding. *Biophysical Journal, 70*, 971–976.

Tang, C. T., Li, S., Long, C., Cha, J., Huang, G., Li, L., et al. (2009). Setting the pace of the Neurospora circadian clock by multiple independent FRQ phosphorylation events. *Proceedings of the National Academy of Sciences of the United States of America, 106*, 10722–10727.

Yang, Y., Cheng, P., He, Q., Wang, L., & Liu, Y. (2003). Phosphorylation of FREQUENCY protein by casein kinase II is necessary for the function of the Neurospora circadian clock. *Molecular and Cellular Biology, 23*, 6221–6228.

Yang, Y., Cheng, P., Zhi, G., & Liu, Y. (2001). Identification of a calcium/calmodulin-dependent protein kinase that phosphorylates the Neurospora circadian clock protein FREQUENCY. *The Journal of Biological Chemistry, 276*, 41064–41072.

Yang, Y., He, Q., Cheng, P., Wrage, P., Yarden, O., & Liu, Y. (2004). Distinct roles for PP1 and PP2A in the Neurospora circadian clock. *Genes and Development, 18*, 255–260.

Young, M. W., & Kay, S. A. (2001). Time zones: A comparative genetics of circadian clocks. *Nature Reviews. Genetics, 2*, 702–715.

Zhang, A., Qi, W., Singh, S., & Fernandez, F. (2011). A new approach to explore the impact of freeze-thaw cycling on protein structure: Hydrogen/deuterium exchange mass spectrometry (HX-MS). *Pharmaceutical Research, 28*, 1179–1193.

Zhou, M., Guo, J., Cha, J., Chae, M., Chen, S., Barral, J. M., et al. (2013). Non-optimal codon usage affects expression, structure and function of clock protein FRQ. *Nature, 495*, 111–115.

Zhou, Z., Liu, X., Hu, Q., Zhang, N., Sun, G., Cha, J., et al. (2013). Suppression of WC-independent frequency transcription by RCO-1 is essential for Neurospora circadian clock. *Proceedings of the National Academy of Sciences of the United States of America, 110*, E4867–E4874.

CHAPTER EIGHT

Detecting KaiC Phosphorylation Rhythms of the Cyanobacterial Circadian Oscillator *In Vitro* and *In Vivo*

Yong-Ick Kim*, Joseph S. Boyd*, Javier Espinosa†, Susan S. Golden*,‡,1

*Center for Circadian Biology, University of California, San Diego, La Jolla, California, USA
†Division of Genetics, University of Alicante, Alicante, Spain
‡Division of Biological Sciences, University of California, San Diego, La Jolla, California, USA
[1]Corresponding author: e-mail address: sgolden@ucsd.edu

Contents

Methods in Enzymology, Volume 551
ISSN 0076-6879
http://dx.doi.org/10.1016/bs.mie.2014.10.003

Abstract

The central oscillator of the cyanobacterial circadian clock is unique in the biochemical simplicity of its components and the robustness of the oscillation. The oscillator is composed of three cyanobacterial proteins: KaiA, KaiB, and KaiC. If very pure preparations of these three proteins are mixed in a test tube in the right proportions and with ATP and $MgCl_2$, the phosphorylation states of KaiC will oscillate with a circadian period, and these states can be analyzed simply by SDS-PAGE. The purity of the proteins is critical for obtaining robust oscillation. Contaminating proteases will destroy oscillation by degradation of Kai proteins, and ATPases will attenuate robustness by consumption of ATP. Here, we provide a detailed protocol to obtain pure recombinant proteins from *Escherichia coli* to construct a robust cyanobacterial circadian oscillator *in vitro*. In addition, we present a protocol that facilitates analysis of phosphorylation states of KaiC and other phosphorylated proteins from *in vivo* samples.

1. THEORY

Circadian clocks are complicated biochemical mechanisms that temporally regulate biological processes and the expression of many genes. Many circadian clock components have been discovered in an array of organisms by *in vivo* experiments. Because of the complexity of circadian systems in eukaryotic organisms, the relatively simple prokaryotic clock of cyanobacteria has been developed as a paradigm for circadian biology, and its molecular mechanism can be studied in exquisite detail. The three-dimensional structures of the key protein components of the central oscillator, KaiA (Ye, Vakonakis, Ioerger, LiWang, & Sacchettini, 2004), KaiB (Villarreal et al., 2013), and KaiC (Pattanayek et al., 2004), have been solved at high resolution. Importantly, the 24-h rhythm of KaiC phosphorylation can be reconstituted by mixing purified KaiA, KaiB, and KaiC with adenosine triphosphate (ATP) *in vitro* (Nakajima et al., 2005).

Before the *in vitro* oscillator was developed, the expected model for circadian oscillation in cyanobacteria was that of a transcription–translation feedback loop mechanism, as is typically observed in eukaryotic systems. Takao Kondo and coworkers (Nagoya University) then discovered that the oscillation of KaiC phosphorylation continues in cyanobacteria even when translation or transcription is blocked (Tomita, Nakajima, Kondo, & Iwasaki, 2005); this finding led them to reconstitute the oscillator *in vitro*. Rhythms of KaiC phosphorylation are readily detected, both *in vivo* and *in vitro*, via sodium dodecyl sulfate polyacrylamide gel electrophoresis (SDS-PAGE) analysis. Use of the *in vitro* oscillator allows researchers to identify specific steps in the biochemical mechanism of the circadian clock at the molecular level.

The purification of the three oscillator proteins is a crucial technique to construct a successful *in vitro* oscillator (Fig. 1). Tiny amounts of impurities, such as proteases and ATPases, will abolish the oscillatory phosphorylation of KaiC. By applying protein-specific purification methods, those three proteins can be purified with sufficiently high purity and yield.

2. EQUIPMENT

Temperature-controlled shaking incubator
Refrigerated centrifuge
French press

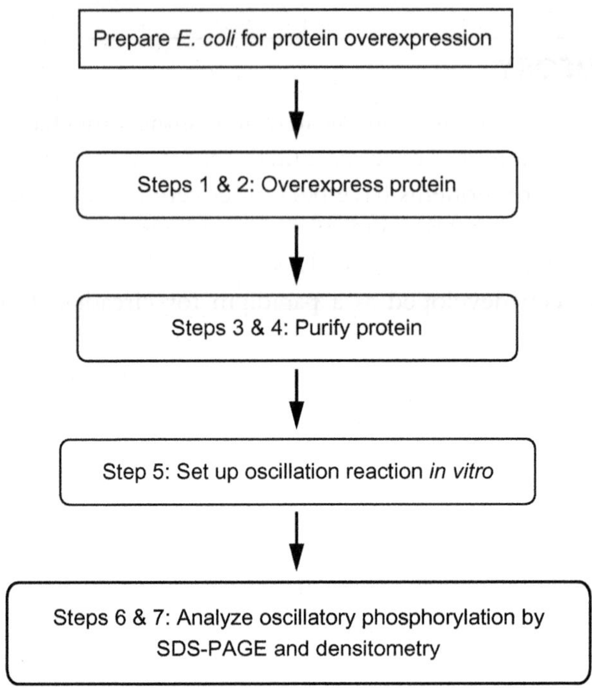

Figure 1 *In vitro* oscillation flowchart.

Fast protein liquid chromatography (FPLC) apparatus
Nickel (Ni) affinity column (such as 5 ml HiTrap Chelating HP prepacked column from GE Healthcare)
Glutathione-*S*-transferase (GST) affinity column (such as 5 ml GSTrap HP prepacked column from GE Healthcare)
Anion exchange column (such as 5 ml HiTrap Q HP prepacked column from GE Healthcare)
Desalting column (such as HiPrap 26/10 desalting column from GE Healthcare)
Membrane protein concentrator (such as an Amicon centrifugal filter, 10 and 30 kDa pore sizes)
Visible spectrophotometer
Sample changer
Heat block
SDS-PAGE system (such as Mini-PROTEAN from BioRad)
Gel documentation system

3. MATERIALS

Sodium chloride (NaCl)
Magnesium chloride ($MgCl_2$)
Hydrogen chloride (HCl)
Soduim hydroxide (NaOH)
Ethylenediaminetetraacetic acid (EDTA)
ATP
Dithiothreitol (DTT)
Imidazole
PreScission™ protease (from GE Healthcare)
Ubiquitin-like-specific protease 1 (Ulp1)
Bradford reagent
Tris
Glycine
SDS
Ammonium persulfate (APS)
Acrylamide (such as 30% acrylamide:bis solution, 29:1 from BioRad)
Bromophenol blue
Glycerol
Guanidinium chloride
Kanamycin
Isopropyl β-D-thiogalactopyranoside (IPTG)
Luria–Bertani broth (LB)
Escherichia coli (BL21DE3)
InstantBlue™ coomassie stain (from Expedeon)

3.1. Solutions and buffers

Ni buffer A

Component	Final concentration (m*M*)	Stock (*M*)	Amount (ml)
Tris–HCl, pH 7.0	20	1	10
NaCl	500	5	50
Imidazole	5	5	0.5

Add water up to 500 ml final volume and pass through a 0.45-μm filter.

Ni buffer B

Component	Final concentration (m*M*)	Stock (*M*)	Amount (ml)
Tris–HCl, pH 7.0	20	1	10
NaCl	500	5	50
Imidazole	500	5	50

Add water up to final volume 500 ml and pass through a 0.45-μm filter.

Desalting buffer

Component	Final concentration (m*M*)	Stock (*M*)	Amount (ml)
Tris–HCl, pH 7.0	20	1	10
NaCl	150	5	15

Add water up to final volume 500 ml and pass through a 0.45-μm filter.

Anion exchange buffer A

Component	Final concentration (m*M*)	Stock (*M*)	Amount (ml)
Tris–HCl, pH 7.0	20	1	10
NaCl	20	5	2

Add water up to final volume 500 ml and pass through a 0.45-μm filter.

Anion exchange buffer B

Component	Final concentration	Stock	Amount
Tris–HCl, pH 7.0	20 m*M*	1 *M*	10 ml
NaCl	1 *M*	5 *M*	100 ml

Add water up to final volume 500 ml and pass through a 0.45-μm filter.

***In vitro* reaction buffer**

Component	Final concentration	Stock	Amount
Tris–HCl, pH 8.0	20 m*M*	1 *M*	5 ml
NaCl	150 m*M*	5 *M*	7.5 ml
$MgCl_2$	5 m*M*	1 *M*	1.25 ml
EDTA	0.5 m*M*	500 m*M*	0.25 ml
ATP	1 m*M*	10 m*M*	25 ml

Add water up to final volume 250 ml and pass through a 0.45-μm filter.

GST buffer A

Component	Final concentration	Stock	Amount
Tris–HCl, pH 7.3	50 m*M*	1 *M*	10 ml
NaCl	150 m*M*	5 *M*	6 ml
$MgCl_2$	5 m*M*	1 *M*	1 ml
EDTA	1 m*M*	500 m*M*	0.4 ml
ATP	5 m*M*	10 m*M*	100 ml

Add water up to final volume 200 ml and pass through a 0.45-μm filter.

GST buffer B

Component	Final concentration	Stock	Amount
Tris–HCl, pH 7.3	50 m*M*	1 *M*	2.5 ml
NaCl	150 m*M*	5 *M*	1.5 ml
$MgCl_2$	5 m*M*	1 *M*	0.25 ml
EDTA	1 m*M*	500 m*M*	0.1 ml
ATP	1 m*M*	10 m*M*	5 ml
Glutathione	6 m*M*	100 m*M*	3 ml

Add water up to final volume 50 ml and pass through a 0.45-μm filter.

GST buffer C

Component	Final concentration	Stock	Amount
Tris–HCl, pH 7.3	50 m*M*	1 *M*	10 ml
NaCl	150 m*M*	5 *M*	6 ml
$MgCl_2$	5 m*M*	1 *M*	1 ml
EDTA	1 m*M*	500 m*M*	0.4 ml
ATP	1 m*M*	10 m*M*	20 ml

Add water up to final volume 200 ml and pass through a 0.45-μm filter.

10× Loading dye

Component	Final concentration	Stock	Amount
Tris–HCl, pH 7.5	100 m*M*	1 *M*	1 ml
DTT	200 m*M*	1 *M*	2 ml

SDS	4%	40%	1 ml
Bromophenol blue	0.2%	2%	1 ml
Glycerol	30%	60%	5 ml

Add water to final volume 10 ml

Tris–Glycine gel buffer

Component	Final concentration	Stock	Amount
Tris–HCl, pH 8.5	25 m*M*	1 *M*	25 ml
Glycine	200 m*M*	2 *M*	100 ml
SDS	1%	40%	25 ml

Add water to final volume 1 l.

Acrylamide running gel

Component	Final concentration	Stock	Amount
Tris–HCl, pH 8.8	400 m*M*	1.5 *M*	8 ml
Acrylamide	6.5%	30%	6.5 ml
SDS	0.1%	40%	0.075 ml
APS	0.1%	10%	0.3 ml
TEMED	0.03%	100%	0.009 ml

Add water to final volume 30 ml and pass through a 0.45-μm filter.

Acrylamide stacking gel

Component	Final concentration	Stock	Amount
Tris–HCl, pH 6.8	60 m*M*	500 m*M*	1.8 ml
Acrylamide	5%	30%	2.5 ml
SDS	0.1%	40%	0.038 ml
APS	0.05%	10%	0.075 ml
TEMED	0.1%	100%	0.015 ml

Add water to final volume 15 ml and pass through a 0.45-μm filter.

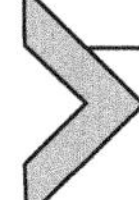

4. PROTOCOL

4.1. Duration

Preparation	3–5 days
Protocol	7–8 days

4.2. Preparation

Kai protein overexpression plasmids were prepared with some modification of pET28a(+) and pET41a(+) vectors. For KaiA and KaiB, sequences that encode SUMO-KaiA or SUMO-KaiB were inserted in between NdeI and HindIII restriction sites on pET28a(+). For KaiC, the PreScission™ protease cutting sequence (LEVLFEQGP) was encoded before the start codon of KaiC and inserted in between the NcoI and HindIII sites on pET41a(+). Each plasmid was confirmed by sequencing and introduced into *E. coli* (BL21DE3) for protein overexpression.

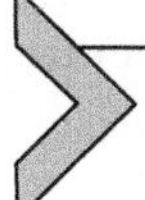

5. STEP 1: EXPRESSION OF KaiA OR KaiB IN *E. COLI*

5.1. Overview

The overexpressed KaiA and KaiB will be prepared for further purification.

5.2. Duration

12 h

1.1 Inoculate 1 l LB (containing the appropriate antibiotic) with transformed *E. coli* (BL21DE3) and grow the culture with vigorous shaking (~250 rpm) at 37 °C until the absorbance reaches $A_{600nm} = 0.7$.

1.2 Add 1 ml of 1 *M* IPTG to induce overexpression of recombinant KaiA or KaiB and incubate an additional 6 h under the same conditions.

1.3 Harvest the cells by centrifugation for 10 min in a refrigerated centrifuge at 5500 × *g*.

1.4 Remove the supernatant fraction and store the pellet at −80 °C.

5.3. Tip

Add three to five drops of Y-30 (Sigma) antifoam to help aeration by preventing foam.

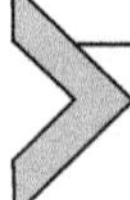

6. STEP 2: EXPRESSION OF KaiC IN *E. COLI*

6.1. Overview

KaiC requires slower induction than KaiA or KaiB to obtain abundant folded, soluble protein. The overexpressed KaiC will be prepared for further purification.

6.2. Duration

12 h

2.1 Inoculate 1 l LB (containing the appropriate antibiotic) with transformed *E. coli* (BL21DE3) and grow the culture at 37 °C until the absorbance reaches $A_{600\text{nm}}=0.6$.

2.2 Cool the culture down to 25 °C.

2.3 Add 1 ml of 100 m*M* IPTG to induce overexpression of recombinant KaiC and incubate the culture with shaking (~150 rpm) at 25 °C for an additional 16 h.

2.4 Harvest the cells by centrifugation for 10 min in a refrigerated centrifuge at 5500 × *g*.

2.5 Remove the supernatant fraction and store the pellet at −80 °C.

6.3. Tip

Room temperature incubation (~23 °C) can be used if temperature control is unavailable. Cool down completely before inducting with IPTG. Incomplete cooling results in a lower yield.

7. STEP 3: PURIFICATION OF KaiA OR KaiB

7.1. Overview

Purify KaiA and KaiB with high purity and yield by using a series of FPLC columns.

7.2. Duration

36 h

3.1 Resuspend KaiA or KaiB pellets (saved in step 1.4) in 60 ml of Ni buffer A.

3.2 Pass the resuspended cells twice through a chilled French press cell at a rate that maintains 16,000 PSI.

3.3 Spin the lysates for 60 min in a refrigerated centrifuge at 20,000 × *g*.

3.4 Filter the supernatant fraction through a 0.45-μm filter.

3.5 Inject the filtered supernatant into a Ni affinity column using FPLC with a 2 ml/min flow rate.

3.6 Wash the column with 25 ml Ni buffer A.

3.7 Elute His-tagged KaiA or KaiB with a 0–80% Ni buffer B gradient over 30 min.

3.8 Identify the protein-positive fractions from step 3.7 by absorbance at 280 nm and combine positive fractions. Run the combined protein sample through a desalting column using desalting buffer with 4 ml/min flow rate.

3.9 Identify the protein-positive fractions from step 3.8 by absorbance at 280 nm and combine positive fractions. Inject the combined protein sample into an anion exchange column using FPLC with a 2 ml/min flow rate.

3.10 Wash the column with 25 ml anion exchange buffer A.

3.11 Elute His-tagged KaiA or KaiB with a 0–80% anion exchange buffer B gradient over 30 min.

3.12 Identify the protein-positive fractions from step 3.11 by absorbance at 280 nm and combine positive fractions. Add 1 ml of 1 *M* NaCl and 0.1 ml of 100 μ*M* Ulp1 per each 10 ml of combined protein-positive sample.

3.13 Incubate the mixture for 16 h at 4 °C.

3.14 Inject the incubated mixture into a Ni affinity column using FPLC with a 2 ml/min flow rate and collect the flow-through. This step will retain the tags on the column and allow untagged KaiA and KaiB to pass through.

3.15 Wash the column with 5 ml Ni buffer A and continue to collect the flow-through.

3.16 Inject the flow-through into a desalting column and wash out continuously using anion exchange buffer A with a 4 ml/min flow rate until all protein-positive fractions are collected. Identify the protein-positive fractions by absorbance at 280 nm and combine positive fractions.

3.17 Inject the combined positive fraction into an anion exchange column using FPLC with 2 ml/min flow rate.

3.18 Wash the column with 25 ml anion exchange buffer A.

3.19 Elute KaiA or KaiB with a 0–80% anion exchange buffer B gradient for 30 min.

3.20 Identify the protein-positive fractions from step 3.19 by absorbance at 280 nm and combine positive fractions. Inject the combined fraction into a desalting column and wash out continuously using *in vitro* reaction buffer with a 4 ml/min flow rate until all protein-positive fractions are collected.

3.21 Concentrate the combined protein-positive fraction from step 3.20 using a membrane protein concentrator (such as an Amicon 10-kDa centrifugal filter). KaiA and KaiB can be concentrated up to 150 and 50 μ*M*, respectively.

3.22 Pass the concentrated protein through a 0.25-μm syringe filter.

3.23 Measure the concentration by Bradford assay.

3.24 Store protein at −80 °C.

7.3. Tip

The purification procedure should be performed at room temperature to obtain high purity.

7.4. Tip

Dialysis can be used instead of a desalting column.

7.5. Tip

The second affinity column (steps 3.14 and 3.15) also helps to remove impurities and is essential for obtaining the purity needed for the *in vitro* oscillator; this step should not be skipped.

8. STEP 4: PURIFICATION OF KaiC

8.1. Overview

Purify KaiC with high purity and yield by using a series of FPLC columns.

8.2. Duration

36 h

4.1 Resuspend KaiC pellets (saved in step 2.5) in 60 ml of GST buffer A.

4.2 Follow steps 3.2–3.4 as for KaiA and KaiB purification.

4.3 Inject the filtered supernatant into a GST affinity column using FPLC with a 1 ml/min flow rate.

4.4 Wash the column with 90 ml GST buffer A.

4.5 Elute GST-tagged KaiC with 12 ml GST buffer B and collect the entire elution volume.

4.6 Add 10 μl (20 U) of PreScission™ protease per each 12 μl of the collected eluent.

4.7 Incubate the KaiC mixture for 16 h at 4 °C.

4.8 Inject the incubated KaiC mixture into a desalting column and wash out using GST buffer C with a 4 ml/min flow rate until all protein-positive fractions are collected.

4.9 Inject the protein-positive fractions into a GST affinity column with a 1 ml/min flow rate and collect the flow-through. This step will retain the tag on the column and allow untagged KaiC to pass through.

4.10 Wash the GST affinity column with 5 ml GST buffer C at a rate of 1 ml/min and continue to collect the flow-through. Combine the flow-through from steps 4.9 to 4.10.

4.11 Inject the combined flow-through from the GST affinity column into a desalting column and wash out using *in vitro* reaction buffer with a 4 ml/min flow rate until all protein-positive fractions are collected.

4.12 Concentrate the combined protein-positive fraction from step 4.11 using a membrane protein concentrator (such as an Amicon 30-kDa centrifugal filter). KaiC can be concentrated to ~15 μ*M*.

4.13 Follow from steps 3.22 to 3.24 as for KaiA and KaiB purification.

8.3. Tip

The purification procedure should be performed at room temperature to obtain high purity.

8.4. Tip

Using a desalting column instead of dialysis may increase yield by preventing precipitation.

9. STEP 5: *IN VITRO* OSCILLATION REACTION

9.1. Overview

The phosphorylation states of KaiC can be observed to oscillate with a 24-h period by mixing KaiA, KaiB, KaiC, and ATP at 30 °C (Fig. 1).

9.2. Duration

76 h

5.1 Mix 1.2 μ*M* KaiA, 3.5 μ*M* KaiB, 3.5 μ*M* KaiC, and Kanamycin (10 μg/ml final concentration) in *in vitro* reaction buffer.

5.2 Incubate the reaction mixture at 30 °C.

5.3 Take 20 μl aliquots of reaction mixture every 2 h for 3 days and add 2 μl 10 × loading dye. Each SDS-PAGE gel will require at least 5 μl of sample, so 20 μl aliquots allow for multiple gels to be run for each experiment.

5.4 Freeze each sample at −20 °C as it is collected until the reaction is complete.

5.5 Keep all samples at −20 °C prior to SDS-PAGE analysis.

9.3. Tip

The total reaction volume can be modified to provide the number of samples desired in step 5.3 to run at least one gel of the timecourse.

10. STEP 6: SDS-PAGE

10.1. Overview

SDS-PAGE (6.5% polyacrylamide gel) is able to separate phosphorylated and unphosphorylated forms of KaiC.

10.2. Duration

2 h

6.1 Prepare eight 15-well polyacrylamide gels with acrylamide running gel and acrylamide stacking gel.

6.2 Mount the gel in the apparatus, place the entire apparatus into a water-tight high-sided tray, and fill the reservoir with Tris–glycine buffer. Remove all bubbles at the bottom of the gel.

6.3 Take out comb and remove all bubbles in the wells. Load the samples (saved in step 5.4) onto the gel.

6.4 Pack ice around the outside of the SDS-PAGE apparatus.

6.5 Run the gel at 60 V for 30 min and at 140 V for 100 min.

6.6 Put the gel in InstantBlue™ coomassie stain solution and shake gently on a platform shaker overnight.

6.7 Destain with water until bands are clearly visible.

10.3. Tip

Load 4–5 μl of sample in each of the 13 interior wells. Gel "smiling" can be avoided by loading dye only, rather than sample, in the outside wells on either end. The resulting straight line of bands will facilitate quantification.

10.4. Tip

The number of gels on step 6.1 and the time on step 6.5 are calculated based on Mini-PROTEAN from BioRad.

11. STEP 7: DENSITOMETRY

11.1. Overview

This is an easy quantification method for determining the phosphorylation level of KaiC in order to graph the oscillation.

11.2. Duration

1 h

7.1 Capture a high-resolution image of the stained gel.

7.2 Measure the density of each band with National Institutes of Health (NIH) ImageJ.

7.3 Refer to Fig. 2 to identify KaiC phosphoforms. Combine values for bands that represent phosphoforms and calculate % phosphorylated KaiC (P-KaiC).

7.4 Draw a graph that plots % P-KaiC versus time.

11.3. Tip

Follow ImageJ user guide on NIH Web site (http://imagej.nih.gov/ij/docs/guide/index.html) to measure the density of each band.

11.4. Tip

PeakFit software can be used instead of NIH ImageJ.

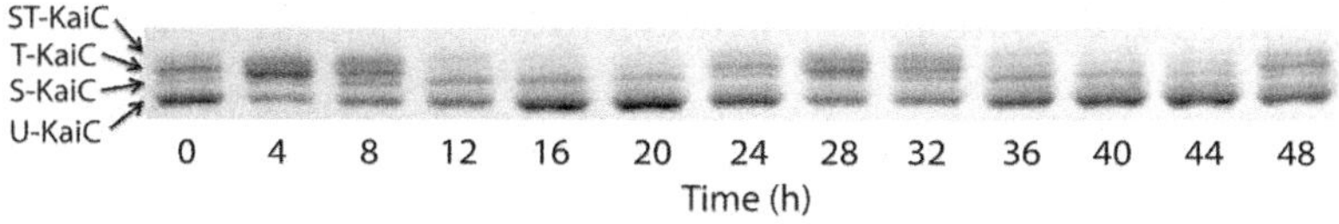

Figure 2 Coomassie-stained SDS-PAGE image of KaiC phosphorylation cycles. The samples were taken every 4 h for 2 days. Four phosphoforms of KaiC can be separated with the method introduced here. ST-KaiC, S431, and T432 KaiC double phosphoform; T-KaiC, T432 KaiC phosphoform; S-KaiC, S431 KaiC phosphoform; and U-KaiC, unphosphorylated KaiC.

12. DETECTION OF PROTEIN PHOSPHORYLATION FORMS FROM *IN VIVO* CELL EXTRACTS

Separation of the phosphoforms of KaiC, which is phosphorylated on Ser/Thr residues, from total soluble cell protein extracts can be facilitated by use of Phos-tag™ reagent in a modified immunoblotting protocol. Phos-tag™ SDS-PAGE allows separation of phospho-states by reagent binding to phosphate groups and retardation of phosphorylated proteins during electrophoresis. This technique can also be applied to phosphorylated histidine kinases and response regulators, for which the inherent lability of phospho-His and phospho-Asp and the lack of phospho-specific antibodies to these residues has hindered *in vivo* detection of the phosphorylated states (Barbieri & Stock, 2008; Moronta-Barrios, Espinosa, & Contreras, 2012). The requirement for specific antibodies can be circumvented by the addition of epitope tags, although this approach has not yet been reported in Phos-tag™ assays. Phos-tag™ acrylamide separations require optimization of gel composition for each protein studied. Ranges of 25–50 μM of Phos-tag™ acrylamide reagent and 50 μM Mn^{2+} are good starting references, but optimal concentrations must be determined empirically to obtain sharp banding. However, even with substantially distorted bands, if sufficient separation of unphosphorylated and phosphorylated protein bands is achieved, quantification of the fraction of phosphorylated proteins can be accurately estimated using image analysis software (Gutu & O'Shea, 2013).

The following protocol has been used to analyze the *in vivo* phosphorylation of KaiC and RpaA in *S. elongatus* cells subjected to 12:12 LD cycles (Fig. 3). It should form a suitable starting protocol that can be optimized for the protein under study.

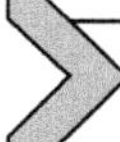

13. EQUIPMENT

Temperature-controlled lighted incubator
Refrigerated centrifuge
Cell homogenizer or benchtop vortexer
Rotary shaker
Power supply
Electrophoretic apparatus for SDS-PAGE
Protein blotting (Wet transfer) apparatus
Standard blotting accessories (filter paper, blotting membrane)
Imaging equipment and software

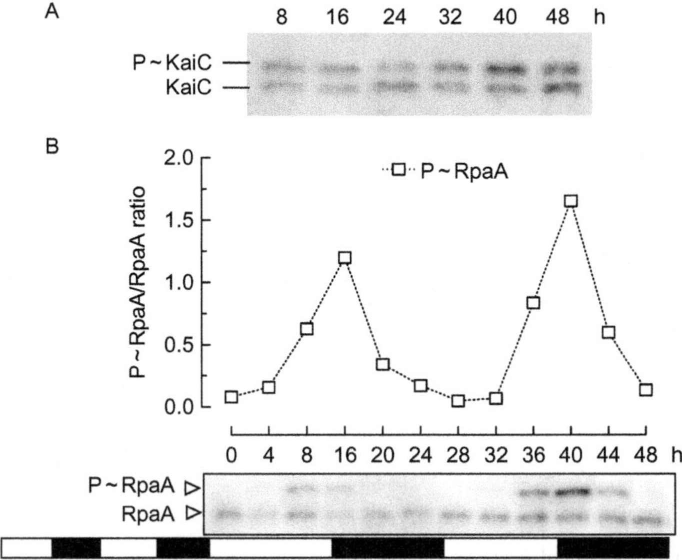

Figure 3 Detection of protein phosphorylation using Phos-tag acrylamide from whole-cell soluble protein extracts. (A) Immunoblot showing separation of phosphorylated and unphosphorylated KaiC after Phos-tag™ SDS-PAGE. Protein samples were collected every 8 h in constant light after entrainment to two 12-h light:dark (LD) cycles. (B) Quantification of P∼RpaA fractions from immunoblot. Specific detection of P∼RpaA and RpaA with anti-RpaA polyclonal antibodies, respectively, is shown. The white and black boxes represent, respectively, light and dark periods. Boxes corresponding to first two LD cycles are not drawn to scale.

14. MATERIALS

29:1 acrylamide: *N*,*N*′-methylene-bis-acrylamide
Deionized distilled water
1.5 *M* Tris–Cl, pH 8.8
0.5 *M* Tris–Cl, pH 6.8
SDS
100% Isopropanol or saturated butanol
APS
TEMED
Phos-tag™ acrylamide AAL-007 (Wako Chemicals)
10 m*M* $MnCl_2$
DTT
Bromophenol blue
Glass beads (100–300 μm diameter)

Tris
Glycine
Methanol
Bradford reagent
EDTA (0.5 *M*)
Antibodies to proteins of interest
Secondary antibodies (for example, horseradish peroxidase, HRP-linked) and chemiluminescent substrate for detection

14.1. Solutions and buffers

TBS-T

100 ml 10 × Tris-buffered saline (TBS)
1 ml Tween-20
Water to 1 l

Lysis Buffer

25 m*M* Tris, pH 8.0
0.5 m*M* EDTA
1 m*M* DTT and bacterial protease inhibitor cocktail (Sigma)

Transfer Buffer

2.2 g CAPS
8.33 μl 10 N NaOH
100 ml Methanol
Deionized water to 1 l

15. PROTOCOL

15.1. Duration

Preparation	3–5 Days, depending on duration of sampling
Protocol	2 Days

16. STEP 1: PREPARATION

16.1. Overview

Cast Phos-tag™ acrylamide gels, entrain cyanobacterial strains, and collect samples

1.1 Phos-tag™acrylamide SDS mini gels are prepared using a discontinuous buffer system (refer to materials Section 3.1 for gel preparation, increasing resolving gel acrylamide concentration to 10% w/v). The resolving gel consists of 10% (w/v) 29:1 acrylamide: *N*,*N*′-methylene-bis-acrylamide containing Tris–HCl, pH 8.8, 0.1% SDS, supplemented with 25 μ*M* of Phos-tag™ acrylamide reagent and 50 μ*M* Mn^{2+}. Overlay the gel with a 2 mm layer of saturated butanol, or layer with isopropanol to exclude oxygen and ensure a flat interface between the resolving and stacking gels. After polymerization, rinse the isopropanol or butanol from the top of the gel with water and drain the water by inverting the gel. Prepare the stacking gel using 4% (w/v) acrylamide:bis-acrylamide, Tris–HCl, pH 6.8, and 0.1% SDS. Prechill acrylamide gels at 4 °C prior to electrophoresis.

1.2 Entrain *S. elongatus* cells for two 12:12 LD cycles, move to constant light, and subsequently sample at chosen time points.

1.3 At each time point, harvest 10–15 ml of cells (grown to an optical density of $A_{750}=0.1$–0.75) by centrifugation at 6000 × *g* for 5 min at 4 °C. Carefully, remove all residual growth medium from cell pellet. Quickly freeze each cell pellet and store at −80 °C until ready to prepare lysates.

1.4 Resuspend the cell pellet in 30–80 μl of lysis buffer (25 m*M* Tris, pH 8.0, 0.5 m*M* EDTA, 1 m*M* DTT, and bacterial protease inhibitor cocktail (Sigma)). Add glass beads (0.1–0.3 μ diameter) equivalent to one-third of the volume. Cell lysis is achieved with 2 cycles of 1 min in a homogenizer (BeadBeaterBioSpec) with 1 min cooling between cycles. Alternatively, cells may be lysed using a benchtop vortexer at maximum setting with 10 cycles alternating 30 s vortexing/30 s cooling on ice.

1.5 Centrifuge cell extract at 10,000 × *g* for 5 min at 4 °C to pellet unbroken cells, cellular debris, and beads. Transfer the blue-green supernatant to a new tube and use an aliquot to quantify protein concentration using a Bradford assay. Samples may be used immediately for gel loading, or stored at −20 °C.

16.2. Tip

For proteins with heat-labile phospho-Asp or phospho-His residues, keep samples on ice after extraction and do not heat prior to gel loading. To obtain optimal levels of phosphorylated protein, proceed directly to electrophoresis and do not subject cell extracts to freeze–thaw cycles.

16.3. Tip

Samples must be prepared in buffer without phosphate. Do not substitute phosphate-buffered saline (PBS) for TBS during resuspension of cell pellets.

17. STEP 2: ELECTROPHORESIS AND BLOTTING

17.1. Overview

Load and run Phos-tag™ acrylamide gels and detect proteins by immunoblotting

2.1 Load 2–5 μg of protein extract in SDS sample buffer per lane (for KaiC detection) and run at 90 V 4 °C for approximately 3 h in a Tris/glycine/SDS buffer.

2.2 Incubate gels twice for 5 min at RT in transfer buffer solution (Tris/glycine/methanol supplemented with 1 m*M* EDTA), then twice more in standard transfer solution without EDTA.

2.3 Transfer proteins from gel to PVDF membrane by wet electroblotting (tank transfer). Transfer at 4 °C and 100 V for at least 3 h is recommended.

2.4 Block the membrane with 2.5% nonfat dry milk (NFDM)/TBS + 0.1% Tween-20 for 1 h at RT and then incubate with Anti-KaiC (1:2000) overnight at 4 °C on a rotary shaker. Then, rinse the membrane twice with TBS and incubate with HRP-linked secondary antibody (1:10,000) in NFDM/TBS-T for 1 h at RT.

2.5 Develop blot using SuperSignal West Femto Maximum Sensitivity Substrate (Thermo Scientific) according to the manufacturer's directions. For examples shown in Fig. 3, chemiluminescence was detected using an Alpha Innotech FluorChem® HD2 Imaging System (Alpha Innotech, San Leandro, CA) (cf. Fig. 3B).

2.6 Perform image analysis to quantify band intensity using NIH ImageJ software (cf. Fig. 3B).

17.2. Tip

Resolution of multiple KaiC phospho-states may be obtained by running gels at lower current for an extended time and/or increasing concentration of Phos-tag™ acrylamide reagent.

ACKNOWLEDGMENTS

This work was supported by National Institutes of Health Grant R01GM062419 (to S. S. G.). We thank Federico Unglaub for careful critique of the methods descriptions.

REFERENCES

Barbieri, C. M., & Stock, A. M. (2008). Universally applicable methods for monitoring response regulator aspartate phosphorylation both in vitro and in vivo using Phos-tag-based reagents. *Analytical Biochemistry*, *376*, 73–82.

Gutu, A., & O'Shea, E. (2013). Two antagonistic clock-regulated histidine kinases time the activation of circadian gene expression. *Molecular Cell*, *50*, 288–294.

Moronta-Barrios, F., Espinosa, J., & Contreras, A. (2012). In vivo features of signal transduction by the essential response regulator RpaB from *Synechococcus elongatus* PCC 7942. *Microbiology*, *158*, 1229–1237.

Nakajima, M., Imai, K., Ito, H., Nishiwaki, T., Murayama, Y., Iwasaki, H., et al. (2005). Reconstitution of circadian oscillation of cyanobacterial KaiC phosphorylation in vitro. *Science*, *308*, 414–415.

Pattanayek, R., Wang, J., Mori, T., Xu, Y., Johnson, C. H., & Egli, M. (2004). Visualizing a circadian clock protein: Crystal structure of KaiC and functional insights. *Molecular Cell*, *15*, 375–388.

Tomita, J., Nakajima, M., Kondo, T., & Iwasaki, H. (2005). No transcription-translation feedback in circadian rhythm of KaiC phosphorylation. *Science*, *307*, 251–254.

Villarreal, S. A., Pattanayek, R., Williams, D. R., Mori, T., Qin, X., Johnson, C. H., et al. (2013). CryoEM and molecular dynamics of the circadian KaiABC complex indicates that KaiB monomers interact with KaiC and block ATP binding clefts. *Journal of Molecular Biology*, *425*, 3311–3324.

Ye, S., Vakonakis, I., Ioerger, T. R., LiWang, A. C., & Sacchettini, J. C. (2004). Crystal structure of circadian clock protein KaiA from *Synechococcus elongatus*. *The Journal of Biological Chemistry*, *279*, 20511–20518.

CHAPTER NINE

The Role of Casein Kinase I in the *Drosophila* Circadian Clock

Jeffrey L. Price[1], Jin-Yuan Fan, Andrew Keightley, John C. Means
Division of Molecular Biology and Biochemistry, School of Biological Sciences, University of Missouri-Kansas City, Kansas City, Missouri, USA
[1]Corresponding author: e-mail address: pricejL@umkc.edu

Contents

Abstract

The circadian clock mechanism in organisms as diverse as cyanobacteria and humans involves both transcriptional and posttranslational regulation of key clock components. One of the roles for the posttranslational regulation is to time the degradation of the targeted clock proteins, so that their oscillation profiles are out of phase with respect to those of the mRNAs from which they are translated. In Drosophila, the circadian transcriptional regulator PERIOD (PER) is targeted for degradation by a kinase (DOUBLETIME or DBT) orthologous to mammalian kinases (CKIε and CKIδ) that also target mammalian PER. Since these kinases are not regulated by second messengers, the mechanism (if any) for their regulation is not known. We are investigating the possibility that regulation of DBT is conferred by other proteins that associate with DBT and PER. In this chapter, the methods we are employing to identify and analyze these factors are discussed. These methods include expression of wild type and mutant proteins with the GAL4/UAS binary expression approach, analysis of DBT in Drosophila S2 cells, *in vitro* kinase assays with DBT isolated from S2 cells, and proteomic analysis of DBT-containing complexes and of DBT phosphorylation with mass spectrometry. The work has led to the discovery of a previously unrecognized circadian rhythm component (Bride of DBT, a non-canonical FK506-binding protein) and the mapping of autophosphorylation sites within the DBT C-terminal domain with potential regulatory roles.

Methods in Enzymology, Volume 551
ISSN 0076-6879
http://dx.doi.org/10.1016/bs.mie.2014.10.012

1. INTRODUCTION

The analysis of circadian rhythms in Drosophila has revealed a requirement for daily changes in the phosphorylation states of key components, and this requirement has also been found in other phyla with largely divergent molecular components of the circadian clock. In Drosophila, posttranscriptional regulatory mechanisms were immediately suggested by the observation that *period* (*per*) mRNA and proteins levels peak at nearly opposite times in the 24-h cycle (Hardin, Hall, & Rosbash, 1990; Zerr, Hall, Rosbash, & Siwicki, 1990). A role for PER phosphorylation was demonstrated by the rhythmic changes in PER phosphorylation state, demonstrated initially by PER electrophoretic mobility shifts on SDS-PAGE gels (Edery, Zwiebel, Dembinska, & Rosbash, 1994) and subsequently by mass spectrometry analysis (Chiu, Vanselow, Kramer, & Edery, 2008; Garbe et al., 2013). The isolation of *doubletime* (*dbt*) mutants that shorten, lengthen, or blunt the oscillations of PER and the molecular cloning of the *dbt* gene, which encodes the Drosophila casein kinase δ/ε ortholog, provided the first insight on a clock component that contributes to this phosphorylation (Kloss et al., 1998; Price et al., 1998). Since then, numerous other kinases, phosphatases, O-GlcNAc transferases, and proteasomal components that contribute to the posttranslational regulation of PER have been identified (Hardin & Panda, 2013), but DBT is still a principal focus of research aimed at the posttranslational part of the clock mechanism because it targets more PER sites than other kinases (Chiu et al., 2008; Muskus, Preuss, Fan, Bjes, & Price, 2007).

DBT interacts directly with PER and regulates its circadian function in a number of ways. With reduced levels of DBT protein or activity, PER does not exhibit circadian changes in level or electrophoretic mobility, and it is expressed at high levels (Muskus et al., 2007; Price et al., 1998). Phosphorylation of PER by DBT has been shown to create a high affinity binding site for the ubiquitin ligase component SLIMB (Chiu et al., 2008). DBT has been shown to move with PER from the cytosol to the nucleus, thereby positioning it to produce both phosphorylation-dependent delays in cytosolic PER accumulation and termination of the nuclear repression phase (Kloss, Rothenfluh, Young, & Saez, 2001); taken together, these effects can explain (at least conceptually) the phase difference between the *per* mRNA and protein profiles as a consequence of DBT-signaled PER degradation in the cytosol and nucleus. DBT has also been suggested to regulate nuclear localization of PER (Bao, Rihel, Bjes, Fan, & Price, 2001; Cyran et al., 2005; Muskus et al., 2007; Nawathean, Stoleru, & Rosbash, 2007), PER's transcriptional activity (Kim, Ko, Yu, Hardin, & Edery, 2007; Nawathean et al., 2007), phosphorylation of CLOCK (CLK)—the transcription factor repressed by PER (Kim & Edery, 2006; Yu, Zheng, Price, & Hardin, 2009)—and the synchrony of the circadian neural network (Zheng, Sowcik, Chen, & Sehgal, 2014).

Since the discovery of its role in the Drosophila clock, CKI has been shown to be a component of most other eukaryotic circadian clocks so far analyzed. The first mammalian clock mutant (the Syrian Hamster *tau* mutant) was shown to carry a mutation in *ckIε* (Lowrey et al., 2000), and subsequent work has shown that both CKIε and CKIδ are involved in mammalian clock function (Lee, Etchegaray, Cagampang, Loudon, & Reppert, 2001; Lee, Chen, Lee, Yoo, & Lee, 2009; Walton et al., 2009; Xu et al., 2005). It was shown that the *tau* mutation produces the same short period in flies as in hamsters, in the context of either the fly or vertebrate enzyme (Fan, Preuss, Muskus, Bjes, & Price, 2009). In addition, casein kinase I is a component of fungal clocks (He et al., 2006). An algal clock, which does not require a transcriptional mechanism, is affected by CKI activity (van Ooijen et al., 2013). Cyanobacteria do not have a casein kinase I ortholog, but rhythmic phosphorylation of clock components is the central feature of the circadian mechanism (Nakajima et al., 2005).

A major conceptual challenge has been to explain the basis for both the short- and long-period *dbt* mutations. *In vitro*, both the short- and long-period *dbt* (or mammalian *ckI*) mutant kinases produce less kinase activity on complex substrates like PER and casein (Fan et al., 2009; Kivimae, Saez, & Young, 2008; Lowrey et al., 2000; Preuss et al., 2004; Suri, Hall, & Rosbash, 2000; Xu et al., 2005). The different mutant effects on circadian period may be explained by different effects of the mutant proteins at different phosphorylation sites within PER. Indeed, the dbt^S and $ckI\varepsilon^{tau}$ mutations have been suggested to phosphorylate some sites more actively than wild type (Gallego, Eide, Woolf, Virshup, & Forger, 2006), and DBT phosphorylates a domain within PER to antagonize its action at the more N-terminal sites that produce SLIMB binding and degradation (Chiu, Ko, & Edery, 2011; Chiu et al., 2008; Kivimae et al., 2008; Yu, Houl, & Hardin, 2011), demonstrating that DBT activity can both decelerate and accelerate the pace of the circadian cycle.

As an alternative or additional mechanism, we are intrigued by the possibility that the short-period mutations affect interactions of the kinase with other circadian complex components to alter circadian period. There are several reasons for this prediction. First, expression of various levels of a dominant negative catalytically inactive $DBT^{K/R}$ protein produces only long-period circadian rhythms, with length proportional to the degree of expression; short periods are not observed, suggesting that they are not produced by reduced kinase activity (Muskus et al., 2007). Likewise, pharmacological inhibitors of CKI activity produce only longer circadian periods (Walton et al., 2009). The effect of DBT on CLK phosphorylation appears to be mediated by its role in protein complexes, since the catalytically inactive $DBT^{K/R}$ can stimulate CLK phosphorylation as does DBT^{WT} (Yu et al., 2009). The effects of the *timeless* (TIM) protein are mediated in part by its suppression of PER phosphorylation when bound with PER; elimination of TIM with light produces rapid phosphorylation of PER, suggesting that TIM at least can suppress DBT activity on PER by associating with PER/DBT (Kloss et al., 2001). Finally, both the *dbt*^{S} and *tau* mutations affect amino acids on the surface of the kinase, where they could affect interaction with another factor that would normally slow the pace of the clock.

For these reasons, we have focused much on our recent efforts on the identification of novel DBT-interacting proteins that may regulate the pace of the circadian cycle. To this end, we have employed both genetic analysis with overexpression and RNAi approaches and proteomic approaches. We are also interested in a possible role for DBT C-terminal domain autophosphorylation in regulation of DBT, since we have demonstrated that DBT rapidly autophosphorylates when phosphatases are inhibited (Fan et al., 2009), and this has previously been suggested to inhibit CKIδ/ε activity (Gietzen & Virshup, 1999; Graves & Roach, 1995). Several methods that have been applied to these analyses are discussed in this chapter. The methods have led to the discovery of a role for the Bride of DBT protein in the circadian mechanism (Fan et al., 2013), as well as several other candidates still under evaluation.

2. EXPRESSION OF MUTANT FORMS OF DBT WITH THE GAL4/UAS BINARY EXPRESSION METHOD

Standard strategies for assessment of mutant protein function typically entail introduction of transgenes expressing the mutant protein into a mutant that is genetically null for the endogenous gene, or alternatively to generate a knock-in mutant in which the transgene replaces the endogenous gene. However, both of these approaches pose problems for analysis of strong loss-of-function DBT proteins, as the null genotype for *dbt* is lethal

(Zilian et al., 1999) and knock-in of transgenes by homologous recombination has not been routinely attainable in Drosophila. Therefore, we have knocked down *dbt* activity specifically in circadian cells to circumvent this lethality.

To target circadian processes, we have employed the GAL4/UAS approach of Brand and Perrimonn (1993) to target expression of the transgenes to circadian cells. A number of circadian transgenic drivers are available in which circadian promoters have been linked to the yeast transcription factor GAL4, which is therefore expressed in the cell types dictated by the promoter (Stoleru, Peng, Agosto, & Rosbash, 2004). Flies containing these transgenes can be crossed to flies containing a *dbt* transgene linked to a yeast upstream activating sequence (UAS). The GAL4 transcription factor binds to the UAS sequence to activate transcription of the *dbt* gene in specific sets of circadian cells. Using this approach with a *tim*GAL4 driver (expressed in all circadian cells), we have shown that expression of a dominant negative catalytically inactive $\mathrm{DBT}^{\mathrm{K/R}}$ protein can antagonize PER phosphorylation, degradation and the pace of the circadian clock, thereby leading to very long periods and arrhythmicity and demonstrating that DBT kinase activity is necessary for the circadian clock period determination and oscillation (Muskus et al., 2007). The degree of period-lengthening is proportional to the amount of expression. Moreover, expression of UAS-dbt^S and -dbt^L transgenes produces flies with the same circadian periods as those of the original short and long *dbt* mutants. Therefore, it has been concluded that high levels of the mutant DBT proteins titrate out endogenous DBT and produce a genotype-specific set point. The approach has also been used to express mutant forms of vertebrate CKIδ in flies; the short-period mutants (including one carrying the mutation causing the Syrian hamster *tau* phenotype) cause similar period-shortening in the context of either the fly or vertebrate enzyme, demonstrating evolutionary conservation of *dbt*'s role in the clock (Fan et al., 2009). Finally, expression of the $\mathrm{DBT}^{\mathrm{K/R}}$ protein supports phosphorylation of the CLK protein, while a lack of DBT association with CLK and PER does not support this phosphorylation, suggesting that DBT is part of a scaffold that recruits another CLK kinase (Yu et al., 2009).

Additional specifically expressed circadian GAL4 drivers have been used with these *dbt* transgenes to drive specific circadian cells at different periods from the rest of the circadian system, demonstrating specific roles for various cell groups (Zhang, Liu, Bilodeau-Wentworth, Hardin, & Emery, 2010) and a limited range of entrainment of more peripheral brain centers to the oscillator of the small ventral lateral neurons (Yao & Shafer, 2014).

Specific expression can be fine-tuned even further by the inclusion of a Gal80 repressor under control of another promoter with expression domains overlapping those of the GAL4 driver, so that GAL4-mediated expression is produced in only part of the driver domain (Stoleru et al., 2004).

Finally, GAL4-mediated expression of UAS-transgenes expressing dsRNAi's has been used to investigate gene products that may interact with DBT in circadian cells. This type of analysis has been facilitated by several collections of fly lines in which double-stranded RNAs for fly genes have been produced on a genome-wide scale and linked to UAS elements (e.g., http://stockcenter.vdrc.at/control/main; http://www.flyrnai.org/TRiP-HOME.html). For instance, expression of dsRNAi for the *bdbt* gene was found to affect DBT autophosphorylation state and lead to constitutively high levels of hypophosphorylated PER (Fan et al., 2013). dsRNAs are designed to target the endogenous single-stranded mRNA for degradation by the RNAi pathway, but they can produce unanticipated side effects by affecting off-target mRNAs. Therefore, it is important to confirm any effects with multiple nonoverlapping dsRNAs, or to rescue the effect with overexpression of the wild-type transgenic protein. In the case of *bdbt* RNAi phenotype, it was possible to suppress the phenotype with a UAS-*dbt* transgene, thereby demonstrating a genetic interaction that confirmed the DBT/BDBT protein interaction and that suggested the *bdbt* phenotype was a consequence of lowered *dbt* activity (Fan et al., 2013).

3. EXPRESSION OF DBT IN DROSOPHILA S2 CELLS FOR ANALYSIS OF DBT KINASE ACTIVITY

In addition to expression of DBT for recovery of protein complexes and analysis of DBT phosphorylation (described below), we commonly employ S2 cells to express mutant and wild-type DBTs for analysis of their activity. To analyze their activity toward PER, transgenic PER is coexpressed with DBT, and the subsequent phosphorylation (as manifested by electrophoretic mobility shifts) and degradation of PER are assayed by immunoblot. This approach was discussed in a previous addition of this series (Ko & Edery, 2005) and will not be described in detail here. We have used this approach to show that the catalytically inactive $DBT^{K/R}$ protein acts as a dominant negative because it antagonizes PER degradation produced by DBT^{WT} (Muskus et al., 2007), to show that C-terminally truncated forms of DBT are more active in triggering PER degradation than full-length DBT^{WT} (Fan et al., 2009), and to show that BDBT enhances

DBT's effects on PER (Fan et al., 2013). S2 cell studies have also been employed to demonstrate autophosphorylation of DBT. Incubation of S2 cells with increasing concentrations of okadaic acid, an inhibitor of serine/threonine phosphatases, produces mobility shifts for DBT^{WT} but not for $DBT^{K/R}$ on a 10% SDS-PAGE gel, demonstrating that they require DBT's kinase activity (Fan et al., 2009). The phosphorylation that produces mobility shifts maps principally to the C-terminal domain because mutants lacking this domain or the phosphorylation sites do not produce this mobility shifts (Fan et al., submitted).

Another application of S2 cells is to express DBT for subsequent purification, so that its kinase activity can be measured. We and others have been unsuccessful in recovering enzymatically active DBT protein from bacteria, although vertebrate CKIδ/ε with site specific mutations found in Drosophila mutants have been used as substitutes for the fly DBTs with bacterial expression. However, Drosophila S2 cells have been used to express various forms of Drosophila DBT, with robust kinase activity recovered from DBT^{WT} and no kinase activity recovered from $DBT^{K/R}$; since $DBT^{K/R}$ can be recovered at the equivalent levels to DBT^{WT}, it serves as a negative control to rule out contaminating kinase activity in the purified material. We have been successful in recovering robust kinase activity in single-step immunoprecipitations like the ones outlined in the next section, with direct *in vitro* assay of kinase activity in the Sepharose beads. Both single condition assays and multiple substrate concentrations (a Michaelis–Menten enzyme kinetic analysis) have been employed (Fan et al., 2009; Muskus et al., 2007; Preuss et al., 2004). Incorporation of ^{32}P into the substrate (dephosphorylated casein or bacterially expressed PER) is measured by phosphorimager quantification of electrophoresed reaction products and is normalized to the amount of DBT detected in the reaction by immunoblot. With this procedure, we have shown that the DBT^{S} and DBT^{L} mutant proteins are both less active on casein and bacterially expressed PER than is DBT^{WT}, but that C-terminally truncated forms of DBT are more active (demonstrating an autoinhibitory role for the DBT C-terminus).

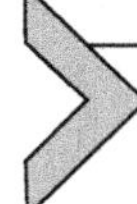

4. PROTEOMIC APPROACHES

4.1. Isolation of DBT-containing complexes

DBT is part of a clock protein complex that includes PER, CLK, CYC, BDBT, and likely other unknown factors. Hence, isolation of DBT-containing complexes and identification of the complex components could identify previously unrecognized circadian factors. We are employing two sorts of proteomic approaches in our lab for identification of DBT interactors. The first approach is the one we have employed most extensively and involves overexpression of MYC-HIS-tagged DBT in Drosophila S2 cells, followed by recovery of DBT with anti-MYC antibody conjugated to Sepharose beads; if additional purity is required, the extract from S2 cells can first be purified on a metal affinity column, eluted with imidazole and then immunoprecipitated with anti-MYC beads. The second approach entails expression of tagged proteins in circadian cells of the fly and recovery of complexes in an *in vivo* context. In both cases, interactors are detected as bands on a gel after electrophoresis of the complexes (SDS-PAGE), subjected to in-gel trypsinization, and then identified by mass spectrometry (nanoLC-tandem-MS; see below).

This first approach allows relatively high levels of DBT expression and tends to produce higher levels of interacting proteins than with methods employing more physiologically normal levels of DBT expression. There are two principal concerns. First, the high levels of expression that are produced from the expression vector may produce artifactual interactions that do not actually occur *in vivo*. Second, S2 cells do not produce circadian rhythms, so some circadian interactors may be missed. The first point can be addressed by follow-up *in vivo* studies that seek to confirm the interaction *in vivo* with co-IP/immunoblot methods and by genetic analysis that alters expression of the candidate interactor to determine if circadian rhythms or DBT state are affected. The second concern is mitigated somewhat by the extent to which S2 cells overexpressing transgenic proteins have been employed in the field to dissect many aspects of circadian function—including progressive PER phosphorylation and degradation (Ko & Edery, 2005), PER's negative feedback on CLK–CYC (Darlington et al., 1998), regulation of PER subcellular localization (Ashmore et al., 2003; Nawathean et al., 2007; Saez & Young, 1996), and the light-dependent interaction of CRY and TIM (Ceriani et al., 1999; Sathyanarayanan et al., 2008). For these responses, S2 cells supply additional components that are required for the response—for instance, the proteasome responsible for PER degradation when DBT and PER are overexpressed, or CYC when transgenic CLK is expressed together with an E-box-containing reporter gene for measurement of CLK/CYC transcriptional activity. Thus, functional complexes involving both transgenic proteins and endogenous S2 cell proteins can be reconstituted in S2 cells.

Using this approach, we have identified a noncanonical FK506-binding protein (BDBT; Fan et al., 2013) and an adaptor protein involved in protein kinase C signaling (RACK1) as DBT interactors. BDBT was found to interact with full-length DBT but not with C-terminally truncated DBT. The bands were then excised, treated with trypsin and subjected to nanoLC-tandem-MS. Multiple peptides from the predicted BDBT protein were identified. The interaction of DBT and BDBT was confirmed with multiple assays, and its role in the clock was further demonstrated with RNAi knock-down techniques that produced circadian phenotypes (long-period and arrhythmic locomotor activity and constitutively high levels of nuclear PER). Moreover, lowered levels of BDBT produced hyperphosphorylated DBT and reduced phosphorylation of PER, while higher levels of BDBT produced increased phosphorylation and degradation of PER in S2 cells. Finally, BDBT was shown to rhythmically accumulate in discrete foci in the fly eye at the time PER transitions from the cytosol to the nucleus. Taken together, the results argue that BDBT stimulates DBT activity toward PER. While the biochemical mechanism for this effect is not currently known, the determination of the X-ray crystal structure showed that BDBT is a noncanonical FK506-binding protein, with extensive structural homology to more canonical members of this protein family, but lacking the residues that bind FK506 or rapamycin and mediate *cis–trans* isomerization of the peptide bond at prolines. Such catalytically inactive FKBPs are thought to contribute to nuclear import of the glucocorticoid receptor and may serve a similar function here.

The analysis of RACK1 has not proceeded as far as that for BDBT, but RNA knock-down with two UAS-dsRNAi lines in circadian cells does dramatically lengthen circadian period (Fig. 1). It is noteworthy that RACK1 was also isolated as a component of mammalian CLOCK-containing complexes (Robles, Boyault, Knutti, Padmanabhan, & Weitz, 2010), of which mammalian CKIε/δ (the DBT orthologs) should also be components. RACK1 has been suggested to mediate protein kinase C signaling.

Recovery of protein complexes from circadian cells of the adult fly is a more daunting task, because most of the clock proteins are expressed at very low levels and fly circadian cells comprise only a subset of the *dbt*-expressing fly tissues—even in the head, where they are most abundant. Recovery of DBT from heads will therefore isolate some DBT complexes from cells that are not circadian, and SDS-PAGE analysis of potential interactors should reveal interactors from many different DBT-containing complexes simultaneously, making it impossible to determine which of these are found in the

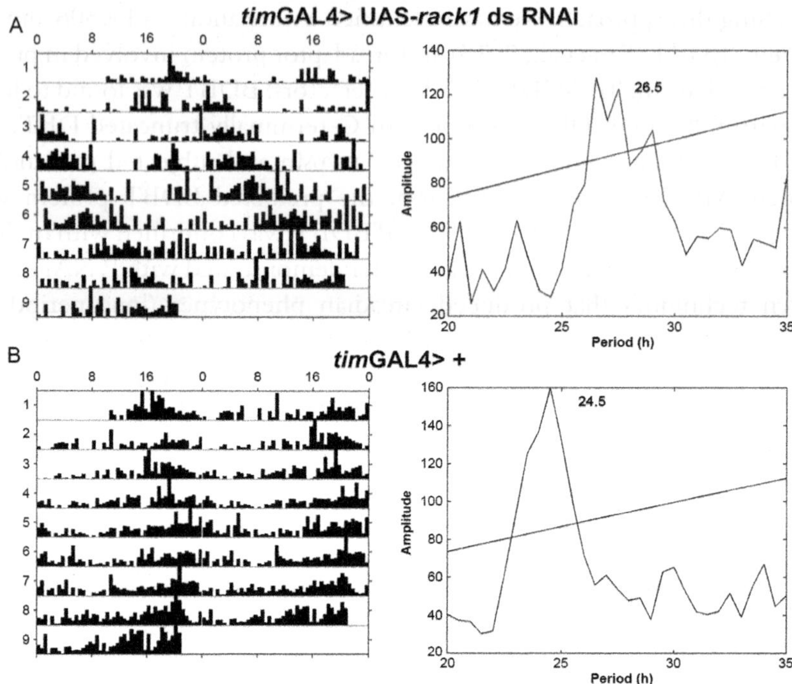

Figure 1 Expression of dsRNAi for *rack1* in circadian cells can lead to long circadian periods of the locomotor activity rhythm. Representative activity records and periodograms from offspring of the cross of UAS-*rack1* dsRNAi males (line 27859 from the Vienna Drosophila RNAi Center, VDRC) to UAS-dcr2; *tim*GAL4 females are shown. The offspring were entrained to three 12 h:12 h light–dark cycles, and the assays started during the first subjective day of DD. The days are double-plotted, and periods above the line in the periodogram analysis are significant with $p < 0.001$. (A) Actogram and periodogram for a fly which inherited the UAS-*rack1* gene. (B) Actogram and periodogram for a fly which inherited a nontransgenic chromosome (a TM3 balancer) from the 27859 stock. The overall results for the three UAS-*rack1* dsRNAi lines from the VDRC and the controls were as follows: 27859 (26.3 ± 0.2 h period, 69% rhythmic, $N = 16$), 27858 (27.6 ± 0.7 h period, 50% rhythmic, $N = 8$), 104470 (24.4 ± 0.2 h period, 65% rhythmic, $N = 17$), and wild-type TM3 controls (24.4 ± 0.2, 94% rhythmic, $N = 16$).

same complexes. Moreover, even circadian cells are not all the same, as analysis linking morning and evening activity peaks to different anatomical regions of the brain has shown (Stoleru et al., 2004). We have made an attempt to restrict our analysis to circadian cells by driving expression of DBT-MYC-HIS or BDBT-FLAG exclusively in circadian cells with the *tim*GAL4 driver, followed by immunoprecipitation of the tagged "bait" from fly heads. Figure 2A demonstrates the recovery of DBT in these

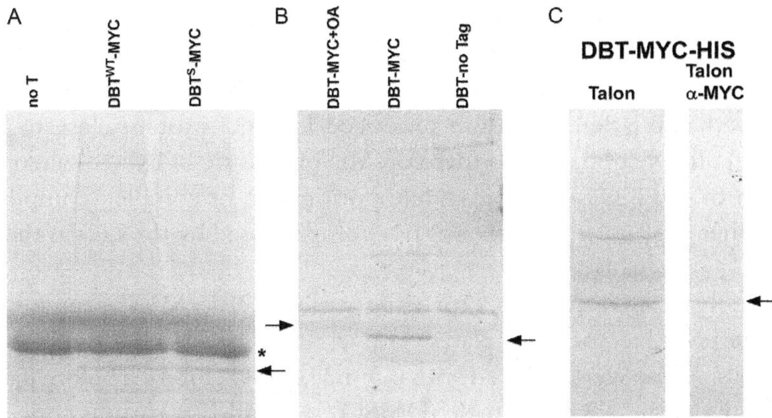

Figure 2 SDS-PAGE analysis of DBT-MYC-HIS purifications from fly heads and S2 cells for proteomic analysis. In all panels, the gels were stained with Coomassie Blue, and the arrows indicate the presence of a protein that was identified by nanoLC-tandem-MS as DBT. (A) DBT-MYC-HIS of the indicated genotype and expressed with the *tim*GAL4 driver was purified by immunoprecipitation from fly heads at ZT1. The star indicates the presence of the antibody protein. "no-T," wild-type flies not expressing any tagged DBT. (B) DBT-MYC-HIS was expressed in Drosophila S2 cells with okadaic acid (+OA) or without it, and the transgenic DBT was immunoprecipitated with anti-MYC bound to Sepharose. Okadaic acid produced a shift in the electrophoretic mobility of DBT indicative of phosphorylation (left arrow), but the levels were too low to detect many DBT peptides by MS, which did detect DBT in the purifications from untreated cells (right arrow). DBT was not detected in control immunoprecipitations from cells expressing untagged DBT (DBT-no tag). (C) DBT-MYC-HIS was purified from S2 cells by Talon resin affinity chromatography (targeting the HIS tag) or by Talon resin affinity chromatography followed by anti-MYC immunoprecipitation.

immunoprecipitates (its identity was supported by mass spectrometry). We hope to scale these recoveries up and perform them at different times of day to determine the circadian changes in the composition of these complexes. If possible, more restrictive drivers (e.g., *pdf*GAL4) may be employed to drive the tagged baits in only one circadian cell type.

4.1.1 Expression and purification of DBT-MYC from S2 cells by immunoprecipitation

We have expressed full-length wild-type DBT-MYC, various mutant DBT-MYCs (including ones truncated within the C-terminal domain of DBT), and untagged DBT as a negative control that will not react with the anti-MYC antibody. Vectors expressing these proteins under control of the metallothionein promoter and S2 cell lines stably transfected with them have been previously described (Fan et al., 2009; Muskus et al., 2007; Preuss et al., 2004). Purification with anti-MYC antibody and gamma-bind Sepharose or with anti-MYC attached to Sepharose directly were both employed; the latter procedure produced less IgG protein close to DBT in molecular weight and was therefore the procedure of choice when purification of DBT was the goal, while purification of complex components with other molecular weights was not compromised by the IgG in the final eluate.

1. Stably transfected lines expressing various DBT-MYC proteins from the pMT vector were plated in five culture wells, each with 1×10^6 cells. These were induced with 0.5 m*M* $CuSO_4$ for 41 h before harvest, and okadaic acid or vehicle (DMSO) was added at various concentrations 17 h before harvest for experiments designed to recover autophosphorylated DBT.
2. The S2 cells were collected and rinsed with PBS in a 15-ml centrifuge tube, and after the second centrifugation (2000 × *g* in a bench top centrifuge), the cells were suspended in 1 ml of 1% NP-40, 50 m*M* Tris–HCl (pH 8.0), 150 m*M* NaCl, 1 m*M* leupeptin, 1 m*M* aprotinin, and 1 m*M* pepstatin A.
3. The cells were placed on ice for 15 min.
4. The cells were homogenized at medium power with eight cycles from a sonicator (10 s sonication/20 s rest) on ice.
5. After sonication, an additional 4 ml of fresh buffer were added into the extract, vortexed briefly, and then centrifuged at 10,000 × *g* at 4 °C for 20 min in a Beckman-Coulter centrifuge.
6. The supernatants were collected and added to a column (Invitrogen; cat# 45-0015), in which 250 μl of anti-MYC bound to Sepharose (Covance; cat# AFC-150p) had been previously equilibrated with the 1% NP-40 buffer with protease inhibitors, and incubated for 4 h or overnight with rocking. Alternatively, the supernatants were rocked with gamma-bind Sepharose beads (GE Healthcare Life Sciences, cat# 17-0885-01) for an hour, and after brief centrifugation the supernatants were then incubated with anti-MYC bound to Sepharose for 4 h at 4 °C.
7. The columns were subjected to brief centrifugation (2000 × *g* for 1 min) to settle the beads, and then the liquid was removed by drainage.
8. Three washes with 0.1% NP-40, 50 m*M* Tris–HCl (pH 8.0), 150 m*M* NaCl followed.
9. The beads were rinsed once with 50 m*M* HEPES, pH 7.5.

10. The beads were resuspended in a 50% slurry with 50 m*M* HEPES, pH 7.5. A 50 μl aliquot of the 50% slurry was pipetted into a new tube, centrifuged briefly, and the supernatant discarded.
11. The beads were resuspended in 25 μl of 2× Laemmli SDS loading buffer, heated to 100 °C for 5 min, and the entire sample was electrophoresed on a 10% SDS-PAGE gel, which was then stained with Coomassie Blue (Thomas Scientific, cat# 24590).

4.1.2 Expression and purification of DBT-MYC from fly heads

We have previously described fly lines with various wild type and mutant forms of MYC-tagged DBT and FLAG-tagged BDBT under the control of the yeast UAS element (Fan et al., 2013, 2009; Muskus et al., 2007). These were crossed to the *tim*GAL4 driver to produce flies expressing the tagged protein specifically in circadian cells. The purifications described below have allowed us to recover the tagged proteins in stainable amounts from fly head circadian cells (Fig. 2A). We are currently optimizing the recoveries (possibly with the addition of a tandem affinity procedure including Talon affinity; see Fig. 2C) to increase the purity, so that complex components can be visualized.

1. DBT-MYC- or BDBT-FLAG-overexpressing fly lines driven by *tim*GAL4 were reared in a 12 h:12 h light–dark cycle for at least 3 days, and then their heads were cut by razor blade at the indicated Zeitgeber times. ~100 Heads were collected and quick-frozen in liquid nitrogen, after which they were stored at −80 °C.
2. The fly heads were suspended in 200 μl of RIPA buffer (50 m*M* Tris–HCl, pH 7.4, 150 m*M* NaCl, 1 m*M* EDTA, 1 m*M* Na_2VO_4, 1 m*M* NaF, 1% NP-40, and 0.5% Na deoxycholate) with protease inhibitors as described previously for 15 min on ice.
3. The heads were then sonicated at medium power eight times with 10 s sonication and 20 s rest on ice.
4. Another 800 μl buffer was added to yield 1 ml total suspension, and the sample was vortexed briefly.
5. The sonicated head extracts were centrifuged at 14,000 × *g* for 10 min at 4 °C.
6. The supernatants were then pipetted into new tubes, and 50 μl of a 50% slurry of gamma-bind Sepharose previously equilibrated in RIPA buffer was added and rocked for 1 h.
7. The solid beads were discarded, and the supernatants were collected into a new tube after 1 min centrifugation at 14,000 × *g*.

8. Either anti-MYC (Covance; cat# MMS-150R; 1:150) or anti-FLAG (Pierce; cat# MA1-91878; 1:150) antibody was added to these supernatants followed by another 50 μl of a 50% slurry of gamma-bind Sepharose. Alternatively, anti-MYC bound to Sepharose was added, as in the previous procedure.
9. The samples were placed in a cold room to rock overnight.
10. Three quick washes/centrifugations in RIPA buffer with protease inhibitors followed (14,000 × *g* for 1 min). After each, the residual wash solution was pipetted off as much as possible.
11. 25 μl of 2 × Laemmli SDS loading buffer was added to each sample to resuspend the beads, mixed by pipette, and heated at 100 °C for 5 min.
12. 30 μl of sample was applied to 10% SDS-PAGE gels to separate all potential interactors. In order to detect interactors, Coomassie Blue (Thermo Scientific; cat# 24590) or silver staining (Bio-Rad; cat# 161-0449) was employed.

4.2. Isolation of DBT for analysis of its autophosphosphorylation

We have shown that DBT (like its vertebrate orthologs) autophosphorylates its C-terminal domain. The phosphorylation can be detected as an electrophoretic mobility shift for DBT in S2 cells when a general phosphatase inhibitor (okadaic acid) is added; an electrophoretic mobility shift is detected for wild-type DBT but not for a catalytically inactive $DBT^{K/R}$, thereby demonstrating that it is due to autophosphorylation (Fan et al., 2009). By analyzing a series of C-terminal truncation mutants expressed in S2 cells with okadaic acid treatment, we have shown that the mobility-shift-inducing phosphorylations occur in the C-terminal domain (Fan et al., submitted). In order to identify the phosphorylation sites, DBT-MYC was purified by sequential Talon affinity chromatography and anti-MYC immunoprecipitation. The initial yield of DBT from the cells untreated with okadaic acid was much higher than from cells treated with okadaic acid (Fig. 2B), so tandem purification targeted DBT-MYC from cells untreated with okadaic acid. Mass spectrometry analysis of the peptides produced by in-gel trypsinization of this DBT identified a phosphorylated residue at Ser408. Since CKI sites often occur in clusters due to processive phosphorylation by CKI, all of the Ser and Thr in the region around this site were mutated to Ala, and this protein did not exhibit okadaic acid induced mobility shifts in S2 cells (Fan et al., submitted). We are currently using this construct to address the role of DBT autophosphorylation in various biological processes.

4.2.1 Tandem affinity expression and purification of DBT-MYC-HIS from Drosophila S2 cells for analysis of its phosphorylation

1. Drosophila S2 cells (1×10^6/ml, 1×10^7 cells) stably transfected with pMT-dbt^{WT}-*myc-his* were induced to express DBT-MYC-HIS by addition of $CuSO_4$ (500 μ*M*) and harvested 48 h later.
2. S2 cells were centrifuged at $1000 \times g$ for 3 min and cell pellets stored at −80 °C for a minimum of 1 day.
3. Cell pellets were resuspended in 20 ml of lysis buffer (200 m*M* Tris–HCl pH 8, 0.4 *M* ammonia sulfate, 10 m*M* $MgCl_2$, 10% glycerol, and 20 m*M* NaF + protease inhibitors) and incubated on ice for 30 min. Resuspended cell pellets were then sonicated for a total of 3 min (15 s bursts, 10 s breaks) on ice.
4. Sonicates were centrifuged at $12{,}000 \times g$ for 5 min to pellet cell debris, and the supernatant was recovered. Aliquots from the supernatants and pellets were saved to determine by immunoblot analysis if DBT was primarily present in the supernatant.
5. To isolate DBT-MYC-HIS, the supernatant was incubated with 200 μl of a 50% slurry of Talon Metal Affinity Resin (Clontech; cat# 635509) for 1 h at 4 °C, with rocking in a 50-ml conical tube.
6. After incubation, the sample was centrifuged at $4500 \times g$ for 5 min and the supernatant removed. An aliquot was saved to determine by immunoblot analysis if DBT was bound to the Talon Metal Affinity Resin.
7. The pellet was then washed with 10 ml of wash buffer (10 m*M* Imidazole, 0.5 *M* NaCl, 100 m*M* KCl, 0.2% Triton X-100, 25 m*M* HEPES, pH 7.6, and 10% glycerol) twice, each time with centrifugation at $4500 \times g$ for 5 min and removal of the supernatant.
8. A total of 4 ml of wash buffer was added to resuspend the pellet and loaded on a column (Invitrogen; cat# 45-0015). DBT was eluted by adding 3 ml of elution buffer (100 m*M* Imidazole, 0.5 *M* NaCl, 100 m*M* KCl, 0.2% Triton X-100, 25 m*M* HEPES, pH 7.6, 10% glycerol, and 100 m*M* EDTA), and 500 μl fractions were collected.
9. Aliquots of the elutions were electrophoresed and Coomassie stained to check for DBT.
10. The fractions containing DBT were pooled, and a second purification was performed using anti-MYC chromatography (9E-10 Monoclonal Antibody, Affinity Matrix from Covance; cat# AFC-150p; 50 μl/1 ml pooled faction). The DBT fractions were incubated with the affinity matrix monoclonal 9E-10 beads overnight at 4 °C, with rocking.
11. The samples were then centrifuged at $4500 \times g$ for 5 min, and the supernatant removed.

12. The pellet was washed with 10 ml of wash buffer (10 m*M* Imidazole, 0.5 *M* NaCl, 100 m*M* KCl, 0.2% Triton X-100, 25 m*M* HEPES, pH 7.6, and 10% glycerol) twice, each time with centrifugation at 4500 × *g* for 5 min and removal of the supernatant.
13. Purified DBT was then eluted with 50 μl of 5 × Laemmli SDS loading buffer and heated for 5 min at 100 °C.
14. The purified DBT (25 μl) was then separated by 10% SDS-PAGE and detected by Coomassie stain (Fig. 2C). DBT was further confirmed by Western blot using either the anti-MYC antibody or an antibody to DBT (anti-DBTC antibody; not shown). The DBT band was then excised from the Coomassie-stained gel and used for mass spectrometry analysis.

4.3. Highly sensitive methods for LC-tandem-MS

The identification of associated proteins carried with an immunoprecipitated bait protein, resolved by SDS-PAGE, and then analyzed by LC–MS is a common strategy for finding biologically relevant binding partners. Procedures described here focus on exploiting nanoLC in capillary columns, eluting directly into a sensitive mass analyzer (tandem MS system) which acquires data dependent MS/MS scans. A comprehensive description can be found elsewhere (Kinter & Sherman, 2000). The analysis methods are now fairly routine, but success depends substantially on steps taken to optimize sensitivity in the data acquisition. The overall process involves excision of gel bands or regions where copurified proteins are suspected to reside, in-gel trypsin digestion, extraction of tryptic peptides, and nanoLC-tandem-MS analysis. We have had success using LC–MS compatible silver-stained gels and colloidal blue staining, or Coomassie-stained gels (avoid aldehyde-containing fixative if possible). If systematic comprehensive excisions are made, there is no need to see bands. Generally speaking, modern mass spectrometers are more sensitive than Coomassie Blue. Make sure the gel is substantially destained and equilibrated in water before beginning.

The immunoprecipitations themselves should include at least one negative control lane using an antibody for a different target or with extracts from similar cells that do not express the tagged bait protein. This helps to distinguish associated protein candidates from nonspecifically associated proteins, including any that may bind to the antibody. Tandem purification methods based on two separate affinity steps can reduce nonspecific associations, but also may suffer loss of legitimate binding partners. Various strategies for tandem purification methods will not be discussed here, but an example is shown above (Fig. 2C). Western analysis to assess yield of the bait protein can be useful in developing the immunoprecipitation before preparing samples for LC–MS.

In-gel trypsin digestion can take advantage of the ability of acetonitrile to extract the aqueous solution (including chemical reagents) from polyacrylamide gels, and the process alternates incubation of gel slices in an aqueous solution and acetonitrile to expose the fixed protein in the gel to a series of conditions. There are many protocols published which describe this process. We use one based on the method described in chapter 6 of Kinter and Sherman (2000). This series of steps results in the reduction and subsequent alkylation of cysteine residues to break and then prevent disulfide formation (critical for identification of peptides containing cysteine), with the equilibration of the gel slice in a volatile buffer which is amenable to MS analysis. Volumes should be kept to a minimum during the process. Peptide mixtures extracted from in-gel digestion are dried down (with a speedvac) and can be stored until analysis.

Once digestions are complete, there are two general goals: (1) the identification of proteins in each sample (including verifying the presence of the bait protein) and optionally, (2) the identification of posttranslational modifications (PTM), such as phosphorylation. The data acquisition mode may depend on what instrumentation is available, and whether any additional steps are taken to enrich for a specific PTM. The most sensitive strategies generally rely on nano-scaled liquid chromatography (150–250 nl/min flow rates), and phosphopeptides can often be detected using reversed phase chromatography with no prior enrichment. Thus, the simultaneous identification of proteins and phosphorylation sites in those proteins is possible. The identification of phosphopeptides in LC–MS data is commonly present in protein/peptide MS analysis packages such as the Trans-Proteomic Pipeline (Seattle Proteome Center), Mascot (Matrix Science, Ltd.), Peaks (Bioinformatics, Inc.), and Proteome Discoverer (Thermo-Fisher). These and other collections of Proteomics analysis tools can accommodate different types of LC-tandem MS data. MS instruments may implement collision-induced diffusion (CID), or may have electron capture/transfer dissociation or high energy collision dissociation for the fragmentation of peptides for MS/MS. These alternative fragmentation modes can result in significant improvement in the detection of PTM (Guthals & Bandeira, 2012).

Enrichment of peptides containing certain PTM can also be applied before data acquisition. There are methods to enrich phosphopeptides (Beltran & Cutillas, 2012; Dunn, Reid, & Bruening, 2010), and these can be done prior to data acquisition. Alternatively, online enrichment conditions have been developed (Pinkse, Uitto, Hilhorst, Ooms, & Heck, 2004; Zhao et al., 2009). However, many phosphopeptides are discovered in samples that are not specifically enriched prior to analysis, as was the case with the samples purified as described above (Fig. 2C).

REFERENCES

Ashmore, L. J., Sathyanarayanan, S., Silvestre, D. W., Emerson, M. M., Schotland, P., & Sehgal, A. (2003). Novel insights into the regulation of the timeless protein. *The Journal of Neuroscience*, *23*, 7810–7819.

Bao, S., Rihel, J., Bjes, E., Fan, J. Y., & Price, J. L. (2001). The Drosophila double-time^{S} mutation delays the nuclear accumulation of period protein and affects the feedback regulation of period mRNA. *The Journal of Neuroscience*, *21*, 7117–7126.

Beltran, L., & Cutillas, P. R. (2012). Advances in phosphopeptide enrichment techniques for phosphoproteomics. *Amino Acids*, *43*, 1009–1024.

Brand, A. H., & Perrimon, N. (1993). Targeted gene expression as a means of altering cell fates and generating dominant phenotypes. *Development*, *118*, 401–415.

Ceriani, M. F., Darlington, T. K., Staknis, D., Mas, P., Petti, A. A., Weitz, C. J., et al. (1999). Light-dependent sequestration of TIMELESS by CRYPTOCHROME. *Science*, *285*, 553–556.

Chiu, J. C., Ko, H. W., & Edery, I. (2011). NEMO/NLK phosphorylates PERIOD to initiate a time-delay phosphorylation circuit that sets circadian clock speed. *Cell*, *145*, 357–370.

Chiu, J. C., Vanselow, J. T., Kramer, A., & Edery, I. (2008). The phospho-occupancy of an atypical SLIMB-binding site on PERIOD that is phosphorylated by DOUBLETIME controls the pace of the clock. *Genes and Development*, *22*, 1758–1772.

Cyran, S. A., Yiannoulos, G., Buchsbaum, A. M., Saez, L., Young, M. W., & Blau, J. (2005). The double-time protein kinase regulates the subcellular localization of the Drosophila clock protein period. *The Journal of Neuroscience*, *25*, 5430–5437.

Darlington, T. K., Wagner-Smith, K., Ceriani, M. F., Staknis, D., Gekakis, N., Steeves, T. D. L., et al. (1998). Closing the circadian loop—CLOCK-induced transcription of its own inhibitors, PER and TIM. *Science*, *280*, 1599–1603.

Dunn, J. D., Reid, G. E., & Bruening, M. L. (2010). Techniques for phosphopeptide enrichment prior to analysis by mass spectrometry. *Mass Spectrometry Reviews*, *29*, 29–54.

Edery, I., Zwiebel, L. J., Dembinska, M. E., & Rosbash, M. (1994). Temporal phosphorylation of the Drosophila period protein. *Proceedings of the National Academy of Sciences of the United States of America*, *91*, 2260–2264.

Fan, J. Y., Agyekum, B., Venkatesan, A., Hall, D. R., Keightley, A., Bjes, E. S., et al. (2013). Noncanonical FK506-binding protein BDBT binds DBT to enhance its circadian function and forms foci at night. *Neuron*, *80*, 984–996.

Fan, J. Y., Preuss, F., Muskus, M. J., Bjes, E. S., & Price, J. L. (2009). Drosophila and vertebrate casein kinase Idelta exhibits evolutionary conservation of circadian function. *Genetics*, *181*, 139–152.

Gallego, M., Eide, E. J., Woolf, M. F., Virshup, D. M., & Forger, D. B. (2006). An opposite role for tau in circadian rhythms revealed by mathematical modeling. *Proceedings of the National Academy of Sciences of the United States of America*, *103*, 10618–10623.

Garbe, D. S., Fang, Y., Zheng, X., Sowcik, M., Anjum, R., Gygi, S. P., et al. (2013). Cooperative interaction between phosphorylation sites on PERIOD maintains circadian period in Drosophila. *PLoS Genetics*, *9*, e1003749.

Gietzen, K. F., & Virshup, D. M. (1999). Identification of inhibitory autophosphorylation sites in casein kinase I epsilon. *The Journal of Biological Chemistry*, *274*, 32063–32070.

Graves, P. R., & Roach, P. J. (1995). Role of COOH-terminal phosphorylation in the regulation of casein kinase I delta. *The Journal of Biological Chemistry*, *270*, 21689–21694.

Guthals, A., & Bandeira, N. (2012). Peptide identification by tandem mass spectrometry with alternate fragmentation modes. *Molecular and Cellular Proteomics*, *11*, 550–557.

Hardin, P. E., Hall, J. C., & Rosbash, M. (1990). Feedback of the Drosophila period gene product on circadian cycling of its messenger RNA levels. *Nature*, *343*, 536–540.

Hardin, P. E., & Panda, S. (2013). Circadian timekeeping and output mechanisms in animals. *Current Opinion in Neurobiology*, *23*, 724–731.

He, Q., Cha, J., He, Q., Lee, H. C., Yang, Y., & Liu, Y. (2006). CKI and CKII mediate the FREQUENCY-dependent phosphorylation of the WHITE COLLAR complex to close the Neurospora circadian negative feedback loop. *Genes and Development*, *20*, 2552–2565.

Kim, E. Y., & Edery, I. (2006). Balance between DBT/CKIepsilon kinase and protein phosphatase activities regulate phosphorylation and stability of Drosophila CLOCK protein. *Proceedings of the National Academy of Sciences of the United States of America*, *103*, 6178–6183.

Kim, E. Y., Ko, H. W., Yu, W., Hardin, P. E., & Edery, I. (2007). A DOUBLETIME kinase binding domain on the Drosophila PERIOD protein is essential for its hyperphosphorylation, transcriptional repression, and circadian clock function. *Molecular and Cellular Biology*, *27*, 5014–5028.

Kinter, M., & Sherman, N. E. (2000). *Protein sequencing and identification using tandem mass spectrometry*. New York: Wiley-Interscience.

Kivimae, S., Saez, L., & Young, M. W. (2008). Activating PER repressor through a DBT-directed phosphorylation switch. *PLoS Biology*, *6*, e183.

Kloss, B., Price, J. L., Saez, L., Blau, J., Rothenfluh, A., Wesley, C. S., et al. (1998). The Drosophila clock gene double-time encodes a protein closely related to human casein kinase Iepsilon. *Cell*, *94*, 97–107.

Kloss, B., Rothenfluh, A., Young, M. W., & Saez, L. (2001). Phosphorylation of period is influenced by cycling physical associations of double-time, period, and timeless in the Drosophila clock. *Neuron*, *30*, 699–706.

Ko, H. W., & Edery, I. (2005). Analyzing the degradation of PERIOD protein by the ubiquitin-proteasome pathway in cultured Drosophila cells. *Methods in Enzymology*, *393*, 394–408.

Lee, H., Chen, R., Lee, Y., Yoo, S., & Lee, C. (2009). Essential roles of CKIdelta and CKIepsilon in the mammalian circadian clock. *Proceedings of the National Academy of Sciences of the United States of America*, *106*, 21359–21364.

Lee, C., Etchegaray, J. P., Cagampang, F. R., Loudon, A. S., & Reppert, S. M. (2001). Posttranslational mechanisms regulate the mammalian circadian clock. *Cell*, *107*, 855–867.

Lowrey, P. L., Shimomura, K., Antoch, M. P., Yamazaki, S., Zemenides, P. D., Ralph, M. R., et al. (2000). Positional syntenic cloning and functional characterization of the mammalian circadian mutation, tau. *Science*, *288*, 483–491.

Muskus, M. J., Preuss, F., Fan, J. Y., Bjes, E. S., & Price, J. L. (2007). Drosophila DBT lacking protein kinase activity produces long-period and arrhythmic circadian behavioral and molecular rhythms. *Molecular and Cellular Biology*, *27*, 8049–8064.

Nakajima, M., Imai, K., Ito, H., Nishiwaki, T., Murayama, Y., Iwasaki, H., et al. (2005). Reconstitution of circadian oscillation of cyanobacterial KaiC phosphorylation in vitro. *Science*, *308*, 414–415.

Nawathean, P., Stoleru, D., & Rosbash, M. (2007). A small conserved domain of Drosophila PERIOD is important for circadian phosphorylation, nuclear localization, and transcriptional repressor activity. *Molecular and Cellular Biology*, *27*, 5002–5013.

Pinkse, M. W., Uitto, P. M., Hilhorst, M. J., Ooms, B., & Heck, A. J. (2004). Selective isolation at the femtomole level of phosphopeptides from proteolytic digests using 2D-nanoLC-ESI-MS/MS and titanium oxide precolumns. *Analytical Chemistry*, *76*, 3935–3943.

Preuss, F., Fan, J. Y., Kalive, M., Bao, S., Schuenemann, E., Bjes, E. S., et al. (2004). Drosophila doubletime mutations which either shorten or lengthen the period of circadian rhythms decrease the protein kinase activity of casein kinase I. *Molecular and Cellular Biology*, *24*, 886–898.

Price, J. L., Blau, J., Rothenfluh, A., Abodeely, M., Kloss, B., & Young, M. W. (1998). *Double-time* is a novel *Drosophila* clock gene that regulates PERIOD protein accumulation. *Cell*, *94*, 83–95.

Robles, M. S., Boyault, C., Knutti, D., Padmanabhan, K., & Weitz, C. J. (2010). Identification of RACK1 and protein kinase Calpha as integral components of the mammalian circadian clock. *Science*, *327*, 463–466.

Saez, L., & Young, M. W. (1996). Regulation of nuclear entry of the Drosophila clock proteins period and timeless. *Neuron*, *17*, 911–920.

Sathyanarayanan, S., Zheng, X., Kumar, S., Chen, C. H., Chen, D., Hay, B., et al. (2008). Identification of novel genes involved in light-dependent CRY degradation through a genome-wide RNAi screen. *Genes and Development*, *22*, 1522–1533.

Stoleru, D., Peng, Y., Agosto, J., & Rosbash, M. (2004). Coupled oscillators control morning and evening locomotor behaviour of Drosophila. *Nature*, *431*, 862–868.

Suri, V., Hall, J. C., & Rosbash, M. (2000). Two novel doubletime mutants alter circadian properties and eliminate the delay between RNA and protein in Drosophila. *The Journal of Neuroscience*, *20*, 7547–7555.

van Ooijen, G., Hindle, M., Martin, S. F., Barrios-Llerena, M., Sanchez, F., Bouget, F. Y., et al. (2013). Functional analysis of Casein Kinase 1 in a minimal circadian system. *PLoS One*, *8*, e70021.

Walton, K. M., Fisher, K., Rubitski, D., Marconi, M., Meng, Q. J., Sladek, M., et al. (2009). Selective inhibition of casein kinase 1 epsilon minimally alters circadian clock period. *The Journal of Pharmacology and Experimental Therapeutics*, *330*, 430–439.

Xu, Y., Padiath, Q. S., Shapiro, R. E., Jones, C. R., Wu, S. C., Saigoh, N., et al. (2005). Functional consequences of a CKIdelta mutation causing familial advanced sleep phase syndrome. *Nature*, *434*, 640–644.

Yao, Z., & Shafer, O. T. (2014). The Drosophila circadian clock is a variably coupled network of multiple peptidergic units. *Science*, *343*, 1516–1520.

Yu, W., Houl, J. H., & Hardin, P. E. (2011). NEMO kinase contributes to core period determination by slowing the pace of the Drosophila circadian oscillator. *Current Biology*, *21*, 756–761.

Yu, W., Zheng, H., Price, J. L., & Hardin, P. E. (2009). DOUBLETIME plays a noncatalytic role to mediate CLOCK phosphorylation and repress CLOCK-dependent transcription within the Drosophila circadian clock. *Molecular and Cellular Biology*, *29*, 1452–1458.

Zerr, D. M., Hall, J. C., Rosbash, M., & Siwicki, K. K. (1990). Circadian fluctuations of period protein immunoreactivity in the CNS and the visual system of Drosophila. *The Journal of Neuroscience*, *10*, 2749–2762.

Zhang, Y., Liu, Y., Bilodeau-Wentworth, D., Hardin, P. E., & Emery, P. (2010). Light and temperature control the contribution of specific DN1 neurons to Drosophila circadian behavior. *Current Biology*, *20*, 600–605.

Zhao, R., Ding, S. J., Shen, Y., Camp, D. G., 2nd., Livesay, E. A., Udseth, H., et al. (2009). Automated metal-free multiple-column nanoLC for improved phosphopeptide analysis sensitivity and throughput. *Journal of Chromatography. B, Analytical Technologies in the Biomedical and Life Sciences*, *877*, 663–670.

Zheng, X., Sowcik, M., Chen, D., & Sehgal, A. (2014). Casein kinase 1 promotes synchrony of the circadian clock network. *Molecular and Cellular Biology*, *34*, 2682–2694.

Zilian, O., Frei, E., Burke, R., Brentrup, D., Gutjahr, T., Bryant, P. J., et al. (1999). double-time is identical to discs overgrown, which is required for cell survival, proliferation and growth arrest in Drosophila imaginal discs. *Development*, *126*, 5409–5420.

CHAPTER TEN

Purification and Analysis of PERIOD Protein Complexes of the Mammalian Circadian Clock

Jin Young Kim, Pieter Bas Kwak, Michael Gebert, Hao A. Duong, Charles J. Weitz[1]

Department of Neurobiology, Harvard Medical School, Boston, Massachusetts, USA
[1]Corresponding author: e-mail address: cweitz@hms.harvard.edu

Contents

Abstract

In mammals, circadian rhythms are generated at least in part by a cell-autonomous transcriptional feedback loop in which the three PERIOD (PER) and two CRYPTOCHROME (CRY) proteins inhibit the activity of the dimeric transcription factor CLOCK-BMAL1, thereby repressing their own expression. Upon nuclear entry, the PER and CRY proteins form a large protein complex (PER complex) that carries out circadian negative feedback by means of at least two basic functions: (1) it brings together multiple effector proteins that repress transcription and (2) it delivers these repressive effectors directly to CLOCK-BMAL1 bound to E-box sequences of circadian target genes. At present, the composition, mechanisms of action, and dynamics of PER complexes in circadian clock negative feedback are incompletely understood. Here, we describe several experimental approaches to the study of PER complexes obtained from mammalian tissues. We focus on the isolation of nuclei from mouse tissues, the extraction of PER complexes from the isolated nuclei, characterization of native PER complexes by gel filtration and blue native polyacrylamide gel electrophoresis, preparative immunoaffinity purification of PER complexes for mass spectrometric identification of constituent proteins, and chromatin immunoprecipitation to monitor the recruitment of PER complex proteins to CLOCK-BMAL1 at E-box sites of clock-regulated genes.

Methods in Enzymology, Volume 551
ISSN 0076-6879
http://dx.doi.org/10.1016/bs.mie.2014.10.013

1. GENERAL STRATEGY

A growing body of evidence indicates that nuclear PERIOD (PER) complexes of >1 mDa (Brown et al., 2005) function in circadian negative feedback by delivering multiple effector proteins with transcriptional inhibitory actions to DNA-bound CLOCK-BMAL1 (Duong, Robles, Knutti, & Weitz, 2011; Duong & Weitz, 2014; Padmanabhan, Robles, Westerling, & Weitz, 2012). The composition of PER complexes, the delivery process, or both is somehow temporally controlled such that some effector proteins are delivered at distinct circadian phases in an ordered manner, ensuring, for example, that successive local chromatin modifications carried out by PER complexes are catalyzed in the correct sequence (Duong & Weitz, 2014). The composition, mechanisms of action, and dynamics of nuclear PER complexes in circadian clock negative feedback are incompletely understood. At present, we do not know if multiple PER complexes of significantly different composition coexist at any given time during the ~14-h duration of the negative feedback phase, if the same PER complexes act on CLOCK-BMAL1 dimers bound to the many different circadian target genes, or if there are tissue-specific variations in PER complex composition or action. Ultimately, it will be important to determine how the activities of many PER complex effector proteins are integrated over time to shape circadian rhythmic gene expression. Biochemical analysis of PER complexes in mammalian tissues will thus be an essential prerequisite for understanding the molecular mechanism of the circadian transcriptional feedback loop itself and of circadian gene expression more generally.

For the characterization of PER complexes in mammalian tissues, we have optimized methods for efficient recovery of large chromatin-associated complexes. In particular, we have found that a high-salt (420 m*M* KCl) extraction procedure solubilizes approximately 95% of the total nuclear PER complex. In contrast, extraction with 150 m*M* KCl recovers only about 25%, most of which is a soluble nucleoplasmic complex, leaving behind the majority of the chromatin-associated PER complex to pellet with the chromatin fraction. In addition, the PER complexes extracted under high-salt conditions are distinctly larger than those obtained by 150 m*M* KCl extraction in gel filtration analysis (Fig. 1). An alternative method for the extraction of chromatin-associated protein complexes is treatment of lysed nuclei with DNase I; we settled instead on the high-salt extraction method, mostly to avoid the possibility of protected DNA

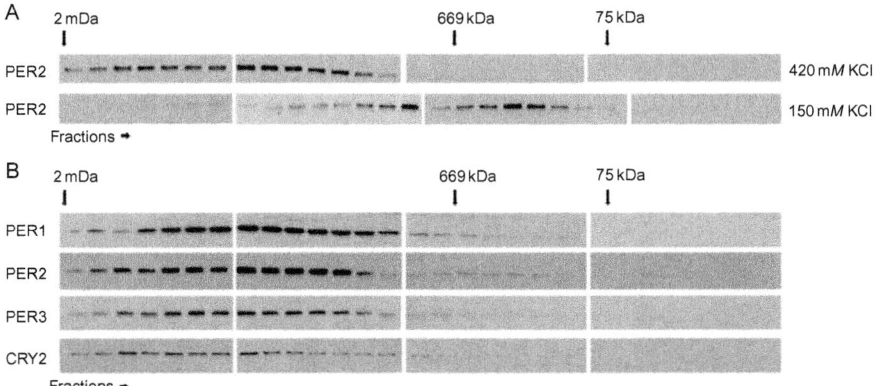

Figure 1 Characterization of nuclear PER complexes by gel filtration chromatography. (A) Immunoblots showing comparison of Sephacryl S-400 chromatography of PER complexes from mouse liver nuclei (circadian time 14 h) extracted with 420 m*M* KCl (top) or 150 m*M* KCl (bottom). Column fractions were run on SDS–PAGE and immunoblotted with an antibody to PER2. Positions of size markers are indicated at the top. The 420 m*M* KCl extraction not only solubilizes PER complexes much more efficiently than the 150 m*M* KCl method (>90% vs. ~25%, not shown), but it solubilizes much more of the very large molecular weight PER complexes (1–2 mDa). The broad peak of PER2-immunoreactive complexes could reflect heterogeneity of PER complexes, but the BN-PAGE profiles shown in Fig. 2 suggest that it may reflect the gradual dissociation of at least some protein subunits from the complex during the course of the gel filtration run. (B) Immunoblots showing Sephacryl S-400 chromatography of PER complexes extracted from liver nuclei under high-salt conditions, as earlier. Different core components of PER complexes exhibit nearly identical gel filtration profiles.

fragments remaining in the complex, complicating or interfering with the analysis.

To identify protein constituents of PER complexes, we developed tools for the preparative purification of PER complexes from tissue extracts. For purification of PER complexes with high-affinity monoclonal antibodies, we generated two mouse lines, one in which endogenous PER1 was replaced by a PER1 fusion protein tagged at the N-terminus with a FLAG–Hemagglutinin tandem epitope tag (FH-PER1) and another in which endogenous PER2 was replaced by a PER2 fusion protein tagged similarly at the C-terminus (PER2-FH; Duong et al., 2011). FH-PER1 or PER2-FH each supports molecular and behavioral circadian rhythms in mice lacking both native PER1 and PER2 (Duong et al., 2011). PER complexes containing PER2-FH or FH-PER1 show essentially the same size distribution on gel filtration or blue native gel electrophoresis as fully wildtype PER complexes (Fig. 2). Because PER1 and PER2 are predominantly present

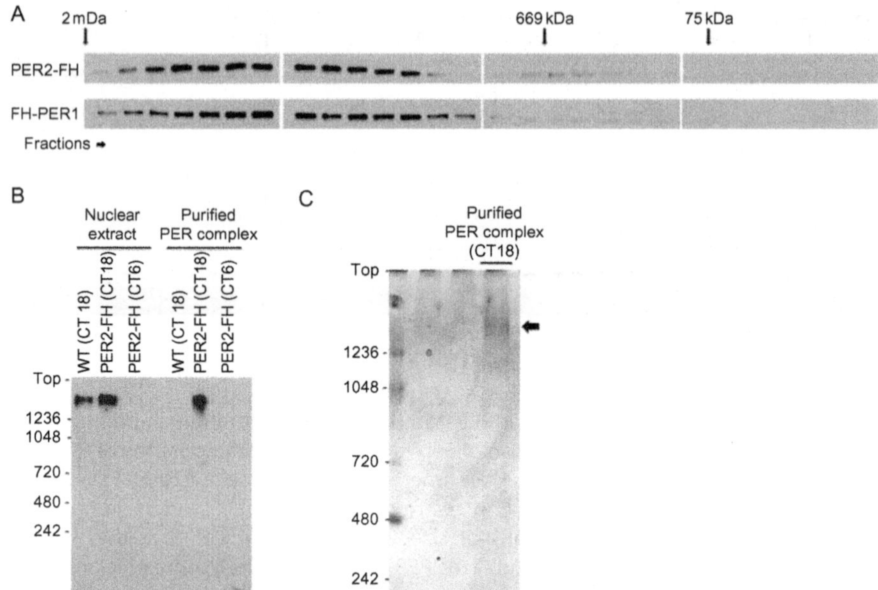

Figure 2 The size distribution of PER complexes containing FH-PER1 or PER2-FH epitope-tagged proteins is essentially identical to native PER complexes. (A) Immunoblots probed with anti-FLAG antibody showing gel filtration analysis (as earlier) of liver nuclear extracts from mice expressing PER2-FH (but no native PER2) or FH-PER1 (but no native PER1). Size distribution of the PER complexes is very similar to that of wildtype mice (see Fig. 1). (B) Immunoblots probed with anti-PER2 antibody showing blue native polyacrylamide gel electrophoresis (BN-PAGE) analysis of PER complexes from mouse liver nuclei. Positions of size markers (in kDa) are indicated on the left. Left lanes (nuclear extract), PER complexes in crude extracts from wildtype (WT) mice or mice in which PER2-FH replaced PER2 show indistinguishable size of ~2000 kDa. As expected, no nuclear PER complex is detected at circadian time (CT) 6 h. Right lanes (purified PER complex), PER complexes from PER2-FH mice immunoaffinity purified with FLAG antibody show a signal of the expected size. WT control mice, lacking a FLAG-tagged PER2 protein, had no detectable immunoaffinity-purified complex, and, as expected, no detectable complex was purified from the PER2-FH mice at CT6. (C) Silver-stained BN-PAGE gel showing immunoaffinity-purified PER complex running at the expected molecular weight of ~2000 kDa (arrow).

in the same complexes, this strategy provided two independent ways of purifying PER complexes from mouse tissues, allowing crossvalidation of results—in earlier work, we obtained nearly identical patterns of copurifying protein bands with FH-PER1 or PER2-FH (Padmanabhan et al., 2012). This approach allows molecular analysis of PER complexes from any desired tissue at any circadian time (CT), as defined precisely by the animal's ongoing circadian rhythm of locomotor behavior.

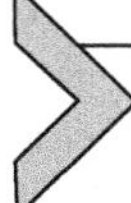

2. EXTRACTION AND CHARACTERIZATION OF PER COMPLEXES FROM MOUSE TISSUES

2.1. Extraction methods

2.1.1 Materials and methods for isolation of cell nuclei from mammalian tissues

- Tissues from 8- to 12-week-old mice entrained to a 12:12-h light–dark cycle for at least 2 weeks.
- 1 × PBS (Ambion).
- Protease inhibitors: complete mini, EDTA-free (Roche)—1 tablet for 10 ml of solution.
- Phosphatase inhibitor cocktails 2 and 3 (Sigma)—10 μl for 1 ml of solution.
- Cushion: 2.05 *M* sucrose, 10 m*M* Hepes, pH 7.6, 15 m*M* KCl, 2 m*M* EDTA at pH 8.0.
- Homogenization buffer: 2.2 *M* sucrose, 10 m*M* Hepes, pH 7.6, 15 m*M* KCl, 2 m*M* EDTA at pH 8.0.
- Nuclei washing buffer: 10 m*M* Hepes, pH 7.6, 100 m*M* KCl, 0.1 m*M* EDTA, pH 8.0, 10% glycerol.

1. Mice kept in constant darkness after appropriate entrainment to light–dark cycles are euthanized at the desired CT-point under infrared light, and tissues are dissected under room light.
2. Dissected tissues are transferred to 25 ml of ice-cold 1 × PBS.
3. Mince tissues with blade and transfer to dounce homogenizer.
4. Add 3 ml of 1 × PBS including freshly added 1 m*M* DTT, 0.15 m*M* spermine, 0.5 m*M* spermidine, 1 m*M* phenylmethylsulfonyl fluoride (PMSF), all inhibitors (protease inhibitor cocktail and phosphatase 2 and 3 inhibitors), to a total volume of ~30 ml, then add 5 ml of homogenization solution.
5. Homogenize tissues 10 times with pestle A, then 15 times with pestle B slowly, so as to avoid bubbles.

   Note: although homogenization 15 times with pestle B breaks some of the nuclei, undeniably reducing the yield, it gives clean nuclei with negligible cytoplasmic contamination.
6. Immediately add homogenized sample to 20 ml of homogenization solution in a 50-ml conical tube, then mix well by inverting.
7. Gently transfer sample from step #6 onto 10 ml of cushion (see earlier) in an ultracentrifuge tube.

8. Pellet down nuclei with ultracentrifugation at 75,000 × *g* at 4 °C for 1 h.
9. Discard supernatant and add 1 ml of nuclei washing buffer including 1 m*M* DTT, 0.15 m*M* spermine, 0.5 m*M* spermidine, 1 m*M* PMSF, and all inhibitors (protease inhibitor cocktail and phosphatase 2, 3 inhibitors).
10. Vortex to resuspend the pellet, then transfer to 1.5-ml Eppendorf tube.
11. Centrifuge at 1500 × *g* at 4 °C for 5 min to pellet nuclei.
12. Repeat steps 9–11 one time, discard supernatant, keep nuclear pellet.

2.1.2 Materials and methods for extraction of nuclear PER complexes

- Isolated nuclei.
- Nuclear lysis buffer (NLB): 10 m*M* Hepes, pH 7.9, 100 m*M* KCl, 3 m*M* $MgCl_2$, 0.1 m*M* EDTA, 20% glycerol.
- Dilution buffer: 10 m*M* Hepes, pH 9.0, 1.5 m*M* $MgCl_2$, 0.25 m*M* EDTA, pH 8.0, 20% glycerol.

1. Resuspend nuclei in 1 ml of NLB with freshly added protease inhibitor cocktail, phosphatase inhibitors 2 and 3, 1 m*M* PMSF, and 1 m*M* DTT.
2. Sonicate nuclei (e.g., with Bioruptor Standard [Diagenode]) at 4 °C for 30 s. Cool the sample for 1 min on ice. Repeat sonication and cooling steps for a total of 15 sonication pulses at setting "high."

   Note: to avoid an undesirable increase in temperature, change the water and ice every five sonications.
3. Transfer the sonicated nuclei to an ultracentrifuge tube, then lyse by adding 3.3 *M* KCl by drop-wise manner to bring the final concentration of KCl to 420 m*M*.

   Note: after adding KCl, the sample becomes very viscous. We suggest that this step be performed with the sample in an ultracentrifuge tube so that no further transfer is necessary prior to ultracentrifugation.
4. Incubate the lysate with rotation at 4 °C for 20 min.
5. Ultracentrifuge at 84,000 × *g* at 4 °C for 1 h.
6. Take supernatant, then add dilution buffer in a drop-wise manner to reduce the final concentration of KCl to 150 m*M*. *Note*: if smaller volume is needed for the next steps, concentrate samples with Centricon Filter Unit (30-kDa cutoff; Millipore).

2.2. Characterizing the size distribution of nuclear PER complexes

An important step in the analysis of PER complexes from a given tissue or CT-point is to obtain an estimate of the molecular mass. For this purpose, we have used two different separation methods, gel filtration chromatography and blue native polyacrylamide gel electrophoresis (BN-PAGE; Wittig & Schägger, 2008), both of which are suitable for the analysis of PER complexes in crude tissue extracts or after preparative immunoaffinity purification (as described later). Of the two, gel filtration chromatography has lower resolution for separating protein complexes on the basis of molecular weight, but it has an advantage in that it easily allows active, native complexes to be recovered from the collected fractions for subsequent analysis, such as assays of enzymatic activity or coimmunoprecipitation, impractical with BN-PAGE. On the other hand, BN-PAGE not only has much greater intrinsic size resolution, but analysis of a particular protein complex can be optimized over a large dynamic range simply by changing the acrylamide concentration or employing acrylamide gradients of varying properties.

2.2.1 Materials and methods for gel filtration chromatography analysis of native PER complexes in nuclear extracts

- Nuclear extracts.
- HiPrep 16/60 Sephacryl S-400 High Resolution (GE Healthcare) gel filtration column.
- Size markers: Blue Dextran (2 mDa, GE Healthcare), Bovine Thyroglobulin (669 kDa, GE Healthcare), Conalbumin (75 kDa, GE Healthcare).
- Gel filtration running buffer: 100 m*M* Hepes, pH 7.9, 150 m*M* NaCl, 1.5 m*M* $MgCl_2$.

1. Equilibrate gel filtration column (e.g., HiPrep 16/60 Sephacryl S-400 High Resolution) with gel filtration running buffer.
2. Load Blue Dextran onto a column to determine the void volume.

   Note that the flow rate is 0.5 ml/min and the fraction size is 1 ml.
3. Run protein molecular weight markers, for example, Bovine Thyroglobulin (669 kDa) and Conalbumin (75 kDa), on the same column and monitor by UV absorbance at 280 nm.
4. After determining the fractions that contain the molecular weight marker proteins, wash column with at least two column volumes of running buffer with the same running conditions.

5. Load samples (volume of 500 μl; if the volume is too large, the peaks will be unnecessarily broadened) onto the same column and run as described earlier.
6. After the void volume (determined by prior running of 2-mDa Blue Dextran), fractions can be loaded on NuPAGE 4–12% Bis-Tris SDS–PAGE gels (Invitrogen) for immunoblot analysis, for example, with anti-PER1 (1:2000, Thermos Scientific), PER2 (1:3000), PER3 (1:1000), or CRYPTOCHROME 2 (CRY2) (1:2000; all from ADI), and FLAG antibodies (1:3000, Abcam) to monitor FH-PER1 or PER2-FH fusion proteins, if appropriate (see Figs. 1 and 2).

2.2.2 Materials and methods for BN-PAGE analysis of PER complexes

- Nuclear extracts from mouse tissues or purified PER complex.
- NativePAGE Novex 3–12% Bis-Tris gels (Invitrogen).
- NativeMark unstained protein standards (Invitrogen).
- Loading buffer (5 × stock): 2.5% Coomassie blue G-250, 50% glycerol, 250 m*M* ε-aminocaproic acid, 50 m*M* Bis-Tris, pH 7.0 (0.22 μm filtered).
- Cathode buffer, compatible with silver staining: 50 m*M* Tricine, 15 m*M* Bis-Tris, 0.004% G-250, pH 7.0 (0.22 μm filtered).
- Anode buffer: 50 m*M* Bis-Tris HCl, pH 7.0.
- Transfer buffer: 25 m*M* Tris, 192 m*M* glycine, 20% methanol, 0.05% sodium dodecyl sulfate (SDS).
- Fixative solution: 40% ethanol, 10% acetic acid.
- Tris-buffered saline (TBS): 20 m*M* Tris, pH 7.4, 150 m*M* NaCl.

1. Place the samples to be analyzed on ice, add loading buffer to a 1 × final concentration, and incubate for at least 5 min prior to loading. Similarly, prepare the molecular weight marker sample (5 μl) using the same buffer as for the sample(s). The NativeMark protein standards (Invitrogen) cover a molecular weight range from 20 to 1236 kDa.
2. Prepare the precast gel for running using cathode and anode buffers that had been precooled to 4 °C and rinse the wells with cathode buffer to remove residual acrylamide monomer and sodium azide from the packaging buffer. Ideally, load equal volumes and fill all empty wells with 1 × loading buffer to prevent disturbances in migration among samples.
3. Perform electrophoresis at 4 °C for 4 h, or overnight, using a voltage limit of 150 V and a current limit of 10 mA.
4. For western blotting after electrophoresis, assemble the transfer sandwich (filter paper, gel, membrane, and filter paper) according to

standard protocols using a polyvinylidene fluoride (PVDF) membrane. Remember to wet the PVDF in methanol prior to immersing it into the transfer buffer.

5. Transfer proteins to the PVDF membrane, for example, with a wet transfer system (BioRad) at 4 °C for 1 h using a voltage limit of 100 V. The transfer buffer contains SDS to facilitate transfer of large protein complexes to the membrane.
6. Following transfer, rinse off the Coomassie stain present on the membrane using pure methanol. As soon as the Coomassie stain is removed (roughly 20–30 s), rinse the membrane using TBS containing 0.05% Tween-20 (TBST) and then block the membrane with 0.5% milk in TBST at room temperature for 1 h.
7. Wash three times with TBST and proceed by incubating with the primary antibody (PER21-A 1:2000, ADI) for 1 h. Subsequently, incubate with secondary antibody (1:5000 antirabbit, conjugated to horseradish peroxidase, GE Healthcare) similarly and perform chemiluminescent detection according to standard protocols.
8. For silver staining after electrophoresis, incubate the gel in fixative solution for 1 h with shaking. Decant and incubate in fixative solution for another 8–16 h to remove any remaining Coomassie stain and to obtain a low background. Continue using the instructions as provided by the silver staining kit (SilverQuest, Invitrogen).

2.3. Preparative purification of PER complexes from mouse tissues

FH-Per1^{+}; *Per1*$^{-/-}$ mice (Duong et al., 2011) express PER1 exclusively in the form of a fusion protein carrying an N-terminal dual epitope tag, FLAG-Hemagglutinin (FH) (Nakatani & Ogryzko, 2003). *Per2-FH*$^{+}$; *Per2*$^{-/-}$ mice (Duong et al., 2011) express PER2 exclusively in the form of a fusion protein carrying the C-terminal FH tag. For simplicity, we refer to these genotypes and the fusion proteins they encode as FH-PER1 and PER2-FH, respectively. These tags allow protein complexes containing PER1 or PER2 to be immunoaffinity purified using high-affinity monoclonal antibodies against FLAG and HA. Either tagged PER protein can rescue molecular and behavioral circadian rhythms of mice lacking both native PER1 and PER2 (Duong et al., 2011), and each is incorporated into one or more multiprotein complexes with a size essentially identical to that of native, wildtype PER proteins (see Fig. 2).

We designed these mouse lines to allow purification of PER complexes sequentially with the two monoclonal antibodies; we have found that a procedure using only the anti-FLAG antibody is usually sufficient to obtain highly pure complexes adequate for detailed analysis, including mass spectrometry. Most of our experience to date is with the purification of PER complexes from mouse liver or lung, tissues that exhibit robust circadian rhythms when explanted in tissue, culture have relatively low cellular heterogeneity, and supply a relatively large amount of protein, such that the tissue obtained from a single mouse is usually adequate for mass spectrometry analysis. Because we are primarily interested in universal or at least common constituents of PER complexes (as opposed to possible tissue-specific constituents), we routinely test by coimmunoprecipitation in wildtype mice to determine whether a PER-associated protein we identified in liver or lung extracts is also associated with PER complex proteins in multiple tissues or cellular sources. (To date, we have not identified any tissue-specific constituents of PER complexes.) Although optimized using liver and lung nuclear extracts, the methods described here should be readily applicable to nuclear extracts from any tissue source expressing PER complexes, given adequate amounts of total protein in the starting samples.

2.3.1 *Materials and methods for preparative purification of PER complexes from tissues of* **FH-Per1** *or* **Per2-FH** *mouse lines*

- Nuclear extracts from tissues of FH-PER1 or PER2-FH mouse lines.
- Mouse monoclonal anti-FLAG M2 affinity gel (Sigma).
- Stringent bead washing buffer: 50 m*M* Hepes, pH 7.4, 1 *M* NaCl, 1% Triton X-100, 1.5 m*M* $MgCl_2$.
- Washing buffer: 50 m*M* Hepes, pH 7.4, 250 m*M* NaCl, 0.2% Triton X-100, 1.5 m*M* $MgCl_2$.

1. Before starting coimmunoprecipitation, wash anti-FLAG M2 affinity gel with 500 μl of stringent bead washing buffer three times for 10 min at 4 °C followed by a rinse with 500 μl of NLB.

   Note: this step is important to remove proteins that contaminate the affinity gels, a potentially troublesome source of background in affinity purifications.
2. Incubate the nuclear extract from one mouse liver with 150 μl of 1:1 slurry of anti-FLAG M2 affinity gel with rotation at 4 °C for 2 h.

   Note: if nuclear extract appears cloudy, add Triton X-100 to 0.1% (final concentration) prior to immunoaffinity purification.
3. Spin down the beads at 100 × *g* on a table-top centrifuge for 1 min, then discard supernatant.

4. Add 1 ml of washing buffer, then incubate at 4 °C for 10 min with rotation.
5. Repeat step 4 three more times.
6. Elute FH-PER1 or PER2-FH complexes with 100 μg/ml of FLAG peptide (Sigma, A2220) in 350 μl of washing buffer at 4 °C for 30 min with rotation.

   Note: the elution efficiency is typically more than 90%, so a second elution is not usually necessary.
7. The eluted, purified complexes can at this point be studied by various biochemical approaches, including BN-PAGE and SDS–PAGE with silver staining or western blotting, and mass spectrometry.

2.4. Chromatin immunoprecipitation analysis of the recruitment of PER complex proteins to circadian target genes

Protein constituents of PER complexes identified by immunoaffinity purification could in principle play any one of several roles in the circadian clock transcriptional feedback loop. They could (1) function as transcriptional effector proteins of the PER complex, recruited to DNA-bound CLOCK-BMAL1 as a part of the negative feedback actions of the PER complex; (2) be associated with CLOCK-BMAL1 during the transcriptional activation phase of the circadian cycle, becoming incorporated into a PER complex only upon interaction of the PER complex with its CLOCK-BMAL1 target; or (3) associate with PER complexes in the nucleoplasm but not with PER complexes that interact with CLOCK-BMAL1 on chromatin. To distinguish among these possibilities for any particular PER-associated protein, we perform chromatin immunoprecipitation (ChIP) studies on samples obtained at time points across the circadian cycle. In the first case, the PER-associated protein will, like PER and CRY proteins, appear at circadian E-box DNA sites only during the transcriptional repression phase of the cycle, approximately CTs 12–24 h in peripheral tissues. In the second case, the PER-associated protein will be present at the E-box site along with CLOCK-BMAL1 during the circadian activation phase when no PER complex is detectable, CTs 4–8 h. In the third case, the PER-associated protein can be immunoprecipitated with PER and CRY proteins from nuclear extracts, but is not detectably recruited to the E-box site during any time point of the circadian cycle. Figure 3 provides an example of such a circadian ChIP experiment.

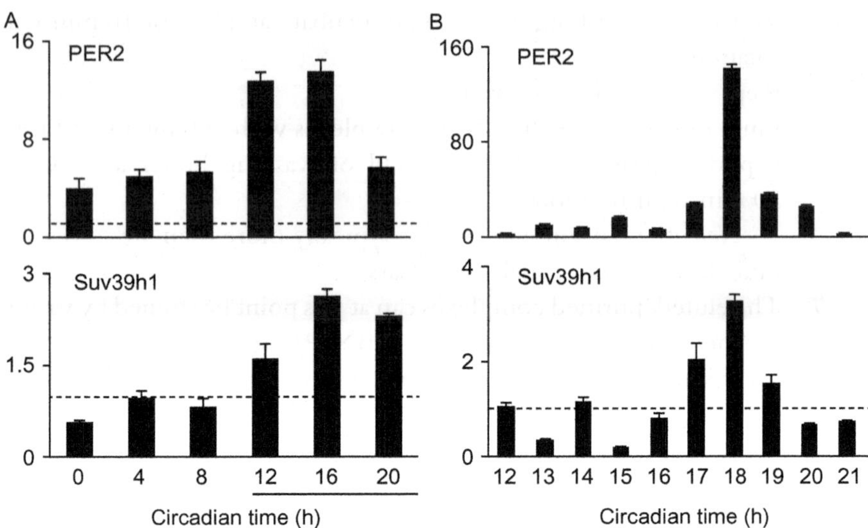

Figure 3 Coordinate circadian rhythms of PER2 and Suv39h at the *Per1* promoter in mouse lung. (A) ChIP assays from mouse lungs sampled across a circadian cycle performed with the antibodies indicated at the top left of each panel. ChIP values are normalized to the signal from a parallel control IgG ChIP assay (dashed line); data are displayed as mean ± SEM of triplicate experiment and are representative of three independent experiments. Underlined segment marks the phase of circadian negative feedback, examined by ChIP at high temporal resolution in panel (B). (B) ChIP assays showing temporal profiles of PER2 and Suv39h1 (as indicated at top left of each panel) at *Per1* E-box site at expanded 1-h time resolution during the phase of circadian feedback transcriptional repression. Shown are mean ± SEM of triplicate experiments, displayed as in (A); representative of three independent ChIP assays. Suv39h1, a histone methyltransferase catalytic subunit, shows a temporal profile at the E-box site that closely parallels PER2, as expected for a PER complex transcriptional effector protein (Duong & Weitz, 2014).

2.4.1 Materials and methods for ChIP of PER complex proteins

Chromatin preparation

- 300 mg (wet weight) of mouse tissue (e.g., liver or lung).
- 7-ml dounce homogenizer.
- Ice-cold PBS.
- 20% formaldehyde (Tousimis).
- 1.25 *M* glycine (10 × stock).
- Crosslinking buffer (XLB): 0.1 *M* NaCl, 1 m*M* EDTA, pH 8.0, 0.5 m*M* EGTA, pH 8.0, 50 m*M* Hepes, pH 8.0, 5 m*M* sodium butyrate.
- Cell lysis buffer (CLB): 5 m*M* PIPES pH 7.5, 85 m*M* KCl, 0.5% NP40, 5 m*M* sodium butyrate.

- ChIP dilution buffer: 0.01% SDS, 1% Triton X-100, 1 m*M* EDTA pH 8.0, 15 m*M* Tris HCl, pH 8.0, 150 m*M* NaCl, 5 m*M* sodium butyrate.
- Nuclear Lysis Buffer (NLB): 1% SDS, 10 m*M* EDTA, 50 m*M* Tris HCl, pH 8.0, 5 m*M* sodium butyrate.
  1. Rinse tissue in 10 ml ice-cold PBS.
  2. Homogenize tissue in 2 ml ice-cold XLB using pestle A and then adjust the volume to 3 ml with XLB.
  3. Add formaldehyde to 1.1% (174 µl of 20% stock) and incubate at room temperature for 10 min with rotation.
  4. Quench the excessive formaldehyde by adding 352 µl of 1.25 *M* glycine and incubate for 5 min at room temperature with rotation.
  5. Collect the cells by centrifugation at 1000 × *g*, 4 °C for 5 min, and rinse cells by adding an excess of ice-cold PBS and repeating centrifugation.
  6. Recover cell pellet and lyse cells in 3.5 ml of ice-cold CLB for 15 min on ice.
  7. Collect nuclei by centrifugation at 1000 × *g* at 4 °C for 5 min and resuspend the nuclei in 1 ml of NLB.
  8. Divide the mixture into two equal aliquots (to reduce the volume) and then sonicate each (for example, with Bioruptor Standard [Diagenode]) at 4 °C for 30 s, followed by cooling for 30 s. Repeat for a total of 10 sonication pulses.
  9. Pellet the debris at 14,000 × *g* at 4 °C for 5 min and collect the supernatant containing the sheared chromatin.
  10. Determine the amount of sheared chromatin in the sample: remove a 50-µl sample of chromatin and, following steps 15–18 below, reverse the crosslinking and purify the DNA. The chromatin concentration is estimated from the amount of purified DNA; the efficiency of the shearing is determined by the size range of DNA fragments resolved on agarose gel electrophoresis, as visualized by staining with ethidium bromide.
  11. Dilute the chromatin 10-fold in ChIP dilution buffer, aliquot (10–15 µg of chromatin per aliquot), and store at −80 °C.

ChIP

- Aliquots of chromatin (as earlier).
- Antibody: 1–3 µg.
- Protein-G magnetic beads (Dynabead, Invitrogen).

- Proteinase K, 20 mg/ml (BioRad).
- Low-salt solution: 0.1% SDS, 1% Triton X-100, 2 m*M* EDTA, pH 8.0, 20 m*M* Tris HCl, pH 8.0, 150 m*M* NaCl, 5 m*M* sodium butyrate.
- High-salt solution: 0.1% SDS, 1% Triton X-100, 2 m*M* EDTA, pH 8.0, 20 m*M* Tris HCl, pH 8.0, 500 m*M* NaCl, 5 m*M* sodium butyrate.
- LiCl solution: 2 m*M* EDTA, pH 8.0, 10 m*M* Tris HCl, pH 8.0, 250 m*M* LiCl, 5 m*M* sodium butyrate.
- Reverse Cross-Link Buffer: 250 m*M* Tris, pH 8.0, 2% SDS.
- Tris-EDTA buffer (TE) (10 m*M* Tris, pH 8.0, 1 m*M* EDTA).

12. Add 1–3 μg of antibody to chromatin aliquot and rotate for 2 h-to-overnight at 4 °C.
13. Add 15 μl of protein G magnetic bead (1:1 slurry) and rotate for 1 h at 4 °C.
14. Wash the chromatin-bead complex sequentially with 1 ml each of low-salt, high-salt, LiCl solution, and TE for 5 min each.
15. Resuspend the chromatin-bead complex in 100 μl Reverse Cross-Link Buffer and incubate for 10 min at 90 °C.
16. Allow the sample to cool to room temperature, add 20 μg of Proteinase K, and incubate at 55 °C for 1 h.
17. Incubate the mixture again for 10 min at 90 °C.
18. Recover the DNA using the Qiagen PCR (Polymerase Chain Reaction) cleaning kit (as recommended in the manufacturer's instructions) and measure the quantity of DNA in the immunoprecipitated chromatin fragments by quantitative PCR (typically in comparison to the amount in the input sample or in a parallel control sample immunoprecipitated by a nonspecific antibody).

REFERENCES

Brown, S. A., Ripperger, J., Kadener, S., Fleury-Olela, F., Vilbois, F., Rosbash, M., et al. (2005). PERIOD1-associated proteins modulate the negative limb of the mammalian circadian oscillator. *Science*, *308*, 693–696.

Duong, H. A., Robles, M. S., Knutti, K., & Weitz, C. J. (2011). A molecular mechanism for circadian clock negative feedback. *Science*, *332*, 1436–1439.

Duong, H. A., & Weitz, C. J. (2014). Temporal orchestration of repressive chromatin modifiers by mammalian circadian clock PERIOD complexes. *Nature Structural & Molecular Biology*, *21*, 126–132.

Nakatani, Y., & Ogryzko, V. (2003). Immunoaffinity purification of mammalian protein complexes. *Methods in Enzymology*, *370*, 430–444.

Padmanabhan, K., Robles, M. S., Westerling, T., & Weitz, C. J. (2012). Feedback regulation of transcriptional termination by the mammalian circadian clock PERIOD complex. *Science*, *337*, 599–602.

Wittig, I., & Schägger, H. (2008). Features and applications of blue-native and clear-native electrophoresis. *Proteomics*, *8*, 3974–3990.

CHAPTER ELEVEN

Best Practices for Fluorescence Microscopy of the Cyanobacterial Circadian Clock

Susan E. Cohen[*,†], **Marcella L. Erb**[†], **Joe Pogliano**[†], **Susan S. Golden**[*,†,1]
[*]Center for Circadian Biology, University of California, San Diego, La Jolla, California, USA
[†]Division of Biological Sciences, University of California, San Diego, La Jolla, California, USA
[1]Corresponding author: e-mail address: sgolden@ucsd.edu

Contents

Abstract

This chapter deals with methods of monitoring the subcellular localization of proteins in single cells in the circadian model system *Synechococcus elongatus* PCC 7942. While genetic, biochemical, and structural insights into the cyanobacterial circadian oscillator have flourished, difficulties in achieving informative subcellular imaging in cyanobacterial cells have delayed progress of the cell biology aspects of the clock. Here, we describe best practices for using fluorescent protein tags to monitor localization. Specifically, we address how to vet fusion proteins and overcome challenges in microscopic imaging of very small autofluorescent cells.

1. INTRODUCTION

Understanding precise protein localization within the cell can reveal valuable insights into its function. It is now appreciated that bacterial cells maintain a high degree of internal architecture. The appropriate spatial organization within the bacterial cell has been demonstrated to be of critical importance for a variety of activities as well the ability to adapt and respond

Methods in Enzymology, Volume 551
ISSN 0076-6879
http://dx.doi.org/10.1016/bs.mie.2014.10.014

to changing environments (reviewed in Rudner & Losick, 2010). In eukaryotic model systems (*Neurospora, Drosophila*, plants, and mammalian cells), changes in the localization of circadian clock proteins, specifically their cycling from cytosolic to nuclear, have been documented; the observed rhythms in nuclear accumulation are an important feature to the timekeeping mechanism (Kondratov et al., 2003; Saez, Meyer, & Young, 2007). Discoveries addressing clock protein localization within the cyanobacterial cell, how spatial distribution changes over the circadian cycle, and how these changes contribute to a robust clock are now beginning to be made (Cohen et al., 2014).

This chapter focuses primarily on using green fluorescent protein (GFP) and other spectral variants to monitor and track the localization of proteins in single cyanobacterial cells. These fluorescent fusion proteins must go through rigorous validation to ensure that phenotypes observed are due to a functional, full-length fusion protein. Immunofluorescence is another method of determining localization in fixed cells without the complication of having to use a tag. However, this technique has been used successfully in cyanobacteria only rarely (Dong et al., 2010; Miyagishima, Wolk, & Osteryoung, 2005), and in our experience, the technical challenges associated with immunofluorescence were too daunting to enable visualization of the clock proteins. Imaging the clock in live cells offers the additional advantage of allowing researchers to ask questions about protein dynamics and how these dynamics are integrated with other cellular functions to contribute to circadian timing.

2. MATERIALS

1. Anti-GFP antibody (AbGENT GFP Tag mouse monoclonal)
2. Anti-FtsZ antibody (Agisera rabbit polyclonal). Antibody produced against FtsZ from *Anabaena* sp. PCC 7120 but has reactivity against FtsZ from *Synechococcus elongatus* PCC 7942
3. Agarose solution (1.2%) in BG-11 medium (Bustos & Golden, 1991)
4. 100 m*M* sodium thiosulfate solution ($Na_2S_2O_3$)
5. BG-11 medium supplemented with sodium bicarbonate ($NaHCO_3$), final concentration 10 m*M*
6. Chamber slides: Single chamber (25 mm × 75 mm Microslide single degression, Erie Scientific) or multichamber slide (Lab-Tek 16-well glass slide, Nunc)
7. Glass slide: 25 mm × 75 mm 1.0-mm thick microslide (VWR)

8. Coverslip: Coverglass for single chamber slide 22mm × 22 mm No. 1.5 cover glass (VWR) Coverglass for 16-well ChamberSlide (Nunc)
9. 1 *M* $NaPO_4$, pH 7.4
10. 16% Paraformaldehyde solution (Electron Microscopy Sciences)

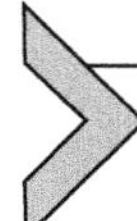

3. METHODS

3.1. Generating fusions to fluorescent proteins

The natural autofluorescence from the photosynthetic thylakoid membranes in cyanobacteria overlaps spectrally with emissions from fluorophores that emit in the red/orange spectrum, including mCherry, precluding their use. In our experience, GFP, ZsGreen, YFP, and ECFP, as well as other variants in the green/yellow color spectrum, are expressed well and easily differentiated from cellular autofluorescence with the appropriate filters. Fluorescent tags are often appended to either the N- or C-terminus of the protein of interest (POI). In some cases, the fluorescent tag may also be inserted into an internal loop, such that each domain is allowed to fold properly and not affect the function of either GFP or the POI. If possible, structural information can be used to make an informed decision about the placement of a fluorescent tag, although even well-guided guesses must be vetted. We have used both N- and C-terminal fusions to KaiC to observe details of subcellular localization. While N-terminal fusions to YFP fully complemented a *kaiC* null stain, C-terminal fusions to either YFP or ECFP display a long-period phenotype, extended by ~5 h (Cohen et al., 2014).

A flexible linker is often introduced between the POI and the fluorescent protein to avoid steric hindrance and allow each domain to fold properly. Glycine, having the smallest side chain, allows for the greatest degree of flexibility (Campbell & Davidson, 2010). We have been successful in using short linkers (2–3 amino acids) composed of either Glycine or Alanine to generate fusions to KaiA. For KaiC, we used a longer linker (17 amino acids) composed of Glycine interspersed with Serine (Cohen et al., 2014) that additionally functions to improve solubility (Campbell & Davidson, 2010). Linkers should be optimized for every application, and in our experience, it is best to initially test multiple fusion proteins to compare N- and C-terminal fusions as well as vary linker lengths and test multiple fluorophores before settling on one fusion protein with which to proceed.

While many exciting discoveries have been made using fluorescent tags, be wary of potential localization artifacts. Examples include clustering artifacts that resulted in ClpX foci that were later found to be not biologically relevant, as well as helical cables observed for MreB that were later found to be an artifact of the high expression of the YFP tag (Landgraf, Okumus, Chien, Baker, & Paulsson, 2012; Margolin, 2012; Swulius & Jensen, 2012).

3.2. Validating fusions

In order to observe subcellular localization patterns that are biologically relevant, the fusion protein must undergo rigorous validation to ensure that the fusion is being expressed as a full-length fusion protein at wild-type (WT) levels and is functional within the context of the cell (Fig. 1). For clock proteins, functionality is easily monitored by measuring clock output activities such as the rhythms of gene expression from luciferase reporter strains (Mackey, Ditty, Clerico, & Golden, 2007). A functional fusion protein will be able to complement a null strain, and in the case of the Kai proteins, will be able to restore rhythmicity, as is the case for fusions 2, 4, and 5 in Fig. 1B. In some cases, the addition of a fluorescent domain may modify your POI in a way that is acceptable; it may be too much to ask for full function after adding a large domain to your POI. Nevertheless, the degree of functionality should be experimentally determined. As an example, we identified a KaiA-GFP fusion that is able to restore rhythmicity to a *kaiA* mutant strain, albeit with an ~2 h period lengthening (Cohen et al., 2014). Although this KaiA-GFP fusion protein did not fully complement a *kaiA* mutant strain, it allowed us to observe KaiA localization under conditions where the clock is running.

Immunoblot analysis to check protein quality is critical to ensure that a full-length fusion protein is being translated and that your fusion is not subject to proteolytic cleavage, resulting in an untagged protein. It is not uncommon to find that your fusion has been cleaved, separating the fluorescent protein from your POI (both of which are functional on their own, but with no relationship to one another), or resulting in a truncated protein fragment as can be observed for fusions 1, 2, and 6 in Fig. 1. Note that while fusion 2 supported WT rhythms of gene expression (Fig. 1B), the fusion is not expressed as a full-length fusion protein but rather as a truncation. Thus, checking both restoration of clock rhythmicity and protein production is critical. In cases where antibodies against the native protein are not available, commercial antibodies against GFP can be used against GFP and some other spectral variants, although the presence of cleaved WT POI would not be detectable. Immunoblots will also inform you about the

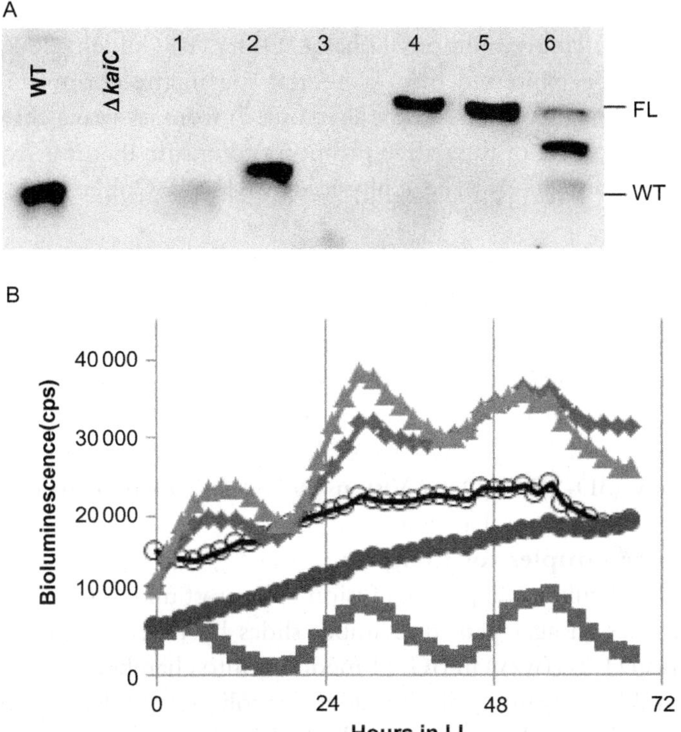

Figure 1 Screening KaiC fusions for quality, quantity, and functionality. (A) Immunoblot of soluble extracts incubated with αKaiC antiserum. Six KaiC fusions (1–6) consisting of N-terminal fusions (1, 3, 4, and 5) and C-terminal fusions (2 and 6) with various linker lengths were tested. WT denotes the expected size for a wild-type untagged KaiC protein and FL denotes the predicted size for a full-length fusion protein. (B) Monitoring rhythms of gene expression from a P_{kaiBC}–*luc* reporter for strains expressing fusions 1–6 as the only copy of *kaiC*. Representative traces for WT (blue squares), Δ*kaiC* (red circles), fusions 1, 3, and 6 (which were indistinguishable, black open circles), fusion 2 (green triangles), and fusions 4 and 5 (which were indistinguishable, purple diamonds). Fusions 4 and 5 produce full-length fusion proteins and complement rhythms of gene expression. In contrast, Fusion 6 produces a full-length fusion protein in addition to truncated products near in size to untagged KaiC, none of which support rhythmicity. Fusion 2 produces near WT rhythms; however, it is not expressed as a full-length fusion protein, and a truncated product near in size to untagged KaiC is observed. (See the color plate.)

quantity of protein produced to ensure that your fusion protein is being produced at appropriate levels. Overexpression can lead to localization artifacts and should be avoided if possible. To ensure proper expression, we have been successful in expressing fusion proteins from their native promoters as well as from a P_{trc} promoter, where low constitutive expression is observed under noninducing conditions (Zhang, Dong, & Golden, 2006). Fusion proteins can be expressed from a neutral site in the chromosome when the endogenous gene has been knocked out, or from its native chromosomal locus under control of the native promoter to ensure that it is expressed in context, and more likely to be at physiological levels (Cohen et al., 2014; Liu et al., 2012).

3.3. Imaging fluorescent fusion proteins

In order to obtain high-resolution images of *S. elongatus* in which details of subcellular localization can be observed, it is best to use a confocal or deconvolution (DeltaVision Core system Applied Precision) microscope (see Note 1 and Fig. 2). We have also used 3D-structured illumination microscopy (3D-SIM) (Delta Vision OMX) as a method to obtain high-quality images in cyanobacteria.

To prepare samples for imaging

1. Grow cell cultures expressing fusion of interest under desired conditions
2. Construct an agar pad in chamber slides by pipetting molten agarose solution (1.2% (w/v) in BG-11 medium) into chamber. Work as quickly as possible so agarose solution does not solidify and flatten immediately by covering with a clean standard glass slide and applying pressure. A completely flat agar pad surface level with the material surrounding the chamber is the desired end result

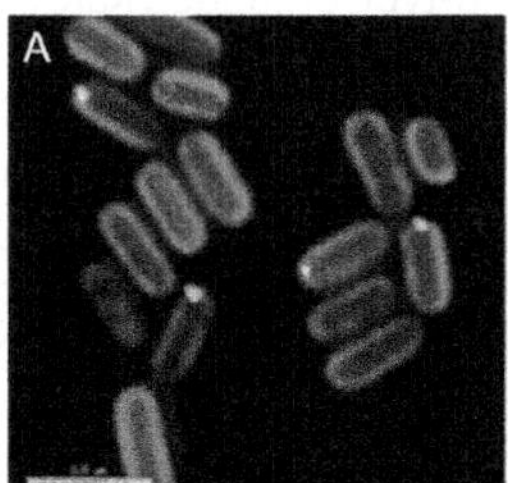

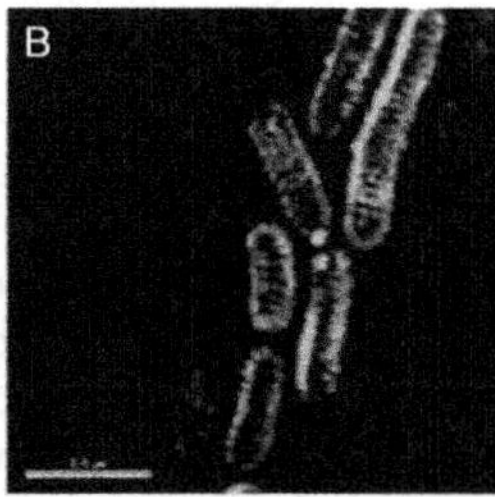

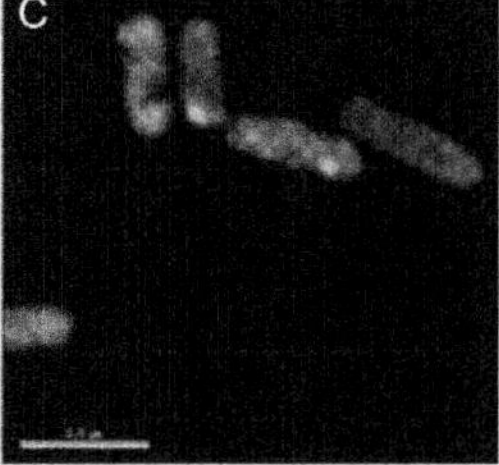

Figure 2 Microscopic images of strains expressing KaiC and KaiA fusion proteins. (A) Deconvolution fluorescence micrograph of cells expressing YFP-KaiC (fusion 4) which is expressed as a full-length fusion, complements rhythms, and appears green, with cell autofluorescence in red; (B and C) KaiA-GFP (green), for which autofluorescence was omitted to improve visualization of the low-abundance KaiA fusion. (B) 3D-SIM micrograph using a FITC filter set where bleed-through from the thylakoid fluorescence is obvious and (C) narrow band-pass GFP filter set on a deconvolution fluorescence microscope to reduce bleed-through from the photosynthetic pigments. Scale bar = 2.5 μm. (See the color plate.)

3. Let solidify (1 min or less) and remove glass slide by sliding off to one side gently without disturbing the agar pad. Try not to touch the surface of the pad and gently wipe away excess agarose mixture from the surrounding glass. Specifically for the use of single chamber glass slides, any residual agarose mixture (or other detritus) on the glass perimeter can prevent good coverslip adherence
4. Add cells to agar pad and let dry before covering with cover slip. For 16-chamber slides 1 μL of moderately dense culture is sufficient (OD_{750} = 0.3–0.5)
5. Use fluorescent microscope to image strains (see Notes 1 and 2)

3.3.1 Image cells over a circadian time course via time-point sampling

Samples can be collected at specified time points and fixed to preserve cellular architectures for imaging at a later time. This method allows you to collect many samples, including different genotypes grown in different conditions, and image at a time that is more convenient. Moreover, this approach allows you to follow how populations of cells are changing over time at a single-cell level.

1. Sample aliquots of cells at designated time points and fix them directly in growth medium (BG-11) by adding a final concentration of 2.4% (v/v) paraformaldehyde (Electron Microscopy Sciences) and 30 m*M* $NaPO_4$ buffer (pH 7.5)
2. Incubate for 20 min at room temperature
3. Samples can be stored at 4 °C and imaged at a later time

3.3.2 Time-lapse imaging of cells

Time-lapse imaging is a powerful tool that allows you to track multiple events including relative circadian phase, protein localization, and cell division in the same subset of cells over time (Yang, Pando, Dong, Golden, & van Oudenaarden, 2010). However, the numbers of different strains and conditions that can be tested are limited by microscope setup and experiment run time.

YFP destabilized by the addition of a C-terminal LVA tag and expressed under the control of a circadian promoter has been used successfully to monitor relative circadian phase in single cells (Dong et al., 2010; Teng, Mukherji, Moffitt, de Buyl, & O'Shea, 2013; Yang et al., 2010). Time-lapse imaging can be achieved by growing cells on an agar pad (see Note 3) or in a microfluidic device. O'Shea and colleagues have successfully tracked *S. elongatus* growth and division in agarose-lined microfluidic chambers, where cells are trapped between a coverglass and a patterned agarose micro-environment (Teng et al., 2013). Microfluidic technology limits cellular crowding and avoids the issue of drying of the agar pad over time, which has previously limited the time course for which cells growing on agar pads could be monitored (Dong et al., 2010; Yang et al., 2010). For time-lapse experiments, an environmental chamber and external light source outfitted to the microscope would be necessary to maintain cells during the experiment.

3.3.3 Investigation of a fluorescent fusion to FtsZ

In addition to KaiA and KaiC, we generated fusions to the bacterial tubulin homolog FtsZ in order to observe clock-controlled dynamics of FtsZ localization in live cells. Our experience with this fusion highlights the iterative process by which we evaluate a fusion construct. FtsZ is conserved in almost all bacteria and is essential for cell division where it assembles into a structure known as the Z-ring at the division site prior to cytokinesis. An N-terminal YFP–FtsZ fusion under the control of the P_{trc} promoter was designed to replace the endogenous *ftsZ*. FtsZ is essential in *S. elongatus* (Jain, Vijayan, & O'Shea, 2012; Miyagishima et al., 2005), and because we observed homogenous segregation of the *yfp–ftsZ* allele in place of the endogenous *ftsZ* (Fig. 3A), we can conclude that FtsZ expressed from this construct is functional to the extent that it supports viability.

Immunoblot analysis demonstrated that this YFP–FtsZ fusion is expressed as a full-length fusion protein, with no obvious truncation products observed (Fig. 3B). However, the abundance of YFP–FtsZ protein is elevated ~55-fold relative to the WT FtsZ. Overexpression of FtsZ in several organisms, including *S. elongatus*, results in cellular filamentation (Mori & Johnson, 2001). However, fluorescence microscopy of our YFP–FtsZ strains indicated that cells are of normal cell length and Z-ring formation can be observed near mid-cell in a subset of cells in an otherwise WT background (Fig. 3C and D). This result suggests that there is no elevation in FtsZ activity when this fusion is present, and it assembles into normal rings. Moreover, when YFP–FtsZ is expressed as the only source of FtsZ in a Δ*cikA* mutant background, elongated cell morphology is observed; FtsZ is also mislocalized in these cells—patchy YFP fluorescence, partial Z-rings, or multiple Z-rings per cell are observed (Fig. 3E). These results are reminiscent of previously reported localization patterns of WT FtsZ in Δ*cikA* backgrounds observed via immunofluorescence (Dong et al., 2010). Taken together, these results suggest that the cell may tolerate such high levels of YFP–FtsZ because this particular

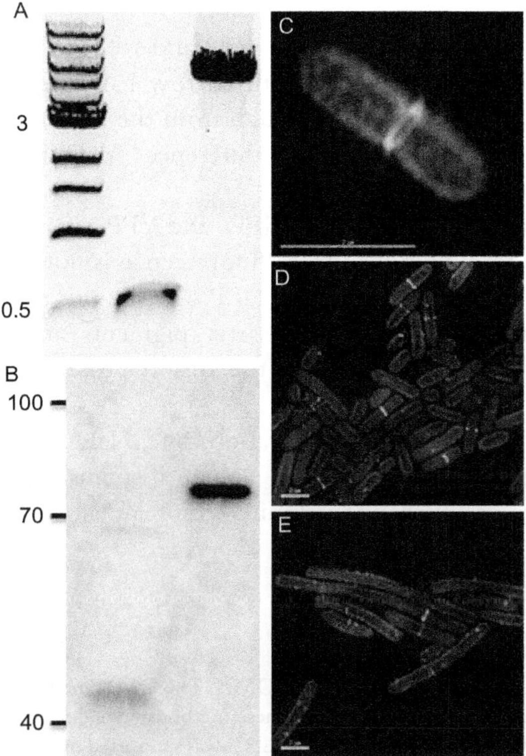

Figure 3 Characterization of a YFP–FtsZ fusion protein. (A) PCR analysis of *ftsZ* locus demonstrates that YFP–FtsZ fusion expressed from the P_{trc} promoter can replace the chromosomal copy of *ftsZ*. Lane 1, 1 kb DNA ladder (NEB); Lane 2, amplification of a 444 bp region of the *ftsZ* locus (from 104 bp upstream to 340 bp into *ftsZ*); Lane 3, amplification of the same chromosomal locus where a construct expressing the spectinomycin resistance cassette-Lac*I*–P_{trc}–YFP–FtsZ has replaced the native *ftsZ*. Homogenous segregation demonstrates that the *yfp–ftsZ* allele can completely replace endogenous *ftsZ*. (B) Immunoblot of soluble extracts incubated with αFtsZ. Lane 1, 13 μg extract from WT; Lane 2, 0.5 μg extract from YFP–FtsZ-expressing cells. YFP–FtsZ is expressed as a full-length fusion protein; however, it is ~55-fold overexpressed compared to the endogenous FtsZ. (C–E) 3D-SIM micrographs of strains expressing YFP–FtsZ as the only source of FtsZ. (C) Representative individual cell in which a Z-ring has formed near mid-cell. Field of cells expressing YFP–FtsZ in a (D) otherwise WT background, normal cell shape, and Z-ring formation is uniformly observed or (E) Δ*cikA* mutant background, where cells are elongated and FtsZ appears mislocalized. Scale bars = 2 μm. (See the color plate.)

fusion protein is not fully functional. The elevated levels of YFP–FtsZ may compensate for decreased functionality of this fusion protein. Thus, despite this overexpression, this fusion accurately reports FtsZ localization patterns in both WT and Δ*cikA* mutant backgrounds.

Notes

1. Changing the GFP filter from the standard FITC (EX 490/20, EM 528/38) to a GFP filter set with narrow bandpass (EX 470/40, EM 515/30) will reduce bleed-through from the photosynthetic pigments. See Fig. 2B and C to observe differences in GFP imaging with the two different filter sets.
2. Exposure times for imaging of GFP- and YFP-expressing strains should be limited to conditions where fluorescence is not observed at all in a WT strain that does not express GFP or YFP; this precaution will limit bleed-through from photosynthetic pigments and ensure that the observed fluorescence is from the fusion protein and not thylakoid fluorescence.
3. $Na_2S_2O_3$ (1 m*M* final concentration) and $NaHCO_3$ (10 m*M*) can be added to the agar pad as a CO_2 source if strains will be grown on an pad, where gas exchange is limited, or maintained for longer periods of time.

ACKNOWLEDGMENTS

We thank Jennifer Santini for help with the OMX. The OMX and Light Microscopy Facility at UCSD is funded by Grant NS047101. These methods were developed with support from the NIH GM100116 to J. P. and S. S. G. S. E. C. was supported by American Cancer Society Postdoctoral Fellowship PF-12-262-01-MPC.

REFERENCES

Bustos, S. A., & Golden, S. S. (1991). Expression of the psbDII gene in *Synechococcus* sp. strain PCC 7942 requires sequences downstream of the transcription start site. *Journal of Bacteriology, 173*(23), 7525–7533.

Campbell, R. E., & Davidson, M. W. (2010). Fluorescent reporter protein. In S. S. Gambhir, & S. S. Yaghoubi (Eds.), *Molecular imaging with reporter genes* (pp. 3–40). New York: Cambridge University Press.

Cohen, S. E., Erb, M. L., Selimkhanov, J., Dong, G., Hasty, J., Pogliano, J., et al. (2014). Dynamic localization of the cyanobacterial circadian clock proteins. *Current Biology, 24*, 1836–1844.

Dong, G., Yang, Q., Wang, Q., Kim, Y. I., Wood, T., Osteryoung, K. W., et al. (2010). Elevated ATPase activity of KaiC constitutes a circadian checkpoint of cell division in *Synechococcus elongatus*. *Cell, 140*(4), 529–539.

Jain, I. H., Vijayan, V., & O'Shea, E. K. (2012). Spatial ordering of chromosomes enhances the fidelity of chromosome partitioning in cyanobacteria. *Proceedings of the National Academy of Sciences of the United States of America, 109*(34), 13638–13643.

Kondratov, R. V., Chernov, M. V., Kondratova, A. A., Gorbacheva, V. Y., Gudkov, A. V., & Antoch, M. P. (2003). BMAL1-dependent circadian oscillation of nuclear CLOCK: Posttranslational events induced by dimerization of transcriptional activators of the mammalian clock system. *Genes & Development, 17*(15), 1921–1932.

Landgraf, D., Okumus, B., Chien, P., Baker, T. A., & Paulsson, J. (2012). Segregation of molecules at cell division reveals native protein localization. *Nature Methods*, *9*(5), 480–482.

Liu, L. N., Bryan, S. J., Huang, F., Yu, J., Nixon, P. J., Rich, P. R., et al. (2012). Control of electron transport routes through redox-regulated redistribution of respiratory complexes. *Proceedings of the National Academy of Sciences of the United States of America*, *109*(28), 11431–11436.

Mackey, S. R., Ditty, J. L., Clerico, E. M., & Golden, S. S. (2007). Detection of rhythmic bioluminescence from luciferase reporters in cyanobacteria. *Methods in Molecular Biology*, *362*, 115–129.

Margolin, W. (2012). The price of tags in protein localization studies. *Journal of Bacteriology*, *194*(23), 6369–6371.

Miyagishima, S. Y., Wolk, C. P., & Osteryoung, K. W. (2005). Identification of cyanobacterial cell division genes by comparative and mutational analyses. *Molecular Microbiology*, *56*(1), 126–143.

Mori, T., & Johnson, C. H. (2001). Independence of circadian timing from cell division in cyanobacteria. *Journal of Bacteriology*, *183*(8), 2439–2444.

Rudner, D. Z., & Losick, R. (2010). Protein subcellular localization in bacteria. *Cold Spring Harbor Perspectives in Biology*, *2*(4), a000307.

Saez, L., Meyer, P., & Young, M. W. (2007). A PER/TIM/DBT interval timer for *Drosophila's* circadian clock. *Cold Spring Harbor Symposia on Quantitative Biology*, *72*, 69–74.

Swulius, M. T., & Jensen, G. J. (2012). The helical MreB cytoskeleton in Escherichia coli MC1000/pLE7 is an artifact of the N-terminal yellow fluorescent protein tag. *Journal of Bacteriology*, *194*(23), 6382–6386.

Teng, S. W., Mukherji, S., Moffitt, J. R., de Buyl, S., & O'Shea, E. K. (2013). Robust circadian oscillations in growing cyanobacteria require transcriptional feedback. *Science*, *340*(6133), 737–740.

Yang, Q., Pando, B. F., Dong, G., Golden, S. S., & van Oudenaarden, A. (2010). Circadian gating of the cell cycle revealed in single cyanobacterial cells. *Science*, *327*(5972), 1522–1526.

Zhang, X., Dong, G., & Golden, S. S. (2006). The pseudo-receiver domain of CikA regulates the cyanobacterial circadian input pathway. *Molecular Microbiology*, *60*(3), 658–668.

CHAPTER TWELVE

Structural and Biophysical Methods to Analyze Clock Function and Mechanism

Martin Egli[1]
Department of Biochemistry, School of Medicine, Vanderbilt University, Nashville, Tennessee, USA
[1]Corresponding author: e-mail address: martin.egli@vanderbilt.edu

Contents

Abstract

Structural approaches have provided insight into the mechanisms of circadian clock oscillators. This review focuses upon the myriad structural methods that have been applied to the molecular architecture of cyanobacterial circadian proteins, their

Methods in Enzymology, Volume 551
ISSN 0076-6879
http://dx.doi.org/10.1016/bs.mie.2014.10.004

interactions with each other, and the mechanism of the KaiABC posttranslational oscillator. X-ray crystallography and solution NMR were deployed to gain an understanding of the three-dimensional structures of the three proteins KaiA, KaiB, and KaiC that make up the inner timer in cyanobacteria. A hybrid structural biology approach including crystallography, electron microscopy, and solution scattering has shed light on the shapes of binary and ternary Kai protein complexes. Structural studies of the cyanobacterial oscillator demonstrate both the strengths and the limitations of the divide-and-conquer strategy. Thus, investigations of complexes involving domains and/or peptides have afforded valuable information into Kai protein interactions. However, high-resolution structural data are still needed at the level of complexes between the 360-kDa KaiC hexamer that forms the heart of the clock and its KaiA and KaiB partners.

1. INTRODUCTION

Cyanobacteria are the simplest organisms known to possess a circadian clock. Initial investigations conducted some 15 years ago focused on a cluster of three genes, *kaiA*, *kaiB*, and *kaiC*, whereby *kaiA* and *kaiBC* messenger RNAs showed circadian cycling (Ishiura et al., 1998). The observations that KaiC overexpression repressed the *kaiBC* promoter and KaiA overexpression enhanced it were consistent with a transcription–translation feedback loop (TTFL) mechanism of the clock, apparently confirming the hypothesis that all biological clocks feature a TTFL at their core. This assumption was toppled by the discovery that the clock in the model organism *Synechococcus elongatus* could be reconstituted *in vitro* by mixing the KaiA, KaiB, and KaiC proteins in the presence of ATP and Mg^{2+} (Nakajima et al., 2005). Obviously, no transcription or translation is occurring in the *in vitro* system, therefore demonstrating that a TTFL as the core mechanism of circadian clocks was not obligatory. Therefore, the three proteins generate a posttranslational oscillator (PTO), with KaiC cycling through hypo and hyperphosphorylated states with a ca. 24-h period. This phosphorylation cycle controls period, formation of heteromultimeric complexes among Kai proteins and clock output signal. The latter involves the histidine kinase SasA that associates with KaiC and phosphorylates the transcription factors RpaA and RpaB that in turn modulate rhythmic expression of cyanobacterial genes in a nonpromoter-specific fashion (Markson, Piechura, Puszynska, & O'Shea, 2013).

The existence of the PTO composed of three proteins in the absence of transcription and translation provides a unique opportunity to dissect a biological clock with biochemical, biophysical, and structural means. Over the course of the last decade, a plethora of approaches have been deployed to gain a better understanding of the structure and function of the Kai proteins and their interactions with each other over the course of the daily cycle. In particular, investigators have focused on the Kai proteins from the mesophilic *S. elongatus* and the thermophilic *Thermosynechococcus elongatus* strains for gaining a better picture of the roles of the three proteins in the PTO. KaiC is the only enzyme in the trio, acting as an ATPase, an autokinase, an autophosphatase, and a phosphotransferase. KaiC undergoes phosphorylation at two sites, Thr-432 and Ser-431 (Nishiwaki et al., 2004; Xu et al., 2004), whereby a strict order is maintained in terms of phosphorylation and dephosphorylation: TS → pTS → pTpS → TpS → TS (Nishiwaki et al., 2007; Rust, Markson, Lane, Fisher, & O'Shea, 2007). KaiA stimulates phosphorylation of KaiC (Williams, Vakonakis, Golden, & LiWang, 2002; Xu, Mori, & Johnson, 2003) and KaiB antagonizes KaiA action (Kitayama, Iwasaki, Nishiwaki, & Kondo, 2003) by sequestering KaiA when KaiC is hyperphosphorylated at the interface of the KaiBC complex toward the end of the clock cycle (Brettschneider et al., 2010; Qin, Byrne, Mori, et al., 2010).

The dynamic nature of the Kai protein associations over the daily clock cycle (Mori et al., 2007) constitutes a particular challenge to the structural characterization of protein–protein interactions in the PTO. However, structural and biophysical techniques that cover the low- to high-resolution range have yielded 3D models of the individual Kai proteins in atomic detail as well as of the binary and ternary complexes at low and medium resolution (Egli, 2014; Egli & Johnson, 2013; Johnson, Egli, & Stewart, 2008; Johnson, Stewart, & Egli, 2011). This chapter provides an overview of the various techniques used to analyze structure and dynamics of Kai proteins and their complexes. Some of these same methods are being applied to the study of structure and function of mammalian circadian clock proteins, but the focus here is upon the cyanobacterial system because it is the best-characterized circadian system from the biochemical/structural perspective, and because it is the area of my expertise. I hope that describing the strengths and limitations of these methods individually will assist other researchers in their application to other clock proteins. Rather than describing individual methods in a recipe-like fashion, I have placed the emphasis on highlighting particular insights into the cyanobacterial clock that were gained from an approach and on a comparison of its advantages and limitations relative to other techniques (Table 1). At least as far as the characterization of the shape and dynamics of the PTO in broad strokes is concerned, a hybrid structural

Table 1 Strengths and limitations of individual techniques

Technique	Strengths	Limitations
PAGE (SDS/native)	SDS: assay of molecular mass, phosphorylation status Native: assay of complexes	SDS: not all proteins show mobility shifts with changes in phosphorylation Native: complexes must be very stable to withstand the long time required for electrophoretic separation
SPR	Assays of protein dynamics and/or protein interactions in solution at physiological concentrations. Can provide K_D values	Potential artifacts due to one or more of the proteins being immobilized on a surface
AUC	Accuracy; not affected by artifacts of gel filtration such as protein sticking to beads	Slow (complexes can dissociate). Sedimentation velocity is a function of both mass and shape. Requires a relatively large amount of protein
DLS	Size distribution profile of particles; determination of quaternary structures	Highly pure samples needed. Difficult to characterize polydisperse samples
TLC	Separation of small molecules such as ATP, ADP, etc	Not very quantitative unless coupled with radioactive labeling
MS	High sensitivity. Information about molecular weight, modification of proteins, e.g., phosphorylation and structure (native MS)	Not very quantitative. Sample heterogeneity sets limitations. Transfer of complexes from solution to gas phase can affect structure
FA/FRET	Assays of protein dynamics and/or protein interactions in solution at physiological concentrations. Can provide K_D values	Labeling problems: sometimes difficult to label protein(s). Protein activity affected, labeled proteins unstable?
Site-directed mutagenesis	Test hypotheses regarding the function of specific residues	Mutant proteins may be more difficult to purify.
EPMR	Information on the environment and dynamics of residues. Provides distance data (DEER)	Labeling problems: sometimes difficult to label protein(s). Protein activity affected, labeled proteins unstable?

Table 1 Strengths and limitations of individual techniques—cont'd

Technique	Strengths	Limitations
EM	Good for visualizing; can actually see proteins and their complexes. Resolutions of cryo-EM structures can extend to better than 4 Å	Static images, preparation (staining, vacuum, etc.) may significantly alter structure and/or formation of complexes
X-ray crystallography	Highest possible resolution. Accurate visualization of active sites can lead to insights into mechanism. Potential insights regarding function from structure	Requires diffraction-quality crystals. Static structure that is relatively uninformative in terms of dynamics. Trapped conformation may not be representative of active state
SAXS	Samples are in solution. Small sample size. Accurate assessment of masses, folding states, volumes, and 3D shapes. Complements crystallography and NMR	Highly sensitive to aggregation. May not provide a unique solution. Overfitting possible as a consequence of the lack of a reliable quality assessment factor
NMR	Provides conformational constraints, information on foldedness, and dynamics of proteins in solution. Chemical shift perturbation assays to detect protein–protein and protein–ligand interactions. Conditions can be readily changed	Relatively small proteins or fragments rather than full length. Results obtained with fragments may be misleading and not reflective of what the full-length protein does. Concentrations typically high, which may produce misleading results
HDX-MS	Insights into protein structural dynamics and conformational changes. Mapping solvent accessibility and protein–protein binding interfaces	Analysis of data is very tedious and time consuming
MD	Key approach for X-ray and NMR refinements, optimization of models built into EM density and SAXS envelopes. Ligand docking, homology modeling, and *ab initio* fold prediction	Simulations of large systems are prohibitively expensive (CPU time) and may limit adequate sampling of conformational states. Current force fields poorly suited to approximate quantum effects
Mathematical modeling	Provide testable predictions based on hypotheses	Sufficiently complicated models can sometimes model any phenomenon without establishing definitive tests

approach has proved to be particularly valuable. However, X-ray crystallography and solution nuclear magnetic resonance (NMR) remain the two most powerful techniques for deriving atomic-resolution structures.

2. KAI PROTEIN OVEREXPRESSION, PURIFICATION, COMPLEX FORMATION, AND ANALYSIS BY DENATURED AND NATIVE POLYACRYLAMIDE GEL ELECTROPHORESIS

2.1. Protein expression and purification

Kai proteins are expressed in *Escherichia coli* as hexahistidine-tagged (Mori et al., 2002), or GST-tagged, or (Mori et al., 2007; Nishiwaki et al., 2004) SUMO-fusion (Kim, Dong, Carruthers, Golden, & LiWang, 2008) proteins (Fig. 1). Purification entails first an affinity chromatography step and is then followed by gel filtration or ion exchange chromatography. Tags or fusions can either be removed by the appropriate proteases (e.g., enterokinase, PreScission, and Ulp1) or be retained for further experiments. In terms of the question of whether to cleave the fusion or proceed with the modified Kai protein for further studies, it is noteworthy that tags or fusion proteins may be helpful or at least not interfere with function. Thus, wild-type and mutant KaiC proteins have thus far all been crystallized with a

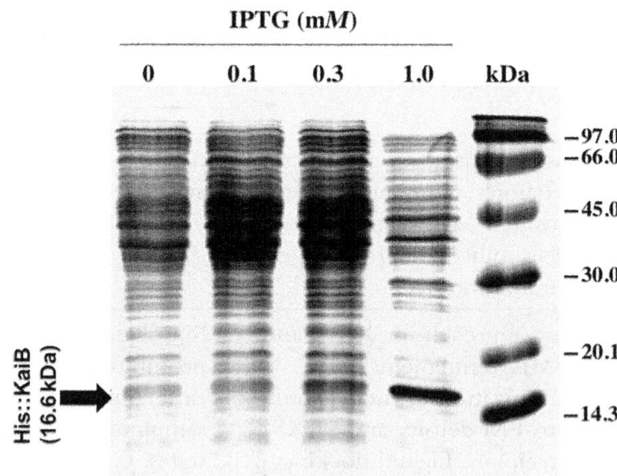

Figure 1 SDS-PAGE assay (20%) of the expression/induction of *S. elongatus* KaiB with an N-terminal $(His)_6$ tag in *E. coli* BL21 (DE3) cells as a function of isopropyl β-D-1-thiogalactopyranoside (IPTG) concentration. The KaiB monomer band is shown with an arrow, and molecular weights of marker bands are indicated on the right.

C-terminal $(His)_6$ tag (Pattanayek et al., 2009, 2004, 2011; Pattanayek, Xu, Lamichhane, Johnson, & Egli, 2014), and the tag appears not to interfere with the *in vitro* cycling reaction according to a recent report (Kitayama, Nishiwaki-Ohkawa, Sugisawa, & Kondo, 2014). In fact, even the bulky Cerulean protein fused to the C-terminal end of KaiC does not appear to distort the rhythm of the *in vitro* PTO (Ma & Ranganathan, 2012). Furthermore, FLAG-tagged Kai protein domains are routinely used for solution NMR studies (Tseng et al., 2014). However, in our hands, the C-terminal $(His)_6$ tag or Cerulean fusion do affect activity *in vitro*. And *in vivo*, there are some reports of poor rhythms with tagged proteins, while more recent papers state the opposite. Hopefully in the near future, these disparate results can be resolved among the various laboratories that study cyanobacterial clock proteins.

The identities of all purified proteins should be established by tryptic digestion in combination with electrospray ionization mass spectrometry (ESI-MS) or matrix-assisted laser desorption ionization time-of-flight mass spectrometry (MALDI-TOF MS). This is particularly important for mutant proteins of KaiC that occasionally copurify with GroEL (the two proteins have similar molecular weights (MWs), ca. 60 kDa, and form oligomers in the presence of ATP). The KaiC proteins from *S. elongatus* and *T. elongatus* also display subtle differences as a result of deviating numbers of basic and acidic residues (Pattanayek et al., 2014). Thus, the *T. elongatus* KaiC protein can be purified in the monomeric state in the absence of ATP (the same applies to a thermophilic KaiC from a source in Yellowstone Park, Mori et al., 2002), whereas KaiC from *S. elongatus* is normally purified as a hexamer with ATP bound to avoid precipitation.

2.2. Denatured and native polyacrylamide gel electrophoresis

The purity of all proteins is checked with denaturing polyacrylamide gel electrophoresis (SDS-PAGE). The KaiB protein forms a stable dimer-of-dimers in solution and in the solid state and at least the dimer band can typically be observed in SDS-PAGE along with the 13-kDa monomer. We commonly use native PAGE to assay complex formation among Kai proteins (Pattanayek et al., 2008, 2006, 2011; Qin, Byrne, Mori, et al., 2010), but have on occasion also relied on fluorescence spectroscopy to probe the interactions between Kai proteins or their peptide fragments (Pattanayek et al., 2008). Kai proteins from mesophilic and thermophilic strains may behave differently in native PAGE assays. Thus, the KaiB proteins from *S. elongatus* and *T. elongatus* display drastically different migrations as a result of deviating numbers of acidic residues in their C-terminal tails (Pattanayek et al., 2008).

Migration in an SDS-PAGE gel cannot be taken as an absolute measure of molar mass because other factors such as charge can influence migration. This phenomenon can be useful; for example, some proteins change their charge sufficiently upon phosphorylation (or other modifications) that their migration in an SDS-PAGE gel is significantly affected. Thus, phosphoforms can be distinguished for some proteins with a simple SDS-PAGE experiment (Fig. 2). This is true for many circadian clock proteins, including KaiC, FRQ, PER, and so forth. Treatment with λ-phosphatase has been helpful with all of these clock proteins, as it converts multiple bands that represent different phosphoforms to a single band of unphosphorylated protein (Fig. 2) (e.g., Hayashi et al., 2004; Pattanayek et al., 2008; Xu et al., 2003).

Two-dimensional native and SDS-PAGE assays have been conducted with Kai clock proteins on multiple occasions (e.g., Hayashi et al., 2004 Mori et al., 2007). Thus, complexes are separated in the first dimension by native PAGE and then analyzed in terms of their composition by SDS-PAGE using staining with Coomassie blue or alternative agents. Such a protocol allows for a time-resolved analysis of the association between Kai proteins over the daily period and demonstrates the formation of KaiA:KaiC complexes during the initial phase with a concomitant increase in KaiC phosphorylation, and formation of binary KaiB:KaiC and ternary KaiA:KaiB:KaiC complexes during the dephosphorylation phase (Mori et al., 2007).

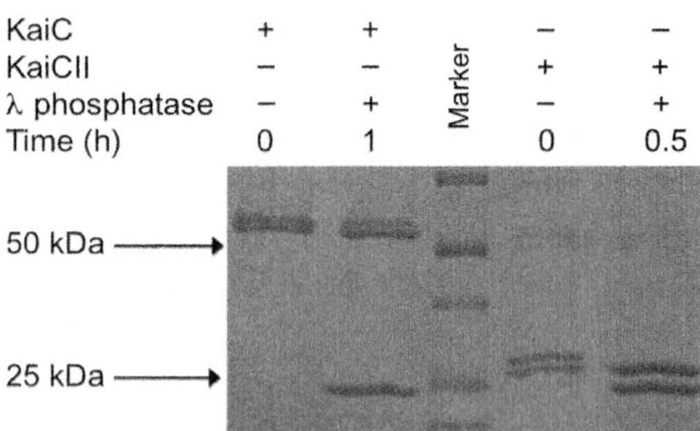

Figure 2 SDS-PAGE of phosphorylated and nonphosphorylated forms of *S. elongatus* KaiC (upper and lower band, respectively, in the 50 kDa range in the two leftmost lanes) and KaiCII, the C-terminal half of KaiC (upper and lower band, respectively, in the 25 kDa range in the two rightmost lanes). KaiC and KaiCII are gradually dephosphorylated by λ phosphatase (bottom band in lanes 2 and 5 from the left).

Related approaches for analyzing association among proteins are gel filtration chromatography in combination with SDS-PAGE or pull-down assays involving affinity tags such as $(His)_6$ or FLAG and SDS-PAGE using either standard staining techniques or immunoblotting with antibodies against individual proteins or particular tags (e.g., Chang, Kuo, Tseng, & LiWang, 2011; Hayashi et al., 2004; Kageyama et al., 2006; Pattanayek et al., 2011; Tseng et al., 2014; Villarreal et al., 2013). Native PAGE analysis of mixtures between two proteins of various ratios provides a means to measure stability parameters of their complexes, such as the dissociation constant K_D. Using this approach, the K_D for the KaiA:KaiC interaction was measured to be 152 ± 26 nM (Hayashi et al., 2004). By plotting the quotient of bound/free KaiA against bound KaiA (pmol) (Scatchard plot) based on native PAGE data, it was determined that two KaiA dimers can bind to a KaiC hexamer, although a single KaiA dimer appears to be sufficient to boost KaiC to the hyperphosphorylated state.

An alternative albeit label-free technique for quantitative measurements of biomolecular interactions is surface plasmon resonance (SPR). In this approach, one protein is immobilized on a biosensor and a solution of the prospective binding partner is channeled over the surface, whereby changes in the refractive index reflected from the biosensor are recorded. In this fashion, rate constants such as k_{on} and k_{off} as well as the dissociation constant K_D can be quantified. Kondo and coworkers used SPR with the Biacore instrument (GE Healthcare) to analyze binding of KaiA and KaiB to KaiC and found that both the association and dissociation rates for the KaiA:KaiC interaction were higher than those for KaiB:KaiC (15-and 4-fold, respectively; Kageyama et al., 2006). The values of K_D for the KaiA:KaiC and KaiB:KaiC interactions were 2.52 ± 0.46 and 8.79 ± 0.57 μM, respectively. The comparison between the affinity constants for the KaiA:KaiC complex based on SPR and native PAGE (see above) indicates that binding parameters may vary widely based on the particular techniques used.

3. ANALYTICAL ULTRACENTRIFUGATION

From the outset of investigations regarding the KaiABC PTO, the quaternary structure of individual proteins was a central concern and the use of analytical ultracentrifugation (AUC) predates by several years the discovery of the PTO. Indeed this issue has remained of crucial importance as the associations among Kai proteins and possibly their quaternary structures change over the daily cycle (Mori et al., 2007). AUC was used in combination with native PAGE and negative stain EM (ns EM) to establish KaiC hexamer formation, and this was the first concrete indication of the homohexamer oligomeric structure of KaiC (Mori et al., 2002). Moreover, this work used AUC and EM to establish KaiC as the first circadian clock protein for which structural information about the full-length protein was visualized. By adjusting the rotor speed, AUC can assess MWs over a size range of about 100 Da–10 GDa. No sizable difference exists in terms of the amount of protein needed to carry out the individual assays. Ideally, one should use various approaches to confirm the oligomeric state of a protein. As we shall see later, the KaiB protein alone exists as a tetramer (dimer-of-dimers) in solution as well as in the crystal (Garces, Wu, Gillon, & Pai, 2004; Hitomi, Oyama, Han, Arvai, & Getzoff, 2005; Iwase et al., 2005; Pattanayek et al., 2008), but is now known to interact with the KaiC hexamer in the monomeric state (Snijder et al., 2014; Villarreal et al., 2013). An alternative method to probe the quaternary structure of proteins is gel filtration chromatography in conjunction with SDS-PAGE, as recently employed in the context of investigations directed at the KaiB:KaiC interaction (Chang, Tseng, Kuo, & LiWang, 2012). AUC is a more reliable indicator than gel filtration for quantifying the MW of a protein complex. However, it is important to note that the sedimentation velocity is a function of both molar mass and shape, and therefore, calculations of molar mass from sedimentation velocity often assume a roughly spherical shape for the protein or complex. Irrespective of the approach one may prefer it is important to bear in mind that quaternary structure can be affected by temperature and protein concentration.

4. DYNAMIC LIGHT SCATTERING

Dynamic light scattering (DLS) is a further method to assay formation of higher order structures and is exquisitely sensitive to aggregation. Parameters that can be extracted from a light scattering experiment include the translational diffusion coefficient (D_T; $[D_T] = cm^2/s$) and the hydrodynamic or Stokes radius (R_H; $[R_H] = nm$). Using DLS, it was established that KaiB forms a tetramer in solution (Hitomi et al., 2005; Fig. 3). However, DLS data that we collected for KaiBs from different cyanobacterial strains or *S. elongatus* KaiB mutants were inconclusive as to the quaternary structure of the proteins. Solutions of wild-type KaiBs from *S. elongatus* and *T. elongatus* are monodisperse and the radii consistent with tetramers that were also observed in the respective crystal structures (Pattanayek et al., 2008; Villarreal et al., 2013).

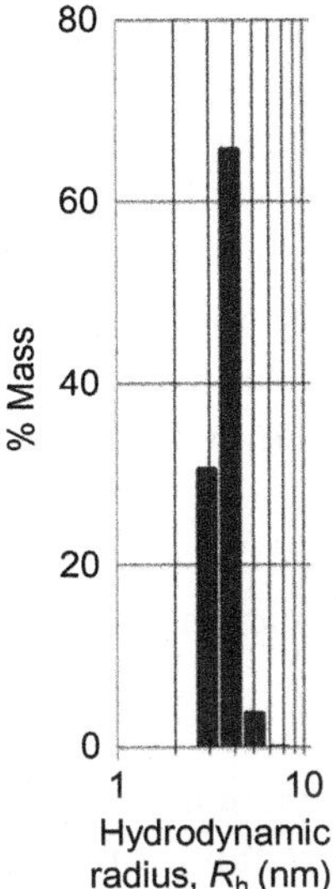

Figure 3 Dynamic light scattering of *S. elongatus* KaiB; the scattering values are the averages of 30 scans, each including 20 different time points. The translational diffusion coefficient D_T is 6.85×10^{-7} cm^{2}s^{-1}, the hydrodynamic radius R_h is 3.54 nm, and the mass is 65 kDa (tetramer).

However, KaiBs with up to three Asp mutations in the hydrophobic dimerization loop intended to shift the equilibrium to the monomeric state exhibit large radii/MWs that are inconsistent with monomer, dimer, or tetramer (R. Pattanayek & M. Egli, unpublished data). DLS proved to be more useful in connection with the question of the effect of oxidized plastiquinone (PQ) on the KaiA and KaiB proteins. As the cellular PQ pool increases at dusk, the compound is being transported to KaiA, causing it to aggregate and thus preventing KaiA from binding to KaiC to stimulate phosphorylation of the latter (Wood et al., 2010). DLS data acquired for solutions of KaiA in the presence of 2,5-dibromo-6-isopropyl-3-methyl-1,4-benzoquinone (DBMIB), a water-soluble PQ analog, are consistent with KaiA aggregation, whereas DBMIB has absolutely no effect on KaiB (Pattanayek, Sidiqi, & Egli, 2012).

5. THIN LAYER CHROMATOGRAPHY

Thin layer chromatography (TLC) remains a useful and inexpensive technique to screen organic chemical reactions, separate mixtures, and check purity and identity of compounds by way of a retardation factor

(R_f). The R_f of each spot can be determined by dividing the distance that a particular compound has traveled by the distance between solvent front and initial spotting site. This parameter is dependent on the TLC plate and the solvent. TLC is of little use to analyze Kai proteins and their interactions. However, a question that had remained unanswered until 2012 concerned the mechanism of the KaiC dephosphorylation reaction. In many publications, KaiC was referred to as an autokinase and autophosphatase. The latter terminology implied that the removal of phosphates from Thr-432 and Ser-431 involved a phosphatase, although the site of this activity in KaiC was unknown. To address the hypothesis that the dephosphorylation reaction possibly proceeded via the formation of ATP, i.e., in a reversal of the kinase reaction and using the same active site at subunit interfaces, we assayed the formation of radioactive ATP from [8-^{14}C]ADP with TLC (Egli et al., 2012). Indeed, polyethyleneimine cellulose TLC with 2 *M* sodium acetate solution revealed the buildup of radioactive ATP, thus supporting an ATP synthase mechanism as the basis for KaiC dephosphorylation. Formation of ATP from ADP and pThr-432/pSer-431 was subsequently also established by others using an alternative approach (Nishiwaki & Kondo, 2012).

6. MASS SPECTROMETRY

In addition to the standard use of mass spectrometry for identifying proteins based on the masses of peptide fragments from protease digests, an MS-based approach is also key for identifying posttranslational modifications in proteins, such as phosphorylation, acetylation, ubiquitination, and so forth. The two phosphorylation sites Thr-432 and Ser-431 in KaiC were recovered by nanoflow liquid chromatography-electrospray tandem MS (MS/MS), following digestion with trypsin and Asp-N (Nishiwaki et al., 2004). We found the same two sites by inspecting around threonine and serine residues difference Fourier electron density maps calculated for a partially refined model of the crystal structure of *S. elongatus* KaiC (Xu et al., 2004). Positive difference density above a certain threshold in the immediate vicinity of the side chain hydroxyl groups of either amino acid was taken as an indication of phosphorylation. Unlike the data from an MS-based analysis, crystallography provides a detailed three-dimensional map of the environment of ATP and phosphorylation sites at the subunit interface. Careful examination of the kinase active site revealed a second threonine, Thr-426, whose side chain was engaged in a H-bond to pSer-431 (Xu et al.,

2004). Subsequent structure and function investigations with KaiCs featuring mutations in the phosphorylation sites provided strong evidence that the residue at position 426 needs to be phosphorylatable and that Thr-426 is a short-lived phosphorylation site that when mutated renders the clock arrhythmic (Pattanayek et al., 2009; Xu et al., 2009). Sequential phosphorylation of KaiC, i.e., first Thr-432 and then Ser-431 followed by dephosphorylation in the same order (TS → pTS → pTpS → TpS → TS), was first established by tracking the *in vitro* cycling reaction and analyzing digested KaiC protein samples mass spectrometrically (Nishiwaki et al., 2007; Rust et al., 2007). However, with the identity of gel bands confirmed, subsequent studies analyzing the effects of mutations on the clock period *in vitro* relied on SDS-PAGE to establish the phosphorylation rhythm.

7. SITE-DIRECTED MUTAGENESIS

Mutagenesis is commonly used to study the function of genes and proteins. In the context of the cyanobacterial clock, phosphorylation site mutants of KaiC as well as KaiCs with mutations in the active site at the subunit interface have been analyzed in detail (Pattanayek et al., 2009, 2011; Xu et al., 2004, 2009). As well, KaiCs with C-terminal deletions were studied in regard to their phosphorylation states and binding to KaiA that is known to contact the C-terminal region of KaiC (Kim et al., 2008; Pattanayek et al., 2006; Vakonakis & LiWang, 2004). Nowadays kits from various manufacturers are used to introduce mutations, insertions, or deletions in a protein. Two primers, with one or both of them carrying the desired mutation(s) are annealed to the plasmid for PCR amplification. Following ligation to circularize the mutated PCR products and transformation of the resulting plasmid into *E. coli*, overexpression of mutant protein proceeds in an analogous fashion to that of wild-type protein. However, depending on the particular changes introduced into the protein, the expression efficiency may vary and it is often necessary to optimize expression parameters such as temperature and induction. Similarly, the purification protocol may have to be adapted as mutations can alter protein stability, structure, and electrostatic surface potential in addition to activity (i.e., ATPase, kinase, and ATP phosphotransferase in the case of KaiC). Whole gene synthesis offers an attractive alternative to the above mutation strategy, particularly in cases where multiple changes are introduced into a protein at distant sites. For example, one may want to produce a KaiC mutant with a C-terminal deletion as well as mutations in the phosphorylation site in the C-terminal half and mutations in the ATPase in the N-terminal half. Rather than going through a stepwise mutation procedure, commercial custom gene synthesis and insertion into the desired expression vector may then offer a cost- and time-efficient alternative.

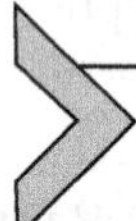

8. FLUORESCENCE TECHNIQUES (LABELED PROTEINS, ANISOTROPY, AND FLUORESCENCE RESONANCE ENERGY TRANSFER)

Fluorescence has become a standard method in many biologists' toolbox, especially in cell biology, where the ability to fuse genetically encoded proteins with fluorescent proteins (e.g., Green Fluorescent Protein) has revolutionized our understanding of where and how proteins act within cells. Fluorescence methods have had a similar impact in biochemistry, and in particular, the biochemistry of circadian clock proteins. One simple fluorescence application has been the labeling of FLAG-tagged KaiC proteins with fluorochromes (e.g., Alexa Fluor 532 and Cy5) and performing *in vitro* pulldowns with an anti-FLAG antibody to confirm that KaiC exchanges monomers among the KaiC hexamers in the population (Ito et al., 2007). Several groups have used the technique of fluorescence anisotropy (FA) to assess the dynamics of Kai protein complex formation and apparent dissociation constants (Qin, Byrne, Mori, et al., 2010; Tseng et al., 2014). The principle of FA in this application is that photons emitted from a fluorophore have a specific polarization with respect to the fluorochrome, and if the fluorochrome is rigidly attached to the protein, the plane of fluorescence emission will then reflect the orientation of the protein. When polarized light is used to excite a group of randomly oriented fluorophore-labeled proteins, most of the excited proteins will be those oriented within a particular range of angles to the applied polarization. If the proteins do not move, the emitted light will also be polarized within a particular range of angles to the applied excitation. If proteins can change their orientation before releasing photons as fluorescence emission (i.e., by tumbling in solution), this will reduce the magnitude of polarization of the emitted light. Larger proteins (or large protein complexes) tumble more slowly than smaller proteins (or small complexes). Therefore, anisotropy of the fluorescence signal can be used as a gauge of individual fluorophore-labeled proteins combining into a larger protein complex. This technique has been applied to the formation of KaiA:KaiC complexes with the fluorophore attached to KaiA (Qin, Byrne, Mori, et al., 2010) and to the binding of fluorophore-labeled KaiB to the CI domain of KaiC (Tseng et al., 2014).

FA is a useful technique if the fluorophores are relatively far apart. If they are very close to one another, they can exchange energy by fluorescence resonance energy transfer (FRET). In this process, one fluorophore (the "donor") transfers its excited-state energy to another fluorophore (the "acceptor") that usually emits fluorescence of a different color. FRET efficiency depends on the spectral overlap, the relative orientation, and the distance between the donor and acceptor fluorophores. Generally, FRET occurs when the donor and acceptor are 10–100 Å apart, so it can be used to assay protein–protein proximity by attaching the donor and acceptor fluorophores to the candidate proteins. Therefore, FRET can be used as a "molecular ruler" to confirm that two proteins are interacting. In the case of circadian clock proteins, FRET was used to confirm the exchange of KaiC monomers among hexamers (Mori et al., 2007; Fig. 4). Kageyama and coworkers had previously used FLAG-tagged KaiC proteins to demonstrate that KaiC monomers appear to exchange between KaiC hexamers (Kageyama et al., 2006). However, pull-down assays of FLAG-tagged

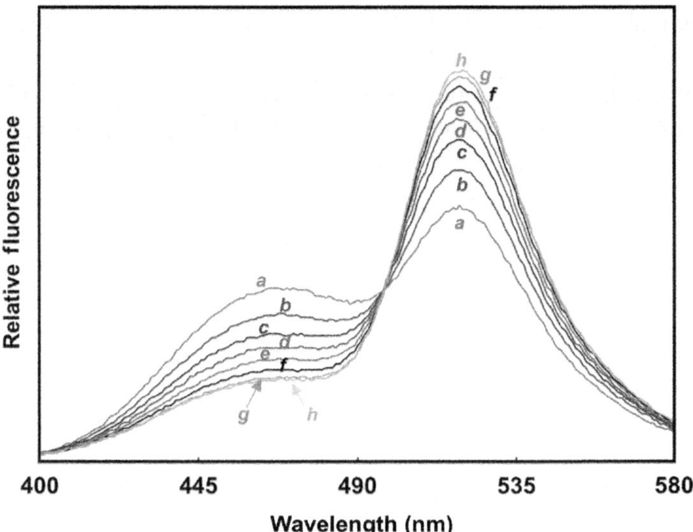

Figure 4 KaiC monomer exchange assayed by FRET. Equal populations of *S. elongatus* KaiC labeled with IAEDANS or MTSF were mixed and emission spectra of KaiC excited at 336 nm recorded at times (a) 0, (b) 0.16, (c) 0.5, (d) 1, (e) 2, (f) 4, (g) 6, and (h) 8 h after mixing at 30 °C. The decrease in fluorescence intensity at 470 nm of IAEDANS-labeled KaiC is indicative of energy transfer due to subunit shuffling between the two KaiC populations. *Adapted from Mori et al. (2007).*

proteins can suffer from differential cross-reactivity and aggregation. Therefore, we used a completely different technique to confirm KaiC monomer exchange, namely FRET (Mori et al., 2007). KaiC has three intrinsic cysteine residues that can be used for labeling with fluorophores. IAEDANS and MTSF, well-characterized FRET fluorophore partners, were used to label KaiC. One group of KaiC hexamers was labeled with IAEDANS and the other with MTSF. These groups were then mixed and incubated while monitoring the time-dependent change in quenching of IAEDANS fluorescence (indicative of FRET) in response to excitation at the excitation maximum for IAEDANS. During the incubation of the two populations of KaiC, the emission at the emission peak for IAEDANS decreased progressively (Mori et al., 2007; Fig. 4). That result indicated that monomer exchange among the two groups of KaiC had occurred, confirming the previous results that utilized pulldowns (Kageyama et al., 2006).

9. ELECTRON MICROSCOPY

Electron microscopy (EM) is a versatile technique that permits visualization of molecules and molecular assemblies with MWs >100 kDa. There is really no upper limit in terms of the size of the molecules that can be studied with EM, and improvements in detector technology now render electron crystallography a viable alternative to X-ray crystallography, as exemplified by recent EM structures of adenovirus (Reddy, Natchiar, Stewart, & Nemerow, 2010) and the ribosome (Amunts et al., 2014). The two most common approaches for analyzing macromolecular samples are ns EM and cryo-EM and both have been used for investigating the architecture of the KaiABC PTO.

9.1. Negative stain EM

For ns EM, samples are applied to a grid and stained with a solution of uranyl formate or acetate. Electron micrographs are then collected and individual particles (raw images), typically 1000s, are selected and classified. The latter procedure can be performed manually or in an automated fashion and basically entails grouping particles into classes of similar orientation by translational and rotational alignment. Averaging of such classes furnishes class sum images that exhibit an improved signal-to-noise ratio compared to raw images. Obviously, the quality of the classification is dependent on how well the alignments are done and alignment and classification steps can be repeated several times to optimize the data analysis. Next the relative orientations of the 2D projections in the class sums have to be determined in order to produce a 3D model, whereby the latter can be used to realign the raw images. Again, this reprojection process can be repeated to optimize the computed 3D electron density map. The resolution of the final reconstruction is assessed with the Fourier shell correlation (FSC) 0.5 threshold criterion. Models of proteins or domains from X-ray or NMR structure determinations can then be built into the EM envelope.

Early studies employed ns EM to confirm the hexameric organization of *S. elongatus* KaiC (Mori et al., 2002; Fig. 5) and the overall shape of *T. elongatus* KaiC at low resolution (Hayashi et al., 2003). By correlating the 3D shape of particles from aliquots removed during the *in vitro* cycling reaction with their composition as analyzed by 2D native blue gel assay, we were able to track the appearance of Kai protein–protein complexes over the daily period (Mori et al., 2007). Overall, four classes of particles can be distinguished: KaiC hexamer, a KaiA:KaiC complex recognizable by a protrusion at one end of the KaiC barrel, a KaiB:KaiC complex with a third layer on top of the two KaiC rings (Fig. 6), and finally a ternary KaiA:KaiB:KaiC

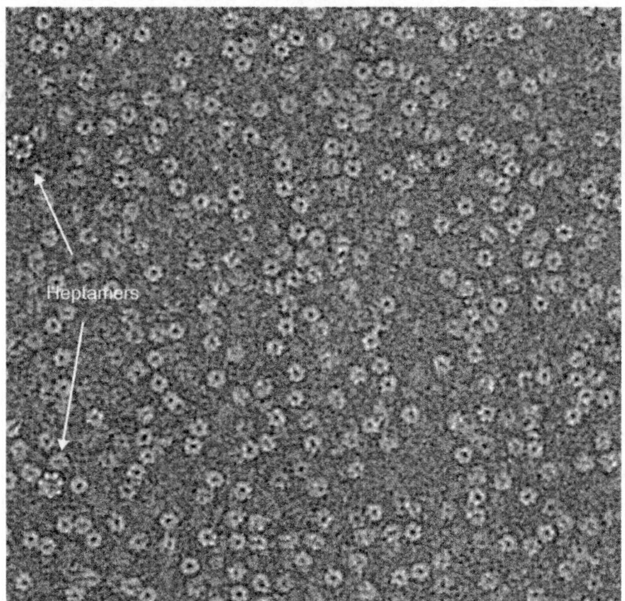

Figure 5 Negative stain electron micrograph of *S. elongatus* KaiC hexamers, viewed mostly along the central channel. Two heptameric particles are indicative of a small population of GroEL. GroEL and KaiC monomers have virtually identical molecular mass, and the former is frequently present in overexpressions of KaiC in *E. coli*.

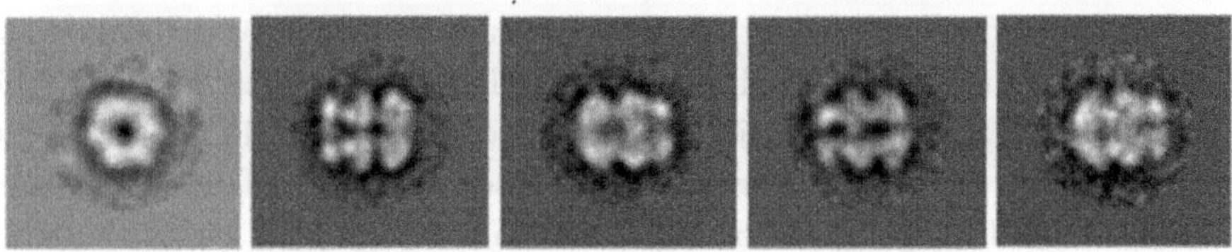

Figure 6 Class sums for the KaiB:KaiC complex from *S. elongatus* based on negative stain EM. Complex particles exhibit a characteristic three-layer shape, with the third thin layer formed by KaiB monomers.

complex with additional protrusions growing sideways from the third layer of the KaiB:KaiC complex. The KaiA:KaiC complexes from both *S. elongatus* and *T. elongatus* were studied in more detail by ns EM, and a detailed 3D reconstruction revealed two orientations of KaiA atop KaiC (Pattanayek et al., 2006). One, referred to as "tethered" exhibits a relatively large spacing between the two proteins, whereas the other, called "engaged" showed the two proteins in tight contact at the C-terminal end of KaiC. Subsequently, a model of the "engaged" complex was built into the EM electron density by taking into account the crystal structure of KaiC (Pattanayek et al., 2004), the complex based on NMR between the C-terminal domains of KaiA and C-terminal KaiC peptide (Vakonakis & LiWang, 2004), and the crystal structure of full-length KaiA (Ye, Vakonakis, Ioerger, LiWang, & Sacchettini, 2004). ns EM also provided the 3D shape of KaiA sequestered at the KaiB:KaiC interface, but the achieved resolution is too low to reveal details of the interactions between the three proteins (Pattanayek et al., 2011).

9.2. Cryo EM

Rather than by staining, samples in cryo-EM are preserved on a holey carbon grid in a frozen-hydrated state by plunging them into ethane slush cooled with liquid nitrogen. This preservation in a vitrified state prevents the collapse of samples due to dehydration upon staining. We used cryo-EM to study the *S. elongatus* KaiB:KaiC complex and an initial model showed two KaiB dimers bound to KaiC (Pattanayek et al., 2008). Because of the overall symmetry of the KaiC hexamer, it is not possible to distinguish between the N- and C-terminal hexameric rings at low resolution. Therefore, we relied on native PAGE to assay binding between separate N- and C-terminal KaiC rings and KaiB. These experiments provided support for the C-terminal but not the N-terminal hexamer interacting with KaiB. In the model of the complex, we therefore assigned KaiB to the C-terminal end of KaiC. Subsequently, we used Ni-NTA nanogold in combination with a KaiC carrying a C-terminal $(His)_6$ tag to demonstrate that gold, His-tag, and KaiB all congregate at the C-terminal end of the KaiC hexamer (Pattanayek, Yadagirib, Ohi, & Egli, 2013). A more recent cryo-EM model of the complex between KaiB and KaiC hexamer that lacks the C-terminal 30 residues per subunit (delta-KaiC) with a resolution of 16 Å (FSC 0.5) shows six KaiB monomers forming a ring on top of the C-terminal KaiC ring, thereby covering the ATP binding clefts (Villarreal et al., 2013). Although the resolution of this $KaiC_6B_6$ model is insufficient to reveal details of the protein interactions, the binding mode suggests that KaiBs can interfere with the active site, thereby potentially limiting the kinase activity and promoting transfer of phosphates from pThr-432 and pSer-431 back to ADP. What is clear is that KaiB does not act as a competitive inhibitor of KaiA (i.e., by binding to C-terminal KaiC tails) and that at the level of the KaiC hexamer, a stable KaiB:KaiC complex exists in the absence of KaiA.

10. X-RAY CRYSTALLOGRAPHY

Single-crystal X-ray crystallography remains the most important approach for generating high-resolution 3D structural information for biomacromolecules, independent of their size. Dramatic improvements in crystallization methodology (robotics, sparse matrix crystallization screens), diffraction data collection (rapid detectors, flash freezing of crystals), phasing approaches (seleno-methionine multi- or single-wavelength anomalous dispersion, Se-Met MAD or SAD, respectively), and refinement strategies (simulated annealing, maximum likelihood analysis) render crystallography the method of choice for structure analysis at atomic resolution of proteins, nucleic acids, multiprotein complexes, and molecular machines.

Crystal structures of all three full-length Kai proteins were determined in 2004: KaiA (Ye et al., 2004) and KaiC (Pattanayek et al., 2004) from *S. elongatus* and KaiB from *Anabaena* (Garces et al., 2004). KaiA forms a domain-swapped dimer and features N-terminal bacterial response regulator-like and C-terminal four-helix bundle domains that are connected by a linker. The bundle domain forms the interface of the dyad-related homodimer and also harbors the KaiC-binding site. KaiB forms a dimer-of-dimers with subunits exhibiting a thioredoxin-like fold. All subsequent crystal structures of KaiB wild-type and mutant proteins confirmed the preference for the tetrameric quaternary structure (Hitomi et al., 2005; Iwase et al., 2005; Pattanayek et al., 2008; Villarreal et al., 2013). A comparison of the surfaces of the KaiA and KaiB dimers revealed a striking similarity between the spacing of arginine pairs from subunits, i.e., Arg-69 residues in KaiA and Arg-23 residues in KaiB (*Anabaena* numbering; Garces et al., 2004). Thus, it seemed possible for KaiA and KaiB to compete for the same binding site on KaiC, consistent with their opposite effect on KaiC phosphorylation (Iwasaki, Nishiwaki, Kitayama, Nakajima, & Kondo, 2002; Kitayama et al., 2003). Although this scenario seemed compelling at the time, subsequent work established that KaiA and KaiB contact KaiC in different locations (Pattanayek et al., 2008, 2006; Vakonakis & LiWang, 2004). Nevertheless, the availability of structures paved the road to an interpretation of clock mutational data by mapping residues in the 3D models (Garces et al., 2004; Ye et al., 2004).

We solved the KaiC crystal structure by SAD using a tantalum bromide derivative (Pattanayek et al., 2004; Fig. 7). Annealing the crystal in the loop, i.e., by briefly diverting the coldstream and reflash cooling (Harp, Hanson, Timm, & Bunick, 1999; Fig. 8) led to a considerable improvement in the resolution limit of the diffraction data to 2.8 Å (Fig. 9). The structure revealed a homohexamer of overall dimensions 100 × 100 × 100 Å, composed of two stacked rings with a constricted waist and a central channel. The N- and C-terminal domains that are the result of a gene duplication adopt a fold similar to that of recombinases (RecA) and helicases (DnaB), as anticipated from sequence considerations (Leipe, Aravind, Grishin, & Koonin, 2000). However, individual KaiC rings display a closer resemblance to F1-ATPase (Pattanayek et al., 2004), although the relationship between the KaiC homohexamer and the ATPase trimer of dimers is not apparent at the sequence level. Conversely, comparison of the KaiC rings with the structures of helicases indicates an inferior correspondence, both in terms of the diameter as well as the locations and orientations of ATP molecules.

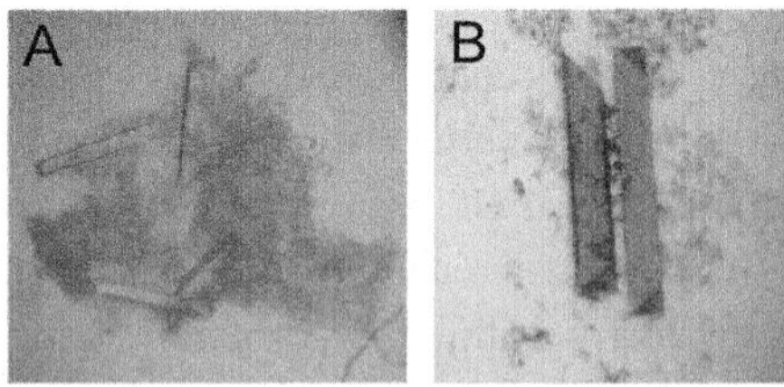

Figure 7 Crystals of (A) full-length *S. elongatus* KaiC (N1-519) with a C-terminal $(His)_6$ tag, and (B) derivatized with a tantalum bromide cluster $Ta_6Br_{12}^{2+}$ compound. (See the color plate.)

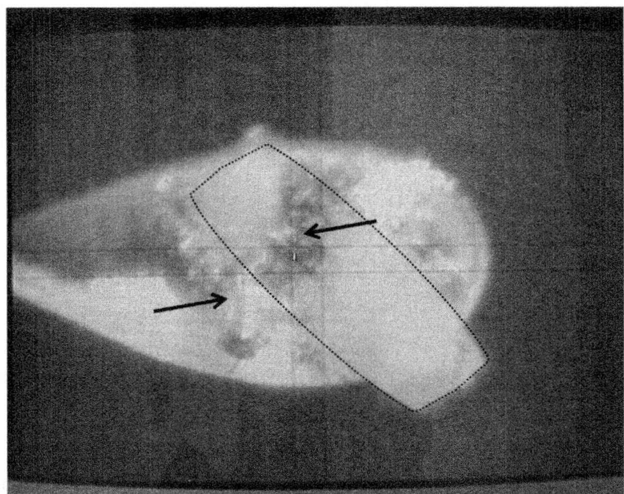

Figure 8 Annealing of a KaiC crystal (outlined with a dashed line) in the cryo-loop. The liquid N_2 stream is temporarily blocked, and the image taken from a monitor at the beam line depicts bubbling mother liquor (arrows) around the crystal as it is warming up. (See the color plate.)

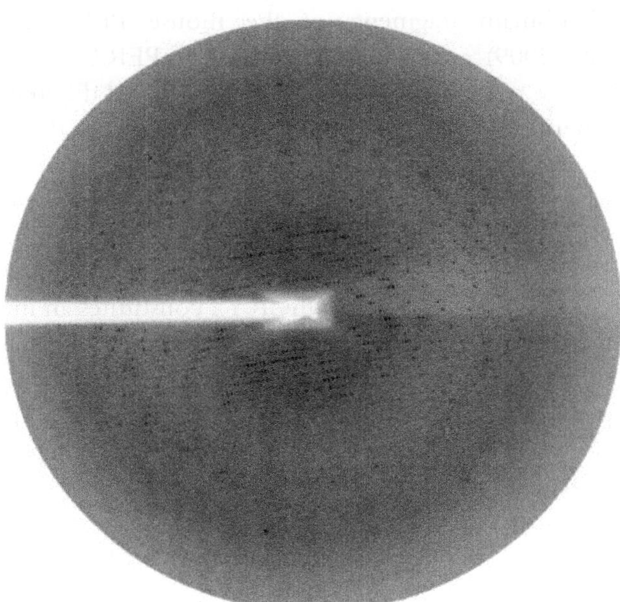

Figure 9 X-ray diffraction pattern from a KaiC crystal; the maximum resolution of the data is ca. 2.8 Å.

KaiC hexamers bind 12 ATPs, 6 each at subunit interfaces in the N- and C-terminal rings. The crystal structures of both *S. elongatus* KaiC (Pattanayek et al., 2004) and *T. elongatus* KaiC (Pattanayek et al., 2014) are trapped in the hyperphosphorylated state, and the former structure initially revealed the Thr-432 and Ser-431 phosphorylation sites in the C-terminal ring. As well, both structures were obtained for constructs with C-terminal $(His)_6$ tags, whereas constructs of full-length KaiC with His-tags or GST fusions cleaved off have thus far resisted all crystallization attempts. Subsequent structure determinations of KaiC proteins with mutations in the phosphorylation sites revealed only minor conformational changes in the overall conformations and at subunit interfaces and allowed us to identify a third short-lived phosphorylation site at Thr-426 (Pattanayek et al., 2009). Careful inspection of electron density maps allowed us to build complete or partial models of the C-terminal tails of KaiC (Pattanayek et al., 2006) that constitute the binding site for C-terminal domains of KaiA (Vakonakis & LiWang, 2004). However, to date, none of the numerous attempts to crystallize the KaiC hexamer with either KaiA or KaiB have been met with success.

X-ray crystallography is increasingly having an impact on our understanding of the mammalian and Drosophila clocks. Thus, the structures of the PAS domain fragments of the mouse PERIOD2 (mPER2; Hennig et al., 2009) and 1 and 3 proteins (mPER1, mPER3; Kucera et al., 2012) were recently reported and have revealed differences between the homodimeric interactions of the three homologues that are likely of importance in terms of their individual clock functions. Crystal structures of *Drosophila* cryptochrome (dCRY, Czarna et al., 2013; Zoltowski et al., 2011) and mouse cryptochrome 1 (mCRY1, Czarna et al., 2013) have also become available. Whereas dCRY serves as an FAD-dependent circadian photoreceptor, mCRY1 along with mCRY2 constitutes an integral part of the clock as they act as repressors of CLOCK/BMAL1-dependent transcription. A crystal structure of the latter complex revealed an asymmetric heterodimer with the basic helix-loop-helix bHLH, PAS-A, and PAS-B domains involved in three distinct dimer interfaces (Huang et al., 2012). Finally, the crystal structure of the complex between a fragment of mCRY1 encompassing the photolyase homology region and a C-terminal mPER2 fragment (Schmalen et al., 2014) provided evidence that the interaction between the two proteins is modulated by zinc binding and mCRY1 disulfide bond formation and might thus be affected by the cell's redox state.

11. SMALL-ANGLE X-RAY AND NEUTRON SCATTERING

Small-angle X-ray scattering (SAXS) has become an integral part of the so-called hybrid structural biology approach in recent years. X-ray crystallography provides a detailed picture of the 3D structure of macromolecules and their complexes, but typically allows only limited insight into dynamic aspects. SAXS affords information that is complementary to X-ray crystallography, such as on the folding and unfolding of macromolecules, aggregation, flexible domains, oligomeric state, complex formation, and 3D shape (Putnam, Hammel, Hura, & Tainer, 2007; Rambo & Tainer, 2013). The amounts of material required for solution scattering experiments are minimal compared with single-crystal X-ray crystallography and NMR solution approaches, and there are no MW limitations like those encountered in EM (lower limit) or NMR (upper limit). Unlike with crystals, the conditions in SAXS (protein concentration, ionic strength, pH, temperature, etc.) can be altered readily, allowing one to optimize various parameters in a relatively short time.

In a SAXS experiment, the scattering intensity $I(q)$ is measured as a function of the resolution or q range that is expressed as $q=(4\pi \sin\theta)/\lambda$, where 2θ is the scattering angle and λ is the radiation wavelength ($[q]=\text{Å}^{-1}$) (Fig. 10). In the so-called Guinier analysis, $\ln\{I(q)\}$ is plotted against q^2 and the resulting curve in the region closest to the zero angle is linear if there is no aggregation or other concentration-dependent phenomenon (Fig. 10). From the slope of the Guinier plot, the radius of gyration, R_G, can be extracted. R_G is the square root of the average squared distance of each scattering atom from the particle center. The Kratky plot $q^2I(q)$ versus q provides insight into the folding and flexibility of a macromolecule or macromolecular assembly. For example, a protein with a globular fold will essentially exhibit a bell-shaped curve, whereas unfolded or partially unfolded species will display a plateau or increasing $q^2I(q)$ values in the upper q range. Finally, the pairwise distribution function $P(r)$ or pair-density distribution function represents the SAXS equivalent of the Patterson function in crystallography. It can be directly computed by Fourier transforming the scattering curve $I(q)$ and provides information about the distances between electrons in the scattering sample. In theory, $P(r)$ is zero at $r=0$ and at $r \geq D_{max}$, the maximum linear dimension of the scattering particle (Fig. 10). One of the most interesting outcomes of a SAXS experiment is the ability to carry out *ab intio* shape calculations. Because scattering curve and 3D shape of particles are related (Volkov & Svergun, 2003), it is possible to generate molecular

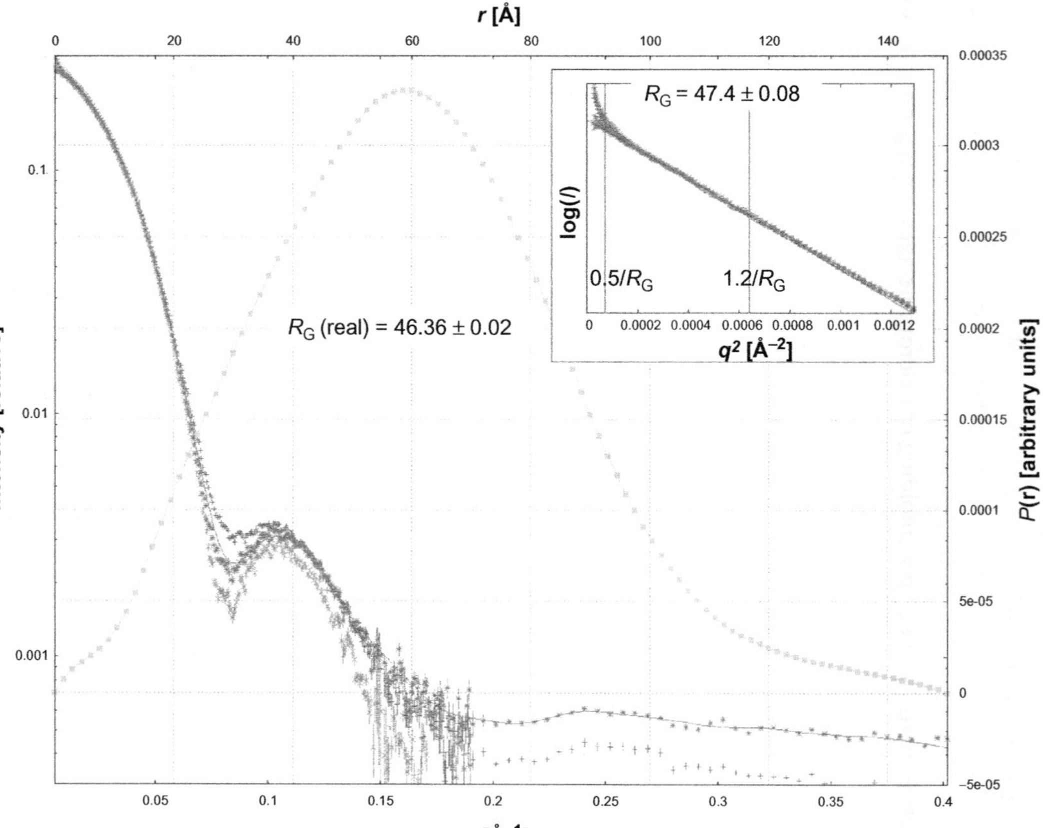

Figure 10 SAXS scattering curves $I(q)$ (red, high concentration; blue, medium concentration, and green, low concentration), pairwise function $P(r)$ (cyan), and Guinier plots (inset) for *S. elongatus* KaiC. *From Pattanayek et al. (2011).* (See the color plate.)

envelopes. Refining these entails calculating a scattering curve based on the initial model that can then be compared to the experimental curve; iterative minimization of the deviations between the experimental and theoretical scattering curves results in an improved envelope. Provided structures of the protein or the components of the macromolecular assembly under investigation are available; these can be built into the envelope and a best fit achieved with rigid-body refinement (Petoukhov & Svergun, 2005) or by applying molecular dynamics (MD) simulations (Fig. 11).

Akiyama and coworkers used SAXS with ternary KaiABC and binary KaiAC and KaiBC mixtures to chart a time course for forward scattering intensity $I(0)$ and R_G over 3 days of *in vitro* cycling (Akiyama, Nohara, Ito, & Maeda, 2008). The oscillation of intensity was in phase with that of the radius of gyration, and the latter indicated the formation of large complexes with $M_r > 470$ kDa at hours 38 and 62 and small complexes at hours 26 and 50. A SAXS-based envelope served as the basis to model KaiA dimer and KaiB tetramer at the C-terminal end of the KaiC hexamer, although no further evidence was provided to support the binding of both proteins to the same half of KaiC. We used SAXS to derive 3D models in solution of KaiA

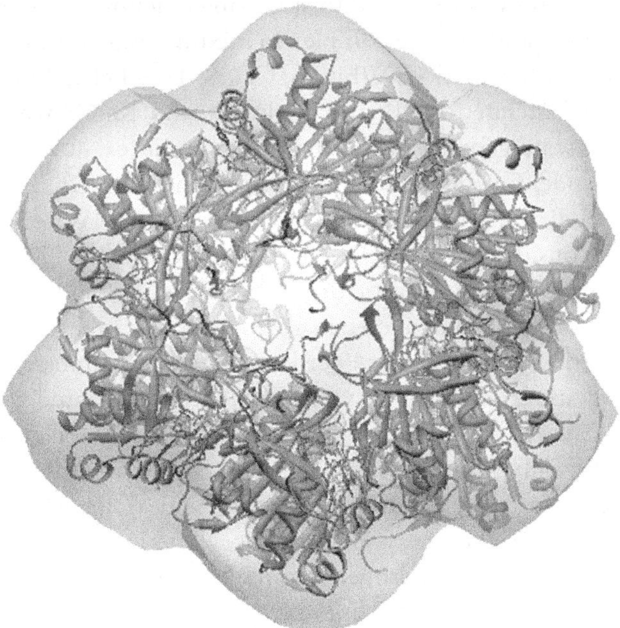

Figure 11 The crystallographic model of *S. elongatus* KaiC hexamer docked into the SAXS-based molecular envelope (using sixfold rotational symmetry constraints).

dimer, KaiB tetramer, and KaiC hexamer as well as the binary KaiAC and KaiBC complexes and, in combination with ns EM, the binary complex between KaiC and the His-kinase SasA (Pattanayek et al., 2011). Unlike EM density at low or medium resolution, the SAXS envelope for KaiC features a protrusion at one end, consistent with the presence of C-terminal tails that emerge from the dome-shaped surface near the central channel (Pattanayek et al., 2006). The SAXS envelope of the KaiBC complex was supportive of KaiB binding to the C-terminal end of KaiC because the protrusion accounting for C-terminal KaiC tails co-locates with that attributed to KaiBs. SAXS was also the key approach for tracking the expansion and contraction of the KaiC C-terminal half over the daily cycle (Murayama et al., 2011). Unlike in a crystal or on an EM grid, environments that can constrain dynamic behavior of a protein or a macromolecular assembly, such constraints are largely absent in solution and scattering provides a means to track volume changes in a particle. Accordingly, the C-terminal half of the KaiC hexamer rhythmically contracts and expands as it proceeds through various phosphorylation states, whereby the change in volume is correlated with the ATPase activity and amounts to maximal 4%.

In an attempt to study the state of KaiB (monomer, dimer, or tetramer) when bound to KaiC, we turned to small-angle neutron scattering (SANS; Jacrot, 1976). By studying the complex in mixtures of H_2O and D_2O of various ratios (contrast variation; Whitten, Cai, & Trewhella, 2008) and working with perdeuterated KaiB and hydrogenated KaiC, we intended to minimize the scatter from KaiC at the match point (ca. 40% D_2O) in order to derive a model for KaiB in combination with SAXS (R. Pattanayek, M. Egli, & W. Heller, unpublished data). However, the MWs of the two proteins differ considerably (13 kDa, KaiB monomer, vs. 360 kDa, KaiC hexamer), and the sensitivity of the approach is probably insufficient for the contribution of KaiB extracted from the overall scatter to be meaningful. Selective, partial, or completely (perdeuterated) recombinant proteins can be produced using expression systems in bacteria adapted to growth in D_2O and relying on deuterated carbon sources (Meilleur, Weiss, & Myles, 2009). The use of fully perdeuterated proteins improves the signal-to-noise ratio in neutron scattering and diffraction experiments and is essential for the study of proteins >40 kDa by solution NMR. The replacement of all hydrogen atoms by deuterium differs from the hydrogen–deuterium exchange approach (see Section 13) that is based on replacement of only a subgroup of H by D to probe protein dynamics and solvent accessibility.

12. NUCLEAR MAGNETIC RESONANCE

NMR spectroscopy in solution provides information that is complementary to that afforded by X-ray crystallography and offers insight into dynamic processes (Keeler, 2005; Wüthrich, 1986). However, compared with crystallography that can be applied to any molecular structure independent of size, NMR is generally limited to structures with MWs below 40 kDa. With specific labeling (e.g., ^{15}N, ^{13}C, ^{2}H) of particular amino acids in combination with transverse relaxation-optimized spectroscopy in two, three, or more dimensions, this limit can be pushed upward, but such experiments can become very time consuming. One of the most common NMR experiments is 2D homonuclear correlation spectroscopy. In the resulting spectra, the diagonal corresponds to the common 1D spectrum and off-diagonal peaks result from the interactions among hydrogen atoms that are relatively closely spaced. A further common type of NMR experiment for proteins is the so-called 2D heteronuclear single-quantum correlation (HSQC) spectroscopy that provides a chemical shift correlation map between directly bonded ^{1}H and nuclei such as ^{15}N or ^{13}C. Thus, each signal in a [^{15}N-^{1}H] HSQC spectrum of a protein represents a single amino acid (Fig. 12). The amount of protein necessary to conduct an NMR investigation is similar to that needed for X-ray crystallography. Unlike with the latter technique, sample conditions (e.g., temperature, and pH) can be changed quickly with NMR. However, artifacts can arise because of the use of isolated fragments or domains in order to keep the size of the molecule within a manageable range. The most time-consuming step of an NMR experiment concerns resonance assignments and the acquisition of data necessary to achieve this can take weeks or months. By comparison, X-ray diffraction data collection and processing of data are now a matter of minutes and advances in derivative preparation and phasing approaches have cut down significantly the time required to determine a crystal structure. For a more detailed comparison between the two techniques, see the overview in Egli (2010).

NMR structures of Kai protein domains emerged early on in the investigation directed at the cyanobacterial circadian clock. An initial structure of the N-terminal domain of *S. elongatus* KaiA revealed a pseudoreceiver-like fold (Williams et al., 2002). Binding assays using KaiA domains and KaiC demonstrated that the contact to the latter is exclusively established via the KaiA C-terminal domain. The NMR structure of the N-terminal

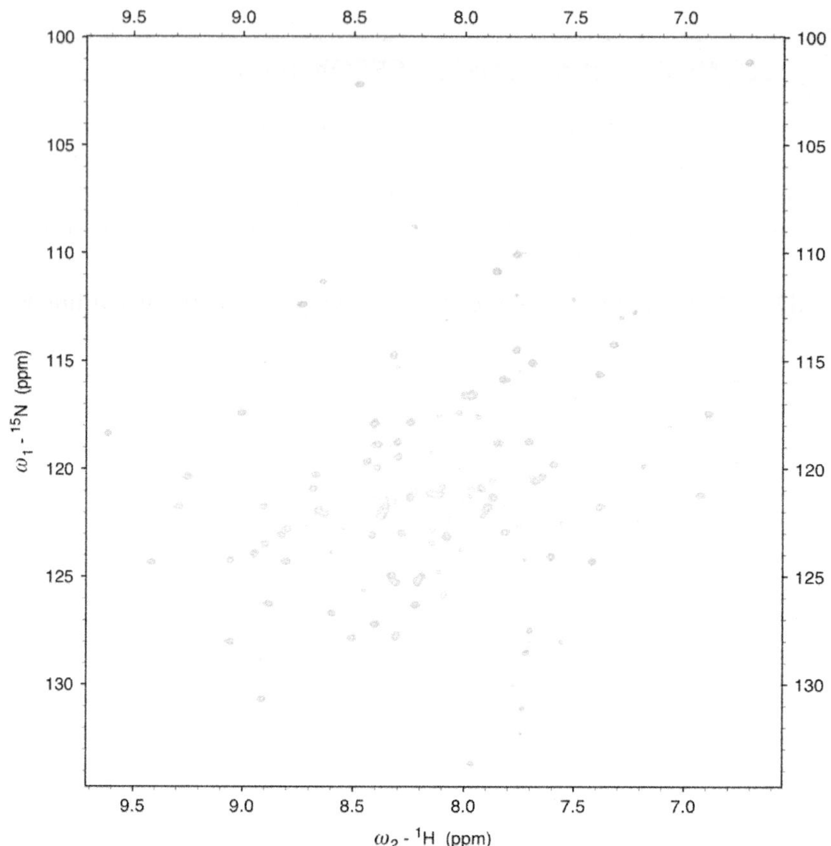

Figure 12 Two-dimensional [^{15}N-^{1}H] HSQC NMR spectrum for *S. elongatus* KaiB recorded on an 800-MHz spectrometer.

domain of the SasA histidine kinase that is involved in the clock output pathway displayed clear differences relative to the crystal structure of KaiB (Vakonakis, Klewer, Williams, Golden, & LiWang, 2004). At the sequence level, the two proteins from *S. elongatus* exhibit 28.6% identity and 55.2% similarity, and it had generally been expected that the structures would be closely related. However, no NMR solution structure of KaiB for any of the cyanobacterial strains has been reported to date. Based on the structural differences established between SasA and KaiB at the time, it appeared unlikely that the two proteins would compete for KaiC binding. More recent studies using native PAGE and ns EM have provided evidence that KaiB and SasA indeed compete for KaiC binding but exhibit divergent affinities for the central cog of the KaiABC clock (Pattanayek et al., 2011).

No consensus exists at the moment as to whether binding between KaiB (or SasA) and KaiC involves the N-terminal or C-terminal half of KaiC or both (Egli, 2014). NMR experiments provided a breakthrough in our understanding of the interaction between KaiA and KaiC. Accordingly, the C-terminal domains of KaiA dimer bind two C-terminal peptides that protrude from the surface of the KaiC hexamer (residues 488–518; *T. elongatus* protein). This binding mode results in the unraveling of so-called A-loops that rim the channel opening on the C-terminal side, a conformational change that can be transmitted to P-loop and ATP on the one hand (Kim et al., 2008) and phosphorylation sites on the other (Egli et al., 2013), consistent with the stimulation of KaiC phosphorylation by KaiA.

More recent NMR investigations of the Kai proteins concern dynamic aspects of the clock. Using separate N- and C-terminal domains from *T. elongatus* KaiC monomer, LiWang and coworkers showed that the flexibility of the C-terminal ring governs the rhythm of KaiC phosphorylation and dephosphorylation (Chang et al., 2011). Over the daily cycle, the C-terminal KaiC ring undergoes transitions from a flexible state (e.g., ST) to more rigid ones (e.g., pTpS, TpS). These changes in flexibility affect not only the interactions with KaiA and KaiB in the PTO but may also control clock output pathways (e.g., via bound SasA). Further NMR experiments with domains of KaiC monomer led to the identification of a KaiB binding site on the N-terminal half of KaiC that is apparently obscured in the KaiC hexamer (Chang et al., 2012). NMR also provides evidence that the complex between a FLAG-tagged, monomeric KaiC N-terminal domain and KaiB* (KaiB lacking its C-terminal tail and with two Tyr residues mutated to Ala in order to destabilize the dimer-of-dimers seen with wt-KaiB) is able to recruit KaiA that lacks the N-terminal domain (ΔN-KaiA) (Tseng et al., 2014). Although the observation of KaiA sequestration by monomeric fragments of KaiB and KaiC that are amenable to detailed NMR investigations alone or in complex is fascinating, it is unclear at the moment whether these structural studies can indeed capture the complexity of the interactions between full-length proteins in the ternary KaiABC complex with a MW of >500 kDa.

13. HYDROGEN–DEUTERIUM EXCHANGE

Replacement of covalently bound hydrogen by deuterium (H/D exchange or HDX) is a common approach to analyze the dynamics of protein folding and to probe solvent accessibility and protein–protein interactions (Englander & Kallenbach, 1983; Katta & Chait, 1991; Konermann, Pan, & Liu, 2011). The exchange is also the basis for neutron crystallography and SANS. HDX combined with tryptic digestion and electrospray ionization mass spectrometry (HDX-MS) permits tracking of conformational changes and mapping of protein–protein interfaces (Mandell, Baerga-Ortiz, Falick, & Komives, 2005; Wales & Engen, 2006). Thus, sites buried upon complex formation will be less or not accessible to HDX, allowing for their subsequent identification among peptides from protein digests. HDX-MS was used to analyze the KaiB:KaiC interaction in conjunction with native MS (van Duijn, 2010) to study the KaiB quaternary structure, i.e., the distribution among monomer, dimer, and tetramer (Snijder et al., 2014). The latter investigation demonstrated that the monomeric state is predominant at lower temperature and protein concentrations. The KaiB tetramer seen in KaiB crystals (Garces et al., 2004; Hitomi et al., 2005; Iwase et al., 2005; Pattanayek et al., 2008; Villarreal et al., 2013) and in solution by SAXS (Pattanayek et al., 2011) and light scattering (Pattanayek et al., 2012) may therefore not be the form that contacts the KaiC hexamer. The HDX-MS analysis revealed limited accessibility to deuterium at subunit interfaces in the C-terminal KaiC half and at the constricted waist, consistent with binding by KaiBs on the same side as KaiA and conformational adjustments upon complex formation at the interface between the N- and C-terminal KaiC rings (Snijder et al., 2014). In combination with computational simulations of complexes with six KaiB monomers bound to either the KaiC surface at the N-or C-terminal ends and comparison of MS cross-collisional section data, it was concluded that a $KaiC_6B_6$ complex with KaiBs bound to the C-terminal ring and covering ATP binding clefts constitutes the most likely scenario. Remarkably, the stoichiometry of the KaiB: KaiC interaction, the state of KaiB as it is bound to the KaiC hexamer, and the location of KaiB binding on KaiC are fully consistent with the results from the cryo-EM analysis that also mapped six KaiB monomers to the C-terminal end of KaiC (Villarreal et al., 2013). The latter study also took into account the consequences for KaiB binding of the KaiC R468C mutation (Xu et al., 2003). This mutation results in higher affinity of KaiB for KaiC and the residue maps to the binding interface in the $KaiC_6B_6$ model proposed based on EM (Villarreal et al., 2013). One difference between the HDX-MS and cryo-EM-derived models of the KaiBC complex concerns the docking approach. The former study relied on the HADDOCK software (Dominguez, Boelens, & Bonvin, 2003) that can make use of a wide variety of information to constrain the docking process for building models of protein–protein complexes (e.g., mutagenesis, NMR shift, and/or HDX data). The cryo-EM study used the Molecular Dynamics Flexible Fitting software (Trabuco, Villa, Schreiner, Harrison, & Schulten, 2009) to evaluate different locations and orientations of KaiB crystal structures on the KaiC hexamer within the EM-based electron density of the complex.

14. MD SIMULATIONS

MD simulations are a key component of many structural and biophysical approaches as indicated at the end of Section 13. Thus, simulated annealing can help overcome local minima in crystallographic refinement. Similarly, constrained refinement of NMR structural ensembles, SAXS-based models and cryo-EM structures often involves MD simulations. MD simulations provide detailed insight into the time-dependent behavior of a molecular system by allowing one to chart atomic fluctuations and conformational changes over a period of time (Durrant & McCammon, 2011; Fig. 13). Whereas X-ray crystallography and high-resolution cryo-EM furnish structural information at atomic resolution, MD simulations shed light on the dynamics of macromolecules and their complexes. Molecular mechanics calculations employ a force field that combines force constants, parameters (bonds, angles, torsion angles, nonbonded distances, etc.), and an energy function that together can be used to calculate the energy of a molecule. By comparison, MD simulates the atomic motions of a molecule by employing classical mechanics (Schwede & Peitsch, 2008). Thus, atoms are modeled as point charges of a certain mass that are under the influence of a force field. In subsequent steps, Newton's equation of motion is integrated at every time interval for each atom in the system to determine all their positions. Molecular motions occur at various time scales, i.e., femtoseconds to seconds (short-range motions, such as bond and angle stretching and side chain motions), nanoseconds to seconds (rigid-body motions, such as domain and subunit motions), and > μs (long-range motions, such as binding events and protein folding and unfolding). To be able to capture the short-range motions, the time step needs to be small enough; typically one time step equals 15 fs of simulation time. Similarly, to be able to observe the slower long-range motions, it is necessary to continue the simulation for as many steps as possible. However, the available computer time often sets limitations on how long a simulation can be continued for. Thus, a typical CPU consumes about 1 s of real time to compute about 1 fs of simulation

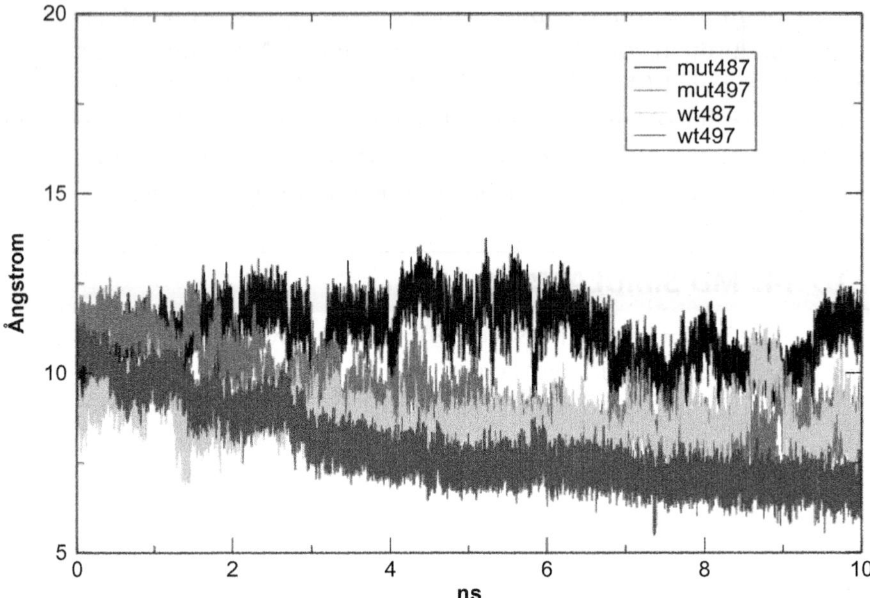

Figure 13 Variations in the distance between the Cα positions of Ser-417 and Gly-421 monitored over 10 ns of MD simulations for four KaiC hexamers: KaiC 1–497, KaiC 1–487 (lacking the so-called A-loop) and the 1–497 and 1–487 KaiCs with an additional A422V mutation. The 422-loop region encompassing residues 417–421 was hypothesized to become more flexible as a result of removal of the A-loop. Indeed, the black and green traces for the 1–487 systems without the A-loop indicate increased distance fluctuations. (See the color plate.)

time for a protein of around 200 amino acids length in a box of solvent molecules. Therefore, it follows that an MD simulations over ca. 5 ns requires about a month of computer time for a typical protein.

We carried out MD simulations for KaiC molecules with or without the A-loop regions (amino acids 487–497) to address the hypothesis that unraveling of the A-loop by KaiA (Vakonakis & LiWang, 2004) results in increased dynamics of selected regions of KaiC, consistent with stimulation of KaiC autophosphorylation by KaiA (Egli et al., 2013). We used the crystal structure of the KaiC hexamer (Pattanayek et al., 2006) as the start coordinates in combination with the AMBER software (Case et al., 2012; Götz et al., 2012), the parameters developed by Meagher and coworkers for the 12 ATP molecules (Meagher, Redman, & Carlson, 2003) and the SPC/E water model (Berendsen, Grigera, & Straatsma, 1987) with compatible ions. KaiC was hydrated with a total of 66,273 water molecules inside an orthorhombic box with an 11 Å clearance to the edges of the box.

Following minimization, the systems were equilibrated over the course of 450 ps under constant-volume conditions and finally the simulations were carried out for 25 ns, with 15 fs steps under constant pressure and no symmetry constraints applied (Egli et al., 2013). The KaiC hexamer, 6 × 497 amino acids (487 in the model without the A-loop) plus 12 ATPs and solvent, represents a rather large system for MD simulations, and we initially attempted to only include three subunits or limit the simulations to the C-terminal hexameric ring to conserve resources. However, neither system behaved in a stable manner and only the full hexameric particle remained stable. This is noteworthy because separate C-terminal domains do not form a stable hexamer (Hayashi, Iwase, Uzumaki, & Ishiura, 2006; Pattanayek et al., 2008). The MD simulations demonstrated that removal of the A-loop increases the root-mean-square deviation (rmsd) from the initial conformation relative to the KaiC hexamer with the A-loop present. In particular, a loop region that interacts with the A-loop and comprises residues 418–425, immediately adjacent to Thr-426, Ser-431, and Thr-432 that have all been shown to be targets of the autokinase, shows increased fluctuations upon A-loop removal (Fig. 13). As well, the absence of the A-loop leads to higher mobility of a neighboring β-strand and indirectly of the P-loop and ATP (Egli et al., 2013). Surprisingly, the lack of the A-loop region does not just cause changes in the mobility of residues that lie in the immediate vicinity of the loop, but altered fluctuations are seen also in the N-terminal ring. Overall, the outcome of the simulations is consistent with a concerted allosteric mechanism of KaiC phosphorylation that is stimulated by KaiA binding to a KaiC C-terminal tail and pulling on the A-loop from a subunit.

15. MODELING THE *IN VITRO* OSCILLATOR

Mathematical modeling can be a powerful method in conjunction with biochemical studies to provide verification of hypotheses and predictions for future experimental tests. Since the publication of the *in vitro* KaiABC rhythm, there have been many attempts to model this oscillator, e.g., reviewed in Johnson et al. (2011), and those presented in references (Byrne, 2009; Clodong et al., 2007; Emberly & Wingreen, 2006; Ito et al., 2007; Kageyama et al., 2006; Kurosawa, Aihara, & Iwasa, 2006; Li, Chen, Wang, & Wang, 2009; Li & Fang, 2007; Ma & Ranganathan, 2012; Mehra et al., 2006; Miyoshi, Nakayama, Kaizu, Iwasaki, & Tomita, 2007; Mori et al., 2007; Nagai, Terada, & Sasai, 2010; Paddock, Boyd, Adin, & Golden, 2013; Phong, Markson, Wilhoite, & Rust, 2013; Qin, Byrne,

Mori, et al., 2010; Qin, Byrne, Xu, Mori, & Johnson, 2010; Rust, Golden, & O'Shea, 2011; Takigawa-Imamura & Mochizuki, 2006; Teng, Mukherij, Moffitt, de Buyl, & O'Shea, 2013; van Zon, Lubensky, Altena, & ten Wolde, 2007; Wang, Xu, & Wang, 2009; Yang, Pando, Dong, Golden, & van Oudenaarden, 2010; Yoda, Eguchi, Terada, & Sasai, 2007; Zwicker, Lubensky, & ten Wolde, 2010). As a representative example, we proposed in 2007, a model that stochastically simulates the kinetics of KaiC hexamers and the degree of phosphorylation of each monomer in every hexamer (Mori et al., 2007). In addition to modeling the general phenomenon, that model addresses the question of how the various KaiC hexamers in a population stay in synch with each other to maintain a robust, high-amplitude oscillation *in vitro* over many days. Inspired by the experimental data of the Kondo lab (Ito et al., 2007; Kageyama et al., 2006), we incorporated phase-dependent KaiC monomer exchange as a mechanism for keeping the phosphorylation state of hexamers synchronized in the population, and our model accurately predicted the observed patterns of *in vitro* KaiC phosphorylation (Mori et al., 2007). Interestingly, monomer exchange was predicted by a modeling study before it was experimentally measured (Emberly & Wingreen, 2006) Other significant models have proposed a different mechanism for KaiC hexamer synchronization, namely synchronization by KaiA sequestration (Brettschneider et al., 2010; Clodong et al., 2007; Rust et al., 2007; van Zon et al., 2007). There is clear experimental evidence for both models (Brettschneider et al., 2010; Ito et al., 2007; Mori et al., 2007; Qin, Byrne, Mori, et al., 2010; Rust et al., 2007), and it is likely that KaiA sequestration may act in concert with monomer exchange to accomplish the synchrony of KaiC phosphorylation that enables the robust high-amplitude rhythms for many cycles *in vitro* (Ito et al., 2007). We have also generated a combined model in which monomer exchange is a mechanism for maintaining phase synchrony among KaiC hexamers while KaiA sequestration is involved in the "switch" from autokinase to autophosphatase mode (Qin, Byrne, Mori, et al., 2010). Finally, modeling studies are beginning to address how the cyanobacterial pacemaker may regulate gene expression (Paddock et al., 2013; Qin, Byrne, Xu, et al., 2010; Teng et al., 2013; Zwicker et al., 2010), metabolism (Hellweger, 2010; Rust et al., 2011), and cell division (Yang et al., 2010).

16. SUMMARY AND OUTLOOK

The KaiABC PTO has been analyzed in significant detail both at the biophysical and structural levels with a wide range of methods as indicated in this chapter. Analyses directed at the association of Kai proteins demonstrate that the cyanobacterial inner timer is a dynamic nanomachine that exhibits changing Kai protein–protein interactions over the daily cycle (Kageyama et al., 2006; Mori et al., 2007). Not only do the compositions of complexes vary (i.e., binary KaiAC and KaiBC complexes and the ternary KaiABC complex), but subunit interactions undergo transformations as well, i.e., KaiC subunit shuffling (Kageyama et al., 2006; Mori et al., 2007) and KaiB tetramer to monomer conversion (Snijder et al., 2014; Villarreal et al., 2013). Moreover, the concentrations of individual Kai proteins and protein complexes oscillate over the daily period (Akiyama et al., 2008; Mori et al., 2007), and the protein–protein interfaces for the KaiA:KaiC interaction (Mori et al., 2007; Pattanayek et al., 2006, 2011; Qin, Byrne, Mori, et al., 2010; Vakonakis & LiWang, 2004) and possibly for the KaiB:KaiC interaction (Chang et al., 2011, 2012; Egli, 2014; Mutoh, Nishimura, Yasui, Onai, & Ishiura, 2013; Pattanayek et al., 2008, 2013; Tseng et al., 2014) undergo changes as KaiC cycles from the hypo to the hyper and back to the hypophosphorylated states. Correlating the dynamic formation of heteromultimeric complexes with models of 3D structure determinations using a hybrid structural approach involving EM, NMR, SAXS, X-ray, and a host of auxiliary approaches has turned out to be a significant challenge. Using ns EM, four particle shapes could be differentiated, i.e., KaiC hexamer and the KaiAC, KaiBC, and KaiABC complexes (Mori et al., 2007). EM and SAXS have provided 3D models of three complexes between full-length Kai proteins, but the resolution limits (maximal 16 Å) have not allowed a detailed visualization of the protein–protein interfaces (Egli, 2014; Pattanayek et al., 2008, 2006, 2011). Cryo-EM has thus far not provided high-resolution images of Kai complexes, and it appears that X-ray crystallography and NMR are the methods of choice for gaining insight into the dynamic protein–protein interactions between KaiA, B, and C at atomic resolution. However, only crystallography may allow visualization at that level of complexes involving the KaiC hexamer, with NMR investigations providing useful insights into interactions between monomeric domains (Chang et al., 2011, 2012; Tseng et al., 2014). Beyond the structural realm, there remain other phenomena governing Kai clock protein interactions whose origin and mechanism remain to be explored, among them temperature compensation (Murakami et al., 2008; Terauchi et al., 2007), a fundamental property of all biological clocks (Dunlap, Loros, & DeCoursey, 2004).

Although I have attempted to be as inclusive as possible in terms of the overview presented here of structural and biophysical approaches deployed to analyze the KaiABC PTO, there remain of course additional techniques that could prove useful in the dissection of clock structure and mechanism. The most quantitative approach for determining the thermodynamic properties of protein–protein interactions is isothermal titration calorimetry (ITC; Pierce, Raman, & Nall, 1999), as it can provide information regarding stoichiometry, enthalpy, entropy, and binding kinetics for interacting proteins in solution without the need to immobilize or label them. We have used ITC in a preliminary fashion to measure changes in stability of the KaiC hexamer as a consequence of point mutations, but ultimately relied on circular dichroism melting experiments to just extract melting temperatures (Egli et al., 2013). Chemical cross-linking and mass spectrometry (XL-MS) is a technique that can help identify proximate amino acids and define the binding interface of protein–protein complexes, but for some reason cross-linking approaches have not been reported for the Kai system. XL-MS is now becoming an integral part of the hybrid structural biology approach for the analysis of protein complexes (Herzog et al., 2012). Provided a suitable cross-linking agent can be found for a particular complex, e.g., by screening of various bifunctional molecules with linkers of various lengths and reactive end groups, SDS-PAGE of cross-linked complexes, followed by proteolytic digestion (trypsin) of products from gel bands, and mass-spectrometric analysis of crosslinked peptides (XL-MS) can then be used to map the protein–protein interface (Leitner et al., 2010; Young et al., 2000).

A further approach that can shed light on protein interactions, the degree of proximity between partners, conformational changes and dynamics is electron paramagnetic resonance spectroscopy (Hubbell, Gross, Langen, & Lietzow, 1998; Mchaourab, Steed, & Kazmier, 2011). The KaiC protein from *S. elongatus* features three cysteins that are all located in the C-terminal half of hexamer subunits. Because they appear to be tucked away and therefore less reactive to labeling with a spin probe, introduction of a Cys mutation could be used to attach a label site specifically, e.g., on the N- or C-terminal surface, to probe the possibility that KaiB can bind on either side of the hexamer (Chang et al., 2012; Egli, 2014; Pattanayek et al., 2008, 2013; Tseng et al., 2014). Further, labeling two Kai proteins could furnish distance information for docking of crystal structures inside EM densities or SAXS envelopes, although this could prove very challenging owing to the hexameric nature of KaiC that will give rise to multiple distance pairs or rather distance distributions because of the intrinsic, environment-dependent mobility of spin labels. A related approach, site-directed spin labeling electron spin resonance analysis, was applied to track down a transient interaction between KaiA and KaiB (Mutoh et al., 2010).

Finally, it could be interesting to apply single-molecule biophysics, e.g., pulling experiments (Aubin-Tam, Olivares, Sauer, Baker, & Lang, 2011), toward a more in-depth investigation of the stability, dynamics, and mechanism of the KaiA:KaiC interaction. Single-molecule approaches can give information about heterogeneous kinetics of molecules in the population of molecules. For example, it is likely that the various KaiC hexamers in the population do not all behave exactly the same, but that is the assumption of most of the biochemical analyses to date. Thus, such experiments may yield the forces necessary to pull out the KaiC C-terminal tail and the unraveling of the A-loop, processes that appear to form the basis for KaiA to stimulate KaiC phosphorylation, and in which different KaiC hexamers (e.g., in different phosphorylation states) may exhibit different kinetics. Another approach that potentially applies single-molecule advantages with the visualization capability of EM is atomic force microscopy (AFM). AFM is high-resolution scanning probe microscopy, with resolutions demonstrated down to subnanometer levels. High-speed AFM has been used to great effect on proteins similar in size and shape to KaiC, e.g., F1-ATPase where rotary catalysis and conformational changes that proceed around the hexameric ring were demonstrated (Uchihashi, Lino, Ando, & Noji, 2011).

ACKNOWLEDGMENTS

This work is supported in part by NIH Grant R01 GM073845. I am grateful to Dr. Carl H. Johnson, Vanderbilt University, for many years of a fruitful collaboration between our labs on research directed at the structure and function of the cyanobacterial circadian clock, insights regarding applications of fluorescence anisotropy and FRET as well as mathematical modeling to studies of the KaiABC oscillator and for helpful comments on the manuscript.

REFERENCES

Akiyama, S., Nohara, A., Ito, K., & Maeda, Y. (2008). Assembly and disassembly dynamics of the cyanobacterial periodosome. *Molecular Cell*, *29*, 703–716.

Amunts, A., Brown, A., Bai, X., Llacer, J. L., Hussain, T., Emsley, P., et al. (2014). Structure of the yeast mitochondrial large ribosomomal subunit. *Nature*, *343*, 1485–1489.

Aubin-Tam, M.-E., Olivares, A. O., Sauer, R. T., Baker, T. A., & Lang, M. J. (2011). Single-molecule protein unfolding and translocation by an ATP-fueled proteolytic machine. *Cell*, *145*, 257–267.

Berendsen, H. J. C., Grigera, J. R., & Straatsma, T. P. (1987). The missing term in effective pair potentials. *Journal of Physical Chemistry*, *91*, 6269–6271.

Brettschneider, C., Rose, R. J., Hertel, S., Axmann, I. M., Heck, A. J., & Kollmann, M. (2010). A sequestration feedback determines dynamics and temperature entrainment of the KaiABC circadian clock. *Molecular Systems Biology, 6*, 389.

Byrne, M. (2009). Mathematical modeling of the in vitro cyanobacterial circadian oscillator. In J. L. Ditty, S. R. Mackey, & C. H. Johnson (Eds.), *Bacterial circadian programs: Vol. 16.* (pp. 283–300). Berlin: Springer Publishers.

Case, D. A., Darden, T. A., Cheatham, T. E., III, Simmerling, C. L., Wang, J., Duke, R. E., et al. (2012). *AMBER 12*. San Francisco: University of California.

Chang, Y. G., Kuo, N. W., Tseng, R., & LiWang, A. (2011). Flexibility of the C-terminal, or CII, ring of KaiC governs the rhythm of the circadian clock of cyanobacteria. *Proceedings of the National Academy of Sciences of the United States of America, 108*, 14431–14436.

Chang, Y. G., Tseng, R., Kuo, N. W., & LiWang, A. (2012). Rhythmic ring-ring stacking drives the circadian oscillator clockwise. *Proceedings of the National Academy of Sciences of the United States of America, 109*, 16847–16851.

Clodong, S., Düring, U., Kronk, L., Axmann, I., Wilde, A., Herzel, H., et al. (2007). Functioning and robustness of a bacterial circadian clock. *Molecular Systems Biology, 3*, 90.

Czarna, A., Berndt, A., Singh, H. R., Grudziecki, A., Ladurner, A. G., Timinszky, G., et al. (2013). Structures of *Drosophila* cryptochrome and mouse cryptochrome1 provide insight into circadian function. *Cell, 153*, 1394–1405.

Dominguez, C., Boelens, R., & Bonvin, A. M. J. J. (2003). HADDOCK: A protein–protein docking approach based on biochemical and/or biophysical information. *Journal of the American Chemical Society, 125*, 1731–1737.

Dunlap, J. C., Loros, J. J., & DeCoursey, P. J. (Eds.). (2004). *Chronobiology: Biological timekeeping*. Sunderland, MA: Sinauer Associates, Inc., Publishers.

Durrant, J. D., & McCammon, J. A. (2011). Molecular dynamics simulations and drug discovery. *BMC Biology, 9*, 71–79.

Egli, M. (2010). Diffraction techniques in structural biology. *Current Protocols in Nucleic Acid Chemistry, 41*, 7.13.1–7.13.35.

Egli, M. (2014). Intricate protein–protein interactions in the cyanobacterial circadian clock. *Journal of Biological Chemistry, 289*, 21267–21275.

Egli, M., & Johnson, C. H. (2013). A circadian clock nanomachine that runs without transcription or translation. *Current Opinion in Neurobiology, 23*, 732–740.

Egli, M., Mori, T., Pattanayek, R., Xu, Y., Qin, X., & Johnson, C. H. (2012). Dephosphorylation of the core clock protein KaiC in the cyanobacterial KaiABC circadian oscillator proceeds via an ATP synthase mechanism. *Biochemistry, 51*, 1547–1558.

Egli, M., Pattanayek, R., Sheehan, J. H., Xu, Y., Mori, T., Smith, J. A., et al. (2013). Loop–loop interactions regulate KaiA-stimulated KaiC phosphorylation in the cyanobacterial KaiABC circadian clock. *Biochemistry, 52*, 1208–1220.

Emberly, E., & Wingreen, N. S. (2006). Hourglass model for a protein-based circadian oscillator. *Physical Reviews Letters, 96*, 0383003.

Englander, S. W., & Kallenbach, N. R. (1983). Hydrogen exchange and structural dynamics of proteins and nucleic acids. *Quarterly Reviews of Biophysics, 16*, 521–655.

Garces, R. G., Wu, N., Gillon, W., & Pai, E. F. (2004). Anabaena circadian clock proteins KaiA and KaiB reveal potential common binding site to their partner KaiC. *EMBO Journal, 23*, 1688–1698.

Götz, A. W., Williamson, M. J., Xu, D., Poole, D., Le Grand, S., & Walker, R. C. (2012). Routine microsecond molecular dynamics simulations with AMBER on GPUs. 1. Generalized born. *Journal of Chemical Theory and Computation, 8*, 1542–1555.

Harp, J. M., Hanson, B. F., Timm, D. E., & Bunick, G. J. (1999). Macromolecular crystal annealing: Evaluation of techniques and variables. *Acta Crystallographica Section D, 55*, 1329–1334.

Hayashi, F., Ito, H., Fujita, M., Iwase, R., Uzumaki, T., & Ishiura, M. (2004). Stoichiometric interactions between cyanobacterial clock proteins KaiA and KaiC. *Biochemical and Biophysical Research Communications*, *316*, 195–202.

Hayashi, F., Iwase, R., Uzumaki, T., & Ishiura, M. (2006). Hexamerization by the N-terminal domain and intersubunit phosphorylation by the C-terminal domain of cyanobacterial circadian clock protein KaiC. *Biochemical and Biophysical Research Communications*, *318*, 864–872.

Hayashi, F., Suzuki, H., Iwase, R., Uzumaki, T., Miyake, A., Shen, J.-R., et al. (2003). ATP-induced hexameric ring structure of the cyanobacterial circadian clock protein KaiC. *Genes to Cells*, *8*, 287–296.

Hellweger, F. L. (2010). Resonating circadian clocks enhance fitness in cyanobacteria *in silico*. *Ecological Modelling*, *221*, 1620–1629.

Hennig, S., Strauss, H. M., Vanselow, K., Yildiz, O., Schulze, S., Arens, J., et al. (2009). Structural and functional analyses of PAS domain interactions of the clock proteins Drosophila PERIOD and mouse PERIOD2. *PLoS Biology*, *7*, e94.

Herzog, F., Kahraman, A., Boehringer, D., Mak, R., Bracher, A., Walzthoeni, T., et al. (2012). Structural probing of a protein phosphatase 2A network by chemical crosslinking and mass spectrometry. *Science*, *337*, 1348–1352.

Hitomi, K., Oyama, T., Han, S., Arvai, A. S., & Getzoff, E. D. (2005). Tetrameric architecture of the circadian clock protein KaiB: A novel interface for intermolecular interactions and its impact on the circadian rhythm. *Journal of Biological Chemistry*, *280*, 18643–18650.

Huang, N., Chelliah, Y., Shan, Y., Taylor, C. A., Yoo, S.-H., Partch, C., et al. (2012). Crystal structure of the heterodimeric CLOCK:BMAL1 transcriptional activator complex. *Science*, *337*, 189–194.

Hubbell, W. L., Gross, A., Langen, R., & Lietzow, M. A. (1998). Recent advances in site-directed spin labeling of protein. *Current Opinion in Structural Biology*, *8*, 649–656.

Ishiura, M., Kutsuna, S., Aoki, S., Iwasaki, H., Andersson, C. R., Tanabe, A., et al. (1998). Expression of a gene cluster *kaiABC* as a circadian feedback process in cyanobacteria. *Science*, *281*, 1519–1523.

Ito, H., Kageyama, H., Mutsuda, M., Nakajima, M., Oyama, T., & Kondo, T. (2007). Autonomous synchronization of the circadian KaiC phosphorylation rhythm. *Nature Structural and Molecular Biology*, *14*, 1084–1088.

Iwasaki, H., Nishiwaki, T., Kitayama, Y., Nakajima, M., & Kondo, T. (2002). KaiA-stimulated KaiC phosphorylation in circadian timing loops in cyanobacteria. *Proceedings of the National Academy of Sciences of the United States of America*, *99*, 15788–15793.

Iwase, R., Imada, K., Hayashi, F., Uzumaki, T., Morishita, M., Onai, K., et al. (2005). Functionally important substructures of circadian clock protein KaiB in a unique tetramer complex. *Journal of Biological Chemistry*, *280*, 43141–43149.

Jacrot, B. (1976). The study of biological structures by neutron scattering from solution. *Reports on Progress in Physics*, *39*, 911–953.

Johnson, C. H., Egli, M., & Stewart, P. L. (2008). Structural insights into a circadian oscillator. *Science*, *322*, 697–701.

Johnson, C. H., Stewart, P. L., & Egli, M. (2011). The cyanobacterial circadian system: From biophysics to bioevolution. *Annual Review of Biophysics*, *40*, 143–167.

Kageyama, H., Nishiwaki, T., Nakajima, M., Iwasaki, H., Oyama, T., & Kondo, T. (2006). Cyanobacterial circadian pacemaker: Kai protein complex dynamics in the KaiC phosphorylation cycle *in vitro*. *Molecular Cell*, *23*, 161–171.

Katta, V., & Chait, B. T. (1991). Conformational changes in proteins probed by hydrogen-exchange electrospray-ionization mass spectrometry. *Rapid Communications in Mass Spectrometry*, *5*, 214–217.

Keeler, J. (2005). *Understanding NMR* (1st ed.). Hoboken, NJ: John Wiley & Sons.

Kim, Y. I., Dong, G., Carruthers, C. W., Jr., Golden, S. S., & LiWang, A. (2008). The day/night switch in KaiC, a central oscillator component of the circadian clock of cyanobacteria. *Proceedings of the National Academy of Sciences of the United States of America, 105*, 12825–12830.

Kitayama, Y., Iwasaki, H., Nishiwaki, T., & Kondo, T. (2003). KaiB functions as an attenuator of KaiC phosphorylation in the cyanobacterial circadian clock system. *EMBO Journal, 22*, 1–8.

Kitayama, K., Nishiwaki-Ohkawa, T., Sugisawa, Y., & Kondo, T. (2014). KaiC intersubunit communication facilitates robustness of circadian rhythms in cyanobacteria. *Nature Communications, 4*, 3897.

Konermann, L., Pan, J., & Liu, Y. H. (2011). Hydrogen exchange mass spectrometry for studying protein structure and dynamics. *Chemical Society Reviews, 40*, 1223–1234.

Kucera, N., Schmalen, I., Hennig, S., Ölliger, R., Strauss, H. M., Grudziecki, A., et al. (2012). Unwinding the differences of the mammalian PERIOD clock proteins from crystal structure to cellular function. *Proceedings of the National Academy of Sciences of the United States of America, 109*, 3311–3316.

Kurosawa, G., Aihara, K., & Iwasa, Y. (2006). A model for the circadian rhythm of cyanobacteria that maintains oscillation without gene expression. *Biophysical Journal, 91*, 2015–2023.

Leipe, D. D., Aravind, L., Grishin, N. V., & Koonin, E. V. (2000). The bacterial replicative helicase DnaB evolved from a RecA duplication. *Genome Research, 10*, 5–16.

Leitner, A., Walzthoeni, T., Kahraman, A., Herzog, F., Rinner, O., Beck, M., et al. (2010). Probing native protein structures by chemical cross-linking, mass spectrometry, and bioinformatics. *Molecular and Cellular Proteomics, 9*, 1634–1649.

Li, C., Chen, X., Wang, P., & Wang, W. (2009). Circadian KaiC phosphorylation: A multilayer network. *PLoS Computational Biology, 5*, e1000568.

Li, S., & Fang, Y. H. (2007). Modelling circadian rhythms of protein KaiA, KaiB and KaiC interactions in cyanobacteria. *Biological Rhythm Research, 38*, 43–53.

Ma, L., & Ranganathan, R. (2012). Quantifying the rhythm of KaiB–C interaction for in vitro cyanobacterial circadian clock. *PLoS One*, 7, e42581.

Mandell, J. G., Baerga-Ortiz, A., Falick, A. M., & Komives, E. A. (2005). Measurement of solvent accessibility at protein–protein interfaces. *Methods in Molecular Biology, 305*, 65–80.

Markson, J. S., Piechura, J. R., Puszynska, A. M., & O'Shea, E. K. (2013). Circadian control of global gene expression by the cyanobacterial master regulator RpaA. *Cell, 155*, 1396–1408.

Mchaourab, H. S., Steed, P. R., & Kazmier, K. (2011). Toward the fourth dimension of membrane protein structure: Insights into dynamics from spin-labeling EPR spectroscopy. *Structure, 19*, 1549–1561.

Meagher, K. L., Redman, L. T., & Carlson, H. A. (2003). Development of polyphosphate parameters for use with the AMBER force field. *Journal of Computational Chemistry, 24*, 1016–1025.

Mehra, A., Hong, C., Shi, M., Loros, J., Dunlap, J., & Ruoff, P. (2006). Circadian rhythmicity by autocatalysis. *PLoS Computational Biology, 2*, e96.

Meilleur, F., Weiss, K. L., & Myles, D. A. A. (2009). Deuterium labeling for neutron structure–function–dynamics analysis. *Methods in Molecular Biology, 544*, 281–292.

Miyoshi, F., Nakayama, Y., Kaizu, K., Iwasaki, H., & Tomita, M. (2007). A mathematical model for the Kai-protein-based chemical oscillator and clock gene expression rhythms in cyanobacteria. *Journal of Biological Rhythms, 22*, 69–80.

Mori, T., Saveliev, S. V., Xu, Y., Stafford, W. F., Cox, M. M., Inman, R. B., et al. (2002). Circadian clock protein KaiC forms ATP-dependent hexameric rings and binds DNA. *Proceedings of the National Academy of Sciences of the United States of America, 99*, 17203–17208.

Mori, T., Williams, D. R., Byrne, M. O., Qin, X., Egli, M., McHaourab, H. S., et al. (2007). Elucidating the ticking of an *in vitro* circadian clockwork. *PLoS Biology, 5*, e93.

Murakami, R., Miyake, A., Iwase, R., Hayashi, F., Uzumaki, T., & Ishiura, M. (2008). ATPase activity and its temperature compensation of the cyanobacterial clock protein KaiC. *Genes to Cells, 13*, 387–395.

Murayama, Y., Mukaiyama, A., Imai, K., Onoue, Y., Tsunoda, A., Nohara, A., et al. (2011). Tracking and visualizing the circadian ticking of the cyanobacterial clock protein KaiC in solution. *EMBO J, 30*, 68–78.

Mutoh, R., Mino, H., Murakami, R., Uzumaki, T., Takabayashi, A., Ishii, K., et al. (2010). Direct interaction between KaiA and KaiB revealed by a site-directed spin labeling electron spin resonance analysis. *Genes to Cells, 15*, 269–280.

Mutoh, R., Nishimura, A., Yasui, S., Onai, K., & Ishiura, M. (2013). The ATP-mediated regulation of KaiB–KaiC interaction in the cyanobacterial circadian clock. *PLoS One, 8*, e80200.

Nagai, T., Terada, T. P., & Sasai, M. (2010). Synchronization of circadian oscillation of phosphorylation level of KaiC in vitro. *Biophysical Journal, 98*, 2469–2477.

Nakajima, M., Imai, K., Ito, H., Nishiwaki, T., Murayama, Y., Iwasaki, H., et al. (2005). Reconstitution of circadian oscillation of cyanobacterial KaiC phosphorylation in vitro. *Science, 308*, 414–415.

Nishiwaki, T., & Kondo, T. (2012). Circadian autodephosphorylation of cyanobacterial clock protein KaiC occurs via formation of ATP as intermediate. *Journal of Biological Chemistry, 287*, 18030–18035.

Nishiwaki, T., Satomi, Y., Kitayama, Y., Terauchi, K., Kiyohara, R., Takao, T., et al. (2007). A sequential program of dual phosphorylation of KaiC as a basis for circadian rhythm in cyanobacteria. *EMBO Journal, 26*, 4029–4037.

Nishiwaki, T., Satomi, Y., Nakajima, M., Lee, C., Kiyohara, R., Kageyama, H., et al. (2004). Role of KaiC phosphorylation in the circadian clock system of Synechococcus elongatus PCC 7942. *Proceedings of the National Academy of Sciences of the United States of America, 101*, 13927–13932.

Paddock, M. L., Boyd, J. S., Adin, D. M., & Golden, S. S. (2013). Active output state of the Synechococcus Kai circadian oscillator. *Proceedings of the National Academy of Sciences of the United States of America, 110*, E3849–E3857.

Pattanayek, R., Mori, T., Xu, Y., Pattanayek, S., Johnson, C. H., & Egli, M. (2009). Structures of KaiC circadian clock mutant proteins: A new phosphorylation site at T426 and mechanisms of kinase, ATPase and phosphatase. *PLoS One, 4*, e7529.

Pattanayek, R., Sidiqi, S. K., & Egli, M. (2012). Crystal structure of the redox-active cofactor DBMIB bound to circadian clock protein KaiA and structural basis for DBMIB's ability to prevent stimulation of KaiC phosphorylation by KaiA. *Biochemistry, 51*, 8050–8052.

Pattanayek, R., Wang, J., Mori, T., Xu, Y., Johnson, C. H., & Egli, M. (2004). Visualizing a circadian clock protein: Crystal structure of KaiC and functional insights. *Molecular Cell, 15*, 375–388.

Pattanayek, R., Williams, D. R., Pattanayek, S., Mori, T., Johnson, C. H., Stewart, P. L., et al. (2008). Structural model of the circadian clock KaiB–KaiC complex and mechanism for modulation of KaiC phosphorylation. *EMBO Journal, 27*, 1767–1778.

Pattanayek, R., Williams, D. R., Pattanayek, S., Xu, Y., Mori, T., Johnson, C. H., et al. (2006). Analysis of KaiA–KaiC protein interactions in the cyano-bacterial circadian clock using hybrid structural methods. *EMBO Journal, 25*, 2017–2028.

Pattanayek, R., Williams, D. R., Rossi, G., Weigand, S., Mori, T., Johnson, C. H., et al. (2011). Combined SAXS/EM based models of the S. elongatus post-translational circadian oscillator and its interactions with the output His-kinase SasA. *PLoS One, 6*, e23697.

Pattanayek, R., Xu, Y., Lamichhane, A., Johnson, C. H., & Egli, M. (2014). An arginine tetrad as mediator of input-dependent and input-independent ATPases in the clock protein KaiC. *Acta Crystallographica Section D*, *70*, 1375–1390.

Pattanayek, R., Yadagirib, K. K., Ohi, M. D., & Egli, M. (2013). Nature of KaiB–KaiC binding in the cyanobacterial circadian oscillator. *Cell Cycle*, *12*, 810–817.

Petoukhov, M. V., & Svergun, D. I. (2005). Global rigid body modeling of macromolecular complexes against small-angle scattering data. *Biophysical Journal*, *89*, 1237–1250.

Phong, C., Markson, J. S., Wilhoite, C. M., & Rust, M. J. (2013). Robust and tunable circadian rhythms from differentially sensitive catalytic domains. *Proceedings of the National Academy of Sciences of the United States of America*, *110*, 1124–1129.

Pierce, M. M., Raman, C. S., & Nall, B. T. (1999). Isothermal titration calorimetry of protein–protein interactions. *Methods*, *19*, 213–221.

Putnam, C. D., Hammel, M., Hura, G. L., & Tainer, J. A. (2007). X-ray solution scattering (SAXS) combined with crystallography and computation: Defining accurate macromolecular structures, conformations and assemblies in solution. *Quarterly Reviews of Biophysics*, *40*, 191–285.

Qin, X., Byrne, M., Mori, T., Zou, P., Williams, D. R., Mchaourab, H., et al. (2010). Intermolecular associations determine the dynamics of the circadian KaiABC oscillator. *Proceedings of the National Academy of Sciences of the United States of America*, *107*, 14805–14810.

Qin, X., Byrne, M., Xu, Y., Mori, T., & Johnson, C. H. (2010). Coupling of a core post-translational pacemaker to a slave transcription/translation feedback loop in a circadian system. *PLoS Biology*, *8*, e1000394.

Rambo, R. P., & Tainer, J. A. (2013). Accurate assessment of mass, models and resolution by small-angle scattering. *Nature*, *496*, 477–481.

Reddy, V. S., Natchiar, S. K., Stewart, P. L., & Nemerow, G. R. (2010). Crystal structure of human adenovirus at 3.5 Å resolution. *Science*, *329*, 1071–1075.

Rust, M. J., Golden, S. S., & O'Shea, E. K. (2011). Light-driven changes in energy metabolism directly entrain the cyanobacterial circadian oscillator. *Science*, *331*, 220–223.

Rust, M. J., Markson, J. S., Lane, W. S., Fisher, D. S., & O'Shea, E. K. (2007). Ordered phosphorylation governs oscillation of a three-protein circadian clock. *Science*, *318*, 809–812.

Schmalen, I., Reischl, S., Wallach, T., Klemz, R., Grudziecki, A., Prabu, J. R., et al. (2014). Interaction of circadian clock proteins CRY1 and PER2 is modulated by zinc binding and disulfide bond formation. *Cell*, *157*, 1203–1215.

Schwede, T., & Peitsch, M. C. (Eds.). (2008). *Computational structural biology: Methods and applications*. Hackensack, NJ: World Scientific Publishing Co.

Snijder, J., Burnley, R. J., Wiegard, A., Melquiond, A. S. J., Bonvin, A. M. J. J., Axmann, I. M., et al. (2014). Insight into cyanobacterial circadian timing from structural details of the KaiB–KaiC interaction. *Proceedings of the National Academy of Sciences of the United States of America*, *111*, 1379–1384.

Takigawa-Imamura, H., & Mochizuki, A. (2006). Predicting regulation of the phosphorylation cycle of KaiC clock protein using mathematical analysis. *Journal of Biological Rhythms*, *21*, 405–416.

Teng, S. W., Mukherij, S., Moffitt, J. R., de Buyl, S., & O'Shea, E. K. (2013). Robust circadian oscillations in growing cyanobacteria require transcriptional feedback. *Science*, *340*, 737–740.

Terauchi, K., Kitayama, Y., Nishiwaki, T., Miwa, K., Murayama, Y., Oyama, T., et al. (2007). The ATPase activity of KaiC determines the basic timing for circadian clock of cyanobacteria. *Proceedings of the National Academy of Sciences of the United States of America*, *104*, 16377–16381.

Trabuco, L. G., Villa, E., Schreiner, E., Harrison, C. B., & Schulten, K. (2009). Molecular dynamics flexible fitting: A practical guide to combine cryo-electron microscopy and X-ray crystallography. *Methods, 49*, 174–180.

Tseng, R., Chang, Y.-G., Bravo, I., Latham, R., Chaudhary, A., Kuo, N.-W., et al. (2014). Cooperative KaiA–KaiB–KaiC interactions affect KaiB/SasA competition in the circadian clock of cyanobacteria. *Journal of Molecular Biology, 426*, 389–402.

Uchihashi, T., Lino, R., Ando, T., & Noji, H. (2011). High-speed atomic force microscopy reveals rotary catalysis of rotorless F_1-ATPase. *Science, 333*, 755–758.

Vakonakis, I., Klewer, D. A., Williams, S. B., Golden, S. S., & LiWang, A. C. (2004). Structure of the N-terminal domain of the circadian clock-associated histidine kinase SasA. *Journal of Molecular Biology, 342*, 9–17.

Vakonakis, I., & LiWang, A. C. (2004). Structure of the C-terminal domain of the clock protein KaiA in complex with a KaiC-derived peptide: Implications for KaiC regulation. *Proceedings of the National Academy of Sciences of the United States of America, 101*, 10925–10930.

van Duijn, E. (2010). Current limitations in native mass spectrometry based structural biology. *Journal of the American Society for Mass Spectrometry, 21*, 971–978.

van Zon, J. S., Lubensky, D. K., Altena, P. R., & ten Wolde, P. R. (2007). An allosteric model of circadian KaiC phosphorylation. *Proceedings of the National Academy of Sciences of the United States of America, 104*, 7420–7425.

Villarreal, S. A., Pattanayek, R., Williams, D. R., Mori, T., Qin, X., Johnson, C. H., et al. (2013). CryoEM and molecular dynamics of the circadian KaiB–KaiC complex indicates KaiB monomers interact with KaiC and block ATP binding clefts. *Journal of Molecular Biology, 425*, 3311–3324.

Volkov, V. V., & Svergun, D. I. (2003). Uniqueness of ab initio shape determination in small-angle scattering. *Journal of Applied Crystallography, 36*, 860–864.

Wales, T. E., & Engen, J. R. (2006). Hydrogen exchange mass spectrometry for the analysis of protein dynamics. *Mass Spectrometry Reviews, 25*, 158–170.

Wang, J., Xu, L., & Wang, E. (2009). Robustness and coherence of a three-protein circadian oscillator: Landscape and flux perspectives. *Biophysical Journal, 97*, 3038–3046.

Whitten, A. E., Cai, S., & Trewhella, J. (2008). MULCh: Modules for the analysis of small-angle neutron contrast variation data from biomolecular assemblies. *Journal of Applied Crystallography, 41*, 222–226.

Williams, S. B., Vakonakis, I., Golden, S. S., & LiWang, A. C. (2002). Structure and function from the circadian clock protein KaiA of *Synechococcus elongatus*: A potential clock input mechanism. *Proceedings of the National Academy of Sciences of the United States of America, 99*, 15357–15362.

Wood, T. L., Bridwell-Rabb, J., Kim, Y.-I., Gao, T., Chang, Y.-G., LiWang, A., et al. (2010). The KaiA protein of the cyanobacterial circadian oscillator is modulated by a redox-active cofactor. *Proceedings of the National Academy of Sciences of the United States of America, 107*, 5804–5809.

Wüthrich, K. (1986). *NMR of proteins and nucleic acids* (1st ed.). New York: John Wiley & Sons.

Xu, Y., Mori, T., & Johnson, C. H. (2003). Cyanobacterial circadian clockwork: Roles of KaiA, KaiB, and the kaiBC promoter in regulating KaiC. *EMBO Journal, 22*, 2117–2126.

Xu, Y., Mori, T., Pattanayek, R., Pattanayek, S., Egli, M., & Johnson, C. H. (2004). Identification of key phosphorylation sites in the circadian clock protein KaiC by crystallographic and mutagenetic analyses. *Proceedings of the National Academy of Sciences of the United States of America, 101*, 13933–13938.

Xu, Y., Mori, T., Qin, X., Yan, H., Egli, M., & Johnson, C. H. (2009). Intramolecular regulation of phosphorylation status of the circadian clock protein KaiC. *PLoS One, 4*, e7509.

Yang, Q., Pando, B. F., Dong, G., Golden, S. S., & van Oudenaarden, A. (2010). Circadian gating of the cell cycle revealed in single cyanobacterial cells. *Science*, *327*, 1522–1526.

Ye, S., Vakonakis, I., Ioerger, T. R., LiWang, A. C., & Sacchettini, J. C. (2004). Crystal structure of circadian clock protein KaiA from *Synechococcus elongatus*. *Journal of Biological Chemistry*, *279*, 20511–20518.

Yoda, M., Eguchi, K., Terada, T. P., & Sasai, M. (2007). Monomer-shuffling and allosteric transition in KaiC circadian oscillation. *PLoS One*, *2*, e408.

Young, M. M., Tang, N., Hempel, J. C., Oshiro, C. M., Taylor, E. W., Kuntz, I. D., et al. (2000). High throughput protein fold identification by using experimental constraints derived from intramolecular cross-links and mass spectrometry. *Proceedings of the National Academy of Sciences of the United States of America*, *97*, 5802–5806.

Zoltowski, B. D., Vaidya, A. T., Top, D., Widom, J., Young, M. W., & Crane, B. R. (2011). Structure of full-length Drosophila cryptochrome. *Nature*, *480*, 396–400.

Zwicker, D., Lubensky, D. K., & ten Wolde, P. R. (2010). Robust circadian clocks from coupled protein-modification and transcription-translation cycles. *Proceedings of the National Academy of Sciences of the United States of America*, *107*, 22540–22545.

CHAPTER THIRTEEN

Identification of Small-Molecule Modulators of the Circadian Clock

Tsuyoshi Hirota[*,†], **Steve A. Kay**[*,†,1]
[*]Molecular and Computational Biology Section, University of Southern California, Los Angeles, California, USA
[†]Institute of Transformative Bio-Molecules, Nagoya University, Nagoya, Japan
[1]Corresponding author: e-mail address: alestell@usc.edu

Contents

Abstract

Chemical biology or chemical genetics has emerged as an interdisciplinary research area applying chemistry to understand biological systems. The development of combinatorial chemistry and high-throughput screening technologies has enabled large-scale investigation of the biological activities of diverse small molecules to discover useful chemical probes. This approach is applicable to the analysis of the circadian clock mechanisms through cell-based assays to monitor circadian rhythms using luciferase reporter genes. We and others have established cell-based high-throughput circadian assays and have identified a variety of novel small-molecule modulators of the circadian clock by phenotype-based screening of hundreds of thousands of compounds. The results demonstrated the effectiveness of chemical biology approaches in clock research field. This technique will become more and more common with propagation of high-throughput screening facilities. This chapter describes assay development, screening setups, and their optimization for successful screening campaigns.

Methods in Enzymology, Volume 551
ISSN 0076-6879
http://dx.doi.org/10.1016/bs.mie.2014.10.015

1. INTRODUCTION

Genetic approaches have been playing a pivotal role in discovering clock genes at the organismal level in a variety of species. In mammals, the clock genes show rhythmic expression not only in the central clock of the hypothalamic suprachiasmatic nucleus but also in peripheral tissues and cultured cell lines (Balsalobre, Damiola, & Schibler, 1998). This finding together with the development of luciferase reporters and recent advances in high-throughput screening technologies enabled us to search for "perturbagens" that alter the cellular clock function in an unbiased manner. Chemical biology approaches use compounds to dissect biological mechanisms and is considered to be effective in the analysis of biological functions by complementing the limitations of conventional genetic approaches caused by lethality, pleiotropy, and functional redundancy. Resulting proof-of-concept probes will provide chemical tools to control target protein functions in a dose-dependent and conditional manner across species, and therefore act as starting points for the development of therapeutics against clock-related disorders. We and other groups have conducted phenotype-based circadian screening of hundreds of thousands of compounds and have identified a number of small molecules that strongly affect the clock function (Chen et al., 2012; Hirota et al., 2008, 2010, 2012; Isojima et al., 2009; Lee et al., 2011). An alternative approach for the identification of clock-modulating compounds is to develop small molecules against the known clock proteins. This target-based method has identified proof-of-concept probes for the protein kinase CKI and the nuclear receptor REV-ERB that modulate circadian properties *in vivo* (Meng et al., 2010; Solt et al., 2012). In addition, chemical screening against CLOCK–BMAL1-mediated activation of E box-containing reporters has identified compounds that indirectly regulate CLOCK–BMAL1 activity (Chun et al., 2014; Hu et al., 2011). In this chapter, we focus on the methods for high-throughput circadian screening in mammalian cells to identify small-molecule modulators of the circadian clock.

2. CELL-BASED CIRCADIAN ASSAY

2.1. Luciferase reporter genes

To monitor circadian rhythms at the cellular level in a noninvasive manner, we visualized the rhythmic expression of clock genes by using reporters.

Firefly luciferase is the most commonly used reporter in the circadian field, because it does not require excitation light that is toxic to the cells for long-term recordings. Several modifications of luciferase have been reported for circadian studies. Ueda and colleagues put a PEST sequence at the C-terminal region of luciferase to make it degradable (dLuc; half-life ~0.5 h) for better reflection of the transcriptional rhythms (Ueda et al., 2002). Yamazaki and colleagues applied green luciferase (Gluc) from the Brazilian click beetle that is 21 times brighter than firefly luciferase and useful for single-cell imaging (Yeom, Pendergast, Ohmiya, & Yamazaki, 2010). Furthermore, Nakajima and colleagues introduced green-emitting luciferase from the Japanese luminous beetle and red-emitting luciferase from the railroad worm, both of which use a common substrate and therefore enable simultaneous recordings of two different reporters (Noguchi, Ikeda, Ohmiya, & Nakajima, 2008). In addition to luciferase, destabilized forms of GFP are used for circadian recordings especially at the single-cell level (Nagoshi et al., 2004; Ohta, Yamazaki, & McMahon, 2005).

Rhythmic expression of the reporter is driven by the promoter region of the clock genes. In most tissues, expression of the clock genes *Bmal1* and *Per2* show robust circadian rhythms with mutually opposite phase. Their promoter region contains key regulatory *cis*-elements (RORE for *Bmal1* and E box for *Per2*), and ~500-bp promoter fragments harboring these elements drive rhythmic transcription (Ueda et al., 2002; Yoo et al., 2005). We employed the dLuc reporter driven by the *Bmal1* or *Per2* promoter (*Bmal1-dLuc* or *Per2-dLuc*) (Liu et al., 2008). Transgenic *Per1-luc* reporter mice have been used for tissue explants by many groups (Asai et al., 2001; Wilsbacher et al., 2002; Yamazaki et al., 2000), but *Per1* exhibits less robust rhythms in cultured cell lines. Other clock gene reporters, *Cry1-luc*, *Dbp-luc*, and *Rev-erbα-luc* are also developed for circadian recordings (Brown et al., 2005; Fustin, O'Neill, Hastings, Hazlerigg, & Dardente, 2009; Stratmann, Stadler, Tamanini, van der Horst, & Ripperger, 2010). Some of the reporters contain the 3′-UTR region of the corresponding gene for mRNA stability control (Brown et al., 2005; Nagoshi et al., 2004), which also plays an important role for rhythmic gene expression. In addition to these promoter-driven reporters, $mPer2^{Luc}$ knock-in reporter mice were generated by Takahashi and colleagues in which luciferase is fused to the C-terminal of PER2 protein at the endogenous *Per2* locus (Yoo et al., 2004). By reflecting not only the transcriptional regulation but also translational and posttranslational regulations of PER2, the reporter shows extremely stable rhythms (Welsh, Yoo, Liu, Takahashi, & Kay, 2004; Yoo et al., 2004).

2.2. Reporter cells

For high-throughput screening, it is important to choose cell lines that show robust circadian rhythms. Schibler and colleagues originally discovered that the rat-1 fibroblast cell line exhibits circadian expression of clock genes (Balsalobre et al., 1998). Reflecting the rhythmic gene expression, the cells harboring luciferase reporters of the clock genes show circadian changes of luminescence (Nagoshi et al., 2004; Ueda et al., 2002; Welsh et al., 2004). Human U2OS osteosarcoma and mouse NIH-3T3 fibroblast cell lines have also been commonly used in circadian reporter assays (Baggs et al., 2009; Hirota et al., 2008; Isojima et al., 2009; Maier et al., 2009; Nagoshi et al., 2004; Vollmers, Panda, & DiTacchio, 2008). These cells survive under confluent conditions and are suitable for long-term recordings. We applied a lentivirus system (pLenti6, Invitrogen) to deliver *Bmal1-dLuc* and *Per2-dLuc* reporter genes into U2OS cells and to establish stable lines by using blasticidin as a selection marker (see Ramanathan, Khan, Kathale, Xu, & Liu, 2012, for details). We then chose a clonal cell line that showed high amplitude, low damping rate, and high intensity rhythms (Fig. 1). Alternatively, plasmid transfection can be used for reporter delivery. In both cases, multiple copies of the reporter gene will be integrated into the genome, and it is necessary to obtain bright cells to measure luminescence rhythms with conventional plate readers.

For $mPer2^{Luc}$ knock-in reporter, spontaneously immortalized fibroblasts can be prepared from embryonic or adult tissues of the homozygote knock-in mice (Welsh et al., 2004). Although they show very robust rhythms, the luminescence intensity is much lower than the U2OS reporter cells (Fig. 1) because of the nature of the knock-in reporter (two copies per cell). An additional SV40 poly(A) signal in the knock-in construct enhances the signal and makes the rhythms detectable with a conventional plate reader (Chen et al., 2012). Application of bright luciferase (Yeom et al., 2010) may also help to obtain higher signal intensity.

3. HIGH-THROUGHPUT SCREENING SYSTEM

3.1. Liquid handling apparatus

Typically, 384 or 1536-well plates are used to achieve both higher throughput and less running cost and time in large-scale screening. Here, we describe experimental setups for a 384-well plate format. There are several types of 384-well tissue culture plates available from various companies. For

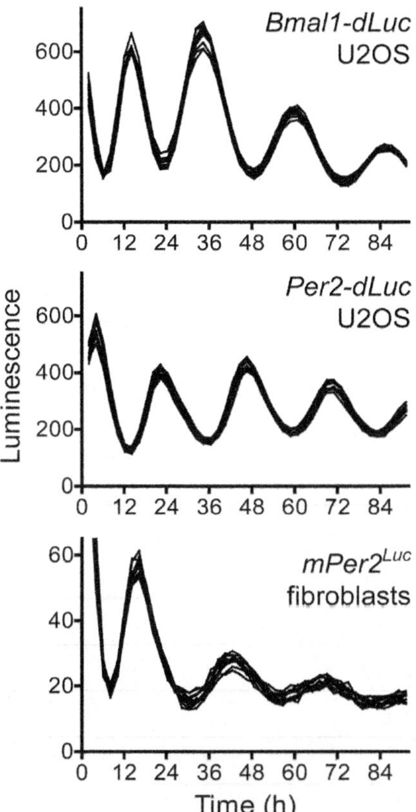

Figure 1 Circadian reporter cells. Luminescence rhythms of *Bmal1-dLuc* U2OS cells (top), *Per2-dLuc* U2OS cells (middle), and *mPer2^{Luc}* knock-in fibroblasts (bottom) are shown ($n = 12$ for each). Note that the luminescence intensity of *mPer2^{Luc}* fibroblasts is more than 10 times lower than U2OS reporter cells.

luminescence recordings, white solid-bottom plates gave a higher signal than others due to light reflection. It is also better to choose plates with lower background counts. In our experiments, we used the plates supplied by Greiner (catalog number 781073 or custom-made thinner plates).

High-throughput screening analyzes tens of thousands of wells at the same time. Therefore, it is essential to handle liquids with high accuracy by applying robotics to obtain consistent and high-quality data. We used a microplate dispenser (MicroFill, BioTek) with a plate stacker (BioStack, BioTek) that automatically dispenses media and cells to multiple plates (up to 50, depending on the stacker size). After trypsinization of the cells, we prepared a batch suspension in the culture medium (Table 1) and plated

Table 1 Composition of the medium

Culture medium
DMEM (11995-073, Gibco)
10% fetal bovine serum
0.29 mg/ml L-glutamine
100 units/ml penicillin
100 μg/ml streptomycin
Explant medium
DMEM (12800–017, Gibco), dissolved in cell culture grade water
2% B-27 supplement (Gibco)
10 m*M* HEPES
0.38 mg/ml sodium bicarbonate
0.29 mg/ml L-glutamine
100 units/ml penicillin
100 μg/ml streptomycin
0.1 mg/ml gentamicin
1 m*M* luciferin
pH 7.2, with NaOH

20 μl containing 2000 cells per well. To avoid precipitation of the cells, the cell suspension was continuously and gently mixed with a magnetic stirrer. It is necessary to optimize the cell number and medium volume for each cell type and culture condition. Due to the small volume of the medium in each well, it is important to minimize evaporation, especially at the edge wells of the plate. Under this condition, the cells reach confluence after 2 days of growth. Cell growth can be checked by plating the cells onto a clear-bottom plate in parallel. To measure luminescence rhythms, we added 50 μl of the explant medium containing HEPES buffer to maintain the pH, a B-27 supplement that synchronizes the cellular clock, and the luciferase substrate luciferin (Table 1). Although we used the medium containing phenol red to visually inspect its pH after the screening, the medium free of phenol red can avoid signal reduction. We then applied 0.5 μl of the compounds (dissolved in DMSO) to cell plates from the compound plates by using a

liquid transfer device (PinTool, GNF Systems). The final concentration of DMSO becomes 0.7% (0.5 μl DMSO in 70 μl medium), and this is tolerated by U2OS cells. The plates were sealed with optically clear films (for example, TempPlate RT, USA Scientific; thickness, 50 μm) and set to a plate reader to measure luminescence rhythms. The film needs to stick firmly to the plate, especially at the edges to avoid evaporation, and a part of the film hanging out from the plate is required to be cut off because it may cause a stuck of the machines. The entire process should be kept as simple as possible to minimize liquid handling steps.

The liquid handling machines are generally equipped with screening facilities. We collaborated with the Genomics Institute of the Novartis Research Foundation (GNF) and operated the liquid handling machines manually (called "off-line" screening) or used a fully automated custom-made robotic system (Melnick et al., 2006), in which a robot arm moves plates among the dispenser, PinTool, plate reader, and incubator (called "on-line" screening). The machines have a barcode reader, and plates (for both compounds and cells) are managed with barcodes. For follow-up studies with small numbers of plates, we manually dispensed liquids by employing electronic multichannel pipettes (such as Matrix Impact 2, Thermo Scientific).

3.2. Plate readers

There are a variety of commercially available plate readers: some are specific for luminescence, others are multi-mode readers also supporting fluorescence and absorbance. They are equipped with a photomultiplier tube (PMT) or a CCD camera as a detector. PMT-based readers are less expensive but take a longer time for plate reading because they scan the wells one by one (for example, 1 s per well takes 7 min per plate). CCD camera-based readers are highly sensitive and fast (about 30 s exposure per plate) but are very expensive. Many readers are compatible with a plate stacker for high-throughput screening. We applied the GNF automated robotic system with a CCD camera-based imager (ViewLux, Perkin Elmer) as described above. The robot arm takes a plate out from the incubator, puts it to the imager, and then brings it back to the incubator. This process takes less than 2 min, and up to 70 plates can be read every 2 h. We monitored luminescence rhythms for 3–4 days to estimate the circadian period. The Chen and Takahashi group employed a PMT-based reader (EnVision, Perkin Elmer) (Chen et al., 2012), and the Ueda group developed a custom-made CCD-Tron system (Isojima et al., 2009).

For single-plate assays in follow-up studies, we used a PMT-based plate reader (Infinite M200, Tecan). We set a longer integration time (14 s per well) to reduce the noise and read each well every 100 min for 5 days. Sealing the plate may cause condensation on the film, but the plate reader we used did not cause condensation, and some other readers can make the top part of the plate warmer than the bottom part to prevent condensation. The distance between the detector and the plate needs to be optimized to minimize crosstalk between the neighboring wells.

3.3. Data analysis software

Reading 70 of the 384-well plates every 2 h for 3.5 days generates more than 1 million data points. Therefore, a specialized algorithm is required to obtain circadian parameters such as period, phase, and amplitude from the large amount of data. By using the R-project computing environment (www.r-project.org), we developed an automated algorithm CellulaRhythm for data analysis and visualization (Hirota et al., 2008). This fits raw luminescence data to a damped cosine curve by using nonlinear least squares and provides period, phase, amplitude, baseline, and damping rate parameters. This algorithm also creates luminescence traces for each well and heat maps for the entire plate, both of which are useful for visual inspection of the validity of the provided parameters. MultiCycle is a more sophisticated software developed by Actimetrics that runs LumiCycle Analysis on a 384-well plate format. Raw luminescence data are detrended with a polynomial curve (usually first-order) and then fitted to a sine curve to provide period, phase, and amplitude parameters. The analytical part is interactive with the visualization part that displays raw or detrended profiles from multiple wells, and also plots calculated phase or period against amplitude. This feature of the software makes analytical optimization and visual inspection more efficient.

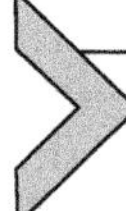

4. CIRCADIAN SCREENING

4.1. Assay optimization and validation

The success of the screening depends on the robustness of the system. For circadian screening, the cellular rhythmicity is a key factor. The above-mentioned points such as reporter gene, cell line, cell number, medium volume, medium composition, compound volume, liquid dispenser, and plate reader need to be optimized for the best results. Another important factor is data analysis to obtain circadian parameters accurately reflecting the raw data. Due to transient changes of luminescence upon the medium change, we excluded the data of the first day from the analysis. Both the start and the end time points used for the curve fitting affect parameter estimation, and they need to be optimized to obtain proper results. After obtaining the circadian parameters, we manually inspected the quality of the curve fitting and filtered out the data with poor fitting that usually arises from low amplitude caused by contamination or by the effect of the compound. Contaminated wells are readily identified with phenol red-containing medium because they become yellow. We also evaluated positional effects of the wells on the rhythms. Edge wells tend to have shifted phase, making phase analysis difficult. In contrast, the period length is consistent, irrespective of the well position. Furthermore, alterations of the period generally arise from deficiencies of the core clock function. Therefore, we focused on the compounds affecting the period length. Although arrhythmicity is a more severe phenotype, it is difficult to discriminate it from cytotoxicity.

We have eventually established a high-throughput circadian assay system by using *Bmal1-dLuc* U2OS cells in which more than 97% of the wells are within ±0.5 h (SD, 0.23 h) from the mean period under control conditions (Fig. 2). By analyzing the effects of positive control compounds such as a CKI inhibitor D4476, we set the final compound concentration for the primary screening at 7 μ*M* (0.5 μl of 1 m*M* of compound in a 70 μl medium). To validate the system setup, we carried out a pilot screening of LOPAC (Library of Pharmacologically Active Compounds, Sigma-Aldrich) that contains 1,280 well-characterized compounds. The library covers a variety

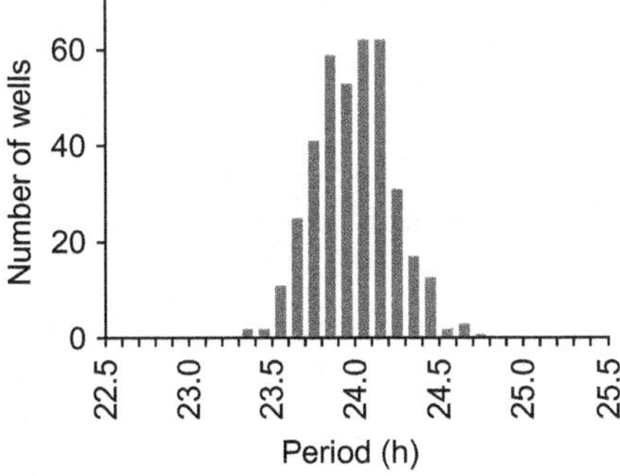

Figure 2 Period distribution of *Bmal1-dLuc* U2OS cells on a 384-well plate-based circadian assay. *Reproduced from Hirota et al. (2008). Copyright (2008) National Academy of Sciences, USA.*

of signaling pathways and drug target classes, such as GPCRs, kinases, ion channels, and transporters. Therefore, it provides a starting point for compound screening campaigns. We repeated the LOPAC screening twice and identified 13 compounds that caused ≥0.5 h period lengthening or shortening in both screens. Among them, the effects of 11 compounds were confirmed by using the original sources. Importantly, many of the "hit" compounds are related to the pathways already known to affect the clock function in other systems, indicating the validity of our system to find clock-modulating compounds (Hirota et al., 2008). Ueda and colleagues also identified a similar set of compounds from the LOPAC screening with NIH-3T3-*mPer2-Luc* and U2OS-*hPer2-Luc* cells (Isojima et al., 2009).

4.2. High-throughput chemical screening

A wide variation of chemical structures has the advantage of probing many classes of potential targets, which may include "undruggable" proteins such as the core clock proteins. We therefore screened a collection of structurally-diverse compounds from commercial sources and academic labs (Ding, Gray, Wu, Ding, & Schultz, 2002; Plouffe et al., 2008). Most compounds are at least 85% pure and have drug-like characteristics with a high observance to Lipinski's rule of five. We analyzed the effects of ~700,000 compounds on the luminescence rhythms of *Bmal1-dLuc* U2OS cells and identified a number of small molecules that strongly changed the period by more than 2 h (Fig. 3). The effects of these "primary hits" were further investigated as 8-point serial dilution series by "hit picking" or "cherry picking" from the original sources (Fig. 3). When the effect is dose-dependent, it is likely that the compound targets key regulators (i.e., rheostats) of the clock. This assay also evaluates the possibility of cross-contamination or decomposition within the compound plates used for the screening. The confirmed hits were further tested with *Per2-dLuc* U2OS cells to investigate reporter-specific effects and with fibroblasts and tissue explants from *mPer2^{Luc}* knock-in mice to evaluate cell-type specific effects.

As subsequent mechanistic studies of the hit compounds requires a lot of effort and time, they need to be prioritized based on the compound's effect, potency, structure, and property. Given that many clock-modulating compounds turned out to affect CKI activity (Chen et al., 2012; Hirota et al., 2008, 2010; Isojima et al., 2009; Lee et al., 2011), analysis of hit compounds by an *in vitro* CKI assay will be useful to reveal CKI-targeting compounds.

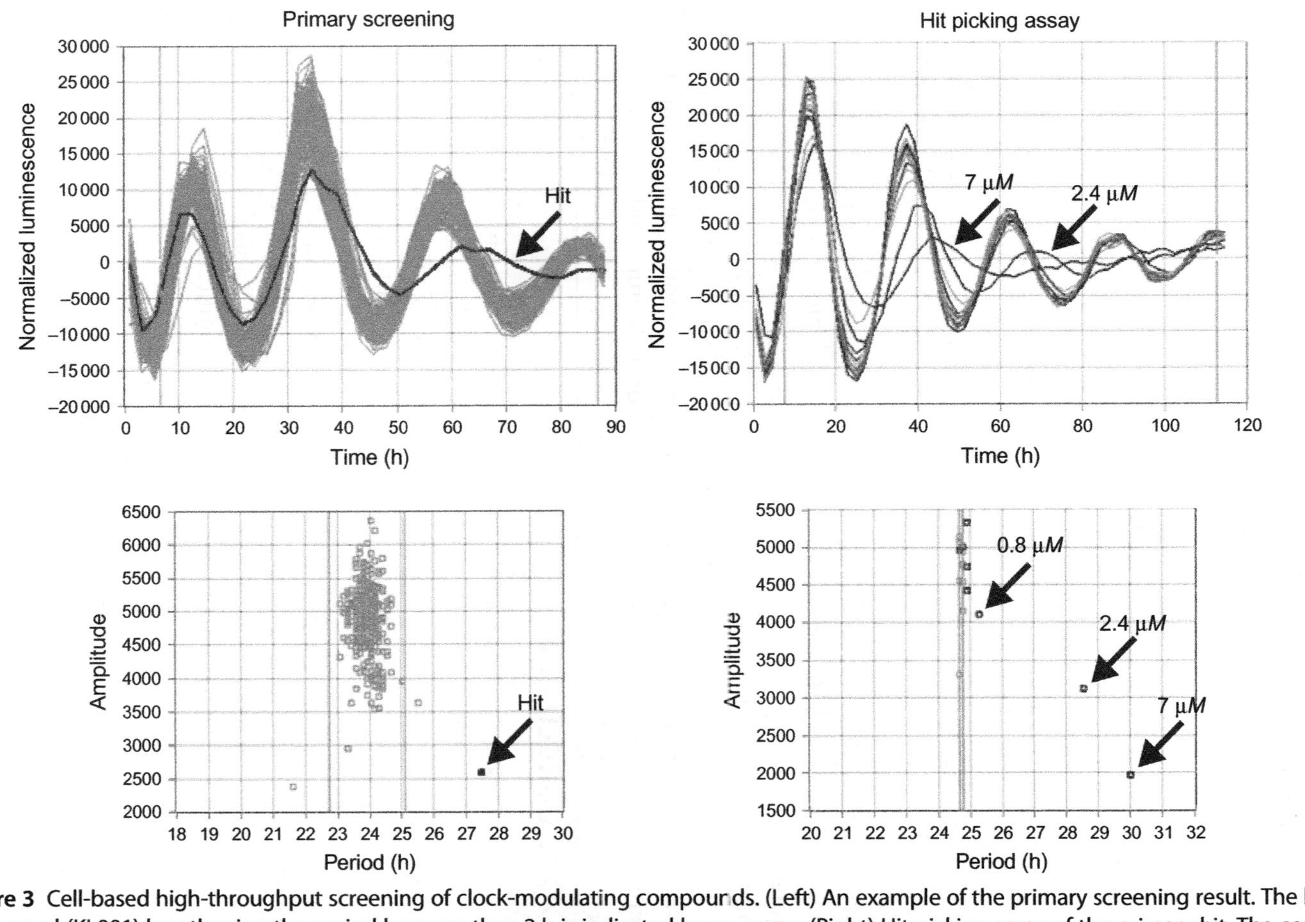

Figure 3 Cell-based high-throughput screening of clock-modulating compounds. (Left) An example of the primary screening result. The hit compound (KL001) lengthening the period by more than 2 h is indicated by an arrow. (Right) Hit picking assay of the primary hit. The compound from the original source was analyzed as 8-point serial dilution series indicated by arrows. Note that the period lengthening effect is dose-dependent.

We took an affinity-based proteomic approach (Rix & Superti-Furga, 2009) to identify molecular targets of hit compounds in an unbiased manner. In collaboration with organic synthetic chemists at the Schultz laboratory of the Scripps Research Institute, we developed affinity probes as well as more active and inactive analogs of the hit compounds. By using these chemical tools, compound-interacting proteins were affinity-purified and then identified by mass spectrometry. With this approach, we successfully discovered the targets of three compounds, longdaysin, LH846, and KL001 (Hirota et al., 2010, 2012; Lee et al., 2011). Further characterization revealed that both longdaysin and LH846 inhibit CKI to stabilize PER and massively lengthen the period. In contrast, KL001 is the first small-molecule targeting CRY and prevents its FBXL3-dependent degradation for period lengthening. Interestingly, although both PER and CRY act as negative regulators in the core clock loop, we found that the effects of longdaysin and KL001 on the *Bmal1* and *Per2* reporters are different: KL001 reduces the basal level and amplitude of the *Per2* rhythm compared with *Bmal1*, while longdaysin has a similar effect on both reporters (Hirota et al., 2012; St John, Hirota, Kay, & Doyle, 2014) (Fig. 4). Therefore, using a different reporter may partially affect the outcome of the screening because of the reporter-specific effects derived from different mechanisms of action of the compounds.

5. CONCLUSION

With the high-throughput screening technology being increasingly common, large-scale identification of clock-modulating compounds is likely to become a more general approach. For the molecular understanding of clock mechanisms, it is essential to identify target proteins of such compounds. Especially, the well-characterized compounds need to be carefully analyzed because of unknown off-target effects. For example, a variety of protein kinase inhibitors such as CDK and MAPK inhibitors have demonstrated to substantially interact with and inhibit CKI (Fabian et al., 2005; Isojima et al., 2009). Therefore, unbiased target identification or confirmation of the phenotype by an alternative approach like RNAi is required. Application of the compounds to *in vivo* studies is also important, because *in vivo* active clock modulators will provide a basis for novel therapies to treat clock-associated diseases. The reported clock-modifying compounds are effective in a variety of cells and tissues including the hypothalamic suprachiasmatic nucleus (Chen et al., 2012; Hirota et al., 2010, 2012; Isojima et al., 2009), indicating the effectiveness of the cell culture model. We

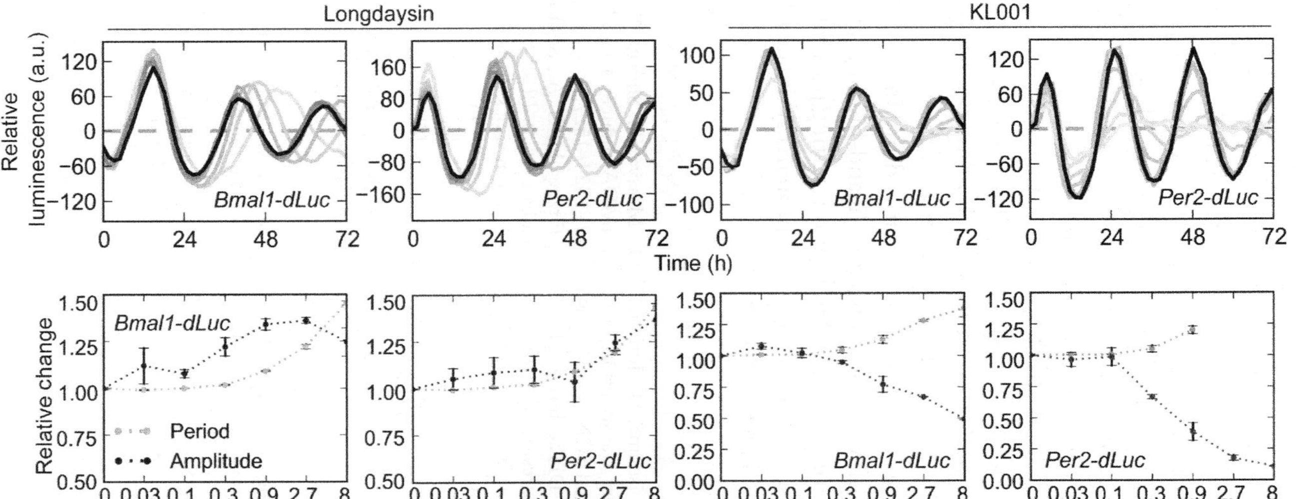

Figure 4 Different amplitude effects of longdaysin and KL001 on *Bmal1-dLuc* and *Per2-dLuc* reporters. (Top) Detrended luminescence profiles with increasing concentrations of the compounds (dark to light colors). (Bottom) Relative changes in period and amplitude. *Reproduced from St John et al. (2014). Copyright (2014) National Academy of Sciences, USA.*

further demonstrated that longdaysin strongly lengthens reporter gene expression rhythms in zebrafish larvae *in vivo* (Hirota et al., 2010; Weger et al., 2013). This 96-well plate-based zebrafish assay will be useful to screen *in vivo* active compounds for clock regulation. Such a chemical screening at the organismal level was achieved using plant seedlings (Toth et al., 2012).

Taken together, chemical biology approaches will play an increasingly important role in dissecting the molecular mechanism of the circadian clock. Discovery of the CRY-targeting compound KL001 (Hirota et al., 2012) demonstrated that the core clock proteins are no longer undruggable. Recent identification of a CRY inhibitor (Chun et al., 2014) supports this idea. CRY is targeted by KL001 at its FAD-binding pocket (Hirota et al., 2012; Nangle, Xing, & Zheng, 2013), and many other core clock proteins also bind with cofactors and ligands (summarized in Hirota & Kay, 2009). Compounds interacting with these cofactor/ligand-binding sites may work as specific modulators of the clock. Another interesting possibility is that the compounds lead to discovery of novel clock protein candidates that have not been implicated in the clock function.

REFERENCES

Asai, M., Yamaguchi, S., Isejima, H., Jonouchi, M., Moriya, T., Shibata, S., et al. (2001). Visualization of mPer1 transcription in vitro: NMDA induces a rapid phase shift of mPer1 gene in cultured SCN. *Current Biology*, *11*, 1524–1527.

Baggs, J. E., Price, T. S., DiTacchio, L., Panda, S., Fitzgerald, G. A., & Hogenesch, J. B. (2009). Network features of the mammalian circadian clock. *PLoS Biology*, 7, e52.

Balsalobre, A., Damiola, F., & Schibler, U. (1998). A serum shock induces circadian gene expression in mammalian tissue culture cells. *Cell*, *93*, 929–937.

Brown, S. A., Ripperger, J., Kadener, S., Fleury-Olela, F., Vilbois, F., Rosbash, M., et al. (2005). PERIOD1-associated proteins modulate the negative limb of the mammalian circadian oscillator. *Science*, *308*, 693–696.

Chen, Z., Yoo, S. H., Park, Y. S., Kim, K. H., Wei, S., Buhr, E., et al. (2012). Identification of diverse modulators of central and peripheral circadian clocks by high-throughput chemical screening. *Proceedings of the National Academy of Sciences of the United States of America*, *109*, 101–106.

Chun, S. K., Jang, J., Chung, S., Yun, H., Kim, N. J., Jung, J. W., et al. (2014). Identification and validation of cryptochrome inhibitors that modulate the molecular circadian clock. *ACS Chemical Biology*, *9*, 703–710.

Ding, S., Gray, N. S., Wu, X., Ding, Q., & Schultz, P. G. (2002). A combinatorial scaffold approach toward kinase-directed heterocycle libraries. *Journal of the American Chemical Society*, *124*, 1594–1596.

Fabian, M. A., Biggs, W. H., 3rd., Treiber, D. K., Atteridge, C. E., Azimioara, M. D., Benedetti, M. G., et al. (2005). A small molecule-kinase interaction map for clinical kinase inhibitors. *Nature Biotechnology*, *23*, 329–336.

Fustin, J. M., O'Neill, J. S., Hastings, M. H., Hazlerigg, D. G., & Dardente, H. (2009). Cry1 circadian phase in vitro: Wrapped up with an E-box. *Journal of Biological Rhythms*, *24*, 16–24.

Hirota, T., & Kay, S. A. (2009). High-throughput screening and chemical biology: New approaches for understanding circadian clock mechanisms. *Chemistry & Biology, 16*, 921–927.

Hirota, T., Lee, J. W., Lewis, W. G., Zhang, E. E., Breton, G., Liu, X., et al. (2010). High-throughput chemical screen identifies a novel potent modulator of cellular circadian rhythms and reveals CKIalpha as a clock regulatory kinase. *PLoS Biology, 8*, e1000559.

Hirota, T., Lee, J. W., St John, P. C., Sawa, M., Iwaisako, K., Noguchi, T., et al. (2012). Identification of small molecule activators of cryptochrome. *Science, 337*, 1094–1097.

Hirota, T., Lewis, W. G., Liu, A. C., Lee, J. W., Schultz, P. G., & Kay, S. A. (2008). A chemical biology approach reveals period shortening of the mammalian circadian clock by specific inhibition of GSK-3beta. *Proceedings of the National Academy of Sciences of the United States of America, 105*, 20746–20751.

Hu, Y., Spengler, M. L., Kuropatwinski, K. K., Comas-Soberats, M., Jackson, M., Chernov, M. V., et al. (2011). Selenium is a modulator of circadian clock that protects mice from the toxicity of a chemotherapeutic drug via upregulation of the core clock protein, BMAL1. *Oncotarget, 2*, 1279–1290.

Isojima, Y., Nakajima, M., Ukai, H., Fujishima, H., Yamada, R. G., Masumoto, K. H., et al. (2009). CKIepsilon/delta-dependent phosphorylation is a temperature-insensitive, period-determining process in the mammalian circadian clock. *Proceedings of the National Academy of Sciences of the United States of America, 106*, 15744–15749.

Lee, J. W., Hirota, T., Peters, E. C., Garcia, M., Gonzalez, R., Cho, C. Y., et al. (2011). A small molecule modulates circadian rhythms through phosphorylation of the period protein. *Angewandte Chemie (International Ed. in English), 50*, 10608–10611.

Liu, A. C., Tran, H. G., Zhang, E. E., Priest, A. A., Welsh, D. K., & Kay, S. A. (2008). Redundant function of REV-ERBalpha and beta and non-essential role for Bmal1 cycling in transcriptional regulation of intracellular circadian rhythms. *PLoS Genetics, 4*, e1000023.

Maier, B., Wendt, S., Vanselow, J. T., Wallach, T., Reischl, S., Oehmke, S., et al. (2009). A large-scale functional RNAi screen reveals a role for CK2 in the mammalian circadian clock. *Genes & Development, 23*, 708–718.

Melnick, J. S., Janes, J., Kim, S., Chang, J. Y., Sipes, D. G., Gunderson, D., et al. (2006). An efficient rapid system for profiling the cellular activities of molecular libraries. *Proceedings of the National Academy of Sciences of the United States of America, 103*, 3153–3158.

Meng, Q. J., Maywood, E. S., Bechtold, D. A., Lu, W. Q., Li, J., Gibbs, J. E., et al. (2010). Entrainment of disrupted circadian behavior through inhibition of casein kinase 1 (CK1) enzymes. *Proceedings of the National Academy of Sciences of the United States of America, 107*, 15240–15245.

Nagoshi, E., Saini, C., Bauer, C., Laroche, T., Naef, F., & Schibler, U. (2004). Circadian gene expression in individual fibroblasts: Cell-autonomous and self-sustained oscillators pass time to daughter cells. *Cell, 119*, 693–705.

Nangle, S., Xing, W., & Zheng, N. (2013). Crystal structure of mammalian cryptochrome in complex with a small molecule competitor of its ubiquitin ligase. *Cell Research, 23*, 1417–1419.

Noguchi, T., Ikeda, M., Ohmiya, Y., & Nakajima, Y. (2008). Simultaneous monitoring of independent gene expression patterns in two types of cocultured fibroblasts with different color-emitting luciferases. *BMC Biotechnology, 8*, 40.

Ohta, H., Yamazaki, S., & McMahon, D. G. (2005). Constant light desynchronizes mammalian clock neurons. *Nature Neuroscience, 8*, 267–269.

Plouffe, D., Brinker, A., McNamara, C., Henson, K., Kato, N., Kuhen, K., et al. (2008). In silico activity profiling reveals the mechanism of action of antimalarials discovered in a high-throughput screen. *Proceedings of the National Academy of Sciences of the United States of America, 105*, 9059–9064.

Ramanathan, C., Khan, S. K., Kathale, N. D., Xu, H., & Liu, A. C. (2012). Monitoring cell-autonomous circadian clock rhythms of gene expression using luciferase bioluminescence reporters. *Journal of Visualized Experiments, 67*, e4234.

Rix, U., & Superti-Furga, G. (2009). Target profiling of small molecules by chemical proteomics. *Nature Chemical Biology, 5*, 616–624.

Solt, L. A., Wang, Y., Banerjee, S., Hughes, T., Kojetin, D. J., Lundasen, T., et al. (2012). Regulation of circadian behaviour and metabolism by synthetic REV-ERB agonists. *Nature, 485*, 62–68.

St John, P. C., Hirota, T., Kay, S. A., & Doyle, F. J., 3rd. (2014). Spatiotemporal separation of PER and CRY posttranslational regulation in the mammalian circadian clock. *Proceedings of the National Academy of Sciences of the United States of America, 111*, 2040–2045.

Stratmann, M., Stadler, F., Tamanini, F., van der Horst, G. T., & Ripperger, J. A. (2010). Flexible phase adjustment of circadian albumin D site-binding protein (DBP) gene expression by CRYPTOCHROME1. *Genes & Development, 24*, 1317–1328.

Toth, R., Gerding-Reimers, C., Deeks, M. J., Menninger, S., Gallegos, R. M., Tonaco, I. A., et al. (2012). Prieurianin/endosidin 1 is an actin-stabilizing small molecule identified from a chemical genetic screen for circadian clock effectors in Arabidopsis thaliana. *The Plant Journal, 71*, 338–352.

Ueda, H. R., Chen, W., Adachi, A., Wakamatsu, H., Hayashi, S., Takasugi, T., et al. (2002). A transcription factor response element for gene expression during circadian night. *Nature, 418*, 534–539.

Vollmers, C., Panda, S., & DiTacchio, L. (2008). A high-throughput assay for siRNA-based circadian screens in human U2OS cells. *PLoS One, 3*, e3457.

Weger, M., Weger, B. D., Diotel, N., Rastegar, S., Hirota, T., Kay, S. A., et al. (2013). Real-time in vivo monitoring of circadian E-box enhancer activity: A robust and sensitive zebrafish reporter line for developmental, chemical and neural biology of the circadian clock. *Developmental Biology, 380*, 259–273.

Welsh, D. K., Yoo, S. H., Liu, A. C., Takahashi, J. S., & Kay, S. A. (2004). Bioluminescence imaging of individual fibroblasts reveals persistent, independently phased circadian rhythms of clock gene expression. *Current Biology, 14*, 2289–2295.

Wilsbacher, L. D., Yamazaki, S., Herzog, E. D., Song, E. J., Radcliffe, L. A., Abe, M., et al. (2002). Photic and circadian expression of luciferase in mPeriod1-luc transgenic mice in vivo. *Proceedings of the National Academy of Sciences of the United States of America, 99*, 489–494.

Yamazaki, S., Numano, R., Abe, M., Hida, A., Takahashi, R., Ueda, M., et al. (2000). Resetting central and peripheral circadian oscillators in transgenic rats. *Science, 288*, 682–685.

Yeom, M., Pendergast, J. S., Ohmiya, Y., & Yamazaki, S. (2010). Circadian-independent cell mitosis in immortalized fibroblasts. *Proceedings of the National Academy of Sciences of the United States of America, 107*, 9665–9670.

Yoo, S. H., Ko, C. H., Lowrey, P. L., Buhr, E. D., Song, E. J., Chang, S., et al. (2005). A noncanonical E-box enhancer drives mouse Period2 circadian oscillations in vivo. *Proceedings of the National Academy of Sciences of the United States of America, 102*, 2608–2613.

Yoo, S. H., Yamazaki, S., Lowrey, P. L., Shimomura, K., Ko, C. H., Buhr, E. D., et al. (2004). PERIOD2::LUCIFERASE real-time reporting of circadian dynamics reveals persistent circadian oscillations in mouse peripheral tissues. *Proceedings of the National Academy of Sciences of the United States of America, 101*, 5339–5346.

PART III

Circadian Regulation of Gene and Protein Expression

CHAPTER FOURTEEN

ChIP-seq and RNA-seq Methods to Study Circadian Control of Transcription in Mammals

Joseph S. Takahashi*,†,1, Vivek Kumar*,†, Prachi Nakashe*, Nobuya Koike*,2, Hung-Chung Huang*,3, Carla B. Green*, Tae-Kyung Kim*

*Department of Neuroscience, University of Texas Southwestern Medical Center, Dallas, Texas, USA
†Howard Hughes Medical Institute, University of Texas Southwestern Medical Center, Dallas, Texas, USA
[1]Corresponding author: e-mail address: joseph.takahashi@utsouthwestern.edu

Contents

[2] Current address: Department of Physiology and Systems Bioscience, Kyoto Prefectural University of Medicine, Kawaramachi-Hirokoji, Kyoto, Japan
[3] Current address: Department of Biology, Jackson State University, Jackson, Mississippi, USA

Methods in Enzymology, Volume 551
ISSN 0076-6879
http://dx.doi.org/10.1016/bs.mie.2014.10.059

Abstract

Genome-wide analyses have revolutionized our ability to study the transcriptional regulation of circadian rhythms. The advent of next-generation sequencing methods has facilitated the use of two such technologies, ChIP-seq and RNA-seq. In this chapter, we describe detailed methods and protocols for these two techniques, with emphasis on their usage in circadian rhythm experiments in the mouse liver, a major target organ of the circadian clock system. Critical factors for these methods are highlighted and issues arising with time series samples for ChIP-seq and RNA-seq are discussed. Finally, detailed protocols for library preparation suitable for Illumina sequencing platforms are presented.

Eukaryotic genomes, the substrate of much of modern biology, are packaged into dynamically regulated units, chromatin, a macromolecular structure consisting of DNA, protein, and RNA. DNA is wrapped around a histone octamer forming a nucleosome, the basic unit of chromatin. Each nucleosome consists of two copies of core histones (H2A, H2B, H3, H4) that assemble when two dimers of H3/H4 form a tetramer and complex with two H2A/H2B dimers. 147 base pairs of DNA are wrapped around each histone octamer and constitute a nucleosome. Histone amino-end tails protrude from the nucleosome core and are extensively and dynamically modified. Arrays of nucleosomes referred to as "beads on a string" are further organized into a 30 n*M* fiber that are packaged in to higher order structures within the nucleus. Transcription, replication, repair, and recombination of DNA have to occur in the context of chromatin. Chromatin is generally thought to act as a physical barrier that must be overcome in order to access DNA and is known to be highly dynamic, with open and closed states. The constituency of chromatin and its interaction with DNA binding factors and cofactors are critical for transcriptional regulation and techniques to dissect these regulatory roles are important for understanding of biological processes including circadian rhythms.

Mammalian circadian rhythms are regulated by a transcription–translation feedback loop in which the bHLH–PAS transcription factors,

CLOCK (and its paralog NPAS2) and BMAL1 (ARNTL) dimerize and activate transcription of the *Period* (*Per1*, *Per2*) and *Cryptochrome* (*Cry1*, *Cry2*) genes (Bunger et al., 2000; Gekakis et al., 1998; King et al., 1997; Kume et al., 1999). As the PER proteins accumulate, they form complexes with the CRY proteins, translocate into the nucleus, and interact with the CLOCK/BMAL1 complex to inhibit their own transcription (Chen et al., 2009; Lee, Etchegaray, Cagampang, Loudon, & Reppert, 2001). As the inhibitory complex turns over and declines, the repression phase ends, and the cycle starts again with a new round of CLOCK/BMAL1-activated transcription.

In many ways, the circadian system is ideally suited to study the many facets of transcription. There exists a well-defined cohort of central regulators with strong genetic and biochemical validity of the system (Lowrey & Takahashi, 2004, 2011). Many mutant alleles of the core components exist that can be exploited. The ~24 h pace of the transcriptional oscillation allows for the study of a naturally occurring endogenous system that is conserved from behavior at the organismal level to the single cells. Large amounts of starting material can be obtained from homogeneous tissues such as liver, at specified times from mice for complex biochemical analysis. At the same time, the circadian field can benefit from the application of modern molecular biology approaches developed by the transcription field. One such approach is Chromatin Immunoprecipitation (ChIP)-seq which we detail here, as well as, RNA-seq, which we describe next. Such genome-wide analyses have recently been published by a number of labs (Hatanaka et al., 2010; Koike et al., 2012; Le Martelot et al., 2012; Menet, Rodriguez, Abruzzi, & Rosbash, 2012; Rey et al., 2011; Vollmers et al., 2012).

ChIP is a powerful technique for detection of protein–DNA interactions and combined with modern next-generation sequencing (NGS), ChIP-seq has revolutionized modern systems-level understanding of the transcriptional landscape. First developed in the late 1970 and early 1980s, ChIP was used to understand the organization of nucleosomes on DNA in its native state (Jackson, 1978; Solomon & Varshavsky, 1985). A variety of reagents were used to cross-link DNA to proteins including formaldehyde, dimethyl sulfate, and UV (Gilmour & Lis, 1984; Karpov, Preobrazhenskaya, & Mirzabekov, 1984; Welsh & Cantor, 1984). The key insight from these pioneering studies was that *in vivo* cross-linking with formaldehyde preserves chromatin structure and the process of cross-linking does not radically alter DNA–histone interactions (Jackson & Chalkley,

1981). Thus these early biochemical and biophysical studies reliably established the technique of cross-linking critical for ChIP. The methods used currently for ChIP were developed in the mid-1990s when antibodies to individual histones and to various modified histones were first made available (Kuo & Allis, 1999). Genome-wide analysis through ChIP was attempted using microarrays in (ChIP–ChIP); however, this technique was not widely adapted due to several factors (Buck & Lieb, 2004; Ren et al., 2000).

The principles of ChIP are fairly straightforward but its successful execution relies on many critical factors that need optimization based on cell type and antibody used and antigen probed. We will attempt to outline the basic steps in ChIP followed by library construction for NGS that has worked well in our laboratory with the caveat that this protocol must be optimized for individual antibodies, factors, and tissue used.

The basic steps in ChIP are outlined in (see accompanying Chapter by Zhou, Yu, & Hardin, 2015). DNA is covalently bound to surrounding proteins, presumably in its native state by using the chemical cross-linker, formaldehyde. The cells are then lysed and the nucleoprotein complex is sheared using either sonication or nuclease, the target cross-linked protein is then immunoprecipitated, and after extensive washing to remove background contaminants, the cross-links are reversed, the DNA isolated, libraries are made for NGS studies.

1. CRITICAL FACTORS

1.1. Antibody

The antibody is the most critical factor in ChIP experiments. An antibody that functions in western or immunohistochemistry may not always perform in ChIP. There is considerable batch to batch variability in polyclonal and monoclonal antibodies from commercial suppliers. We routinely purchase large batches of a particular "working" antibody from commercial vendors. When obtaining a new antibody it is critical to confirm its usefulness with a known positive and negative target. In this protocol, we detail the use of the following antibodies.

Antibodies against PER1, PER2, CLOCK, and BMAL1 were made as described previously (Lee et al., 2001). CRY1 antibody was made as described (Lee, Weaver, & Reppert, 2004). CRY2 (epitope: residues 514–592) and p300 (epitope; residues 60–242 of human p300) antibodies were generated using guinea pigs (Cocalico Biological) and serum was affinity purified using the same protein used to raise antibody. NPAS2 antibody (Reick, Garcia, Dudley, & McKnight, 2001) was a kind gift from Dr. Steven McKnight (UT Southwestern Medical Center). RNAPII-8WG16 (MMS-126R) antibody (Jones et al., 2004) was purchased from Covance. RNAPII-Ser5P (clone 3E8, 04-1572) antibody (Chapman et al., 2007) was purchased from Millipore and RNAPII-Ser5P (ab5131) antibody (Rahl et al., 2010) was purchased from Abcam. H3K4me1 (ab8895), H3K4me3 (ab1012), H3K9ac (ab4441), H3K27ac (ac4729), H3K36me3 (ab9050), and H3K79me2 (ab3594) antibodies were purchased from Abcam. CBP antibody was monoclonal AC238 culture supernatant (Eckner et al., 1996).

1.2. Cross-linking/fixation

Formaldehyde covalently links peptide side-chain nitrogens of lysines, arginine, histidine as well as the α-amino groups of all amino acids to exocyclic amino groups and the endocyclic imino groups of DNA bases (Chaw, Crane, Lange, & Shapiro, 1980; McGhee & von Hippel, 1975a,1975b). Because of its ease of use, fast-acting nature and cross-link reversibility it is the most commonly used cross-linker for ChIP. Formaldehyde, which cross-links reactive groups within a 2 Å distance, is best suited for studying direct protein–DNA interactions. Formaldehyde cross-linking can be optimized by varying the time of fixation, temperature, and concentration. Typically short times are required for immunoprecipitating with core histones and DNA binding factors; however, extended times are required for cofactors that indirectly bind DNA. When fixation is too short, stable DNA–protein complexes that can be pulled down with the antibody will not form. When samples are over fixed, sonication, pulldown, and reversing cross-links will be inefficient, leading to lowered yield. In order to study cofactors that bind several layers away in the sandwich, dual cross-linkers or increased length of cross-linking with formaldehyde can be used.

We have used two cross-linking methods, depending on antigen targeted. If the protein of interest is a DNA binding protein, 1% formaldehyde works well in most cases. However, formaldehyde has a short cross-linking spacer arm and is not efficient to examine the proteins indirectly associated with DNA, such as PERs and CRYs. Dual cross-linking using a protein–protein cross-linker and formaldehyde works better in these cases (Koike et al., 2012; Nowak, Tian, & Brasier, 2005; Zeng, Vakoc, Chen, Blobel, & Berger, 2006).

1.3. Sonication

Too much sonication can disrupt the protein–DNA complex or cause damage to DNA and lead to low levels of immunoprecipitated DNA. Low levels of sonication will lead to large DNA fragment length and low resolution of the genomic region that is immunoprecipitated. Sonication is strongly affected by the type and concentration of detergent used and length of fixation.

1.4. Detergents

Detergents such as SDS or sarcosyl have multiple functions, they lyse cross-linked cells, expose and solubilize the antigenic complex, are important for proper sonication, and decrease background binding. But they can also denature the antigen and disrupt the antigen–antibody interaction surface of some antibodies, lowering yield. In many cases, gentler detergents such as Triton-X100 must be used, but this will lead to decreases in sonication efficiency. Thus it is important to characterize each antibody with a range of detergent concentrations and types.

1.5. Bioinformatics

Computational analysis of ChIP-seq data varies between labs and can be a source of irreproducibility. Even when the software that is used is consistent, parameters used should be properly documented. Circadian data is further complicated by the cyclical nature of interactions that we are interested in detecting. We analyze cycling using three independent programs, COSOPT (Panda et al., 2002), JTK cycle (Hughes, Hogenesch, & Kornacker, 2010), and ARSER (Yang & Su, 2010). For example, in one study (Koike et al., 2012), in order for a gene to be considered cycling, it was scored as cycling if two out of the three software programs detected it.

2. ChIP-SEQ METHOD FOR MOUSE LIVER

2.1. Tissue sampling

Male C57BL/6J at 8–12 weeks of age are housed in light–tight boxes and entrained to LD 12:12 conditions for minimum of 7 days. Thirty-six hours after mice are transferred to constant darkness, liver samples are collected every 4 h. We dissect the entire liver.

2.2. ChIP-seq

A. 1% formaldehyde cross-linking

1. Homogenize mouse livers immediately in 4 mL per liver of 1 × PBS containing 1% formaldehyde.
   - **a.** Wash liver with PBS by soaking.
   - **b.** Mince liver with a razor blade into small pieces.
   - **c.** Add liver pieces to 4 mL (per liver) of PBS containing 1% formaldehyde.
   - **d.** Homogenize with a Dounce homogenizer (seven strokes each with A(loose) and B(tight) pestle).
2. Incubate for 8 min at room temperature.
3. Add 250 μL of 2.5 *M* glycine to stop the reaction on ice.

B. Dual cross-linking

1. Homogenize mouse livers immediately in 4 mL per liver of 1 × PBS containing 2 m*M* EGS (ethylene glycol bis[succinimidylsuccinate]).
2. Incubate for 20 min at room temperature.
3. Add formaldehyde to final concentration of 1%.
4. Incubate for 8 min at room temperature.
5. Add 250 μL of 2.5 *M* glycine to stop the reaction on ice.

C. Nuclei isolation

1. Add 10 mL of ice-cold 2.3 *M* sucrose containing 150 m*M* glycine, 10 m*M* HEPES pH 7.6, 15 m*M* KCl, 2 m*M* EDTA, 0.15 m*M* spermine, 0.5 m*M* spermidine, 0.5 m*M* DTT, and 0.5 m*M* PMSF to the homogenate.
2. Layer the homogenate on the top of a 3 mL cushion of 1.85 *M* sucrose (containing the same ingredients and including 10% glycerol).
3. Centrifuge for 1 h at 24,000 rpm at 4 °C in a Beckman SW32.1 rotor.
4. Wash the precipitated nuclei with 1 mL of 10 m*M* Tris pH 7.5, 150 m*M* NaCl, 2 m*M* EDTA, and transfer to a 1.5 mL microfuge tube.
5. Centrifuge for 3 min at 3000 rpm at 4 °C and washed again.
6. Stored at −80 °C until use.

D. Chromatin sonication

We used two different sonicators (Covaris S2 and Misonix S-4000) for chromatin shearing and four different buffers (1% SDS, 0.5% Sarkosyl, 1% Triton-X100, or MNase digestion buffers) depending on the antibody. As previously stated, this should be optimized.

1. BMAL1 and RNAPII-8WG16 antibodies
   a. Resuspend the formaldehyde-cross-linked nuclei in 0.8 mL per liver of lysis buffer (50 m*M* Tris pH 7.5, 10 m*M* EDTA, 1% SDS, 1 m*M* PMSF and Roche complete EDTA free protease inhibitor cocktail).
   b. Sonicate 10 × for 30 s at 4 °C using a Covaris S2 ultrasonicator.
   c. Dilute 10-fold with IP buffer (10 m*M* Tris pH 7.5, 150 m*M* NaCl, 1 m*M* EDTA, 1% Triton X-100, 0.1% sodium deoxycholate, 1 m*M* PMSF and protease inhibitor cocktail).
2. CLOCK and NPAS2 antibodies
   a. Resuspend the dual cross-linked nuclei in 0.8 mL per liver of Sarkosyl lysis buffer (50 m*M* Tris pH 7.5, 10 m*M* EDTA, 0.5% *N*-lauroylsarcosine, 1 m*M* PMSF, and Roche complete EDTA free protease inhibitor cocktail).
   b. Sonicate 6 × for 30 s at 4 °C using a Covaris S2 ultrasonicator.
   c. Dilute 10-fold with IP buffer.
3. PER1, PER2, CRY1, CRY2, CBP, and p300 antibodies
   a. Resuspend the dual-cross-linked nuclei in 3 mL per liver of IP buffer.
   b. Sonicate 48 × for 5 s on ice using a Misonix S-4000 sonicator.
4. RNAPII-Ser5P antibody
   a. Resuspend formaldehyde-cross-linked nuclei in 0.8 mL per liver of Sarkosyl lysis buffer.
   b. Sonicate 6 × for 30 at 4 °C using a Covaris S2 ultrasonicator
   c. Dilute 10-fold with IP buffer.
5. H3K4me1, H3K4me3, H3K9ac, H3K27ac, H3K36me3, and H3K79me2 antibodies
   a. Resuspend the formaldehyde-cross-linked nuclei in 0.7 mL per liver of MNase Buffer (10 m*M* Tris pH 7.5 and 100 m*M* NaCl).
   b. Sonicate 5 × for 30 s at 4 °C using a Covaris S2 ultrasonicator.
   c. Incubate for 40 min at 37 °C with 200 kunitz units of Micrococcal Nuclease and 2 m*M* CaCl2.
   d. Stop the reaction with 10 m*M* EGTA and 1% SDS.
   e. Dilute 10-fold with IP buffer.

E. ChIP

We used approximately 120 μg (for transcription factors) or 80 μg (for histones) of fragmented chromatin for ChIP-seq.

1. Preclear sonicated nuclear lysates. Add 40 μL (final bed volume) of protein A beads (preblocked with PBS containing 5 mg/mL BSA) to each lysate and incubate for 2 h in rotation at 4 °C.
2. Centrifuge the beads at 14,000 rpm for 10 min at 4 °C.
3. Carefully take out supernatant. Remove 1/10 to 1/20 of the lysate and save as INPUT (go to Step F for reverse cross-linking).
4. The amount of antibody per IP should be determined by careful titration.
5. Add antibody to the precleared chromatin and incubate overnight at 4 °C on a rotating wheel.
6. Add 10 μL (final bed volume) of Protein A/G Plus-agarose (Santa Cruz, sc-2003) and incubated for 1.5 h at 4 °C.
7. Centrifuge the beads at 5000 rpm for 1 min at 4 °C.
8. Remove the supernatant.
9. Wash twice with IP buffer.
10. Wash twice with high salt wash buffer (20 m*M* Tris pH 7.5, 500 m*M* NaCl, 2 m*M* EDTA, 1% Triton X-100, 1 m*M* PMSF).
11. Wash twice with LiCl wash buffer (20 m*M* Tris pH 7.5, 250 m*M* LiCl, 2 m*M* EDTA, 0.5% Igepal CA-630, 1% sodium deoxycholate, 1 m*M* PMSF).
12. Wash once with TE.
13. Carefully remove residual TE.
14. Add 50 μL of Elution Buffer (20 m*M* Tris pH 7.5, 5 m*M* EDTA, 0.5% SDS).
15. Place tubes in 65 °C heat block for 10 min. Gently vortex.
16. Centrifuge the tube at 5000 rpm for 1 min.
17. Transfer the supernatant to new tube.
18. Repeat the elution one more time. Final elution volume is 100 μL.

F. Reverse cross-linking

1. Incubate the eluted chromatin at 65 °C for 5–8 h or up to 12 h to reverse the cross-linking.
2. Add 10 μg of RNaseA and incubate for 30 min at 37 °C.
3. Add 160 μg of proteinase K and incubate for 30 min at 55 °C.
4. Purify DNA using a Qiaquick PCR purification Kit (Qiagen).

2.3. Library preparation for ChIP-seq

This protocol uses previously isolated ChIP DNA and converts it into DNA libraries suitable for subsequent cluster generation and sequencing. The protocol is based on the Illumina workflow and is comparable to the Illumina® TruSeq® ChIP Sample Preparation Kit that has been used as a reference.

2.4. Equipment and reagents needed

1. 1.5 mL nuclease-free tubes
2. 96 well PCR plate, non-skirted
3. Adhesive PCR plate seal
4. 2, 10, 20, 200, and 1000 μL pipettes and 200 μL multichannel pipette
5. PCR machine
6. Magnetic stand-96
7. 0.2 and 1.5 mL nuclease-free tubes
8. Agencourt Ampure XP beads (Beckman Coulter)
9. Bioanalyzer
10. Qubit
11. Qubit dsDNA BR assay kit

2.5. Buffers and enzyme mixes recipes

Ligase storage buffer

Component	Final concentration	Stock solution	10 mL
Water			4.39 mL
Tris–HCl, pH 7.4	10 m*M*	1 *M* pH 7.4 at RT	100 μL
EDTA, pH 8.0	0.1 m*M*	500 m*M* pH 8.0 at RT	2 μL
DTT	1 m*M*	1 *M*	10 μL
KCl	50 m*M*	1 *M*	500 μL
Glycerol	50%	100%	5 mL

Store at −20 °C

End-repair buffer

Component	Volume/reaction (μL)	Vendor	Catalog number
10 m*M* dNTPs	2.5	Enzymatics	N2050-10-L
10× End-repair buffer	4.5	Enzymatics	B9140

Store at −20 °C

End-repair enzyme

Component	Volume/reaction (μL)	Vendor	Catalog number
End-repair mix (low concentration)	3	Enzymatics	Y9140-LC-L

Store at −20 °C

A-tailing mix

Component	Volume/reaction (μL)	Vendor	Catalog number
10 m*M* dATP	1	Enzymatics	N2010-A-L
10 × Blue buffer	2	Enzymatics	B0110
Klenow (3′-5′ exo-) (Low concentration)	0.5	Enzymatics	P7010-LC-L

Store at −20 °C

Ligation mix

Component	Volume/reaction (μL)	Vendor	Catalog number
2 × Ligase buffer	25	Enzymatics	B1010L
Ligase storage buffer	2	–	–
T4 DNA ligase (Rapid)	1	Enzymatics	L6030-HC-L

Store at −20 °C

PCR amplification mix

Component	Volume/reaction (μL)	Vendor	Catalog number
Kapa dNTP Mix	1	Kapa Biosystems	KK2101
KAPA HiFi Fidelity Buffer (5 ×)	10	Kapa Biosystems	KK2101
KAPA HiFi DNA Polymerase (1 U/μL)	1	Kapa Biosystems	KK2101

Store at −20 °C

Library dilution buffer

Component	Final concentration	Stock solution	100 mL
Tris–HCl, pH 8.0	10 m*M*	1 *M*	1 mL
Tween-20	0.05%	100%	50 μL

Store at room temperature

Library normalization buffer

Component	Final concentration	Stock solution	100 mL
Tris–HCl, pH 8.5	10 m*M*	1 *M*	1 mL
Tween-20	0.1%	100%	100 μL

2.6. Adapters and primers

1. The barcoded Y-shaped adapters are ordered from Bioo Scientific (Catalog # 514123). They are stored at −20 °C.
2. The PCR primers are ordered from Integrated DNA Systems and subsequently reconstituted at 100 μ*M* and then diluted to 25 μ*M* each and mixed in equal volume to make a 12.5 μ*M* PCR primer mix. Store at −20 °C.
   PCR primer 1: 5′AATGATACGGCGACCACCGAGATCTACAC
   PCR primer 2: 5′CAAGCAGAAGACGGCATACGAGAT

2.7. Detailed protocol

2.7.1 End repair

The end-repair step converts DNA with overhangs to 5′ phosphorylated, blunt-ended DNA that can be subsequently used for adapter ligation. The conversion of fragmented DNA to blunt-ended is carried out by the 3′ to 5′ and 5′ to 3′ exonuclease activities of T4 DNA polymerase and the 5′ phosphorylation is carried out by the T4 Polynucleotide Kinase in the enzyme mix.

Perform the following reaction in a 96 well plate. Mix,

40 μL	ChIP DNA
7 μL	End-repair buffer
3 μL	End-repair enzyme

Incubate at 25 °C for 30 min.

2.7.2 Bead based size selection

Bead based size selection is based on removing large DNA fragments first by binding them on the beads and doing a supernatant transfer and then subsequently binding all the DNA on the beads except small fragments (less than 100 bp) and then eluting the DNA of interest from the beads.

1. For performing a 150 bp size selection, add 60 μL of well-mixed AMPure XP Beads to each sample and mix thoroughly by pipetting.
2. Incubate the plate for 5 min at room temperature.
3. Place the plate on the magnetic stand for 5 min at room temperature or until the liquid appears completely clear.
4. DO NOT discard the supernatant. Gently transfer 108 μL of the supernatant to a fresh well from the plate without disturbing the beads.
5. Add 40 μL of well-mixed AMPure XP Beads to each sample and mix thoroughly by pipetting.
6. Incubate the plate for 5 min at room temperature.
7. Place the plate on the magnetic stand for 5 min at room temperature or until the liquid appears completely clear.
8. Remove and discard all of the supernatant from the plate taking care not to disturb the beads.
9. With plate on stand, add 200 μL of freshly prepared 80% ethanol to each well without disturbing the beads and incubate the plate for at least 30 s at room temperature. Carefully remove and discard all the supernatant.
10. Repeat Step 9, for a total of two ethanol washes. Ensure the ethanol has been removed.
11. Remove the plate from the magnetic stand and let it dry at room temperature for 2 min.
12. Resuspend dried beads in 18 μL Resuspension Buffer. Gently, pipette the entire volume up and down to mix thoroughly. Ensure that the beads are completely rehydrated and resuspended.
13. Incubate resuspended beads at room temperature for 2 min.
14. Place the plate on the magnetic stand for 5 min at room temperature or until the supernatant appears completely clear.
15. Gently transfer 17 μL of the clear supernatant to a fresh well.
16. The procedure may be stopped at this point and the reactions stored at −20 °C.

Tip: Use multichannel pipette for performing bead cleanups to ensure consistency in processing across samples. Ensure beads don not crack when drying. Complete resuspension of beads will maximize recovery.

2.7.3 A-tailing

A-tailing is performed by utilizing the polyermase activity of Klenow (3′-5′ exo-) in presence of dATP to add a single "A" to the 3′ end of a blunt, double-stranded DNA. A-tailing prevents the blunt fragments from self ligating during the adapter ligation step.

1. For each reaction, mix:

17 μL	End-repaired DNA
3.5 μL	A-tailing mix

2. Mix well by pipetting and then incubate at 37 °C for 30 min followed by 70° for 5 min. Immediately proceed to adapter ligation.

2.7.4 Y-shaped adapter ligation

The ligation step ligates barcoded Y-shaped adapters to the ends of A-tailed DNA fragments. The adapters have a "T" overhang, which is complementary to the adenylated DNA. The ligation step prepares the DNA fragments for subsequent hybridization onto the flow cells.

1. For each reaction, mix:

20.5 μL	Adenylated DNA
2 μL	NEXTflex™ Barcoded Adapter (0.6 μ*M*)
28 μL	Ligation mix

2. Mix well by pipetting and then incubate at 22 °C for 15 min.

2.7.5 Double-bead cleanup

Double-bead cleanup is performed at the end of ligation to remove any excess adapters that might have been self-ligated or be free floating and prevent them from getting amplified during PCR.

1. Add 50.5 μL of well-mixed AMPure XP Beads to each sample and mix thoroughly by pipetting.
2. Incubate the plate for 5 min at room temperature.
3. Place the plate on the magnetic stand for 5 min at room temperature or until the liquid appears completely clear.
4. Remove and discard all of the supernatant from the plate taking care not to disturb the beads.

5. With plate on stand, add 200 μL of freshly prepared 80% ethanol to each well without disturbing the beads and incubate the plate for at least 30 s at room temperature. Carefully, remove and discard the supernatant.
6. Repeat Step 5, for a total of two ethanol washes. Ensure the ethanol has been removed.
7. Remove the plate from the magnetic stand and let dry at room temperature for 2 min.
8. Resuspend dried beads in 51 μL Resuspension Buffer. Gently, pipette the entire volume up and down to mix thoroughly. Ensure that the beads are completely rehydrated and resuspended.
9. Incubate resuspended beads at room temperature for 2 min.
10. Place the plate on the magnetic stand for 5 min at room temperature or until the supernatant appears completely clear.
11. Gently transfer 50 μL of the clear supernatant to a fresh well.
12. Add 50 μL of well-mixed AMPure XP Beads to each well containing sample and mix thoroughly by pipetting.
13. Incubate the plate for 5 min at room temperature.
14. Place the plate on the magnetic stand for 5 min at room temperature or until the liquid appears completely clear.
15. Remove and discard all of the supernatant from the plate taking care not to disturb the beads.
16. With plate on stand, add 200 μL of freshly prepared 80% ethanol to each well without disturbing the beads and incubate the plate for at least 30 s at room temperature. Carefully, remove and discard the supernatant.
17. Repeat Step 16, for a total of two ethanol washes. Ensure the ethanol has been removed.
18. Remove the plate from the magnetic stand and let dry at room temperature for 2 min.
19. Resuspend dried beads in 36 μL Resuspension Buffer. Gently, pipette the entire volume up and down to mix thoroughly. Ensure that the beads are completely rehydrated and resuspended.
20. Incubate resuspended beads at room temperature for 2 min.
21. Place the plate on the magnetic stand for 5 min at room temperature or until the supernatant appears completely clear.
22. Gently transfer 35 μL of the clear supernatant to a fresh well.
23. The procedure may be stopped at this point and the reactions stored at −20 °C.

2.7.6 PCR amplification

PCR amplification is performed to selectively amplify the DNA fragments that have adapters bound to them. The PCR primers anneal in part to the adapter sequences.

1. For each reaction, mix:

35 μL	Adapter ligated DNA
12 μL	PCR amplification mix
2 μL	PCR primer mix (12.5 μ*M*)

2. Mix well by pipetting.

3. PCR cycling:

98 °C 2 min
98 °C 30 s
65 °C 30 s (repeat for 12–20 cycles)
72 °C 60 s
72 °C 4 min

Tip: Always do the minimum number of PCR cycles possible.

2.7.7 Double-bead cleanup

Post PCR amplification a double-bead cleanup is performed to get rid of excess primer and primer dimers.

1. Add 50 μL of well-mixed AMPure XP Beads to each sample and mix thoroughly by pipetting.
2. Incubate the plate for 5 min at room temperature.
3. Place the plate on the magnetic stand for 5 min at room temperature or until the liquid appears completely clear.
4. Remove and discard all of the supernatant from the plate taking care not to disturb the beads.
5. With plate on stand, add 200 μL of freshly prepared 80% ethanol to each well without disturbing the beads and incubate the plate for at least 30 s at room temperature. Carefully, remove and discard the supernatant.
6. Repeat Step 5, for a total of two ethanol washes. Ensure the ethanol has been removed.
7. Remove the plate from the magnetic stand and let dry at room temperature for 2 min.

8. Resuspend dried beads in 51 μL Resuspension Buffer. Gently, pipette the entire volume up and down to mix thoroughly. Ensure that the beads are completely rehydrated and resuspended.
9. Incubate resuspended beads at room temperature for 2 min.
10. Place the plate on the magnetic stand for 5 min at room temperature or until the supernatant appears completely clear.
11. Gently transfer 50 μL of the clear supernatant to a fresh well.
12. Add 50 μL of well-mixed AMPure XP Beads to each well containing sample and mix thoroughly by pipetting.
13. Incubate the plate for 5 min at room temperature.
14. Place the plate on the magnetic stand for 5 min at room temperature or until the liquid appears completely clear.
15. Remove and discard all of the supernatant from the plate taking care not to disturb the beads.
16. With plate on stand, add 200 μL of freshly prepared 80% ethanol to each well without disturbing the beads and incubate the plate for at least 30 s at room temperature. Carefully, remove and discard the supernatant.
17. Repeat Step 16, for a total of two ethanol washes. Ensure the ethanol has been removed.
18. Remove the plate from the magnetic stand and let dry at room temperature for 2 min.
19. Resuspend dried beads in 32 μL Resuspension Buffer. Gently, pipette the entire volume up and down to mix thoroughly. Ensure that the beads are completely rehydrated and resuspended.
20. Incubate resuspended beads at room temperature for 2 min.
21. Place the plate on the magnetic stand for 5 min at room temperature or until the supernatant appears completely clear.
22. Gently transfer 30 μL of the clear supernatant to a fresh well.
23. The procedure may be stopped at this point and the libraries stored at −20 °C until they are validated for quality and quantified for sequencing.

2.8. Quality control

Check the size and quality of the library by running it on a Bioanalyzer using the High Sensitivity DNA assay. If on the Bioanalyzer trace there are two bands, one of expected size and one of higher molecular weight, it's indicative of a bubble product. This double product will not affect the outcome to the sequencing run as double-stranded product is denatured prior to sequencing. As an extra verification step, a portion of this product (1–2 μL) can be denatured manually by heating the sample to 95 °C for 5 min and then placing it on ice and subsequently be run on a Bioanalyzer RNA Pico 6000 Chip Kit. The denatured product should appear as a single band on a Pico 6000 chip.

2.9. Quantification of libraries

In order to get consistent number of reads across different samples it is important to accurately quantify the DNA library templates and then normalize all the samples before sequencing. To get the best sequencing results, it is important to get optimum cluster densities across every lane on every flow cell and this also makes quantification an important step. Perform a Qubit based assay for the quantification of the double stranded libraries using the Qubit dsDNA BR assay kit as per the guidelines provided by the kit. Alternatively, a qPCR based quantification can be performed for quantifying libraries. In our experience the Qubit based quantification is more accurate and reliable.

2.10. Normalizing and pooling libraries for sequencing

1. If you have barcoded libraries, follow Bioo Scientific's guidelines in the barcode manual for pooling normalized samples for sequencing.
2. Normalize the concentration of each library to 20 n*M* using Library normalization buffer and then pool samples at equimolar concentration.
3. Based on the coverage you want you can determine how many samples to pool per lane as 50 bp single end sequencing.

2.11. Data analysis for ChIP-seq

Data analysis

1. Reads are trimmed using fastq-mcf (https://code.google.com/p/ea-utils/wiki/FastqMcf), reads less than 35 bp after trimming are discarded.
2. Bowtie2 is used with the program default values for these two relevant parameters –end-to-end & –sensitive (Langmead & Salzberg, 2012). After mapping, we removed reads of quality less than Q10 using SAMtools (Li et al., 2009).
3. Perform random sampling. In order to normalized for differences in sequencing depth among timed ChIP-seq samples, the sequence reads are "down sampled" to the lowest number of the uniquely mapped reads with duplicates among the time points for each ChIP factor.

4. Remove duplicates using Picard MarkDuplicates (http://picard.sourceforge.net).
5. The peaks are identified from uniquely mapped reads without duplicates using MACS with following parameters: genomic size = mm (1.87 Gb), shift = 60 and input chromatin samples as control data (Zhang et al., 2008). We use a *p*-value threshold of 10^{-5} (default) and a ratio between the ChIP-seq tag count and λlocal of 10 (fold_enrichment threshold). The false peaks called by MACS that repeatedly emerged from low complexity sequence are removed from further analysis.
6. The peaks are then subdivided by PeakSplitter (Salmon-Divon, Dvinge, Tammoja, & Bertone, 2010) with options of –valley 0.7 and –cutoff 7. To construct a master peak list from the six time points, the peaks with summit height more than 6 obtained after PeakSplitter are merged, compared for overlaps and the peak with the highest summit value is chosen if the summit coordinates are within 120 bp. Figure 1 illustrates the master peak process in which MACS peaks are called, then subdivided with PeakSplitter and then compared for overlap and summit height (Koike et al., 2012).The ChIP-seq peak overlaps (peak summit ±120 bp) from the master peak lists are determined using HOMER (Heinz et al., 2010).
7. Results are analyzed using HOMER (Heinz et al., 2010). A tool, "makeTagDirectory," creates Tag Directories for each samples. The numbers of mapped reads in each peak can be quantified by HOMER using a Perl script, "annotatePeaks.pl." HOMER also provides a tool, "makeUCSCfile," to create UCSC visualization file, which can be uploaded as a custom track to UCSC genome browser. We normalize genome browser views to display uniquely mapped reads per 10 million uniquely mapped reads with duplicates.

3. RNA-SEQ METHOD FOR MOUSE LIVER

3.1. Overview of RNA-seq strategy

1. Isolate total RNAs from mouse livers using Trizol reagent (Life Technologies).
2. Determine the quality of isolated total RNAs by Agilent 2011 Bioanalyzer. We usually use total RNA with RIN values of more than 8.
3. For whole transcriptome (WT) RNA-seq, deplete ribosomal RNAs in 10 μg of total RNA pooled from three mice using Ribo-Zero Gold kit

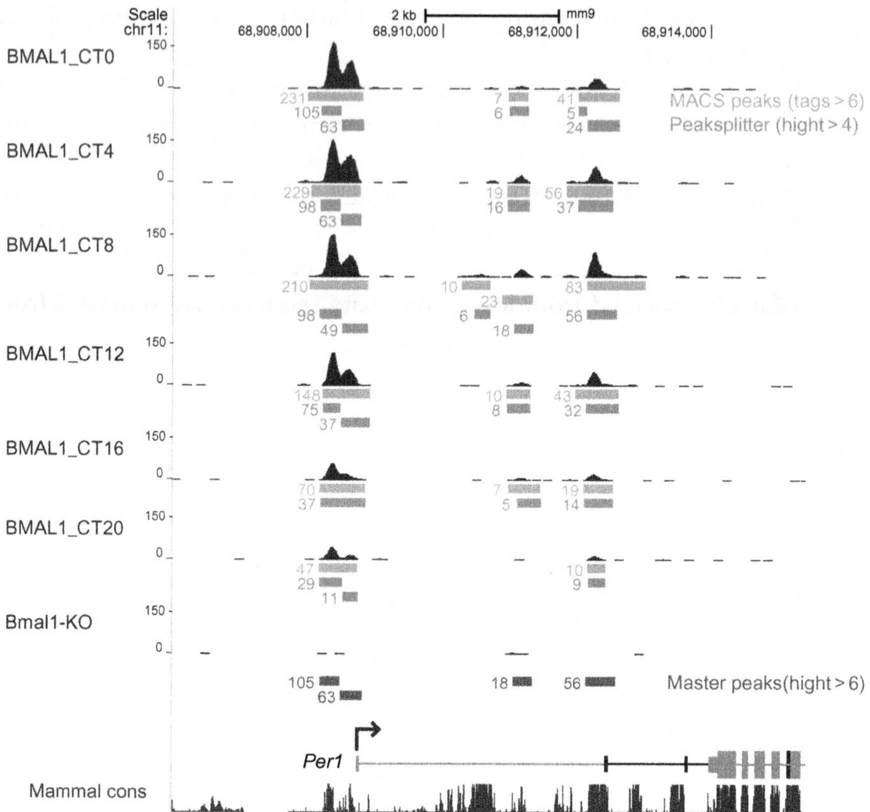

Figure 1 UCSC genome browser view of MACS peak calls for six-timed BMAL1 ChIP-seq occupancy at the *Per1* gene. Orange bars indicate the MACS peak calls and green bars indicate the peak after using Peaksplitter. The numbers to the left of each bar refer to the peak height. Red bars at the bottom show the final consolidated peaks used to construct the master peak list. The Peaksplitter peak with the largest peak height in the region of overlap of peaks is chosen to represent this peak in the master peak list. *Data adapted from Koike et al. (2012).* (See the color plate.)

for Human/Mouse/Rat (Illumina) using the manufacturer's instructions. For mRNA-seq, follow the detailed protocol below.

4. Construct strand-specific RNA-seq libraries using the detailed protocol below to make sequencing libraries for the Illumina HiSeq 2500 platform.
5. For Illumina platforms, we use 50 bp single end reads or 100 bp × 100 bp paired-end reads for WT RNA-seq. The samples can be multiplexed using barcode primers (below). The ability to detect low copy number transcripts depends on the depth of sequencing. A minimum depth of

100 million reads is currently recommended for a typical mammalian tissue, according to Standards, Guidelines, and Best Practices for RNA-Seq from the ENCODE consortium (https://genome.ucsc.edu/ENCODE/protocols/dataStandards/RNA_standards_v1_2011_May.pdf). We recommend at least 100 million reads for WT RNA-seq and at least 30 million reads for mRNA-seq.

3.2. Library preparation for RNA-Seq

This protocol uses total RNA and provides instructions on enriching mRNA that can be subsequently converted into DNA libraries retaining strand origin information. This library can then be used for cluster generation and DNA sequencing. The protocol is based on the Illumina workflow and is comparable to the Illumina® TruSeq® Stranded mRNA Sample Preparation Kit which has been used as a reference.

3.3. Equipment and reagents needed

NEXTflex™ Poly(A) Beads
1.5 mL nuclease-free tubes
96 well PCR plate, non-skirted
Adhesive PCR plate seal
2, 10, 20, 200, and 1000 μL pipettes and 200 μL multichannel pipette
PCR machine
Magnetic stand-96
0.2 and 1.5 mL nuclease-free tubes
Agencourt Ampure XP beads (Beckman Coulter)
Bioanalyzer
Qubit
Qubit dsDNA BR assay kit

3.4. Buffers and enzyme mixes recipes

Actinomycin D (1 mg/mL)

Component	Vendor	Catalog number	5 mL
Actinomycin D	Sigma Aldrich	A1410	5 mg
100% ethanol			5 mL

Actinomycin D stock is stored at −80 °C as aliquots in lightsafe microcentrifuge tubes

120 ng/μL Actinomycin D

Component	Final concentration	Stock solution	1 mL
Actinomycin D	120 ng/μL	1 mg/mL	120 μL
100% Ethanol			880 μL

Store at −80 °C in light safe micro-centrifuge tubes

Ligase storage buffer

Component	Final concentration	Stock solution	10 mL
Water			4.39 mL
Tris–HCl, pH 7.4	10 m*M*	1 *M* pH 7.4 at RT	100 μL
EDTA, pH 8.0	0.1 m*M*	500 m*M* pH 8.0 at RT	2 μL
DTT	1 m*M*	1 *M*	10 μL
KCl	50 mMa	1 *M*	500 μL
Glycerol	50%	100%	5 mL

Store at −20 °C

RNA fragmentation buffer

Component	Final concentration	Stock solution	10 mL
Tris–HCl, pH 8.3	250 m*M*	1 *M*	2.5 mL
KCl	375 m*M*	1 *M*	3.75 mL
$MgCl_2$	10 m*M*	1 *M*	100 μL

Store at −20 °C

First strand synthesis buffer (stranded)

Component	Volume/ reaction (μL)	Vendor	Catalog number
100 m*M* DTT	2	Enzymatics	Supplied with EnzScript™
10 m*M* dNTPs	1	Enzymatics	N2050-10-L
120 ng/μL Antinomycin D	0.5	–	–
RNase Inhibitor	0.5	Enzymatics	Y9240L

Store at −20 °C

EnzScript™

Component	Volume/ reaction (μL)	Vendor	Catalog number
EnzScript™ (M-MLV reverse transcriptase RNase H minus)	0.5	Enzymatics	P7600L

Store at −20 °C

Second strand synthesis mix (stranded)

Component	Volume/ reaction (μL)	Vendor	Catalog number
10 × Blue buffer	3	Enzymatics	Supplied with DNA Polymerase I
dNTP/dUTP mix(1:1:1:2)	1	Enzymatics	–
RNase H	0.5	Enzymatics	Y9220L
DNA Polymerase I	1	Enzymatics	P7050L

Store at −20 °C

A-tailing mix

Component	Volume/reaction (μL)	Vendor	Catalog number
10 m*M* dATP	1	Enzymatics	N2010-A-L
10 × Blue buffer	2	Enzymatics	B0110
Klenow (3′–5′ exo-) (low concentration)	0.5	Enzymatics	P7010-LC-L

Store at −20 °C

Ligation mix

Component	Volume/reaction (μL)	Vendor	Catalog number
2 × Ligase buffer	25	Enzymatics	B1010L
Ligase storage buffer	2	–	–
T4 DNA ligase (Rapid)	1	Enzymatics	L6030-HC-L

Store at −20 °C

PCR amplification mix

Component	Volume/ reaction (μL)	Vendor	Catalog number
Kapa dNTP Mix	1	Kapa Biosystems	KK2101
KAPA HiFi Fidelity Buffer (5 ×)	10	Kapa Biosystems	KK2101
KAPA HiFi DNA Polymerase (1 U/μL)	1	Kapa Biosystems	KK2101

Store at −20 °C

Uracil DNA glycosylase

Component	Volume/reaction (μL)	Vendor	Catalog number
Uracil DNA Glycosylase	1	Enzymatics	G0505L

Store at −20 °C

Library dilution buffer

Component	Final concentration	Stock solution	100 mL
Tris–HCl, pH 8.0	10 m*M*	1 *M*	1 mL
Tween-20	0.05%	100%	50 μL

Store at room temperature

Library normalization buffer

Component	Final concentration	Stock solution	100 mL
Tris–HCl, pH 8.5	10 m*M*	1 *M*	1 mL
Tween-20	0.1%	100%	100 μL

Store at room temperature

3.5. Adapters and primers

3. Random hexamer (NNNNNN) was ordered as a ReadyMade™Primer from Integrated DNA Systems and reconstituted to 100 μ*M* and then diluted to 50 μ*M* to use for first strand synthesis.

4. The barcoded Y-shaped adapters are ordered from Bioo Scientific (Catalog # 512914). They are stored at −20 °C.
5. The PCR primers are ordered from Integrated DNA Systems and subsequently reconstituted at 100 μ*M* and then diluted to 25 μ*M* each and mixed in equal volume to make a 12.5 μ*M* PCR primer mix. Store at −20 °C.

   PCR primer 1: 5′AATGATACGGCGACCACCGAGATCTACAC
   PCR primer 2: 5′CAAGCAGAAGACGGCATACGAGAT

3.6. Detailed protocol

3.6.1 mRNA isolation from total RNA

This step is performed to pull down the mRNA from the total RNA samples using magnetic beads that have oligo(dT) to select for poly(A) mRNA.

Bead washing

This procedure takes approximately 10 min and should be carried out before starting mRNA purification to remove sodium azide in which the beads are stored.

1. Resuspend the magnetic beads thoroughly in the vial to obtain a uniform suspension.
2. Transfer 20 μL of NEXTflex™ Poly(A) Beads to a fresh tube.
3. Place the tube on a DynaMag™-2 Magnet (Life Technologies Cat # 123-21D)/or/similar for 2 min.
4. Remove and discard the supernatant while the tube remains on the magnet.
5. Remove the tube from the magnet and add 100 μL of NEXTflex™ Poly(A) Binding Buffer to the tube, resuspending the beads thoroughly.
6. Place the tube on the magnet for 2 min.
7. Remove and discard the supernatant while the tube remains on the magnet.
8. Resuspend the beads in 100 μL of NEXTflex™ Poly(A) Binding Buffer.

mRNA pulldown

1. If your total RNA sample (1–10 μg) is below 100 μL in volume, adjust its volume to 98 μL using nuclease-free water. Add recommended dilution of ERCC Spike-in mix 1 and then add 100 μL of NEXTflex™ Poly(A) Binding Buffer.

   For example: If using 5 μg of RNA, adjust its volume to 99 μL using nuclease-free water. Add 1 μL of 1:10 dilution of ERCC Spike-in mix

1 and then add 100 μL of NEXTflex™ Poly(A) Binding Buffer. For samples greater than 100 μL in starting volume, add an equal volume of NEXTflex™ Poly(A) Binding Buffer.

2. Heat the total RNA sample to 65 °C for 2 min to disrupt secondary RNA structures. Immediately place on ice.
3. Add your total RNA to the 100 μL of washed beads (as previously described).
4. Mix thoroughly by rotating continuously on a Tube Rotator-unit for 5 min at room temperature.
5. Place the tube on the magnet for 2 min then carefully remove and discard the clear supernatant.
6. Separately aliquot 100 μL of NEXTflex™ Poly(A) Binding Buffer to a fresh 1.5 mL tube.
7. Remove the tube from the magnet, add 200 μL NEXTflex™ Poly(A) Washing Buffer and mix by pipetting. Place the tube on the magnet. Once the beads have pelleted, remove and discard the clear supernatant.
8. Repeat Step 7 for a total of two bead washes.
9. Resuspend the bead pellet with 50 μL of NEXTflex™ Poly(A) Elution Buffer.
10. Heat at 80 °C for 2 min and place the tube immediately on the magnet. After the bead pellet forms, transfer the clear supernatant to the tube prepared in Step 6. Do not discard the used bead pellet.
11. Heat the supernatant sample to 65 °C for 2 min to disrupt secondary structures. Immediately place on ice.
12. Add 200 μL of NEXTflex™ Poly(A) Washing Buffer to the bead pellet from Step 10. Mix by pipetting. Place the tube on the magnet. Once the beads have pelleted, remove and discard the clear supernatant.
13. Repeat Step 12 for a total of two bead washes.
14. Add the RNA sample from Step 11 to the washed beads from Step 13.
15. Mix thoroughly by rotating continuously on a tube rotator for 5 min at room temperature.
16. Place the tube on the magnet for 1–2 min then carefully remove and discard the clear supernatant.
17. Remove the tube from the magnet, add 200 μL NEXTflex™ Poly(A) Washing Buffer and mix by pipetting. Place the tube on the magnet. Once the beads have pelleted, remove and discard the clear supernatant.

18. Repeat Step 17 for a total of two bead washes.
19. Resuspend the bead pellet with 20 μL of NEXTflex™ Poly(A) Elution Buffer.
20. Heat the resuspended pellet to 80 °C for 2 min, then place the tube immediately on the magnet. Transfer the mRNA to a fresh tube or plate. If needed, use 1 μL of eluted mRNA for quantification using a nanodrop or Qubit.

3.6.2 RNA fragmentation

This step is performed to fragment the mRNA to smaller fragments for cDNA synthesis and subsequent library preparation steps.

1. For each reaction, mix in a PCR plate:

14 μL	mRNA (10–100 μg)
5 μL	RNA fragmentation buffer

2. Mix well by pipetting and then incubate at 95 °C for 10 min and then immediately place on ice.

3.6.3 Directional first strand synthesis

This step is performed to synthesize cDNA from the mRNA using random hexamer primers and reverse transcriptase enzyme.

1. For each reaction, add 1 μL random hexamer primer to the fragmented RNA (from Step 2)
2. Incubate at 65 °C for 5 min, and then immediately place on ice.
3. Add 0.5 μL of EnzScript™ per reaction to 4 μL of First strand synthesis buffer (stranded). Add this mix to each reaction, mix gently and spin down.
4. Incubate at 25 °C for 10 min, followed by 42 °C for 50 min and then 70 °C for 15 min.

3.6.4 Directional second strand synthesis

This step is performed to remove the RNA strand and synthesis the second DNA strand using the cDNA strand as a template while incorporating dUTP in the place of dTTP. The dUTP incorporation quenches second strand during amplification because the polymerase does not incorporate past it.

For each reaction, mix:

24.5 μL	First strand synthesis product (from Step 3)
5.5 μL	Second strand synthesis mix (stranded)

Mix well by pipetting and incubate at 16 °C for 1 h.

3.6.5 Bead cleanup

1. Add 54 μL of well-mixed AMPure XP Beads to each sample and mix thoroughly by pipetting.
2. Incubate the plate for 5 min at room temperature.
3. Place the plate on the magnetic stand for 5 min at room temperature or until the liquid appears completely clear.
4. Remove and discard all of the supernatant from the plate taking care not to disturb the beads.
5. With plate on stand, add 200 μL of freshly prepared 80% ethanol to each well without disturbing the beads and incubate the plate for at least 30 s at room temperature. Carefully, remove and discard the supernatant.
6. Repeat Step 5, for a total of two ethanol washes. Ensure the ethanol has been removed.
7. Remove the plate from the magnetic stand and let dry at room temperature for 2 min.
8. Resuspend dried beads in 18 μL Resuspension Buffer. Gently, pipette the entire volume up and down to mix thoroughly. Ensure that the beads are completely rehydrated and resuspended.
9. Incubate resuspended beads at room temperature for 2 min.
10. Place the plate on the magnetic stand for 5 min at room temperature or until the supernatant appears completely clear.
11. Gently transfer 17 μL of the clear supernatant to a fresh well. The procedure can be safely stopped at this point and the samples stored at −80 °C.

3.6.6 A-tailing

A-tailing is performed by utilizing the polyermase activity of Klenow (3′–5′ exo-) in presence of dATP to add a single "A" to the 3′ end of a blunt, double-stranded DNA. A-tailing prevents the blunt fragments from self ligating during the adapter ligation step.

For each reaction, mix:

17 μL	End-repaired DNA
3.5 μL	A-tailing mix

Mix well by pipetting and then incubate at 37 °C for 30 min followed by 70° for 5 min. Immediately proceed to adapter ligation

3.6.7 Y-shaped adapter ligation

The ligation step ligates barcoded Y-shaped adapters to the ends of A-tailed DNA fragments. The adapters have a "T" overhang, which is complementary to the adenylated DNA. The ligation step prepares the DNA fragments for subsequent hybridization onto the flow cells.

For each reaction, mix:

20.5 μL	Adenylated DNA
2 μL	NEXTflex™ barcoded adapter (0.6 μ*M*)
28 μL	Ligation mix

Mix well by pipetting and then incubate at 22 °C for 15 min.

3.6.8 Double-bead cleanup

Double-bead cleanup is performed at the end of ligation to remove any excess adapters that might have been self-ligated or be free floating and prevent them for getting amplified during PCR.

1. Add 50.5 μL of well-mixed AMPure XP Beads to each sample and mix thoroughly by pipetting.
2. Incubate the plate for 5 min at room temperature.
3. Place the plate on the magnetic stand for 5 min at room temperature or until the liquid appears completely clear.
4. Remove and discard all of the supernatant from the plate taking care not to disturb the beads.
5. With plate on stand, add 200 μL of freshly prepared 80% ethanol to each well without disturbing the beads and incubate the plate for at least 30 s at room temperature. Carefully, remove and discard the supernatant.

6. Repeat Step 5, for a total of two ethanol washes. Ensure the ethanol has been removed.
7. Remove the plate from the magnetic stand and let dry at room temperature for 2 min.
8. Resuspend dried beads in 51 μL Resuspension Buffer. Gently, pipette the entire volume up and down to mix thoroughly. Ensure that the beads are completely rehydrated and resuspended.
9. Incubate resuspended beads at room temperature for 2 min.
10. Place the plate on the magnetic stand for 5 min at room temperature or until the supernatant appears completely clear.
11. Gently transfer 50 μL of the clear supernatant to a fresh well.
12. Add 50 μL of well-mixed AMPure XP Beads to each well containing sample and mix thoroughly by pipetting.
13. Incubate the plate for 5 min at room temperature.
14. Place the plate on the magnetic stand for 5 min at room temperature or until the liquid appears completely clear.
15. Remove and discard all of the supernatant from the plate taking care not to disturb the beads.
16. With plate on stand, add 200 μL of freshly prepared 80% ethanol to each well without disturbing the beads and incubate the plate for at least 30 s at room temperature. Carefully, remove and discard the supernatant.
17. Repeat Step 16, for a total of two ethanol washes. Ensure the ethanol has been removed.
18. Remove the plate from the magnetic stand and let dry at room temperature for 2 min.
19. Resuspend dried beads in 36 μL resuspension buffer. Gently, pipette the entire volume up and down to mix thoroughly. Ensure that the beads are completely rehydrated and resuspended.
20. Incubate resuspended beads at room temperature for 2 min.
21. Place the plate on the magnetic stand for 5 min at room temperature or until the supernatant appears completely clear.
22. Gently transfer 35 μL of the clear supernatant to a fresh well.
23. The procedure may be stopped at this point and the reactions stored at −20 °C.

3.6.9 Uracil-DNA Glycosylase treatment and PCR amplification

In this step, the Uracil DNA Glycosylase (UDG) hydrolyzes the *N*-glycosylic bond between uracil and sugar in DNA, selectively degrading the dUTP marked strand and therefore the remaining strand is amplified to generate directional cDNA library. The PCR primers anneal in part to the adapter sequences.

For each reaction, mix:

35 μL	Adapter ligated DNA
1 μL	Uracil DNA Glycosylase
12 μL	PCR amplification mix
2 μL	PCR primer mix (12.5 μ*M*)

Mix well by pipetting.

PCR cycling:

37 °C 2 min
98 °C 2 min
98 °C 30 s
65 °C 30 s (repeat for 12–20 cycles)
72 °C 60 s
72 °C 4 min

Tip: Always do the minimum number of PCR cycles possible.

3.6.10 Double-bead cleanup

Post PCR amplification a double-bead cleanup is performed to get rid of excess primer and primer dimers.

1. Add 50 μL of well-mixed AMPure XP Beads to each sample and mix thoroughly by pipetting.
2. Incubate the plate for 5 min at room temperature.
3. Place the plate on the magnetic stand for 5 min at room temperature or until the liquid appears completely clear.
4. Remove and discard all of the supernatant from the plate taking care not to disturb the beads.
5. With plate on stand, add 200 μL of freshly prepared 80% ethanol to each well without disturbing the beads and incubate the plate for at least 30 s at room temperature. Carefully, remove and discard the supernatant.
6. Repeat Step 5, for a total of two ethanol washes. Ensure the ethanol has been removed.
7. Remove the plate from the magnetic stand and let dry at room temperature for 2 min.

8. Resuspend dried beads in 51 μL Resuspension Buffer. Gently, pipette the entire volume up and down to mix thoroughly. Ensure that the beads are completely rehydrated and resuspended.
9. Incubate resuspended beads at room temperature for 2 min.
10. Place the plate on the magnetic stand for 5 min at room temperature or until the supernatant appears completely clear.
11. Gently transfer 50 μL of the clear supernatant to a fresh well.
12. Add 50 μL of well-mixed AMPure XP Beads to each well containing sample and mix thoroughly by pipetting.
13. Incubate the plate for 5 min at room temperature.
14. Place the plate on the magnetic stand for 5 min at room temperature or until the liquid appears completely clear.
15. Remove and discard all of the supernatant from the plate taking care not to disturb the beads.
16. With plate on stand, add 200 μL of freshly prepared 80% ethanol to each well without disturbing the beads and incubate the plate for at least 30 s at room temperature. Carefully, remove and discard the supernatant.
17. Repeat Step 16, for a total of two ethanol washes. Ensure the ethanol has been removed.
18. Remove the plate from the magnetic stand and let dry at room temperature for 2 min.
19. Resuspend dried beads in 32 μL Resuspension Buffer. Gently, pipette the entire volume up and down to mix thoroughly. Ensure that the beads are completely rehydrated and resuspended.
20. Incubate resuspended beads at room temperature for 2 min.
21. Place the plate on the magnetic stand for 5 min at room temperature or until the supernatant appears completely clear.
22. Gently transfer 30 μL of the clear supernatant to a fresh well.
23. The procedure may be stopped at this point and the libraries stored at −20 °C until they are validated for quality and quantified for sequencing.

3.7. Quality control

Check the size and quality of the library by running it on a Bioanalyzer using the High Sensitivity DNA assay. If on the Bioanalyzer trace there are two bands, one of expected size and one of higher molecular weight, it's indicative of a bubble product. This double product will not affect the outcome to the sequencing run as double-stranded product is denatured prior to sequencing. As an extra verification step, a portion of this product (1–2 μL) can be denatured manually by heating the sample to 95 °C for 5 min and then placing it on ice and subsequently be run on a Bioanalyzer RNA Pico 6000 Chip Kit. The denatured product should appear as a single band on a Pico 6000 chip.

3.8. Quantification of libraries

In order to get consistent number of reads across different samples, it is important to accurately quantify the DNA library templates and then normalize all the samples before sequencing. To get the best sequencing results, it is important to get optimum cluster densities across every lane on every flow cell and this also makes quantification an important step. Perform a Qubit based assay for the quantification of the double stranded libraries using the Qubit dsDNA BR assay kit as per the guidelines provided by the kit. Alternatively, a qPCR based quantification can be performed for quantifying libraries. In our experience the Qubit based quantification is more accurate and reliable.

3.9. Normalizing and pooling libraries for sequencing

1. If you have barcoded libraries, follow Bioo Scientific's guidelines in the barcode manual for pooling normalized samples for sequencing
2. Normalize the concentration of each library to 20 n*M* using Library normalization buffer and then pool samples at equimolar concentration
3. Based on the coverage you want you can determine how many samples to pool per lane for 50 bp single end sequencing.

3.10. Data analysis of RNA-seq data

1. Map the sequence reads to the mouse genome with Tophat (Kim et al., 2013). For strand-specific RNA-seq data, "–library-type fr-firststrand" is the parameter to use for Illumina.
2. Remove the reads mapped with low mapping quality (<5) using SAMtools (Li et al., 2009) to get rid of the reads mapped to multiple locations.
3. Use Homer (Heinz et al., 2010) to process the mapped reads. Homer includes tools to analyze RNA-seq, ChIP-seq, etc. First, tag directories for each sample will be created by a tool "makeTagDirectory." Then, RNA expression is quantified using Perl scripts "analyzeRNA.pl" or "analyzeRepeats.pl." The scripts have options to count reads mapped

to intron, exon, or gene body for each gene. For WT RNA-seq data, we interpret the intron signal as a representation of pre-mRNA expression or nascent transcription (Ameur et al., 2011) and the exon signal as representation of mRNA expression. The expression levels are normalized as reads per kilobase per million mapped reads (RPKM), because longer genes have chance to be mapped more reads. For gene annotation, we used the UCSC known canonical gene set to eliminate transcript variants. For gene annotation, Homer can use GTF files, which can be downloaded from UCSC Table browser.

3.11. Time series analysis for circadian cycling

Time series analysis of very short data sets is nontrivial. Ideally if one were to use Fourier Transform methods to assess the frequency and amplitude of time series data as in the case of locomotor activity data (Takahashi & Menaker, 1982), it would be necessary to analyze at least 10 cycles of the target periodicity at a sampling resolution that matches the Nyquist frequency ($f_s/2$, where f_s = sample rate). Because of the expense of ChIP-seq and RNA-seq samples, obtaining 10 cycles of molecular data is highly unlikely, and it is customary to assay only two cycles of circadian time series. Indeed analysis of only one cycle of data cannot reliably estimate period. A number of algorithms have been developed to estimate period and amplitude of short time series. Most use some type of fitting procedure to either sinusoidal or prespecified waveforms. Significance thresholds are usually then estimated using bootstrap methods. We have used three different programs for RNA cycling, COSOPT (Panda et al., 2002), JTK Cycle (Hughes et al., 2010), and ARSER (Yang & Su, 2010). COSOPT runs on Microsoft Windows, JTK requires R packages, and ARSER is implemented by a Python program calling some R functions. For COSOPT and JTK Cycle analyses, data is detrended by linear regression. Previously (Koike et al., 2012) we considered a cycling gene if two out of three programs detected cycling with threshold of $p < 0.05$. The period and phase from ARSER was used for further analysis. For ChIP-seq peak analysis, two cycles were concatenated and the cycling was analyzed with ARSER ($p < 0.05$). Because COSOPT is slow and much less sensitive at detection of cycling transcripts compared to JTK Cycle and ARSER, we now routinely use the latter two programs for circadian cycling detection. While these programs are adequate, each has its propensity for false negative and false positive detection of cycling, and each is very sensitive to the details of sampling interval, number of replicates, and time series duration. We find that the sets of cycling genes detected by these programs is variable across experiments and can be discordant between programs because of the specific waveform features of the time series. Thus, the analysis of short circadian time series is clearly an area for future development.

REFERENCES

Ameur, A., Zaghlool, A., Halvardson, J., Wetterbom, A., Gyllensten, U., Cavelier, L., et al. (2011). Total RNA sequencing reveals nascent transcription and widespread co-transcriptional splicing in the human brain. *Nature Structural & Molecular Biology, 18*, 1435–1440.

Buck, M. J., & Lieb, J. D. (2004). ChIP-chip: Considerations for the design, analysis, and application of genome-wide chromatin immunoprecipitation experiments. *Genomics, 83*, 349–360.

Bunger, M. K., Wilsbacher, L. D., Moran, S. M., Clendenin, C., Radcliffe, L. A., Hogenesch, J. B., et al. (2000). Mop3 is an essential component of the master circadian pacemaker in mammals. *Cell, 103*, 1009–1017.

Chapman, R. D., Heidemann, M., Albert, T. K., Mailhammer, R., Flatley, A., Meisterernst, M., et al. (2007). Transcribing RNA polymerase II is phosphorylated at CTD residue serine-7. *Science, 318*, 1780–1782.

Chaw, Y. F., Crane, L. E., Lange, P., & Shapiro, R. (1980). Isolation and identification of cross-links from formaldehyde-treated nucleic acids. *Biochemistry, 19*, 5525–5531.

Chen, R., Schirmer, A., Lee, Y., Lee, H., Kumar, V., Yoo, S. H., et al. (2009). Rhythmic PER abundance defines a critical nodal point for negative feedback within the circadian clock mechanism. *Molecular Cell, 36*, 417–430.

Eckner, R., Ludlow, J. W., Lill, N. L., Oldread, E., Arany, Z., Modjtahedi, N., et al. (1996). Association of p300 and CBP with simian virus 40 large T antigen. *Molecular and Cellular Biology, 16*, 3454–3464.

Gekakis, N., Staknis, D., Nguyen, H. B., Davis, F. C., Wilsbacher, L. D., King, D. P., et al. (1998). Role of the CLOCK protein in the mammalian circadian mechanism. *Science, 280*, 1564–1569.

Gilmour, D. S., & Lis, J. T. (1984). Detecting protein-DNA interactions in vivo: Distribution of RNA polymerase on specific bacterial genes. *Proceedings of the National Academy of Sciences of the United States of America, 81*, 4275–4279.

Hatanaka, F., Matsubara, C., Myung, J., Yoritaka, T., Kamimura, N., Tsutsumi, S., et al. (2010). Genome-wide profiling of the core clock protein BMAL1 targets reveals a strict relationship with metabolism. *Molecular and Cellular Biology, 30*, 5636–5648.

Heinz, S., Benner, C., Spann, N., Bertolino, E., Lin, Y. C., Laslo, P., et al. (2010). Simple combinations of lineage-determining transcription factors prime cis-regulatory elements required for macrophage and B cell identities. *Molecular Cell, 38*, 576–589.

Hughes, M. E., Hogenesch, J. B., & Kornacker, K. (2010). JTK_CYCLE: An efficient nonparametric algorithm for detecting rhythmic components in genome-scale data sets. *Journal of Biological Rhythms, 25*, 372–380.

Jackson, V. (1978). Studies on histone organization in the nucleosome using formaldehyde as a reversible cross-linking agent. *Cell, 15*, 945–954.

Jackson, V., & Chalkley, R. (1981). A new method for the isolation of replicative chromatin: Selective deposition of histone on both new and old DNA. *Cell, 23*, 121–134.

Jones, J. C., Phatnani, H. P., Haystead, T. A., MacDonald, J. A., Alam, S. M., & Greenleaf, A. L. (2004). C-terminal repeat domain kinase I phosphorylates Ser2 and Ser5 of RNA polymerase II C-terminal domain repeats. *The Journal of Biological Chemistry*, *279*, 24957–24964.

Karpov, V. L., Preobrazhenskaya, O. V., & Mirzabekov, A. D. (1984). Chromatin structure of hsp 70 genes, activated by heat shock: Selective removal of histones from the coding region and their absence from the 5' region. *Cell*, *36*, 423–431.

Kim, D., Pertea, G., Trapnell, C., Pimentel, H., Kelley, R., & Salzberg, S. L. (2013). TopHat2: Accurate alignment of transcriptomes in the presence of insertions, deletions and gene fusions. *Genome Biology*, *14*, R36.

King, D. P., Zhao, Y., Sangoram, A. M., Wilsbacher, L. D., Tanaka, M., Antoch, M. P., et al. (1997). Positional cloning of the mouse circadian clock gene. *Cell*, *89*, 641–653.

Koike, N., Yoo, S. H., Huang, H. C., Kumar, V., Lee, C., Kim, T. K., et al. (2012). Transcriptional architecture and chromatin landscape of the core circadian clock in mammals. *Science*, *338*, 349–354.

Kume, K., Zylka, M. J., Sriram, S., Shearman, L. P., Weaver, D. R., Jin, X., et al. (1999). mCRY1 and mCRY2 are essential components of the negative limb of the circadian clock feedback loop. *Cell*, *98*, 193–205.

Kuo, M. H., & Allis, C. D. (1999). In vivo cross-linking and immunoprecipitation for studying dynamic Protein:DNA associations in a chromatin environment. *Methods*, *19*, 425–433.

Langmead, B., & Salzberg, S. L. (2012). Fast gapped-read alignment with Bowtie 2. *Nature Methods*, *9*, 357–359.

Lee, C., Etchegaray, J. P., Cagampang, F. R., Loudon, A. S., & Reppert, S. M. (2001). Posttranslational mechanisms regulate the mammalian circadian clock. *Cell*, *107*, 855–867.

Lee, C., Weaver, D. R., & Reppert, S. M. (2004). Direct association between mouse PERIOD and CKIε is critical for a functioning circadian clock. *Molecular and Cellular Biology*, *24*, 584–594.

Le Martelot, G., Canella, D., Symul, L., Migliavacca, E., Gilardi, F., Liechti, R., et al. (2012). Genome-wide RNA polymerase II profiles and RNA accumulation reveal kinetics of transcription and associated epigenetic changes during diurnal cycles. *PLoS Biology*, *10*.

Li, H., Handsaker, B., Wysoker, A., Fennell, T., Ruan, J., Homer, N., et al. (2009). The sequence alignment/map format and SAMtools. *Bioinformatics*, *25*, 2078–2079.

Lowrey, P. L., & Takahashi, J. S. (2004). Mammalian circadian biology: Elucidating genome-wide levels of temporal organization. *Annual Review of Genomics and Human Genetics*, *5*, 407–441.

Lowrey, P. L., & Takahashi, J. S. (2011). Genetics of circadian rhythms in Mammalian model organisms. *Advances in Genetics*, *74*, 175–230.

McGhee, J. D., & von Hippel, P. H. (1975a). Formaldehyde as a probe of DNA structure. I. Reaction with exocyclic amino groups of DNA bases. *Biochemistry*, *14*, 1281–1296.

McGhee, J. D., & von Hippel, P. H. (1975b). Formaldehyde as a probe of DNA structure. II. Reaction with endocyclic imino groups of DNA bases. *Biochemistry*, *14*, 1297–1303.

Menet, J. S., Rodriguez, J., Abruzzi, K. C., & Rosbash, M. (2012). Nascent-Seq reveals novel features of mouse circadian transcriptional regulation. *Elife*, *1*, e00011.

Nowak, D. E., Tian, B., & Brasier, A. R. (2005). Two-step cross-linking method for identification of NF-kappaB gene network by chromatin immunoprecipitation. *Biotechniques*, *39*, 715–725.

Panda, S., Antoch, M. P., Miller, B. H., Su, A. I., Schook, A. B., Straume, M., et al. (2002). Coordinated transcription of key pathways in the mouse by the circadian clock. *Cell*, *109*, 307–320.

Rahl, P. B., Lin, C. Y., Seila, A. C., Flynn, R. A., McCuine, S., Burge, C. B., et al. (2010). c-Myc regulates transcriptional pause release. *Cell*, *141*, 432–445.

Reick, M., Garcia, J. A., Dudley, C., & McKnight, S. L. (2001). NPAS2: An analog of clock operative in the mammalian forebrain. *Science (New York, NY)*, *293*, 506–509.
Ren, B., Robert, F., Wyrick, J. J., Aparicio, O., Jennings, E. G., Simon, I., et al. (2000). Genome-wide location and function of DNA binding proteins. *Science*, *290*, 2306–2309.
Rey, G., Cesbron, F., Rougemont, J., Reinke, H., Brunner, M., & Naef, F. (2011). Genome-wide and phase-specific DNA-binding rhythms of BMAL1 control circadian output functions in mouse liver. *PLoS Biology*, *9*, e1000595.
Salmon-Divon, M., Dvinge, H., Tammoja, K., & Bertone, P. (2010). PeakAnalyzer: Genome-wide annotation of chromatin binding and modification loci. *BMC Bioinformatics*, *11*, 415.
Solomon, M. J., & Varshavsky, A. (1985). Formaldehyde-mediated DNA-protein crosslinking: A probe for in vivo chromatin structures. *Proceedings of the National Academy of Sciences of the United States of America*, *82*, 6470–6474.
Takahashi, J. S., & Menaker, M. (1982). Role of the suprachiasmatic nuclei in the circadian system of the house sparrow, Passer domesticus. *The Journal of Neuroscience*, *2*, 815–828.
Vollmers, C., Schmitz, R., Nathanson, J., Yeo, G., Ecker, J., & Panda, S. (2012). Circadian oscillations of protein-coding and regulatory RNAs in a highly dynamic mammalian liver epigenome. *Cell Metabolism*, *16*, 833–845.
Welsh, J., & Cantor, C. R. (1984). Protein-DNA cross-linking. *Trends in Biochemical Sciences*, *9*, 505–508.
Yang, R., & Su, Z. (2010). Analyzing circadian expression data by harmonic regression based on autoregressive spectral estimation. *Bioinformatics*, *26*, i168–i174.
Zeng, P. Y., Vakoc, C. R., Chen, Z. C., Blobel, G. A., & Berger, S. L. (2006). In vivo dual cross-linking for identification of indirect DNA-associated proteins by chromatin immunoprecipitation. *Biotechniques*, *41*, 694, 696, 698.
Zhang, Y., Liu, T., Meyer, C. A., Eeckhoute, J., Johnson, D. S., Bernstein, B. E., et al. (2008). Model-based analysis of ChIP-Seq (MACS). *Genome Biology*, *9*, R137.
Zhou, J., Yu, W., & Hardin, P. E. (2015). ChIPing away at the Drosophila clock. *Methods in Enzymology*, (in press).

CHAPTER FIFTEEN

ChIPping Away at the *Drosophila* Clock

Jian Zhou, Wangjie Yu, Paul E. Hardin[1]
Department of Biology and Center for Biological Clocks Research, Texas A&M University, College Station, Texas, USA
[1]Corresponding author: e-mail address: phardin@bio.tamu.edu

Contents

Abstract

In eukaryotes, the circadian clock controls 24 h rhythms in physiology, metabolism, and behavior via cell autonomous transcriptional feedback loops. These feedback loops keep circadian time and control rhythmic outputs by driving rhythms in transcription; thus, it is important to determine when clock transcription factors bind their target sequences *in vivo* to promote or repress transcription. Interactions between proteins and DNA can be measured in cells, tissue, or whole organisms using a technique called chromatin immunoprecipitation (ChIP). The principle underlying ChIP is that protein is cross-linked to associated chromatin to form a protein–DNA complex, the DNA is then sheared, and the protein of interest is immunoprecipitated. The cross-links are then removed from the antibody–protein–DNA complex, and the associated DNA fragments are purified. The DNA is then used to quantify specific targets by real-time quantitative PCR or to generate libraries for global analysis of protein target sites by high-throughput sequencing (ChIP-seq). ChIP has been widely used in circadian biology to assess rhythmic binding of clock components, RNA polymerase II, and rhythms in chromatin modifications such as histone acetylation and methylation. Here, we present a detailed method for ChIP analysis in Drosophila that can be used to assess protein–DNA-binding rhythms at specific genomic target sites. With minor modifications, this technique can

Methods in Enzymology, Volume 551
ISSN 0076-6879
http://dx.doi.org/10.1016/bs.mie.2014.10.019

be used to assess protein–DNA-binding rhythms at all target sites via ChIP-seq. ChIP analysis has revealed the relationship between clock factor binding, transcription, and chromatin modifications and promises to reveal circadian transcription networks that control phase and tissue specificity.

1. INTRODUCTION

The circadian timekeeping mechanism in eukaryotes is comprised of one or more transcriptional feedback loops. The conserved molecular architecture of these loops includes positive factors that activate transcription of negative factor genes, feedback of negative factors to inhibit positive factors, and removal of negative factors, thereby allowing the positive factors to initiate the next round of transcription (Bell-Pedersen et al., 2005; Dunlap, 1999; Young & Kay, 2001). Although the 24-h periodicity of this feedback loop is largely regulated posttranscriptionally, the transcriptional activation and repression that forms the basis of this feedback loop also drives transcription of output genes that mediate overt rhythms in metabolism, physiology, and behavior (Hogenesch, Panda, Kay, & Takahashi, 2003; Vitalini, de Paula, Park, & Bell-Pedersen, 2006).

Among the best-characterized feedback loops is that in Drosophila, where the basic helix-loop-helix transcription factors CLOCK (CLK) and CYCLE (CYC) heterodimerize to form the positive factor, and PERIOD (PER) and TIMELESS (TIM) heterodimerize to form the negative factor (Hardin, 2011). The timing of PER–TIM repression is controlled posttranscriptionally at different steps including the accumulation of PER in the cytoplasm, the movement of PER–TIM into the nucleus to inhibit CLK–CYC, and the degradation of PER–TIM in the nucleus to release transcriptional repression (Hardin, 2011). These positive and negative factors are largely conserved in mammals: two CLK orthologs, CLK and NPAS2, form heterodimers with the CYC ortholog BMAL1 to form positive factors, whereas PER orthologs, PER1 and PER2, form a heterodimer with CRY1 or CRY2 rather than TIM to form the negative factor (Lowrey & Takahashi, 2011). As in Drosophila, posttranscriptional regulation of negative factor accumulation, nuclear localization, and degradation plays a primary role in controlling when and how long transcription is repressed (Lowrey & Takahashi, 2011).

In Drosophila and mammals, positive factors bind CACGTG or related E-box sequences to activate transcription of target genes, which include negative factors and many output genes. Given that the timing of transcription by positive factors is determined by negative factors, it was important to understand how negative factors effect transcriptional inhibition. Negative factors directly interact with positive factors when transcription is repressed (Lee, Bae, & Edery, 1998; Lee, Etchegaray, Cagampang, Loudon, & Reppert, 2001), suggesting that negative factors repress transcription by inactivating positive factors or removing them from E-box sequences altogether. Pioneering studies by Reppert and colleagues used chromatin immunoprecipitation (ChIP) assays to distinguish between these possibilities in mice. These studies showed that CLK and BMAL1 were bound to DNA at all times of the circadian cycle (Etchegaray, Lee, Wade, & Reppert, 2003; Lee et al., 2001), yet rhythms in histone acetylation accompanied transcriptional activation (Etchegaray et al., 2003), suggesting that PER–CRY complexes rhythmically inactivate E-box bound CLK–BMAL1.

In Drosophila, *in vitro* DNA-binding experiments showed that CLK and CYC in fly head extracts made at any time of the circadian cycle was capable of binding E-boxes (Wang et al., 2001), though previous western analysis showed that CLK levels were low when transcription was high, and high when transcription was low (Lee et al., 1998). To understand the relationship between CLK levels, E-box binding, and transcriptional activation, we employed a new CLK antibody that was useful for immunohistochemistry and western analysis (Houl, Ng, Taylor, & Hardin, 2008). This antibody detected rhythms in CLK levels similar to those seen previously on westerns (i.e., antiphase to transcription), yet CLK immunoreactivity was present at constant levels in clock cells (Houl et al., 2008). ChIP analysis using this antibody and an existing CYC antibody showed that CLK and CYC rhythmically bound to E-box sequences upstream of the *per* and *tim* genes coincident with transcriptional activation (Yu & Hardin, 2006). Western analysis of head extracts made using stringent procedures (i.e., similar to those used in ChIP but without cross-linking) showed that CLK levels were similar at all times of day, but that CLK was hypophosphorylated during times of transcriptional activation (Yu & Hardin, 2006). These results suggested that stringent extraction procedures were necessary to remove hypophosphorylated CLK from E-boxes during times of transcriptional activation. Although rhythmic CLK and CYC binding in Drosophila contrasted with the constant CLK and BMAL1 binding seen in mice, ChIP analysis by Schibler and colleagues using different antibodies and procedures showed that CLK and BMAL1 rhythmically bind E-boxes coincident with transcriptional activity (Ripperger & Schibler, 2006). Subsequent ChIP analysis in Drosophila and mice support the general concept that positive factors rhythmically bind target gene E-boxes, where binding coincides with times when negative factors are absent (Abruzzi et al., 2011; Koike et al., 2012; Menet, Abruzzi,

Desrochers, Rodriguez, & Rosbash, 2010; Menet, Rodriguez, Abruzzi, & Rosbash, 2012; Rey et al., 2011; Taylor & Hardin, 2008).

Here, we describe the ChIP protocol we currently use to assess transcription factor binding in fly heads. For those new to ChIP, we begin by outlining the basic steps needed to carry out ChIP analysis and the principle behind each step (Fig. 1). Unlike cultured cells and soft tissues from mammals, fly tissues are covered with cuticle that make it more difficult to carry out ChIP, especially for steps such as cross-linking and extracting nuclei. As part of our description, we provide tips for avoiding mistakes that reduce sample quantity and/or quality and indicate points when samples can be safely stored without compromising subsequent steps. Importantly, this protocol can be used to characterize rhythms in transcription activity and chromatin state by determining when core transcription factors (e.g., RNA polymerase II) bind to target genes and the presence of chromatin modifications (e.g., histone modifications or isoforms) associated with transcriptional activity or repression (Taylor & Hardin, 2008). Although our description is limited to the analysis of factor binding to specific genomic target sites, we discuss how samples generated using our protocol can be used for whole-genome analysis of transcription factor-binding sites or chromatin modifications when combined with high-throughput sequencing (ChIP-seq).

2. EQUIPMENT

Sonicator (Microson or Diagenode Bioruptor)
3/32 in. microprobe for the Microson sonicator
Diagenode Bioruptor sonication tubes (necessary for Diagenode Bioruptor)
Agarose gel electrophoresis system
−80 °C freezer
−20 °C freezer
4 °C refrigerator
Gel documentation system
Micropipettors
Micropipettor tips
Centrifuge
1.5 ml Eppendorf (EP) tubes
Vortex mixer
15 and 50-ml Falcon tubes
Spectrophotometer
#25 and #40 sieves with collection pan
Tube rotator

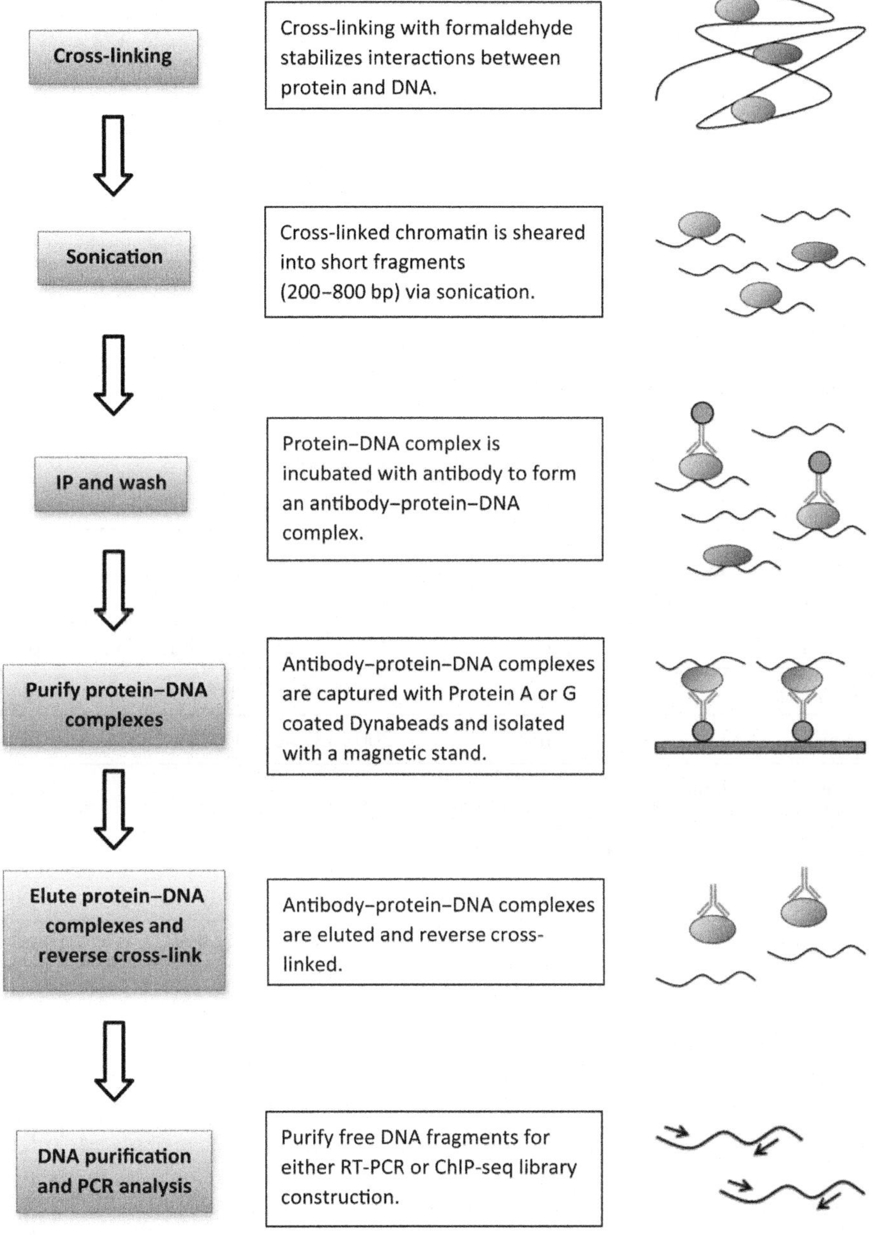

Figure 1 ChIP principle and workflow. The basic steps for ChIP (gray boxes) are outlined in the workflow shown on the left and with a brief description of the principle behind each step (white boxes). On the right, a series of molecular models depict the state of the sample at each step of process. Proteins, green (black in the print version) and brown (dark gray in the print version) ovals; DNA, black lines; Dynabeads, gray circles; antibodies, blue (light gray in the print version) structures; magnetic stand, gray bar; PCR primers, arrows.

7-ml homogenizer with loose pestle (two sets for each sample)
Heat blocks (at 37 and 65 °C)
100 μ*m* nylon mesh
Magnetic stand
Paint brushes (both soft/flexible and hard/stiff)
60 and 80-mm funnels

3. SOLUTIONS

XIP homogenization buffer (HB)[a]

Component	Final concentration	Stock solution	10 ml
Water			4.58 ml
2 × XIP-HB-HSEEIT	1 ×	2 ×	5 ml
Formaldehyde (HCHO)	1%	37%	270.2 μl
PMSF[b]	1 m*M*	200 m*M* in ethanol at 4 °C	50 μl
Na_3VO_4	1 m*M*	200 m*M* at −20 °C	50 μl
NaF	1 m*M*	200 m*M* at RT	50 μl

[a]Make fresh each time.
[b]PMSF is not stable in aqueous solution; add PMSF to the solution just before use.

XIP homogenize dilution buffer (HDB)[a]

Component	Final concentration	Stock solution	10 ml
Water			4.84 ml
2 × XIP-HB-HSEEIT	1 ×	2 ×	5 ml
PMSF[b]	0.5 m*M*	200 m*M* at 4 °C	25 μl
Aprotinin	10 μg/ml	10 μg/μl at 4 °C	10 μl
Leupeptin	10 μg/ml	10 μg/μl at −20 °C	10 μl
Pepstatin A	2 μg/ml	1 μg/μl at −20 °C	20 μl
Na_3VO_4	1 m*M*	200 m*M* at −20 °C	50 μl
NaF	1 m*M*	200 m*M* at RT	50 μl

[a]Make fresh each time.
[b]PMSF is not stable in aqueous solution; add PMSF to the solution just before use.

2× XIP-HB-HSEEIT[a]

Component	Final concentration	Stock solution	50 ml
Water			35.88 ml
HEPES·K (pH 8.0)	100 m*M*	1 *M* at 4 °C	5 ml
NaCl	280 m*M*	5 *M* at RT	2.8 ml
EDTA (pH 8.5)	2 m*M*	0.5 *M* at 4 °C	200 μl
EGTA (pH 8.0)	1 m*M*	0.4 *M* at 4 °C	125 μl
Igpel CA-630	0.8%	10%	4 ml
Triton X-100	0.4%	10%	2 ml

[a]This solution can be stored temporary at 4 °C, or long-term at −20 °C.

XIP nuclei wash buffer[a]

Component	Final concentration	Stock solution	10 ml
Water			8.835 ml
10× XIP-TSEE	1×	10×	1 ml
PMSF[b]	0.5 m*M*	200 m*M* at 4 °C	25 μl
Aprotinin	10 μg/ml	10 μg/μl at 4 °C	10 μl
Leupeptin	10 μg/ml	10 μg/μl at −20 °C	10 μl
Pepstatin A	2 μg/ml	1 μg/μl at −20 °C	20 μl
Na_3VO_4	1 m*M*	200 m*M* at −20 °C	50 μl
NaF	1 m*M*	200 m*M* at RT	50 μl

[a]Make fresh each time.
[b]PMSF is not stable in aqueous solution; add PMSF to the solution just before use.

10× XIP-TSEE[a]

Component	Final concentration	Stock	50 ml
Water			23.375 ml
Tris Cl (pH 7.5)	200 m*M*	1 *M*, pH 7.5 at RT	10 ml
NaCl	1.5 *M*	5 *M* at RT	15 ml
EDTA (pH 8.5)	10 m*M*	0.5 *M* at 4 °C	1 ml
EGTA (pH 8.0)	5 m*M*	0.4 *M* at 4 °C	625 μl

[a]This solution can be stored temporary at 4 °C, or long-term at −20 °C.

XIP-SonicBuffer[a]

Component	Final concentration	Stock	5 ml
Water			2.425 ml
2 × XIP-SonicBuf-GTDSTSEE	1 ×	2 ×	2.5 ml
PMSF[b]	0.5 m*M*	200 m*M* at 4 °C	12.5 μl
Aprotinin	10 μg/ml	10 μg/μl at 4 °C	5 μl
Leupeptin	10 μg/ml	10 μg/μl at −20 °C	5 μl
Pepstatin A	2 μg/ml	1 μg/μl at −20 °C	10 μl
Na_3VO_4	1 m*M*	200 m*M* at −20 °C	25 μl
NaF	1 m*M*	200 m*M* at RT	25 μl

[a]Make fresh each time.
[b]PMSF is not stable in aqueous solution; add PMSF to the solution just before use.

2 × XIP-SonicBuf-GTDSTSEE[a]

Component	Final concentration	Stock solution	30 ml
Water			9 ml
Glycerol	20%	100%	6 ml
Triton X-100	2%	10%	6 ml
DOC	0.8%	10%	2.4 ml
SDS	0.2%	10%	600 μl
10 × TSEE	2 ×	10 ×	6 ml

[a]This solution can be stored temporary at 4 °C, or long-term at −20 °C.

Bioruptor sonication buffer[a]

Component	Final concentration	Stock	10 ml
Water			8.48 ml
HEPES–Na (pH 7.5)	20 m*M*	1 *M*	200 μl
EDTA (pH 8.0)	2 m*M*	500 m*M*	40 μl
SDS	1%	10%	1000 μl
Triton X-100	0.2%	10%	200 μl
Spermidine	0.5 m*M*	0.5 *M*	10 μl
Spermine	0.15 m*M*	0.15 *M*	10 μl

PMSF[b]	0.5 m*M*	200 m*M* at 4 °C	25 μl
Aprotinin	10 μg/ml	10 μg/μl at 4 °C	10 μl
Leupeptin	10 μg/ml	10 μg/μl at −20 °C	10 μl
Pepstatin A	2 μg/ml	1 μg/μl at −20 °C	20 μl

[a]Make fresh each time.
[b]PMSF is not stable in aqueous solution; add PMSF to the solution just before use.

XIP-IPBuffer[a]

Component	Final concentration	Stock	10 ml
Water			4.84 ml
2× XIP-IPBuf-TSTSEE	1×	2×	5 ml
PMSF[b]	0.5 m*M*	200 m*M* at 4 °C	25 μl
Aprotinin	10 μg/ml	10 μg/μl at 4 °C	10 μl
Leupeptin	10 μg/ml	10 μg/μl at −20 °C	10 μl
Pepstatin A	2 μg/ml	1 μg/μl at −20 °C	20 μl
Na_3VO_4	1 m*M*	200 m*M* at −20 °C	50 μl
NaF	1 m*M*	200 m*M* at RT	50 μl

[a]Make fresh each time.
[b]PMSF is not stable in aqueous solution; add PMSF to the solution just before use.

Blocking buffer

Component	Final concentration	Stock solution	1 ml
XIP-IPBuffer			850 μl
Sonicated salmon sperm DNA[a]	0.1 μg/μl	1 μg/μl	100 μl
BSA	5 μg/μl	100 μg/μl	50 μl

[a]For ChIP-seq, add 0.1 μg/μl yeast tRNA instead.

2× XIP-IPBuf-TSTSEE[a]

Component	Final concentration	Stock solution	50 ml
Water			29.9 ml
Triton X-100	2%	10%	10 ml
SDS	0.02%	10%	100 μl
10× TSEE	2×	10×	10 ml

[a]This solution can be stored temporary at 4 °C, or long-term at −20 °C.

XIP-HiSalt Buffer[a]

Component	Final concentration	Stock solution	10 ml
Water			3.835 ml
2× XIP-HiLoSalt-TTEE	1×	2×	5 ml
PMSF[b]	0.5 m*M*	200 m*M* at 4 °C	25 μl
Aprotinin	10 μg/ml	10 μg/μl at 4 °C	10 μl
Leupeptin	10 μg/ml	10 μg/μl at −20 °C	10 μl
Pepstatin A	2 μg/ml	1 μg/μl at −20 °C	20 μl
Na_3VO_4	1 m*M*	200 m*M* at −20 °C	50 μl
NaF	1 m*M*	200 m*M* at RT	50 μl
NaCl	500 m*M*	5 *M*	1.0 ml

[a]Make fresh each time.
[b]PMSF is not stable in aqueous solution; add PMSF to the solution just before use.

XIP-LowSalt Buffer[a]

Component	Final concentration	Stock solution	10 ml
Water			3.835 ml
2× XIP-HiLoSalt-TTEE	1×	2×	5 ml
PMSF[b]	0.5 m*M*	200 m*M* at 4 °C	25 μl
Aprotinin	10 μg/ml	10 μg/μl at 4 °C	10 μl
Leupeptin	10 μg/ml	10 μg/μl at −20 °C	10 μl
Pepstatin A	2 μg/ml	1 μg/μl at −20 °C	20 μl
Na_3VO_4	1 m*M*	200 m*M* at −20 °C	50 μl
NaF	1 m*M*	200 m*M* at RT	50 μl
Water			1.0 ml

[a]Make fresh each time.
[b]PMSF is not stable in aqueous solution; add PMSF to the solution just before use.

2× XIP-HiLoSalt-TTEE[a]

Component	Final concentration	Stock solution	50 ml
Water			37.68 ml
Tris Cl pH 7.5	40 m*M*	1 *M* pH 7.5 at RT	2.0 ml
Triton X-100	2%	10%	10 ml

EDTA pH 8.5	2 m*M*	0.5 *M* at 4 °C	200 μl
EGTA pH 8.0	1 m*M*	0.4 *M* at 4 °C	125 μl

[a]This solution can be stored temporary at 4 °C, or long-term at −20 °C.

Li Buffer[a]

Component	Final concentration	Stock solution	10 ml
Water			7.4 ml
Tris Cl pH 7.5	10 m*M*	1 *M* pH 7.5 at RT	100 μl
Igpel CA-630	1%	10%	1.0 ml
DOC[b]	1%	10%	1.0 ml
LiCl	250 m*M*	8 *M*	312.5 μl
EDTA	1 m*M*	500 m*M*	20 μl
PMSF[c]	0.5 m*M*	200 m*M* at 4 °C	25 μl
Aprotinin	10 μg/ml	10 μg/μl at 4 °C	10 μl
Leupeptin	10 μg/ml	10 μg/μl at −20 °C	10 μl
Pepstatin A	2 μg/ml	1 μg/μl at −20 °C	20 μl
Na_3VO_4	1 m*M*	200 m*M* at −20 °C	50 μl
NaF	1 m*M*	200 m*M* at RT	50 μl

[a]Make fresh each time.
[b]Warm up 10% DOC at 25 °C before use since it will precipitate at room temperature (RT).
[c]PMSF is not stable in aqueous solution; add PMSF to the solution just before use.

Elution buffer[a]

Component	Final concentration	Stock solution	1 ml
Water			0.8 ml
SDS	1%	10%	0.1 ml
$NaHCO_3$	100 m*M*	1 *M*[b]	0.1 ml

[a]Make fresh each time.
[b]Make fresh before use.

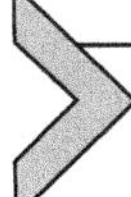

4. PROTOCOL

4.1. Step 1. Isolating fly heads

In this step, the procedure for isolating fly heads is described. Fly heads are typically used to study molecular clock mechanisms in Drosophila because they are highly enriched for clock cells. The vast majority of clock cells in fly heads are photoreceptors, which show rhythms in clock protein expression similar to brain pacemaker neurons and peripheral tissues (Glossop & Hardin, 2002). Although it would be useful to assess DNA-binding profiles of feedback loop components in individual tissues of flies, it is not practical to purify large quantities of fly tissues in contrast to the situation in mammals.

Fly collection

1. Place 50 ml Falcon tubes and a 80-mm funnel on dry ice for cooling.
2. Collect approximately 20 ml of flies into a Falcon tube on dry ice using the funnel.
3. Freeze flies at −80 °C for at least 3 h before collecting heads.
4. You may keep flies frozen at −80 °C for many weeks or months if necessary before collecting fly heads.

Tips: 20 standard 6 oz fly bottles will yield ~20 ml of flies. When collecting flies from bottles, be careful not to hit the bottles too forcefully on the funnel as this can cause food to fall into and clog the funnel.

Head collection

1. Place a #25 and a #40 sieve both with the collection holders and a 60-mm funnel on dry ice for cooling.
2. Vortex each Falcon tube for 20 s twice, shake the tube vigorously for 20 s, and then keep the tube on dry ice for approximately 1–2 min in between vortexing and shaking.
3. Stack the #40 sieve on top of the collection pan and pour the entire 20 ml of flies from the Falcon tube onto the #40 sieve. Using a soft paint brush, brush the flies so that the wings and legs fall through the #40 sieve. Fly heads and bodies will be left on the #40 sieve.
4. Stack the #25 sieve on top of another collection pan and pour the fly heads and bodies from the #40 sieve onto the #25 sieve, brush the heads and bodies on the #25 sieve with a hard paint brush until all the heads fall through and into the collection pan.
5. Transfer fly heads from the collection pan into an EP tube that was precooled on dry ice using the 60-mm funnel. Label each EP tube and keep at −80 °C.
6. The procedure can be stopped here when the samples are frozen at −80 °C, or continue on to the next step.

Tips: One sieve of each size is sufficient to process multiple samples. Clean the sieve with a hard brush between processing each sample. Always keep whole flies and fly heads on dry ice and never let flies or fly heads thaw. 20 ml of flies should yield ~1 ml of fly heads.

4.2. Step 2. X-Nuclei preparation

This step describes how to prepare cross-linked chromatin–protein complexes. Because fly heads are covered with cuticle, it is important to first grind the heads using a homogenizer to break apart the cuticle so that nuclei can be efficiently released by HB. During the homogenization process, any protein–DNA complexes present in cell nuclei will be stabilized by formaldehyde cross-linking.

Preparing cross-linked chromatin–protein complexes

1. Measure 1 ml fly heads.
2. Warm up 2 × XIP-HB-HSEEIT, 10 × XIP-TSEE, and 1.4 *M* glycine solutions at 37 °C water bath.
3. Prepare 5 ml of HB, 7 ml of HDB, and 8 ml wash buffer for each sample. Keep at RT.
4. Place a 7-ml homogenizer with loose pestle and a 60-mm funnel on dry ice.
5. Place 1 ml frozen fly heads into a 7-ml homogenizer and grind them on dry ice for 80 strokes.
6. Pour ground heads into another 7-ml homogenizer at RT or a 25 °C water bath using the dry ice-cooled funnel.
7. Immediately add 5 ml of HB, gently homogenize for 10 min at RT with occasional vortexing.
8. Add 570 μl of 1.4 *M* glycine, homogenize gently for 5 min at RT with occasional vortexing.
9. Filter the homogenate through 100 μm nylon mesh into a 50-ml Falcon tube by placing the mesh on top of the Falcon tube and pouring the homogenate onto the mesh.
10. Wash the homogenizer using 7 ml of HDB by homogenizing for 3–5 strokes, then filter the homogenate through the nylon mesh.
11. Pellet X-Nuclei
    - **(a)** Spin the 50-ml Falcon tube at 1500 × *g* for 5 min, carefully remove the supernatant, do not touch the pellet at the bottom of the tube.
    - **(b)** Resuspend the pellet with 1 ml wash buffer and transfer the suspension into three EP tubes evenly.
    - **(c)** Centrifuge the EP tubes at 950 × *g* for 5 min, remove supernatant.
    - **(d)** Wash the pellets twice with 1 ml wash buffer, each time spin the tube at 950 × *g* for 5 min.
    - **(e)** Transfer all nuclei suspension into one EP tube.
    - **(f)** Spin at 950 × *g* for 5 min, remove the supernatant.

(g) Label the tube containing X-Nuclei, store at −80 °C.
(h) The procedure can be stopped at this point, if necessary, once the sample is frozen at −80 °C.

Tips: A sample prepared from 1 ml fly heads should contain enough material for two ChIP assays. It is easiest to process large amounts of sample at one time. People with experience performing ChIP assays can try starting with fewer fly heads for one ChIP. Timing is critical for the cross-linking step; homogenizing longer than 10 min will make it much more difficult to shear chromatin to a consistent and reproducible length.

4.3. Step 3. Sonication

In this step, the DNA in cross-linked protein–DNA complexes will be sheared into short fragments. Chromatin shearing is a critical step because DNA fragment size will significantly affect IP efficiency and background. DNA-shearing protocols differ depending on the type of sonicator that is used. Here, we provide two protocols for DNA shearing using a standard Microson sonicator or a Diagenode Bioruptor sonicator. Since the volume to be sonicated is low (∼400 μl), a 3/32 in. microprobe should be used in the Microson sonicator. After the DNA is sonicated, the size of sheared DNA is measured (Fig. 2) and should ideally fall into the 200–800 bp size range.

Microson sonication

1. Thaw X-Nuclei on ice. Add 3× volume of XIP-SonicBuffer to X-Nuclei. Typically, the volume of nuclei is ∼100 μl; thus, the sonication volume will be ∼400 μl.
2. Set the Microson XL 2000 sonicator output at 4–5 W on the display.
3. Sonicate for 10 s × 15 times in a cold ethanol bath on crushed ice, 150 s in total. Wait 50 s between each sonication.
4. Centrifuge at 25,000 × *g* for 10 min at 4 °C to remove debris (most of the debris is cuticle) and save supernatant as X-Nuclear extract (SXN).
5. Estimate the concentration of SXN.
   (a) Make a series of standards by diluting bovine serum albumin (BSA) into XIP-SonicBuffer.
   (b) Dilute 2 ml of Bio-Rad Protein Assay Dye Reagent concentrate (Cat. #500-0006) with 8 ml of H_2O to make protein dye mix.
   (c) Mix 5 μl of each standard or sample with 1 ml of protein dye mix by vortexing.

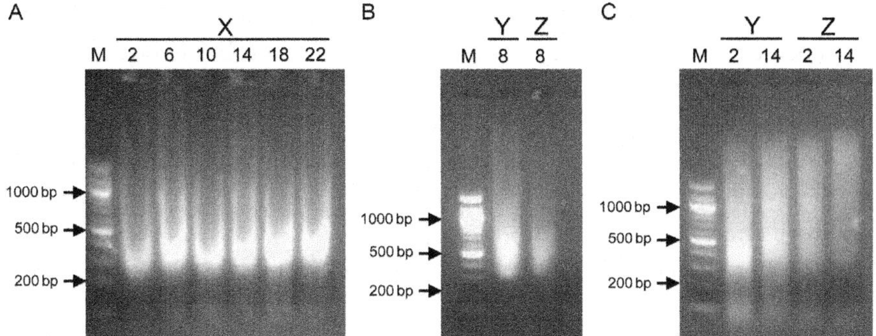

Figure 2 Sizing sheared genomic DNA. (A) ChIP samples prepared from the heads of fly strain X collected at the indicated Zeitgeber Times were sheared in a Diagenode Bioruptor, resulting in 250–500 bp (average size of ~350 bp) DNA fragments that are suitable for qPCR or sequencing analysis. (B) ChIP samples prepared from the heads of fly strains Y and Z collected at ZT8 were sheared in a Microson sonicator, resulting in 300–700 bp (average size of ~500 bp) DNA fragments that are suitable for qPCR or sequencing analysis. (C) ChIP samples prepared from the heads of fly strains Y and Z collected at the indicated Zeitgeber Times were sheared in a Microson sonicator, resulting in long 400 to >3000 bp (~1000 bp average) DNA fragments that are not suitable for RT-PCR or sequencing analysis.

(d) Measure each standard in a spectrophotometer and generate a standard curve.
(e) Measure each SXN sample in a spectrophotometer, calculate the protein concentration of each sample according to the standard curve.

6. Calculate the volume of SXN needed for ChIP (500 μg protein) and input (50 μg protein) and store aliquots at −80 °C.
7. Stop here if necessary with the sample frozen at −80 °C, or continue to the next step.

Tips: The cold ethanol bath is for cooling the samples since their temperature will rise during sonication. However, make sure the sample does not freeze in the cold ethanol bath. During sonication, avoid foaming of the sample. If bubbles are produced during sonication, centrifuge the tube at 25,000 × *g* for 5 min at 4 °C to remove the bubbles, then continue sonication. If 1 ml of fly heads is used as starting material, the final protein concentration of the SXN is typically 2.5–3.0 μg/μl.

Diagenode bioruptor sonication

1. Resuspend X-Nuclei (~100 μl) in 400 μl of Bioruptor sonication buffer.
2. Split the sample into two Bioruptor tubes (250 μl in each).

3. Sonicate using the following program: A cycle of 30 s on and 30 s off is repeated for a total of 20–30 cycles. Vortex the tubes every 10 cycles.
4. Centrifuge at 25,000 × *g* for 10 min at 4 °C to remove debris and save supernatant as Bioruptor X-Nuclear extract (BXN).
5. Estimate concentration and calculate the volume for ChIP using the same method as described in steps 5 and 6 for the Microson sonicator. Store aliquots at −80 °C.
6. Stop here if necessary with the sample frozen at −80 °C, or continue to the next step.

Tips: An advantage of using the Diagenode Bioruptor is that multiple samples can be sonicated in parallel.

Determine DNA fragment size

1. Add 2 μl of 5 *M* NaCl to a 10–20 μl aliquot of the SXN or BXN, incubate in a 65 °C water bath for 6 h to overnight (or incubate in a boiling water bath for 15 min) to reverse the cross-links, and centrifuge at max speed for 5 min.
2. Run the supernatant on a 2% agarose gel to determine the fragment size (Fig. 2).

Tips: The number of sonication cycles or the length of each cycle may need to be adjusted to generate DNA fragments within the 200–800 bp range. DNA fragments longer than 800 bp or shorter than 200 bp may increase background or decrease IP efficiency, respectively. Once you have identified conditions that produce DNA fragments in the desired range, it is not necessary to determine fragment size for each experiment. Reversing cross-links by boiling is fast, but is not as efficient as incubation at 65 °C for 6 h to overnight in our hands.

4.4. Step 4. IP and washes

In this step, the protein–DNA complexes will be incubated with an antibody raised against the protein of interest to form antibody–protein–DNA complex. The antibody–protein–DNA complex will be isolated using Dynabeads coupled to either Protein A or Protein G, which are bacterial proteins with high affinity for immunoglobulins. Once the antibody–protein–DNA complexes are bound to Dynabeads, nonspecific protein–DNA complexes, and free DNA fragments are removed during a series of washes, leaving specific antibody–protein–DNA complexes bound to the Dynabeads.

Isolating antibody–protein–DNA complexes

1. Preincubate SXN or BXN with antibody.
   - (a) Add 3 μl antibody to the SXN or BXN (500 μg), then add IP buffer to SXN samples to bring the total volume to ~800 μl or add IP buffer to BXN samples to dilute the sample 10-fold (~2.5 ml).
   - (b) Add 10% NaN_3 to a final concentration of 0.025%.
   - (c) Place the sample in a tube rotator, set at ~10 rpm overnight at 4 °C.
2. Blocking beads (prepare the same day).
   - (a) Use 30–50 μl Dynabeads for each sample. Wash Dynabeads with 1 ml IP buffer. For this and subsequent washes, capture the beads using a magnetic stand, add the solution to resuspend the beads, and then recapture beads.
   - (b) Wash the beads with 1 ml blocking solution twice.
   - (c) Resuspend the beads with blocking buffer and add NaN_3 to a final concentration of 0.025%.
   - (d) Rotate the beads at ~10 rpm overnight at 4 ° C.

Tips: The antibody volume added to the SXN or BXN depends on the antibody being used, but 3 μl is typically sufficient. If the sample was sonicated using a Bioruptor, a large dilution of the sonicated material is critical because the Bioruptor sonication buffer has 1% SDS, and the SDS must be diluted before the IP so the antibody will not be denatured and inactivated. In our experience, IP efficiency is compromised to some extent even after a 10× dilution of Bioruptor sonication buffer.

3. Incubate immunocomplexes with Dynabeads.
   - (a) Capture blocked beads, wash beads with 1 ml IP buffer.
   - (b) Resuspend the beads with the preincubated SXN or BXN with antibody.
   - (c) Rotate at ~10 rpm at 4 °C for 2 h.
4. Wash (rotate at ~10 rpm for 5 min at 4 °C for each wash).
   - (a) Wash with 1 ml of IP buffer twice.
   - (b) Wash with 1 ml of LowSalt buffer twice.
   - (c) Wash with 1 ml of HiSalt buffer twice.
   - (d) Wash with 1 ml of Li buffer twice.
   - (e) Wash with 1 ml of TE buffer twice.

Tips: Washing additional times, or for a longer time, should reduce the background but may also decrease the specific binding signal.

Likewise, fewer washes, or washing for a shorter time, should increase the specific binding signal but may also increase background. Keep the sample on ice or at 4 °C as much as possible during incubation and washes.

4.5. Step 5. Elution and DNA extraction

In this step, the antibody–protein–DNA complex is eluted from the beads, treated to remove RNA and proteins, and then reverse cross-linked. DNA fragments are then purified for real-time quantitative PCR (qPCR) or sequencing analysis.

Isolating DNA from immunoprecipitates

1. Elution
   - **(a)** Add 50 μl of elution buffer to the washed beads and incubate at 65 °C for 15 min (vortex every 2–3 min), then move the supernatant (50 μl) to a new tube.
   - **(b)** Repeat the elution on the same beads using another 50 μl elution buffer, combine the supernatants together to give a total volume of 100 μl.
   - **(c)** Stop here if necessary with the sample frozen at −80 °C, or continue to the next step.

Tips: Move the washed beads into a new EP tube before adding elution buffer. This will reduce the background by eliminating the unspecific protein attached to the tube wall during incubation and washes.

2. DNA extraction:
   - **(a)** Add one volume (100 μl) of 2 × TE buffer to the eluates. For input samples, add 1 × TE buffer to total volume of 200 μl.
   - **(b)** Add RNase A to a final concentration of 50 μg/ml and incubate at 37 °C for 30 min.
   - **(c)** Adjust SDS in the input sample to 0.5%, then add proteinase K to all the samples to a final concentration of 1 μg/μl. Incubate at 37 °C for 6 h or overnight.
   - **(d)** Add 5 *M* NaCl to final concentration of 0.3 *M*, reverse cross-link at 65 °C for 6 h to overnight.
   - **(e)** Add 300 μl of phenol–chloroform for extraction, vortex thoroughly, centrifuge at 14,000 rpm for 5 min at 4 °C. Remove and save the upper layer into another labeled EP tube. Add 50 μl of

1 × TE to the lower layer, vortex thoroughly, and centrifuge at 14,000 rpm for 5 min at 4 °C. Remove the upper layer (~50 μl) and combine with the previously extracted upper layer.

(f) To prepare DNA for qPCR analysis, add 650 ml of ethanol, glycogen (0.5 μl of 20 μg/μl), and 0.5 μg sonicated salmon sperm DNA to the IP sample (do not add salmon sperm DNA to the input sample, only glycogen) and precipitate at −20 °C overnight. To prepare DNA for ChIP-seq, do not add salmon sperm DNA, add a total of 1 μg glycogen instead.

(g) Centrifuge at 14,000 rpm for 30 min at 4 °C, remove the supernatant. Add 1 ml of cold 70% ethanol, mix thoroughly, centrifuge at 14,000 rpm for 10 min at 4 °C. Discard supernatant and air-dry pellet for 10 min, then resuspend in 50 μl of 1 × TE for qPCR or ChIP-seq.

(h) Stop here if necessary with the sample frozen at −80 °C, or continue to the next step.

Tips: Before phenol–chloroform extraction, remove the sample to a new EP tube to prevent leakage when vortexing (65 °C heat overnight will weaken the seal of the tube cap). DNA purification can also be done using QIAquick PCR purification kit; however, the yield could be less compared to phenol–chloroform extraction and ethanol precipitation. Also keep in mind that the QIAquick PCR purification kit can only purify up to 10 μg of DNA for each tube.

4.6. Step 6. qPCR analysis

In this step, qPCR is carried out to quantify the amount of DNA that was immunoprecipitated at a particular target site in each sample (Fig. 3), which is a measurement of the affinity of the protein for that target site. This method quantifies binding at a known or hypothesized site as a percentage of this site in input DNA (% of input) minus the % of input value for a negative control site that shows no binding. Alternatively, samples can be used to prepare libraries for ChIP-seq analysis, which will be discussed in more general terms below.

Quantifying immunoprecipitated DNA

1. Design two sets of primers: One set for amplifying a 100–200-bp DNA fragment containing a target-binding site, the other set for amplifying an

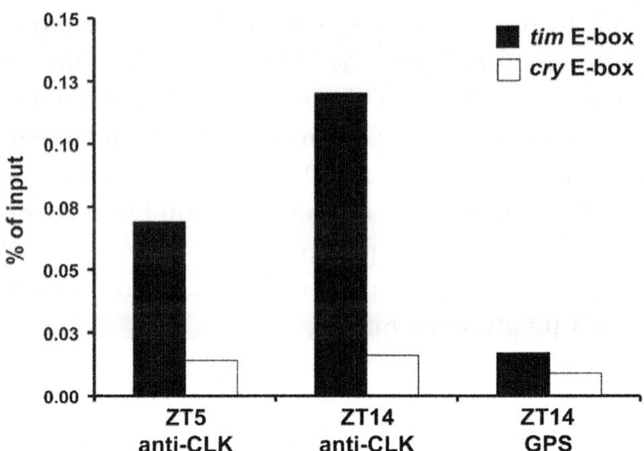

Figure 3 ChIP data quantification. CLK antiserum (anti-CLK) and Guinea pig serum (GPS) were used to prepare ChIP samples from the heads of wild-type flies collected at the indicated Zeitgeber Times. The relative level of CLK and GPS binding (% of input) to the upstream *tim* E-box (black bar) and *cry* E-box (white bar) was determined by qPCR analysis. The *tim* E-box is specifically bound by CLK, and the *cry* E-box serves as a negative control for nonspecific binding because it is not bound by CLK antibody. The GPS data also serve as control for nonspecific antibody binding, but are not necessary for ChIP quantification.

untranslated region or other genomic region that is known not to be bound by the protein of interest as a background control.

2. There are several methods to quantify DNA levels using qPCR. Here, we will use the standard curve method. To make a standard curve for the qPCR, a series of dilutions for one of the inputs needs to be generated using TE buffer. Set the standard quantity as follows:

Dilution of the input sample	Arbitrary quantity of standard
1:100,000	1
1:20,000	5
1:10,000	10
1:2000	50
1:1000	100
1:200	500
1:100	1000

3. Dilute each input sample into 1:1000 and each IP sample into 1:10 with TE buffer to use as DNA templates.
4. Reaction setup in 96-well PCR plates (Cat. # MLL9601, BIO-RAD):

7.5 μl	Ssofast EvaGreen Supermix (Cat. #172-5201, BIO-RAD)
1.25 μl	10 μ*M* primer forward
1.25 μl	10 μ*M* primer reverse
5 μl	Diluted DNA template

5. Centrifuge the 96-well plates in 1500 rpm for 2 min, then put the plate in a BIO-RAD CFX96 real-time PCR machine, design the qPCR program according to the fragment length and primer annealing temperature. Save the data file containing the qPCR results and analyze the data with CFX manager software according to the manufacturer's instructions.
6. A standard curve will be automatically generated from the dilutions of the input sample (see step 2 in this section) by the CFX manager software. The qPCR starting quantity (SQ) for each IP sample and input will then be calculated by the software based on the standard curve that was produced. Because the IP samples and the inputs were diluted before qPCR, the original quantity of IP samples and inputs is then calculated: IP quantity = IP SQ × 10 and input quantity = input SQ × 1000 (the 10 and 1000 multipliers come from the dilution fold in step 3). If different IP and input dilutions were used, the equation should be adjusted accordingly.
7. Relative ChIP abundance is represented as the % of input, which is the proportion of DNA fragments that are enriched from the starting material (input). The % of input is calculated as follows: % of input = (IP quantity)/(input quantity × 10) × 100% (the 10 multiplier comes from the 500 μg used for IP vs. the 50 μg for input). The same calculation is used for negative control data.
8. Correction for nonspecific binding = % of input for binding site − % of input for negative control.
   Tips: Adjust dilution concentration of the input and IP samples based on the efficiency of IP. IgG or serum can also be used as control, but is not required using this quantification method. There are short protocols online for using the CFX96 real-time PCR machine.

5. DISCUSSION

ChIP is a powerful technique for studying protein–DNA interactions. By applying this technique to circadian biology, important principles underlying rhythmic transcription in animal model systems have been derived. For instance, positive factors bind E-box sequences when transcription is high and are released from E-boxes when negative factors interact with positive factors (Ripperger & Schibler, 2006; Yu & Hardin, 2006). This principle extends beyond animals to fungi, where the positive factor WHITE COLLAR 2 (WC2) rhythmically forms a complex with WHITE COLLAR 1 (WC1) on "clock-box" DNA regulatory elements when transcription is high, and WC2 is released after FREQUENCY (FRQ) interacts with this complex to repress transcription (Belden, Loros, & Dunlap, 2007). Additionally, histone modifications that correspond to transcriptionally active or repressed states are rhythmic for genes activated by positive circadian factors in Drosophila, mice, and Neurospora (Belden et al., 2007; Etchegaray et al., 2003; Ripperger & Schibler, 2006; Yu & Hardin, 2006). Despite these advances, traditional ChIP that measures binding at specific target sites has its limitations. For example, a "ChIP-grade" antibody that can efficiently immunoprecipitate cross-linked binding protein is required and target sequences must be known. In other words, this method is used to test a hypothesized protein–DNA interaction; therefore, one should have candidates for both the binding protein and the genomic DNA target region. Consequently, traditional ChIP is not suitable for identifying genes that are targeted by the binding protein of interest.

With the advent of high-throughput sequencing technology, the identity of all genomic-binding sites for a transcription factor can now be achieved using ChIP-seq analysis (Barski et al., 2007; Johnson, Mortazavi, Myers, & Wold, 2007; Robertson et al., 2007). For ChIP-seq analysis, instead of designing primers to amplify a specific genomic region, the immunoprecipitated DNA fragments are processed for making DNA libraries that are then sequenced using high-throughput sequencing technology. The procedure for preparing immunoprecipitated DNA fragments for ChIP-qPCR and ChIP-seq is the same. Although this chapter does not include protocols for making DNA libraries, sequencing, and bioinformatic analysis, these procedures have been documented (Bailey et al., 2013; Landt et al., 2012). ChIP-seq has been applied to circadian biology in several model systems. These studies have documented thousands of rhythmic (and some nonrhythmic)-binding sites for positive factors in Drosophila, mice, and Neurospora (Abruzzi et al., 2011; Koike et al., 2012; Menet et al., 2012; Mireles-Filho, Bardet, Yanez-Cuna, Stampfel, & Stark, 2013; Rey et al., 2011; Smith et al., 2010), described the relationship between RNA polymerase II binding, nascent transcripts, and rhythmic transcription in mouse liver (Koike et al., 2012; Le Martelot et al., 2012; Menet et al., 2012), and identified chromatin marks that occur during rhythmic transcription in mouse liver (Koike et al., 2012; Vollmers et al., 2012). Principals emerging from these studies include tissue-specific binding of positive factors to drive output gene transcription (Mireles-Filho et al., 2013), rhythmic expression of many transcripts is regulated at the posttranscriptional level (Koike et al., 2012; Menet et al., 2012), and the phase of transcript rhythms is not generally determined by the phase of positive factor binding (Koike et al., 2012; Menet et al., 2012). Future use of ChIP-seq in circadian systems may reveal circadian transcription networks that control the phase of rhythmic output genes and transcription factors that determine target gene selection in different tissues.

REFERENCES

Abruzzi, K. C., Rodriguez, J., Menet, J. S., Desrochers, J., Zadina, A., Luo, W., et al. (2011). Drosophila CLOCK target gene characterization: Implications for circadian tissue-specific gene expression. *Genes & Development, 25*(22), 2374–2386.

Bailey, T., Krajewski, P., Ladunga, I., Lefebvre, C., Li, Q., Liu, T., et al. (2013). Practical guidelines for the comprehensive analysis of ChIP-seq data. *PLoS Computational Biology, 9*(11), e1003326.

Barski, A., Cuddapah, S., Cui, K., Roh, T. Y., Schones, D. E., Wang, Z., et al. (2007). High-resolution profiling of histone methylations in the human genome. *Cell, 129*(4), 823–837.

Belden, W. J., Loros, J. J., & Dunlap, J. C. (2007). Execution of the circadian negative feedback loop in Neurospora requires the ATP-dependent chromatin-remodeling enzyme CLOCKSWITCH. *Molecular Cell, 25*(4), 587–600.

Bell-Pedersen, D., Cassone, V. M., Earnest, D. J., Golden, S. S., Hardin, P. E., Thomas, T. L., et al. (2005). Circadian rhythms from multiple oscillators: Lessons from diverse organisms. *Nature Reviews. Genetics, 6*(7), 544–556.

Dunlap, J. C. (1999). Molecular bases for circadian clocks. *Cell, 96*(2), 271–290.

Etchegaray, J. P., Lee, C., Wade, P. A., & Reppert, S. M. (2003). Rhythmic histone acetylation underlies transcription in the mammalian circadian clock. *Nature, 421*(6919), 177–182.

Glossop, N. R., & Hardin, P. E. (2002). Central and peripheral circadian oscillator mechanisms in flies and mammals. *Journal of Cell Science, 115*(Pt 17), 3369–3377.

Hardin, P. E. (2011). Molecular genetic analysis of circadian timekeeping in Drosophila. *Advances in Genetics, 74*, 141–173.

Hogenesch, J. B., Panda, S., Kay, S., & Takahashi, J. S. (2003). Circadian transcriptional output in the SCN and liver of the mouse. *Novartis Foundation Symposium, 253*, 171–180, discussion 152–175, 102–179, 180–173 passim.

Houl, J. H., Ng, F., Taylor, P., & Hardin, P. E. (2008). CLOCK expression identifies developing circadian oscillator neurons in the brains of Drosophila embryos. *BMC Neuroscience*, *9*, 119.

Johnson, D. S., Mortazavi, A., Myers, R. M., & Wold, B. (2007). Genome-wide mapping of in vivo protein-DNA interactions. *Science*, *316*(5830), 1497–1502.

Koike, N., Yoo, S. H., Huang, H. C., Kumar, V., Lee, C., Kim, T. K., et al. (2012). Transcriptional architecture and chromatin landscape of the core circadian clock in mammals. *Science*, *338*(6105), 349–354.

Landt, S. G., Marinov, G. K., Kundaje, A., Kheradpour, P., Pauli, F., Batzoglou, S., et al. (2012). ChIP-seq guidelines and practices of the ENCODE and modENCODE consortia. *Genome Research*, *22*(9), 1813–1831.

Le Martelot, G., Canella, D., Symul, L., Migliavacca, E., Gilardi, F., Liechti, R., et al. (2012). Genome-wide RNA polymerase II profiles and RNA accumulation reveal kinetics of transcription and associated epigenetic changes during diurnal cycles. *PLoS Biology*, *10*(11), e1001442.

Lee, C., Bae, K., & Edery, I. (1998). The *Drosophila* CLOCK protein undergoes daily rhythms in abundance, phosphorylation, and interactions with the PER-TIM complex. *Neuron*, *21*(4), 857–867.

Lee, C., Etchegaray, J. P., Cagampang, F. R., Loudon, A. S., & Reppert, S. M. (2001). Posttranslational mechanisms regulate the mammalian circadian clock. *Cell*, *107*(7), 855–867.

Lowrey, P. L., & Takahashi, J. S. (2011). Genetics of circadian rhythms in mammalian model organisms. *Advances in Genetics*, *74*, 175–230.

Menet, J. S., Abruzzi, K. C., Desrochers, J., Rodriguez, J., & Rosbash, M. (2010). Dynamic PER repression mechanisms in the Drosophila circadian clock: From on-DNA to off-DNA. *Genes & Development*, *24*(4), 358–367.

Menet, J. S., Rodriguez, J., Abruzzi, K. C., & Rosbash, M. (2012). Nascent-Seq reveals novel features of mouse circadian transcriptional regulation. *Elife*, *1*, e00011.

Mireles-Filho, A. C. A., Bardet, A. F., Yanez-Cuna, J. O., Stampfel, G., & Stark, A. (2013). Cis-regulatory requirements for tissue-specific programs of the circadian clock. *Current Biology*, *24*, 1–10.

Rey, G., Cesbron, F., Rougemont, J., Reinke, H., Brunner, M., & Naef, F. (2011). Genome-wide and phase-specific DNA-binding rhythms of BMAL1 control circadian output functions in mouse liver. *PLoS Biology*, *9*(2), e1000595.

Ripperger, J. A., & Schibler, U. (2006). Rhythmic CLOCK-BMAL1 binding to multiple E-box motifs drives circadian Dbp transcription and chromatin transitions. *Nature Genetics*, *38*(3), 369–374.

Robertson, G., Hirst, M., Bainbridge, M., Bilenky, M., Zhao, Y., Zeng, T., et al. (2007). Genome-wide profiles of STAT1 DNA association using chromatin immunoprecipitation and massively parallel sequencing. *Nature Methods*, *4*(8), 651–657.

Smith, K. M., Sancar, G., Dekhang, R., Sullivan, C. M., Li, S., Tag, A. G., et al. (2010). Transcription factors in light and circadian clock signaling networks revealed by genomewide mapping of direct targets for neurospora white collar complex. *Eukaryotic Cell*, *9*(10), 1549–1556.

Taylor, P., & Hardin, P. E. (2008). Rhythmic E-box binding by CLK-CYC controls daily cycles in per and tim transcription and chromatin modifications. *Molecular and Cellular Biology*, *28*(14), 4642–4652.

Vitalini, M. W., de Paula, R. M., Park, W. D., & Bell-Pedersen, D. (2006). The rhythms of life: Circadian output pathways in Neurospora. *Journal of Biological Rhythms*, *21*(6), 432–444.

Vollmers, C., Schmitz, R. J., Nathanson, J., Yeo, G., Ecker, J. R., & Panda, S. (2012). Circadian oscillations of protein-coding and regulatory RNAs in a highly dynamic mammalian liver epigenome. *Cell Metabolism*, *16*(6), 833–845.

Wang, G. K., Ousley, A., Darlington, T. K., Chen, D., Chen, Y., Fu, W., et al. (2001). Regulation of the cycling of *timeless* (*tim*) RNA. *Journal of Neurobiology*, *47*(3), 161–175.

Young, M. W., & Kay, S. A. (2001). Time zones: A comparative genetics of circadian clocks. *Nature Reviews. Genetics*, *2*(9), 702–715.

Yu, W., & Hardin, P. E. (2006). Circadian oscillators of Drosophila and mammals. *Journal of Cell Science*, *119*(Pt 23), 4793–4795.

CHAPTER SIXTEEN

Considerations for RNA-seq Analysis of Circadian Rhythms

Jiajia Li*, Gregory R. Grant[†,‡], John B. Hogenesch[§], Michael E. Hughes*[,1]

*Department of Biology, University of Missouri-St. Louis, St. Louis, Missouri, USA
[†]Department of Genetics, University of Pennsylvania, Philadelphia, Pennsylvania, USA
[‡]Penn Center for Bioinformatics, University of Pennsylvania, Philadelphia, Pennsylvania, USA
[§]Department of Pharmacology, Institute for Translational Medicine and Therapeutics, University of Pennsylvania School of Medicine, Philadelphia, Pennsylvania, USA
[1]Corresponding author: e-mail address: hughesmi@umsl.edu

Contents

Abstract

Circadian rhythms are daily endogenous oscillations of behavior, metabolism, and physiology. At a molecular level, these oscillations are generated by transcriptional–translational feedback loops composed of core clock genes. In turn, core clock genes drive the rhythmic accumulation of downstream outputs—termed clock-controlled genes (CCGs)—whose rhythmic translation and function ultimately underlie daily oscillations at a cellular and organismal level. Given the circadian clock's profound influence on human health and behavior, considerable efforts have been made to systematically identify CCGs. The recent development of next-generation sequencing has dramatically expanded our ability to study the expression, processing, and stability of rhythmically expressed mRNAs. Nevertheless, like any new technology, there are many technical issues to be addressed. Here, we discuss considerations for studying circadian rhythms

Methods in Enzymology, Volume 551
ISSN 0076-6879
http://dx.doi.org/10.1016/bs.mie.2014.10.020

using genome scale transcriptional profiling, with a particular emphasis on RNA sequencing. We make a number of practical recommendations—including the choice of sampling density, read depth, alignment algorithms, read-depth normalization, and cycling detection algorithms—based on computational simulations and our experience from previous studies. We believe that these results will be of interest to the circadian field and help investigators design experiments to derive most values from these large and complex data sets.

1. INTRODUCTION

Circadian rhythms are daily endogenous oscillations of behavior, physiology, and metabolism that allow organisms to anticipate and respond to predictable environmental changes. In animals, these oscillations are governed by a dedicated timing system composed in large part by transcriptional–translational feedback loops of core clock genes (Ko & Takahashi, 2006). At an organismal level, circadian rhythms have profound influence over normal physiological rhythms such as sleep–wake cycles, while disruption of the clock contributes to many human disorders, including cardiovascular disease, neurodegenerative disease, obesity, diabetes, and cancer (Halberg et al., 2006; Klerman, 2005; Levi & Schibler, 2007).

In both mammals and insects, the principal circadian oscillator resides in a small number of neurons in the central nervous system (Nitabach & Taghert, 2008; Slat, Freeman, & Herzog, 2013). The molecular circadian clock in these neurons is entrained by external stimuli, ultimately synchronizing organismal rhythms. In mammals, this central clock is located in the suprachiasmatic nuclei (SCN) of the hypothalamus (Hastings, Reddy, & Maywood, 2003; Stratmann & Schibler, 2006). SCN neurons receive both photic and nonphotic information from the environment and coordinate behavioral rhythms in locomotion, feeding, and sleep–wake cycles.

Through both direct and indirect mechanisms, the SCN also synchronizes downstream molecular circadian clocks in the brain and in peripheral tissues throughout the body. Peripheral clocks are typically phase-delayed from the SCN by 4–6 h (Panda et al., 2002), but otherwise have many of the same genetic and biochemical properties of clocks in the central oscillator. Notably, peripheral oscillations are endogenous and self-sustaining, persisting for days or even weeks *in vitro* (Yoo et al., 2004). Even cultured cell lines that have been maintained *in vitro* for many years maintain endogenous circadian oscillators that can be synchronized by a variety of stimuli

(Balsalobre, Marcacci, & Schibler, 2000; Nagoshi et al., 2004). The discovery of circadian rhythms in tissue culture has had an enormous impact on the field, as these cellular circadian models have proved to be a fruitful resource for investigating core clock mechanisms (Baggs et al., 2009; Zhang et al., 2009).

In both central and peripheral oscillators, core clock proteins drive the rhythmic expression of downstream targets, which are termed "clock-controlled genes" (CCGs). These output genes do not participate directly in the mechanism of the circadian timekeeper, but instead are translated and ultimately impose rhythmicity on downstream cellular and physiological functions (Hastings et al., 2003). Many CCGs regulate the rate-limiting steps of metabolic and genetic pathways, indicating that they play a key role in temporally compartmentalizing cellular functions (Panda et al., 2002). Although a systematic review of every CCG with an established molecular function is beyond the scope of this chapter, it is worth emphasizing that maintaining appropriate rhythmic expression of single genes can be a matter of life and death. For example, a number of key ion channels are under circadian control in cardiomyocytes, and dysregulation of their rhythmic expression has profound consequences for the physiology of the heart while predisposing animals to fatal arrhythmia (Jeyaraj et al., 2012; Schroder et al., 2013).

The total number of cycling transcripts in any given tissue is difficult to ascertain and depends on many assumptions, but we can be confident that it ranges from a few hundred to several thousand transcripts, depending on the tissue (Hughes et al., 2007, 2009). Notably, although the core clock machinery is largely conserved in different tissues, circadian output genes are highly tissue-specific (Ceriani et al., 2002; Hughes et al., 2009; Panda et al., 2002; Storch et al., 2002). This observation makes intuitive sense, as the physiological demands on the liver, for example, are substantially different from those on neural tissues. But the diversity of CCGs complicates matters for investigators studying the molecular mechanisms of circadian clock output, and it provides strong motivation for experiments aimed at systematically identifying CCGs in different tissues and species.

To identify CCGs and understand the mechanism of their regulation, microarrays have been used to profile rhythmic gene expression systematically in cyanobacteria, plants, insects, fungi, mice, and cellular models (Atwood et al., 2011; Covington, Maloof, Straume, Kay, & Harmer, 2008; Hughes et al., 2009; Hughes, Hong, et al., 2012; Keegan, Pradhan, Wang, & Allada, 2007; Kornmann, Schaad, Bujard, Takahashi, &

Schibler, 2007; McGlincy et al., 2012; Menger et al., 2007; Rund, Hou, Ward, Collins, & Duffield, 2011; Vollmers et al., 2009; Xu, DiAngelo, Hughes, Hogenesch, & Sehgal, 2011). These studies have contributed significantly to our understanding of circadian output in *wild-type* animals, and over time they have matured into investigations of more focused tissues and cell types (Collins, Kane, Reeves, Akabas, & Blau, 2012; Kula-Eversole et al., 2010). Most of these data sets are freely available in online and provide a powerful resource for researchers interested in visualizing the expression of multiple genes in many human and mouse tissues (Pizarro, Hayer, Lahens, & Hogenesch, 2013; Zhang, Lahens, Ballance, Hughes, & Hogenesch, 2014). Moreover, these data have contributed significantly to computational modeling studies of the molecular mechanism of circadian rhythms (Anafi et al., 2014; Bozek et al., 2009).

Besides simply cataloging CCGs, microarray profiling of rhythmic gene expression has also been instrumental in elucidating the mechanism of circadian output pathways. For example, a pair of recent studies used microarrays in conjunction with tissue-specific manipulation of clock genes to explore the relationship between central and peripheral oscillators (Hughes, Hong, et al., 2012; Kornmann et al., 2007). Both studies show that the peripheral circadian clock is essential for normal CCG expression and identified a number of candidate genes that may coordinate the synchronization of peripheral rhythms. A similar approach has characterized the fundamental role feeding cues have in driving CCG expression in peripheral tissues. By simply manipulating the time of day at which mice were allowed to feed, the phase of most CCGs in the liver was dramatically changed, underscoring the complexity of CCG regulation in the periphery (Vollmers et al., 2009). Finally, a recent study of rhythmic gene expression in *Dicer*-mutant fruit flies has shown the key role that miRNAs have in regulating circadian rhythms and transcriptional output (Kadener et al., 2009). The role of miRNAs in clock regulation has been confirmed in mouse (Chen, D'Alessandro, & Lee, 2013), thus motivating follow-up experiments aimed at understanding in much greater detail the interplay between ncRNAs and mRNAs in circadian output.

The development of next-generation sequencing (NGS) has accelerated studies into the global regulation of gene expression, and these technical advances offer significant opportunities for the circadian field. Early studies using RNA sequencing (RNA-seq) to profile circadian gene expression have demonstrated the potential of these approaches (Du, Arpat, De Matos, & Gatfield, 2014; Filichkin & Mockler, 2012; Hughes, Grant,

Paquin, Qian, & Nitabach, 2012; Menet, Rodriguez, Abruzzi, & Rosbash, 2012). Moreover, the single base pair resolution of these data enables the detection of new cycling transcripts, as well as measuring alternative splice forms, RNA editing, and other forms of RNA processing. Besides RNA-seq, there are other emerging NGS technologies that have begun to influence circadian research. For example, nascent RNA-seq (Menet et al., 2012; Rodriguez et al., 2013) provides information about the transcriptional and posttranscriptional regulation of cycling mRNAs. ChIP-seq enables the characterization of how these output rhythms are regulated by transcription factor binding and chromatin regulation (Bugge et al., 2012; Koike et al., 2012; Meireles-Filho, Bardet, Yanez-Cuna, Stampfel, & Stark, 2014; Menet, Pescatore, & Rosbash, 2014; Rey et al., 2011).

Despite the impact that NGS has had on circadian research, many technical challenges await investigators conducting these experiments. Several of these challenges involving experimental and statistical design are common to all global gene expression studies, including those using microarrays. We point the interested reader to several excellent articles that have discussed these issues in detail (Deckard, Anafi, Hogenesch, Haase, & Harer, 2013; Hsu & Harmer, 2014; Walker & Hogenesch, 2005; Wijnen, Naef, & Young, 2005). In addition, the use of RNA-seq introduces a number of technical issues that have not been satisfactorily addressed by the circadian field, such as the depth of sequencing coverage, read-depth normalization, and choice of alignment algorithm. Here, we present recommendations for future work using RNA-seq to explore circadian mRNA rhythms, with a focus on the computational and statistical approaches necessary for data interpretation.

2. RESULTS

2.1. Overview

Conceptually, the systematic identification of CCGs is remarkably straightforward. Tissue samples are collected at regular intervals, mRNA expression is measured on a global scale for each of these samples, and appropriate statistical tests are used to identify rhythmic components of the data. Typically, tissue collections are performed in constant darkness in order to isolate rhythms driven by the circadian clock, but many valuable studies have also been performed under LD conditions or in the presence of different zeitgebers. The key considerations when designing circadian RNA-seq experiments include: (1) number of time points and replicates, (2) choice of alignment algorithm, (3) method of read-depth normalization, (4) number of reads per sample, and (5) choice of statistical analyses and interpretation of the results.

To explore these issues, we have performed computational simulations on RNA-seq data collected over two consecutive days from the brain of fruit flies (*Drosophila melanogaster*) (Hughes, Grant et al., 2012) and the liver of mice (*Mus musculus*) (Zhang et al., 2014). Unlike microarrays, RNA-seq data are inherently discrete, i.e., expression values are calculated from the total number of reads in a sample that align to a given transcript. The fly data we use include a total number of 15–21 million reads per sample, with an average depth of 18 million reads; the mouse data include a total of 30–46 million reads per sample with an average depth of 36 million reads. Alignment of raw sequenced reads to the genome/transcriptome and the calculation of transcript expression levels were performed using RNA-seq unified mapper (RUM) (Grant et al., 2011). Cycling transcripts in each data set were identified using JTK_Cycle (Hughes, Hogenesch, & Kornacker, 2010). Using these data, we have assessed the accuracy of cycling transcript identification using a variety of different sample density and read-depth combinations.

2.2. Sample density

Statistical tests for rhythmicity are extremely sensitive to the frequency of sampling (Atwood et al., 2011; Hughes et al., 2007, 2009). However, given the expense and relative novelty of RNA-seq, there are presently no circadian RNA-seq studies using very dense sampling schemes. To simulate higher sampling densities using the available data, we randomly combined subsets of reads from neighboring time points. These synthetic data points were then used to calculate expression levels at intermediate data points and thus gain a measure of the relationship between sampling density and cycling identification. *We emphasize that this computational approach is an expedient to generate synthetic test data*, rather than an approach to identify bona fide cycling transcripts. Nevertheless, we note that over half of the top cycling transcripts in these data also cycle in previous microarray studies (data not shown) (Hughes et al., 2009; Hughes, Grant, et al., 2012). The total number of cycling transcripts and their distribution of phases and amplitudes in these data were also consistent with previous studies. Most importantly, the number of uniquely aligned reads and the dynamic range of transcript expression were both within normal ranges, indicating that our synthetic data realistically model the properties of a circadian RNA-seq experiment.

We found that 2-h sampling resolution over two consecutive days dramatically increases the number of identified cycling transcripts relative to 4- and 6-h sampling schemes (Fig. 1A and B). Moreover, the identification of cycling transcripts at 2-h resolution yielded considerably fewer false positives, which we determined by comparing to cycling transcripts identified in *period*-null fruit flies (data not shown). These results agree with the previous circadian microarray studies mentioned above, and based on their

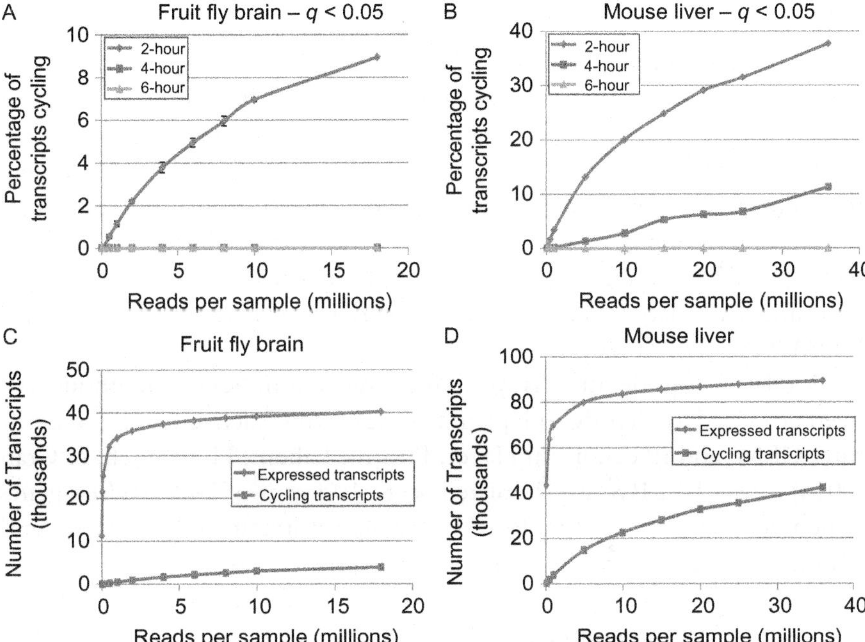

Figure 1 The discovery of cycling transcripts depends on sampling density and read depth. To assess the relationship among sampling density, read depth, and the identification of cycling transcripts, subsets of raw reads were randomly selected from legacy data sets and used to measure gene expression. Two-hour sampling resolution was simulated from these data by randomly pooling reads from neighboring time points. The discovery cycling transcripts in the fruit fly brain (A) and the mouse liver (B) showed a clear positive dependence on total read depth and sampling density. The total number of expressed transcripts (>10 uniquely aligned reads across the entire data set) is plotted as a function of read depth per sample for fruit fly brain (C) and mouse liver (D). Note that the blue traces (dark gray in the print version) in A and B have been replotted in C and D for the sake of clarity. In both data sets, the total number of cycling transcripts does not plateau, even at maximum read depths. Similarly, although the total number of expressed transcripts begins to plateau (~2.5 million reads per sample for flies; ~5 million reads per sample for mice), expressed transcripts continue to be identified even at maximal read depths.

consensus, we strongly recommend using 2-h sampling over two consecutive days. The approximately twofold increase in cost is more than compensated by dramatic improvements in the accuracy and reproducibility of cycling identification.

2.3. Alignment algorithm and splice form detection

Many different algorithms have been developed to align raw RNA-seq reads to a reference genome/transcriptome and calculate expression values. These algorithms, including but not limited to Bowtie, Tophat, RUM, Star, and GSNAP have unique strengths and weaknesses based on speed, memory footprint, sensitivity, and accuracy (Dobin et al., 2012; Grant et al., 2011; Langmead, Trapnell, Pop, & Salzberg, 2009; Trapnell, Pachter, & Salzberg, 2009; Trapnell et al., 2010; Wu & Nacu, 2010). We recommend RUM because it is robust, user-friendly, and exceptionally good at mapping reads to exon–exon junctions. However, we note that other algorithms each have specific advantages. Star, for example, is particularly effective when working with large data sets, being orders of magnitude faster than RUM, GSNAP, or Tophat.

A related consideration is splice form detection. Several methods have been proposed to identify and quantify splice forms, such as Cufflinks, Scripture, CEM, and iReckon. In Hayer, Pizarro, Lahens, Hogenesch, & Grant (2014), using BEERS, we simulated up to 10 forms of 5000 Refseq genes. When detecting one or two forms, most algorithms are able to detect with reasonable accuracy the internal gene structures of these models. However, when three or more forms are included, both the false discovery and false-negative rates become substantial (≥50%). Put simply, when you need these algorithms to detect multiple forms, they fail more often than not, even using 100 bp paired-end reads. Because of their ineffectiveness in this regard, it is difficult to evaluate their quantification properties. While increasing read length may improve splice form detection (250 bp paired-end reads are now possible), at this point, this is an unresolved problem. Alternatively, cycling analysis may be performed at an exon-level to thereby avoid the difficulties of accurately detecting alternative splice forms.

2.4. Read-depth normalization

Read depth determines signal to noise in detection and, consequently, the ability to detect cycling. Unfortunately, with RNA-seq it is impossible to get the same number of reads for multiple samples. This is in contrast to arrays, where normalization methods such as RMA and GCRMA are robust for small changes in overall signal between arrays. If the number of reads per sample is roughly equal, downsampling can be an attractive option. In this way, you "fix" the read depth to the sample for which you have the fewest mapped reads. For example, in the above fly samples with 15–21 m reads, we could sample 15 m reads from each time point. This has the unattractive property of throwing away data, but it is probably the best strategy to normalize read depth between samples.

2.5. Read depth

Read depth is a key factor that determines the accuracy of measurements made by high-throughput sequencing (Hart, Therneau, Zhang, Poland, & Kocher, 2013; Jung et al., 2014; Liu et al., 2013). More reads increase the statistical power of gene expression measurements and allow the detection of rare transcripts. But unlike microarrays whose cost is driven by the number of samples, the cost of RNA-seq is largely determined by the number of sequenced reads. Therefore, care must be taken to avoid underpowered experiments on one hand and wasted resources on the other.

As seen in Fig. 1, the number of identified cycling transcripts depends on read depth. Notably, the number of cyclers does not plateau, even at the maximum read depths available in this study. For comparison, we plotted the total number of *detectable* transcripts in these data, defined as 10 uniquely aligned reads across the entire data set (Fig. 1C and D). Unlike cycling transcripts, expressed transcripts begin to plateau in these data, but some expressed transcripts are only detectable at very high read depths. Consistent with this, our unpublished data indicate that even one billion reads per sample may be insufficient to detect every expressed transcript in mouse tissues. It stands to reason that determining whether a transcript cycles requires considerably more reads than merely detecting its expression, and we speculate that hundreds of millions of reads per sample (at 2-h resolution or greater) will be required to identify every cycling transcript in any given tissue.

On the other hand, if the goal of an experiment is to assess changes in the circadian transcriptome rather than catalog every cycling transcript, far fewer reads per sample may be necessary. This distinction is akin to the difference between *de novo* sequencing of a genome versus resequencing to identify allelic variants. To explore this relationship, we compared the expression profiles of 25 known cycling transcripts in the fly brain at different read depths (Fig. 2A). We found that the overall rhythmic pattern of these

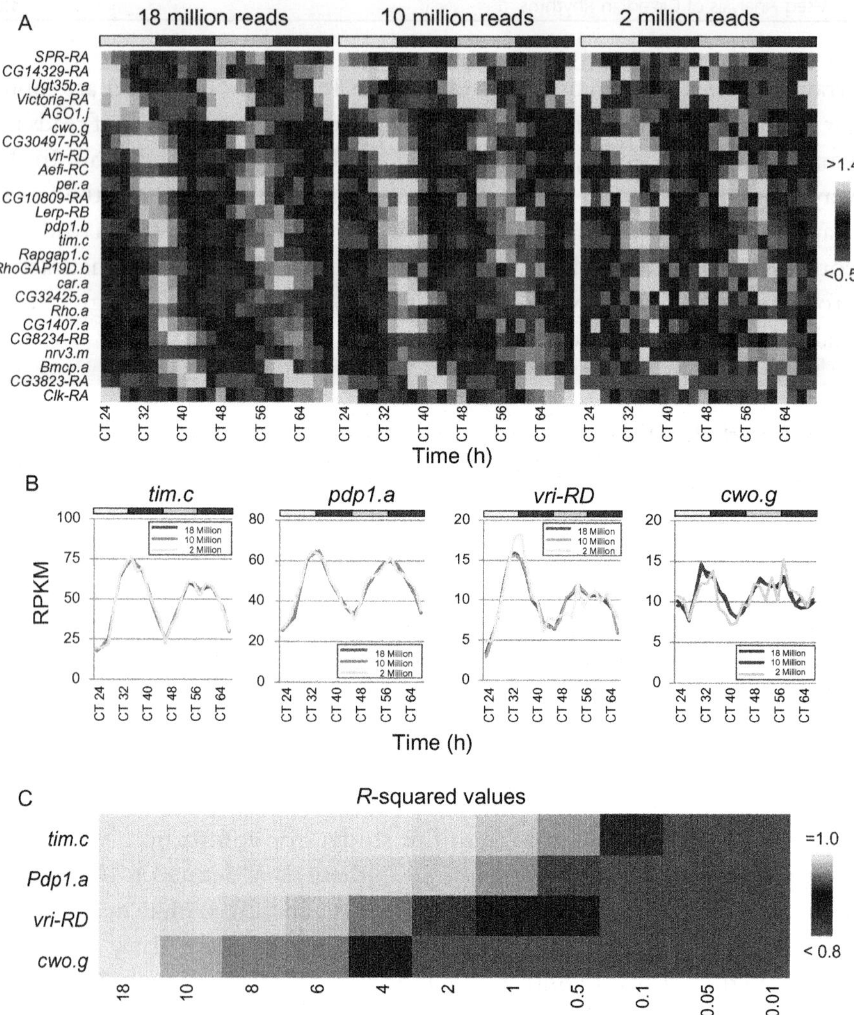

Figure 2 Many cycling transcripts are detectable with far fewer reads per sample than seen in legacy data sets. (A) The expression pattern of 25 cycling transcripts in the fruit fly brain at different read depths are shown as heat maps. The top gray and black bars represent subjective day and night, respectively. Every transcript's expression profile has been median-normalized. Expression levels greater than 1.4-fold are shown as bright yellow; expression levels less than 0.5-fold are shown as bright blue (legend on the far right). (B) The rhythmic expression pattern of four representative transcripts, *timless.c*, *pdp1.a*, *vrille-RD*, and *cwo.g* are shown at different read depths. The top gray and black bars represent subjective day and night, respectively. The vertical axis represents RPKM values; the horizontal axis indicates time points which are marked on the bottom. (C) Pearson correlation coefficients between maximal read depth and subsampled data of expression values of the four transcripts in (B). The horizontal axis indicates different read depth. Correlations equal to 1.0 are shown as bright yellow; correlations less than 0.8 are shown as bright blue (legend on the far right). (See the color plate.)

transcripts is maintained at ~50% the maximal read depth, and much of the rhythmic signal persists even at ~10% of the maximal read depth. To illustrate this observation, we plotted the expression pattern of four clock genes, *timeless*, *vrille*, *Pdp1*, and *cwo*, whose rhythmicity is evident even at low read depths (Fig. 2B). In fact, the correlation between downsampled expression profiles and the "true" expression profile is maintained at surprisingly low read depths. For example, as little as 500 thousand reads per sample are sufficient to detect rhythms in *timeless* and *Pdp1* (Fig. 2C).

We broadened this observation to the whole transcriptome by determining how many cycling transcripts (defined as cycling at maximum reads depths) are also identified at lower read depths (Fig. 3). As one might expect, in both fruit flies and mice, high-amplitude cyclers were detectable at relatively low read depths. Similarly, highly expressed transcripts are more likely to be identified as cycling at lower read depths (data not shown). We used these observations to calculate a rough estimate of the number of reads per sample necessary to identify 50%, 75%, or 87.5% of the total circadian transcriptome (Fig. 4). Since the maximal read depths available in this study are insufficient to identify every cycling transcript as discussed in detail above, we caution the reader that these figures represent low-end estimates. With that caveat in mind, however, these data indicate that 10–15 million reads per sample in flies and 20–25 million reads per sample in mice may be

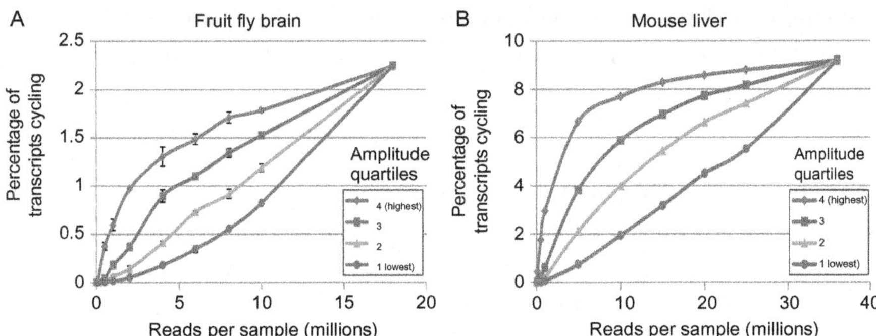

Figure 3 Detection of cycling transcripts at lower read depth depends on amplitude. Cycling transcripts were divided in quartiles based on the magnitude of the amplitude. The percentage of cycling transcripts from each quartile is plotted as a function of read depth in fruit fly brain (A) and mouse liver (B) data sets. Transcripts with higher amplitudes are more likely to be detected at a lower read depth, while the detection of low-amplitude transcripts requires increasingly large read depths. The vertical axis refers to the percentage of transcripts cycling (e.g., since ~8% of transcripts in the fruit fly brain cycle, each quartile contains ~2% of the transcriptome cycling, at max).

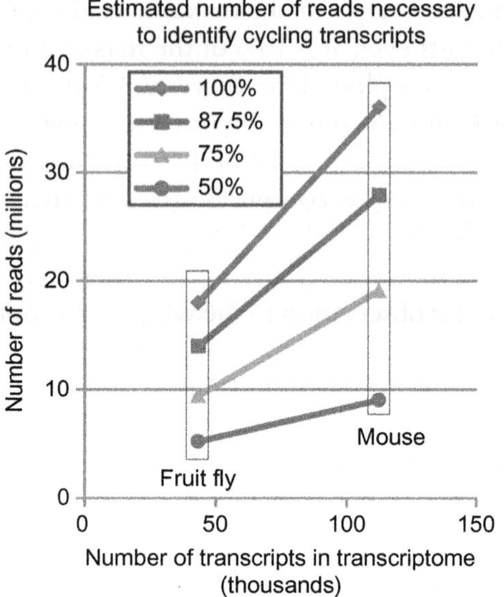

Figure 4 Estimated number of reads necessary to detect subsets of the circadian transcriptome. The number of reads per sample necessary to detect 50%, 75%, 87.5%, and 100% of cycling transcripts in the fruit fly and mouse transcriptome are shown here. The vertical axis indicates read depth per sample, and the horizontal axis indicates the total number of transcripts within the transcriptomes of fruit fly and mice.

sufficient to characterize the majority of the circadian transcriptome, especially if the investigator is content to measure the highest amplitude and most highly expressed rhythmic genes.

2.6. Cycling detection algorithms

Many statistical tests have been developed to detect cycling components of complex data sets, including but not limited to COSPOT, Fisher's *G* test, F24, Haystack, Arser, JTK_Cycle, Lomb Scargle, and Delichtenberg (Claridge-Chang et al., 2001; de Lichtenberg et al., 2005; Glynn, Chen, & Mushegian, 2006; Hughes et al., 2010; Mockler et al., 2007; Straume, 2004; Wichert, Fokianos, & Strimmer, 2004; Yang & Su, 2010). They are based on different mathematical assumptions, and their relative strengths and weaknesses have been compared in several recent benchmarking studies (Deckard et al., 2013; Hughes et al., 2010). In our on-going experiments, we favor using JTK_Cycle as it has proved to be powerful, accurate, and efficient, while also being relatively user-friendly.

But we note that many of the related approaches have been used to great effect in previous cycling studies.

2.7. False discovery correction

Accurately estimating the rate of false discoveries is a well-recognized problem in any high-throughput analysis (Macarthur, 2012). For circadian experiments, the *p*-value is calculated independently for every transcript, and it reflects the probability that a given expression profile occurred by chance alone. However, this measure does not account for the enormous size of genomics experiments, which may include tens of thousands or even hundreds of thousands of expression profiles. Consequently, even extremely unlikely occurrences (i.e., transcripts with low *p*-values) are expected to be present with some frequency. Therefore, the false discovery rate (FDR or *q*-value) is calculated for the experiment as a whole, and it reflects the probability that transcripts at a given statistical threshold are in actuality false discoveries. The rank order of *p*- and *q*-values will always be the same, and the *q*-value will inevitably be greater than or equal to the *p*-value (i.e., more conservative). Although the choice of statistical threshold depends on the aims of the investigation and the tolerance for false discoveries, we strongly recommend the inclusion of an explicit false discovery correction to account for the enormous number of comparisons being made.

2.8. Validation and follow-up

Like any high-throughput assay, several steps should be taken to validate the accuracy gene expression measurements. Fortunately, circadian profiling experiments have excellent internal controls built-in, as the expression pattern of many core clock genes are known or predicted from previous work. Moreover, independent biological samples should be collected, and novel candidate cyclers should be verified using independent technical approaches such as quantitative PCR or *in situ* hybridization. Finally, comparisons with legacy data sets—both expression profiling and ChIP-seq—can be used to effectively strengthen the confidence that a novel cycling gene is worth pursuing in downstream functional experiments.

3. CONCLUSIONS

Circadian rhythms exert an enormous influence on normal and pathological physiology. Nevertheless, the pathways by which the core circadian oscillator drives rhythms in peripheral tissues and the mechanisms through which peripheral clocks influence human health are incompletely understood. RNA-seq analysis of the circadian transcriptome can be used to build a comprehensive list of candidate CCGs as well as explore the underlying molecular biology through which these rhythms are generated. We expect that both applications will have a major impact on the circadian field in the years ahead. Care must be taken to use appropriate experimental designs and statistical approaches, particularly with respect to the sampling density and the number of reads per sample. Using experimental designs appropriate to the goals of an investigation will ultimately increase the value and staying power of the data collected.

4. METHODS

Data processing: Previously published circadian RNA-seq data from fruit fly brain (Hughes, Grant, et al., 2012) were downloaded from GEO (GSE36108). Mouse liver RNA-seq data were obtained in advance of publication (Zhang et al., 2014). To simulate 2-h sampling in both data sets, randomly selected reads from neighboring time points were merged together using custom-built Ruby scripts; three independent replicates were generated for each data set in the fruit fly data.

Alignment: RUM (Grant et al., 2011) was used to align all reads (75 bp, paired-end) to the genome and transcriptome of either *D. melanogaster* (build dm3) or *M. musculus* (build mm9), using the following parameters: "–bowtie-nu-limit 10 –nu-limit 10." For the fruit fly data set, 52–59% of reads were mapped uniquely to the genome and transcriptome. For mouse data sets, 77–82% of reads were mapped uniquely to the genome and transcriptome. Reads per kilobase per million reads (RPKMs) for each transcript were calculated by RUM from uniquely mapped reads.

Circadian analysis: Detection of cycling was performed using either JTK_Cycle or JTK_Cycle_v2 (Hughes et al., 2010; Miyazaki et al., 2011) implemented in R (64-bit, version 2.12.1). Benjamini–Hochberg corrected q-values of <0.05 were generally used as statistical threshold. To mitigate the effect of false discovery in the fruit fly data set, per^0 data (i.e., circadian mutant) were used as a negative control. Cycling transcripts found in per^0 samples were considered to be false discoveries. In the analysis of RPKM or amplitude, all the cycling transcripts in legacy data set were divided into four groups according to either mean RPKM or amplitude values, with the same numbers of transcripts in each group.

RUM can be downloaded and installed from the following: http://www.cbil.upenn.edu/RUM/userguide.php. JTK_Cycle is available from: http://openwetware.org/wiki/HughesLab:JTK_Cycle. All Ruby scripts and data sets are available on demand.

ACKNOWLEDGMENTS

We thank members of the Hughes and Hogenesch labs for helpful comments during the preparation of this chapter. This work is supported by the National Institute of Neurological Disorders and Stroke (1R01NS054794-06 to J. B. H.), the Defense Advanced Research Projects Agency (DARPA-D12AP00025, to John Harer, Duke University), and by the Penn Genome Frontiers Institute under an HRFF grant with the Pennsylvania Department of Health. MEH is supported by University of Missouri-St. Louis and College of Arts and Sciences research awards.

REFERENCES

Anafi, R. C., Lee, Y., Sato, T. K., Venkataraman, A., Ramanathan, C., Kavakli, I. H., et al. (2014). Machine learning helps identify CHRONO as a circadian clock component. *PLoS Biology*, *12*(4), e1001840. http://dx.doi.org/10.1371/journal.pbio.1001840.

Atwood, A., DeConde, R., Wang, S. S., Mockler, T. C., Sabir, J. S. M., Ideker, T., et al. (2011). Cell-autonomous circadian clock of hepatocytes drives rhythms in transcription and polyamine synthesis. *Proceedings of the National Academy of Sciences of the United States of America*, *108*(45), 18560–18565. http://dx.doi.org/10.1073/pnas.1115753108.

Baggs, J. E., Price, T. S., DiTacchio, L., Panda, S., Fitzgerald, G. A., & Hogenesch, J. B. (2009). Network features of the mammalian circadian clock. *PLoS Biology*, 7(3), e52. http://dx.doi.org/10.1371/journal.pbio.1000052.

Balsalobre, A., Marcacci, L., & Schibler, U. (2000). Multiple signaling pathways elicit circadian gene expression in cultured Rat-1 fibroblasts. *Current Biology*, *10*(20), 1291–1294. http://dx.doi.org/10.1016/S0960-9822(00)00758-2.

Bozek, K., Relogio, A., Kielbasa, S. M., Heine, M., Dame, C., Kramer, A., et al. (2009). Regulation of clock-controlled genes in mammals. *PLoS One*, *4*(3), e4882. http://dx.doi.org/10.1371/journal.pone.0004882.

Bugge, A., Feng, D., Everett, L. J., Briggs, E. R., Mullican, S. E., Wang, F., et al. (2012). Rev-erbα and Rev-erbβ coordinately protect the circadian clock and normal metabolic function. *Genes & Development*, *26*(7), 657–667. http://dx.doi.org/10.1101/gad.186858.112.

Ceriani, M. F., Hogenesch, J. B., Yanovsky, M., Panda, S., Straume, M., & Kay, S. A. (2002). Genome-wide expression analysis in Drosophila reveals genes controlling circadian behavior. *The Journal of Neuroscience*, *22*(21), 9305–9319.

Chen, R., D'Alessandro, M., & Lee, C. (2013). miRNAs are required for generating a time delay critical for the circadian oscillator. *Current Biology*, *23*(20), 1959–1968. http://dx.doi.org/10.1016/j.cub.2013.08.005.

Claridge-Chang, A., Wijnen, H., Naef, F., Boothroyd, C., Rajewsky, N., & Young, M. W. (2001). Circadian regulation of gene expression systems in the Drosophila head. *Neuron*, *32*(4), 657–671. http://dx.doi.org/10.1016/S0896-6273(01)00515-3.

Collins, B., Kane, E. A., Reeves, D. C., Akabas, M. H., & Blau, J. (2012). Balance of activity between LN(v)s and glutamatergic dorsal clock neurons promotes robust circadian rhythms in Drosophila. *Neuron*, *74*(4), 706–718. http://dx.doi.org/10.1016/j.neuron.2012.02.034.

Covington, M. F., Maloof, J. N., Straume, M., Kay, S. A., & Harmer, S. L. (2008). Global transcriptome analysis reveals circadian regulation of key pathways in plant growth and development. *Genome Biology*, *9*(8), R130. http://dx.doi.org/10.1186/gb-2008-9-8-r130.

de Lichtenberg, U., Jensen, L. J., Fausboll, A., Jensen, T. S., Bork, P., & Brunak, S. (2005). Comparison of computational methods for the identification of cell cycle-regulated genes. *Bioinformatics*, *21*(7), 1164–1171. http://dx.doi.org/10.1093/bioinformatics/bti093.

Deckard, A., Anafi, R. C., Hogenesch, J. B., Haase, S. B., & Harer, J. (2013). Design and analysis of large-scale biological rhythm studies: A comparison of algorithms for detecting periodic signals in biological data. *Bioinformatics*, *29*(24), 3174–3180. http://dx.doi.org/10.1093/bioinformatics/btt541.

Dobin, A., Davis, C. A., Schlesinger, F., Drenkow, J., Zaleski, C., Jha, S., et al. (2012). STAR: Ultrafast universal RNA-seq aligner. *Bioinformatics*, *29*(1), 15–21. http://dx.doi.org/10.1093/bioinformatics/bts635.

Du, N. H., Arpat, A. B., De Matos, M., & Gatfield, D. (2014). MicroRNAs shape circadian hepatic gene expression on a transcriptome-wide scale. *elife*, *3*, e02510. http://dx.doi.org/10.7554/eLife.02510.

Filichkin, S., & Mockler, T. (2012). Unproductive alternative splicing and nonsense mRNAs: A widespread phenomenon among plant circadian clock genes. *Biology Direct*, 7(1), 1–15. http://dx.doi.org/10.1186/1745-6150-7-20.

Glynn, E. F., Chen, J., & Mushegian, A. R. (2006). Detecting periodic patterns in unevenly spaced gene expression time series using Lomb-Scargle periodograms. *Bioinformatics*, *22*(3), 310–316. http://dx.doi.org/10.1093/bioinformatics/bti789.

Grant, G. R., Farkas, M. H., Pizarro, A. D., Lahens, N. F., Schug, J., Brunk, B. P., et al. (2011). Comparative analysis of RNA-Seq alignment algorithms and the RNA-Seq unified mapper (RUM). *Bioinformatics*, *27*(18), 2518–2528. http://dx.doi.org/10.1093/bioinformatics/btr427.

Halberg, F., Cornelissen, G., Ulmer, W., Blank, M., Hrushesky, W., Wood, P., et al. (2006). Cancer chronomics III. Chronomics for cancer, aging, melatonin and experimental therapeutics researchers. *Journal of Experimental Therapeutics & Oncology*, *6*(1), 73–84.

Hart, S. N., Therneau, T. M., Zhang, Y., Poland, G. A., & Kocher, J. P. (2013). Calculating sample size estimates for RNA sequencing data. *Journal of Computational Biology*, *20*(12), 970–978. http://dx.doi.org/10.1089/cmb.2012.0283.

Hastings, M. H., Reddy, A. B., & Maywood, E. S. (2003). A clockwork web: Circadian timing in brain and periphery, in health and disease. *Nature Reviews. Neuroscience*, *4*(8), 649–661. http://dx.doi.org/10.1038/nrn1177.

Hayer, K., Pizarro, P., Lahens, N. L., Hogenesch, J. B., & Grant, G. R. (2014). Benchmark analysis for determining and quantifying full-length mRNA splice forms from RNA-seq data. *bioRxiv*, 007088. http://dx.doi.org/10.1101/007088.

Hsu, P. Y., & Harmer, S. L. (2014). Global profiling of the circadian transcriptome using microarrays. *Methods in Molecular Biology*, *1158*, 45–56. http://dx.doi.org/10.1007/978-1-4939-0700-7_3.

Hughes, M., Deharo, L., Pulivarthy, S. R., Gu, J., Hayes, K., Panda, S., et al. (2007). High-resolution time course analysis of gene expression from pituitary. *Cold Spring Harbor Symposia on Quantitative Biology*, *72*, 381–386. http://dx.doi.org/10.1101/sqb.2007.72.011.

Hughes, M. E., DiTacchio, L., Hayes, K. R., Vollmers, C., Pulivarthy, S., Baggs, J. E., et al. (2009). Harmonics of circadian gene transcription in mammals. *PLoS Genetics*, *5*(4), e1000442. http://dx.doi.org/10.1371/journal.pgen.1000442.

Hughes, M. E., Grant, G. R., Paquin, C., Qian, J., & Nitabach, M. N. (2012). Deep sequencing the circadian and diurnal transcriptome of Drosophila brain. *Genome Research*, *22*(7), 1266–1281. http://dx.doi.org/10.1101/gr.128876.111.

Hughes, M. E., Hogenesch, J. B., & Kornacker, K. (2010). JTK_CYCLE: An efficient nonparametric algorithm for detecting rhythmic components in genome-scale data sets. *Journal of Biological Rhythms*, *25*(5), 372–380. http://dx.doi.org/10.1177/0748730410379711.

Hughes, M. E., Hong, H. K., Chong, J. L., Indacochea, A. A., Lee, S. S., Han, M., et al. (2012). Brain-specific rescue of Clock reveals system-driven transcriptional rhythms in peripheral tissue. *PLoS Genetics*, *8*(7), e1002835. http://dx.doi.org/10.1371/journal.pgen.1002835.

Jeyaraj, D., Haldar, S. M., Wan, X., McCauley, M. D., Ripperger, J. A., Hu, K., et al. (2012). Circadian rhythms govern cardiac repolarization and arrhythmogenesis. *Nature*, *483*(7387), 96–99. http://dx.doi.org/10.1038/nature10852.

Jung, Y. L., Luquette, L. J., Ho, J. W., Ferrari, F., Tolstorukov, M., Minoda, A., et al. (2014). Impact of sequencing depth in ChIP-seq experiments. *Nucleic Acids Research*, *42*(9), e74. http://dx.doi.org/10.1093/nar/gku178.

Kadener, S., Menet, J. S., Sugino, K., Horwich, M. D., Weissbein, U., Nawathean, P., et al. (2009). A role for microRNAs in the Drosophila circadian clock. *Genes & Development*, *23*(18), 2179–2191. http://dx.doi.org/10.1101/gad.1819509.

Keegan, K. P., Pradhan, S., Wang, J. P., & Allada, R. (2007). Meta-analysis of Drosophila circadian microarray studies identifies a novel set of rhythmically expressed genes. *PLoS Computational Biology*, *3*(11), e208. http://dx.doi.org/10.1371/journal.pcbi.0030208.

Klerman, E. B. (2005). Clinical aspects of human circadian rhythms. *Journal of Biological Rhythms*, *20*(4), 375–386. http://dx.doi.org/10.1177/0748730405278353.

Ko, C. H., & Takahashi, J. S. (2006). Molecular components of the mammalian circadian clock. *Human Molecular Genetics*, *15*(Spec No 2), R271–R277. http://dx.doi.org/10.1093/hmg/ddl207.

Koike, N., Yoo, S.-H., Huang, H.-C., Kumar, V., Lee, C., Kim, T.-K., et al. (2012). Transcriptional architecture and chromatin landscape of the core circadian clock in mammals. *Science*, *338*(6105), 349–354. http://dx.doi.org/10.1126/science.1226339.

Kornmann, B., Schaad, O., Bujard, H., Takahashi, J. S., & Schibler, U. (2007). System-driven and oscillator-dependent circadian transcription in mice with a conditionally active liver clock. *PLoS Biology*, *5*(2), e34. http://dx.doi.org/10.1371/journal.pbio.0050034.

Kula-Eversole, E., Nagoshi, E., Shang, Y., Rodriguez, J., Allada, R., & Rosbash, M. (2010). Surprising gene expression patterns within and between PDF-containing circadian neurons in Drosophila. *Proceedings of the National Academy of Sciences of the United States of America*, *107*(30), 13497–13502. http://dx.doi.org/10.1073/pnas.1002081107.

Langmead, B., Trapnell, C., Pop, M., & Salzberg, S. L. (2009). Ultrafast and memory-efficient alignment of short DNA sequences to the human genome. *Genome Biology*, *10*(3), R25. http://dx.doi.org/10.1186/gb-2009-10-3-r25.

Levi, F., & Schibler, U. (2007). Circadian rhythms: Mechanisms and therapeutic implications. *Annual Review of Pharmacology and Toxicology*, *47*, 593–628. http://dx.doi.org/10.1146/annurev.pharmtox.47.120505.105208.

Liu, Y., Ferguson, J. F., Xue, C., Silverman, I. M., Gregory, B., Reilly, M. P., et al. (2013). Evaluating the impact of sequencing depth on transcriptome profiling in human adipose. *PLoS One*, *8*(6), e66883. http://dx.doi.org/10.1371/journal.pone.0066883.

Macarthur, D. (2012). Methods: Face up to false positives. *Nature*, *487*(7408), 427–428. http://dx.doi.org/10.1038/487427a.

McGlincy, N. J., Valomon, A., Chesham, J. E., Maywood, E. S., Hastings, M. H., & Ule, J. (2012). Regulation of alternative splicing by the circadian clock and food related cues. *Genome Biology*, *13*(6), R54. http://dx.doi.org/10.1186/gb-2012-13-6-r54.

Meireles-Filho, A. C., Bardet, A. F., Yanez-Cuna, J. O., Stampfel, G., & Stark, A. (2014). cis-Regulatory requirements for tissue-specific programs of the circadian clock. *Current Biology*, *24*(1), 1–10. http://dx.doi.org/10.1016/j.cub.2013.11.017.

Menet, J. S., Pescatore, S., & Rosbash, M. (2014). CLOCK: BMAL1 is a pioneer-like transcription factor. *Genes & Development*, *28*(1), 8–13. http://dx.doi.org/10.1101/gad.228536.113.

Menet, J. S., Rodriguez, J., Abruzzi, K. C., & Rosbash, M. (2012). Nascent-Seq reveals novel features of mouse circadian transcriptional regulation. *eLife*, *1*, e00011. http://dx.doi.org/10.7554/eLife.00011.

Menger, G. J., Allen, G. C., Neuendorff, N., Nahm, S.-S., Thomas, T. L., Cassone, V. M., et al. (2007). Circadian profiling of the transcriptome in NIH/3T3 fibroblasts: Comparison with rhythmic gene expression in SCN2.2 cells and the rat SCN. *Physiological Genomics*, *29*, 280–289.

Miyazaki, M., Schroder, E., Edelmann, S. E., Hughes, M. E., Kornacker, K., Balke, C. W., et al. (2011). Age-associated disruption of molecular clock expression in skeletal muscle of the spontaneously hypertensive rat. *PLoS One*, *6*(11), e27168. http://dx.doi.org/10.1371/journal.pone.0027168.

Mockler, T. C., Michael, T. P., Priest, H. D., Shen, R., Sullivan, C. M., Givan, S. A., et al. (2007). The DIURNAL project: DIURNAL and circadian expression profiling, model-based pattern matching, and promoter analysis. *Cold Spring Harbor Symposia on Quantitative Biology*, *72*, 353–363. http://dx.doi.org/10.1101/sqb.2007.72.006.

Nagoshi, E., Saini, C., Bauer, C., Laroche, T., Naef, F., & Schibler, U. (2004). Circadian gene expression in individual fibroblasts: Cell-autonomous and self-sustained oscillators pass time to daughter cells. *Cell*, *119*(5), 693–705. http://dx.doi.org/10.1016/j.cell.2004.11.015.

Nitabach, M. N., & Taghert, P. H. (2008). Organization of the Drosophila circadian control circuit. *Current Biology*, *18*(2), R84–R93. http://dx.doi.org/10.1016/j.cub.2007.11.061.

Panda, S., Antoch, M. P., Miller, B. H., Su, A. I., Schook, A. B., Straume, M., et al. (2002). Coordinated transcription of key pathways in the mouse by the circadian clock. *Cell*, *109*(3), 307–320.

Pizarro, A., Hayer, K., Lahens, N. F., & Hogenesch, J. B. (2013). CircaDB: A database of mammalian circadian gene expression profiles. *Nucleic Acids Research*, *41*(Database issue), D1009–D1013. http://dx.doi.org/10.1093/nar/gks1161.

Rey, G., Cesbron, F., Rougemont, J., Reinke, H., Brunner, M., & Naef, F. (2011). Genome-wide and phase-specific DNA-binding rhythms of BMAL1 control circadian output functions in mouse liver. *PLoS Biology*, *9*(2), e1000595. http://dx.doi.org/10.1371/journal.pbio.1000595.

Rodriguez, J., Tang, C.-H. A., Khodor, Y. L., Vodala, S., Menet, J. S., & Rosbash, M. (2013). Nascent-Seq analysis of Drosophila cycling gene expression. *Proceedings of the National Academy of Sciences*, *110*(4), 275–284. http://dx.doi.org/10.1073/pnas.1219969110.

Rund, S. S. C., Hou, T. Y., Ward, S. M., Collins, F. H., & Duffield, G. E. (2011). Genome-wide profiling of diel and circadian gene expression in the malaria vector Anopheles gambiae. *Proceedings of the National Academy of Sciences*, *108*(32), 421–430. http://dx.doi.org/10.1073/pnas.1100584108.

Schroder, E. A., Lefta, M., Zhang, X., Bartos, D. C., Feng, H. Z., Zhao, Y., et al. (2013). The cardiomyocyte molecular clock, regulation of Scn5a, and arrhythmia susceptibility. *American Journal of Physiology. Cell Physiology*. *304*(10), C954–C965. http://dx.doi.org/10.1152/ajpcell.00383.2012.

Slat, E., Freeman, G. M., Jr., & Herzog, E. (2013). The clock in the brain: Neurons, glia, and networks in daily rhythms. In A. Kramer, & M. Merrow (Eds.), *Circadian clocks: Vol. 217* (pp. 105–123). Berlin Heidelberg: Springer.

Storch, K. F., Lipan, O., Leykin, I., Viswanathan, N., Davis, F. C., Wong, W. H., et al. (2002). Extensive and divergent circadian gene expression in liver and heart. *Nature*, *417*(6884), 78–83. http://dx.doi.org/10.1038/nature744.

Stratmann, M., & Schibler, U. (2006). Properties, entrainment, and physiological functions of mammalian peripheral oscillators. *Journal of Biological Rhythms*, *21*(6), 494–506. http://dx.doi.org/10.1177/0748730406293889.

Straume, M. (2004). DNA microarray time series analysis: Automated statistical assessment of circadian rhythms in gene expression patterning. *Methods in Enzymology*, *383*, 149–166. http://dx.doi.org/10.1016/s0076-6879(04)83007-6.

Trapnell, C., Pachter, L., & Salzberg, S. L. (2009). TopHat: Discovering splice junctions with RNA-Seq. *Bioinformatics*, *25*(9), 1105–1111. http://dx.doi.org/10.1093/bioinformatics/btp120.

Trapnell, C., Williams, B. A., Pertea, G., Mortazavi, A., Kwan, G., van Baren, M. J., et al. (2010). Transcript assembly and quantification by RNA-Seq reveals unannotated transcripts and isoform switching during cell differentiation. *Nature Biotechnology*, *28*(5), 511–515. http://dx.doi.org/10.1038/nbt.1621.

Vollmers, C., Gill, S., DiTacchio, L., Pulivarthy, S. R., Le, H. D., & Panda, S. (2009). Time of feeding and the intrinsic circadian clock drive rhythms in hepatic gene expression. *Proceedings of the National Academy of Sciences of the United States of America*, *106*(50), 21453–21458. http://dx.doi.org/10.1073/pnas.0909591106.

Walker, J. R., & Hogenesch, J. B. (2005). RNA profiling in circadian biology. In M. W. Young (Ed.), *Methods in enzymology: Vol. 393*. (pp. 366–376). San Diego: Elsevier Academic Press.

Wichert, S., Fokianos, K., & Strimmer, K. (2004). Identifying periodically expressed transcripts in microarray time series data. *Bioinformatics*, *20*(1), 5–20.

Wijnen, H., Naef, F., & Young, M. W. (2005). Molecular and statistical tools for circadian transcript profiling. In M. W. Young (Ed.), *Methods in enzymology: Vol. 393*. (pp. 341–365). San Diego: Elsevier Academic Press.

Wu, T. D., & Nacu, S. (2010). Fast and SNP-tolerant detection of complex variants and splicing in short reads. *Bioinformatics*, *26*(7), 873–881. http://dx.doi.org/10.1093/bioinformatics/btq057.

Xu, K., DiAngelo, J. R., Hughes, M. E., Hogenesch, J. B., & Sehgal, A. (2011). The circadian clock interacts with metabolic physiology to influence reproductive fitness. *Cell Metabolism*, *13*(6), 639–654. http://dx.doi.org/http://dx.doi.org/10.1016/j.cmet.2011.05.001.

Yang, R., & Su, Z. (2010). Analyzing circadian expression data by harmonic regression based on autoregressive spectral estimation. *Bioinformatics*, *26*(12), i168–i174. http://dx.doi.org/10.1093/bioinformatics/btq189.

Yoo, S. H., Yamazaki, S., Lowrey, P. L., Shimomura, K., Ko, C. H., Buhr, E. D., et al. (2004). PERIOD2::LUCIFERASE real-time reporting of circadian dynamics reveals persistent circadian oscillations in mouse peripheral tissues. *Proceedings of the National Academy of Sciences of the United States of America*, *101*(15), 5339–5346. http://dx.doi.org/10.1073/pnas.0308709101.

Zhang, E. E., Liu, A. C., Hirota, T., Miraglia, L. J., Welch, G., Pongsawakul, P. Y., et al. (2009). A genome-wide RNAi screen for modifiers of the circadian clock in human cells. *Cell*, *139*(1), 199–210. http://dx.doi.org/10.1016/j.cell.2009.08.031.

Zhang, R., Lahens, N. F., Ballance, H. I., Hughes, M. E., & Hogenesch, J. B. (2014). A circadian gene expression atlas in mammals: Implications for biology and medicine. *Proceedings of the National Academy of Sciences of the United States of America*, http://dx.doi.org/10.1073/pnas.1408886111.

CHAPTER SEVENTEEN

RNA-seq Profiling of Small Numbers of *Drosophila* Neurons

Katharine Abruzzi*, Xiao Chen*, Emi Nagoshi†, Abby Zadina*, Michael Rosbash*,[1]

*Department of Biology, Howard Hughes Medical Institute and National Center for Behavioral Genomics, Brandeis University, Waltham, Massachusetts, USA

†Department of Genetics and Evolution, University of Geneva, Geneva, Switzerland

[1]Corresponding author: e-mail address: rosbash@brandeis.edu

Contents

Abstract

Drosophila melanogaster has a robust circadian clock, which drives a rhythmic behavior pattern: locomotor activity increases in the morning shortly before lights on (M peak) and in the evening shortly before lights off (E peak). This pattern is controlled by ~75 pairs of circadian neurons in the *Drosophila* brain. One key group of neurons is the M-cells (PDF^+ large and small LN_vs), which control the M peak. A second key group is the E-cells, consisting of four LN_ds and the fifth small LN_v, which control the E peak. Recent studies show that the M-cells have a second role in addition to controlling the M peak; they communicate with the E-cells (as well as DN_1s) to affect their timing, probably as a function of environmental conditions (Guo, Cerullo, Chen, & Rosbash, 2014).

To learn about molecules within the M-cells important for their functional roles, we have adapted methods to manually sort fluorescent protein-expressing neurons of interest from dissociated *Drosophila* brains. We isolated mRNA and miRNA from sorted M-cells and amplified the resulting DNAs to create deep-sequencing libraries. Visual inspection of the libraries illustrates that they are specific to a particular neuronal subgroup; M-cell libraries contain timeless and dopaminergic cell libraries contain ple/TH. Using these data, it is possible to identify cycling transcripts as well as many mRNAs and miRNAs specific to or enriched in particular groups of neurons.

Methods in Enzymology, Volume 551
ISSN 0076-6879
http://dx.doi.org/10.1016/bs.mie.2014.10.025

1. INTRODUCTION

Circadian clocks allow organisms to predict and respond to daily fluctuations in their environments. In most organisms, these clocks oscillate with an ~24-h period and are entrained by environmental cues such as light. In *Drosophila*, the clock is driven by a several well-defined transcriptional feedback loops, one of which is focused on a heterodimer of the transcriptional factors CLK and CYC. CLK/CYC drives the transcription of the repressors PER and TIM in the early evening. PER and TIM levels accumulate and repress CLK/CYC-driven transcription in the late night. This negative feedback loop contributes to oscillating gene expression, which has a major impact on the circadian outputs including locomotor activity rhythms.

In *Drosophila*, there are ~75 pairs of neurons in the brain that express high levels of these clock components (CLK, CYC, PER, TIM) and are therefore considered circadian neurons. They have been divided into two main subgroups: dorsal neurons (DNs) and the lateral neurons (LNs). DNs are further subdivided into four groups based primarily on their location in the brain: DN_{1a}, DN_{1p}, DN_2, and DN_3. LNs have been divided into two main groups based on their expression of the neuropeptide pigment-dispersing factor (PDF; Helfrich-Forster, 1995): LPNs, LN_ds, and the fifth small LN_v are PDF^-, whereas the large and four of the five small LN_vs are PDF^+. Further experiments have shown that the PDF^+ s- and l-LN_vs are critical for driving the morning activity period in *Drosophila* and are known as the morning cells (M-cells). The PDF^- LN_ds and the 5th small LN_v are important for driving evening behavior and are known as the evening cells (E-cells; Grima, Chelot, Xia, & Rouyer, 2004; Stoleru, Peng, Agosto, & Rosbash, 2004). Immunostaining studies have begun to suggest the function(s) of these different groups, by revealing different expression patterns. For example, the circadian photoreceptor Cryptochrome (CRY) and different neuropeptides that impact the circadian system are differentially expressed within the circadian network (reviewed in Yoshii, Rieger, & Helfrich-Forster, 2012). In addition, different cell-specific drivers from the GAL4/UAS system have been used to manipulate subsets of these neurons with different UAS proteins to determine changes in circadian behavior. For example, electrical silencing of the M-cells causes a severe deficit in free-running locomotor rhythms (Depetris-Chauvin et al., 2011; Nitabach, Blau, & Holmes, 2002).

Studies in the last decade have provided further evidence that the control of circadian rhythms is not a simple case of attributing a specific task to a single group of neurons. Evidence indicates that interactions between different circadian neurons are necessary to achieve the complex regulation that drives circadian behaviors. M-cells are considered the master pacemakers since they can keep time in constant darkness, but they communicate with both the DN_1s and the E-cells via PDF signaling (Guo et al., 2014; Zhang, Chung, et al., 2010; Zhang, Liu, Bilodeau-Wentworth, Hardin, & Emery, 2010). Moreover, manipulating the E-cells as well as the DN_1s impacts the morning peak (Guo et al., 2014; Zhang, Chung, et al., 2010; Zhang, Liu, et al., 2010). The data indicate that the DN_1s and E-cells are downstream of PDF signaling, but they may also feed back to influence the M-cells.

This complexity indicates that it will be important to characterize subgroups of neurons and eventually single neurons. Although immunostaining has been a valuable tool to start to decipher the unique expression patterns of circadian neurons, a genome-wide view of differential gene expression patterns would greatly expand our vision. To this end, we and others manually sorted subgroups of neurons from dissociated *Drosophila* brains and used microarrays to assay neuron-specific gene expression (Kula-Eversole et al., 2010; Mizrak et al., 2012; Nagoshi et al., 2010). We have now used deep-sequencing technologies to sequence the mRNA and miRNA populations of these M-cells. We also show here a bit of data from a large group of noncircadian neurons (e.g., dopaminergic: ~130 neurons per brain; Mao & Davis, 2009) as well as the smaller group of M neurons (M-cells; l- and s-LN_vs; ~14 neurons/brain). By identifying mRNAs that are enriched and/or undergo cycling in particular groups of neurons, we hope to learn more about the roles of these neurons in contributing to particular aspects of circadian rhythms.

2. RESULTS/METHODS

2.1. Isolating neurons of interest

To isolate neurons of interest, we express a fluorescent protein in a specific subset of neurons and then manually sort these fluorescent neurons from dissociated brains. One of the key steps of this procedure is ensuring that the fluorescent protein (1) is sufficiently bright to make cell sorting possible and (2) has no leaky expression outside of the neurons of interest. Although we have used UAS-MCD8-GFP (Lee & Luo, 1999; Bloomington Stock Center #56168) and UAS-EGFP (Bloomington stock center #1522) effectively, EGFP is generally brighter. We have used this strategy to successfully sort a variety of neurons. They include large subsets of the brain such as all ELAV-expressing cells (*elav*-GAL4, UAS-EGFP flies) or dopaminergic cells (*TH*-GAL4, UAS-EGFP) as well as smaller subsets of neurons such as M-cells (*pdf*-GAL4, UAS-MCD8GFP) and E-cells (*Dv-pdf*-GAL4, UAS-EGP, *pdf*-RFP).

Neurons are isolated from adult fly brains at different circadian times essentially as described previously by Nagoshi et al. (2010). Young adult flies are entrained for 4 days in 12:12 LD cycles and harvested every 4 h to collect six timepoints throughout the day. Flies are chilled on ice and ~100 brains (for M-cells; fewer brains needed if cells are abundant, e.g., ~10 brains are used for *elav*-GAL4, UAS-EGFP) are dissected in cold dissecting saline (9.9 m*M* HEPES-KOH buffer, 137 m*M* NaCl, 5.4 m*M* KCl, 0.17 m*M* NaH_2PO_4, 0.22 m*M* KH_2PO_4, 3.3 m*M* glucose, 43.8 m*M* sucrose, pH 7.4) containing 20 μ*M* 6,7-dinitroquinoxaline-2,3-dione (DNQX; Sigma), 50 μ*M* D(−)-2-amino-5-phosphonovaleric acid (AP5/APV; Sigma), and 0.1 μ*M* tetrodotoxin (Tocris). Brains are collected in cold SM-active Bis–Tris media (SM^{active} media plus the same drugs included in the saline) and are centrifuged at room temperature for 2 min at 1000 rpm, the supernatant is removed, and the brains are washed in 300 μl cold dissecting saline. Approximately 2 μl/brain of L-cysteine-activated papain (50 units/ml, Worthington; heat activated at 37 °C for 10 min) is added, and samples are incubated at room temperature for 20–30 min with occasional mixing using a 20-μl pipette tip. A fivefold volume of cold SM-active Bis–Tris media is added to quench the digestion. Brains are centrifuged for 2 min at room temperature and resuspended in 7 μl of cold SM-active Bis–Tris media per brain (with a minimum volume of 400 μl to prevent difficulties in trituration).

To break apart the digested brains, we make a collection of flame-rounded 1-ml filter tips with either large, medium, or small orifices left at the tip after flaming. Brains are triturated by pipetting up and down 30× with a large flame-rounded tip, 20× with the medium flame-rounded tip, and 10× with the small flame-rounded tip. To ensure that the samples stay chilled, samples are placed on ice after every 10 iterations of pipetting up and down. The amount to triturate is not exact; it will differ depending on the number of brains. The small tip should be used until the liquid goes through easily without getting stuck. Do not over triturate since it can cause cells to burst.

Once cells are dissociated, they are placed in Sylgard plates (Sylgard 184 Elastomer Base and Curing Agent; Dow Corning) for sorting under the fluorescence scope (Leica M165 FC). Cold SM-active Bis–Tris media are added to Sylgard plates: 2.5 ml into two 6-cm Sylgard plates and 1.5 ml into two 3.5-cm plates. Half of the cell suspension is added to the center of each of the 6-cm plates. The plates are incubated on ice for 20–30 min so that cells can settle. Cells are sorted using micropipettes pulled from capillary tubes (World Precision Glass Capillaries #1B100-4) using a micropipette puller (Sutter Instrument Company). Before use, the micropipette tips are broken by punching them through a Kimwipe until they are ~1 cell wide. Cells are identified under the scope and then a cell aspirator (described in Hempel, Sugino, & Nelson, 2007) is used to move them to a Sylgard plate filled with fresh buffer. Since there is some probability that a nonfluorescent cell is accidently aspirated with a fluorescent cell, the cells undergo three rounds of sorting before being placed in a 0.2-ml tube. Approximately 60–100 cells are sorted for each timepoint and cell type of interest. To isolate microRNAs from specific neurons, at least 100 cells are placed in a 0.2-ml tube with 100 μl TRIzol (Invitrogen) and frozen at −80 °C. To isolate and amplify mRNA from neurons of interest, the isolated neurons were placed in 50 μl of lysis/binding buffer (Invitrogen; Dynabeads mRNA direct kit) and frozen at −80 °C.

2.2. Amplification of mRNA

From ~100 manually sorted neurons, we can isolate approximately 200–500 pg of total RNA or ~5 pg of mRNA. In order to generate libraries for RNA deep sequencing (Illumina HiSeq 2000), this RNA needs to be amplified >2000-fold to generate between 10 and 100 ng of mRNA (recommended starting amount Illumina RNA Tru-Seq v2). To this end, we have modified a linear amplification method traditionally used to amplify mRNA for microarray analysis (Kula-Eversole et al., 2010; Nagoshi et al., 2010; Sugino et al., 2006) and currently utilized in single-cell sequencing approaches (Hashimshony, Wagner, Sher, & Yanai, 2012). In this method, small amounts of RNA are reverse transcribed using dT primers containing a T7 promoter to make a cDNA template that can be used for multiple rounds of linear amplification using *in vitro* transcription. Although this strategy works well, it only amplifies the 3′-most end of the mRNA. To overcome this limitation, we isolated the mRNA population on oligo-dT beads and then made the cDNA template for *in vitro* transcription using both oligo-dT T7 and random-T7 primers (random deoxyribonucleotides combined with a T7 promoter sequence) in an attempt to circumvent the 3′ bias. We settled on this strategy after trying dT-T7 priming alone, random-T7 priming alone, and the mixture of the two. Indeed, we found that dT-T7 priming typically led to strongly 3′-biased libraries in which the 5′-ends of mRNAs are not represented (Fig. 1; *Act5c*). Random-T7 priming alone led to the underrepresentation of short mRNA molecules and the 3′-ends of longer mRNAs (Fig. 1; *RpS26* and 3′-end of *Act5c*). A mixture of random-T7 and dT-T7 did the best job at balancing the 5′- and 3′-ends, and we therefore moved forward with this approach (Fig. 1, note the presence of 5′-most exon of *Act5c*).

Poly-A RNA is isolated from ~100 neurons using Dynabeads oligo-dT using a scaled-down version of the manufacturer's instructions (Dynabeads

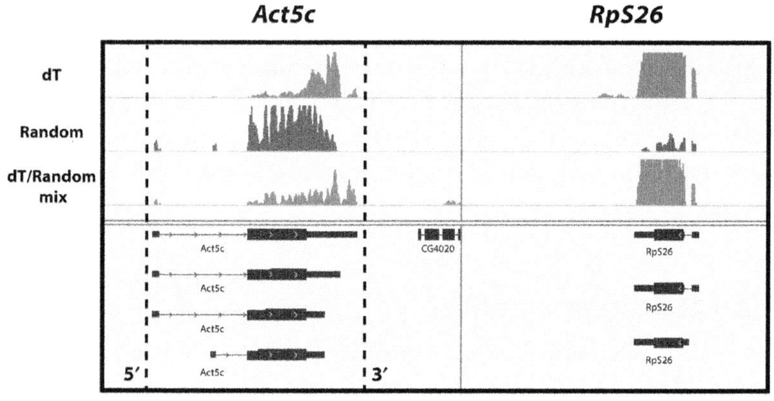

Figure 1 RNA is amplified using a mix of random and dT-T7 primers. Fluorescent cells were sorted from dissociated brains of *elav*-GAL4, UAS-EGFP flies, and mRNA was isolated from ~100 cells. The mRNA was reverse transcribed into cDNA using dT-T7 primers (top), random-T7 primers (middle), or a mix of both dT- and random-T7 primers (bottom). The resulting cDNA was used as a template for *in vitro* transcription and sequencing libraries were made. The resulting data for *Act5c* and *RpS26* are shown here using the Integrative Genomics Viewer (IGV; Robinson et al., 2011; Thorvaldsdottir et al., 2013). Solid blue boxes (dark gray in the print version) indicate exons and the intervening lines represent introns. *Act5c* is on the top strand and transcription is going from left to right; the 3′- and 5′-ends of the *Act5c* mRNA are denoted with dashed lines. *RpS26* is on the bottom strand and transcription is going from right to left. Priming with dT-T7 yields 3′-biased libraries that lack signal in the 5′-most exon of *Act5c*. Priming with random-T7 primer shows good coverage of the 5′-most exon of *Act5c*, but the 3′-end of the mRNA is not represented. Small mRNAs such as RpS26 are underrepresented with random priming. A mixture of dT and random priming leads to a decrease in 3′-bias and the coverage of both 3′-ends of mRNAs and shorter mRNAs.

mRNA direct kit; Invitrogen). The frozen cell suspension is thawed on ice, brought to 100 μl using lysis/binding buffer, and mixed well by pipetting. The lysate is transferred to 20 μl of washed Dynabeads (Dynabeads are prewashed 1 × in lysis/binding buffer and then resuspended in 20 μl of lysis/binding buffer), mixed gently by pipetting up and down, and then rotated at room temperature for 5 min. Samples are placed on the magnet (Invitrogen; DynaMag-2 Magnet) and the supernatant discarded. Beads are washed 2 × with 100 μl of wash buffer A, resuspended in wash buffer B, and transferred to a new tube. Samples are then washed 1 × in ice-cold 10 m*M* Tris–HCl before being resuspended in 10 μl of 10 m*M* Tris–HCl and incubated in a heat block at 70 °C for 2 min. The poly-A mRNA-containing supernatant is moved to a 0.2-ml tube and concentrated to 1 μl using a Speed-Vac (RC1010; Jouan, Winchester, VA). It is important not to let the samples dry out.

The resulting mRNA is reverse transcribed using a mixture of dT and random primers fused to a T7 promoter to generate a double-stranded cDNA to be used as a template for *in vitro* transcription. dT-T7 (GGCCAGTGAATTGTAATACGACTCACTATAGGGAGGCGGT(24)) and random-T7 (GGCCAGTGAATTGTAATACGACTCACTATAGGG AGGCGGN(20)) were added to mRNA at a concentration of 1.25 μ*M* each, and SuperScript III (Invitrogen) was used to generate double-stranded cDNA according to the manufacturer's guidelines. The resulting double-stranded cDNA was ethanol-precipitated overnight and washed 2 × in 75% ethanol. The cDNA is resuspended in 5 μl of RNase-free H_2O.

To validate the cell sorting, 1 μl of the cDNA can be diluted 1:30 and used as a template for q-PCR to determine whether the sample is enriched for genes of interest and/or shows cycling gene expression. For example, when cDNA is made from six timepoints of M-cells (*pdf*-GAL4, UAS-MCD8 GFP), PDF is greatly enriched compared to a similar experiment done with a more heterogeneous group of neurons isolated using the *elav*-GAL4, UAS-GFP line (Fig. 2A). Indeed, q-PCR from the same M-cell samples shows the circadian oscillation of the *tim* mRNA whose expression peaks in the evening (ZT14; Fig. 2B; there are six timepoints, so the data are double-plotted).

The remaining cDNA is used as a template for *in vitro* transcription using T7 polymerase (MEGAscript kit; Ambion) in a 10-μl reaction. Samples are incubated overnight at 37 °C. The resulting mRNA is isolated using RNA MinElute columns (Qiagen) and eluted in 14 μl of RNase-free H_2O. This typically yields between 10 and 50 ng/μl when quantified using

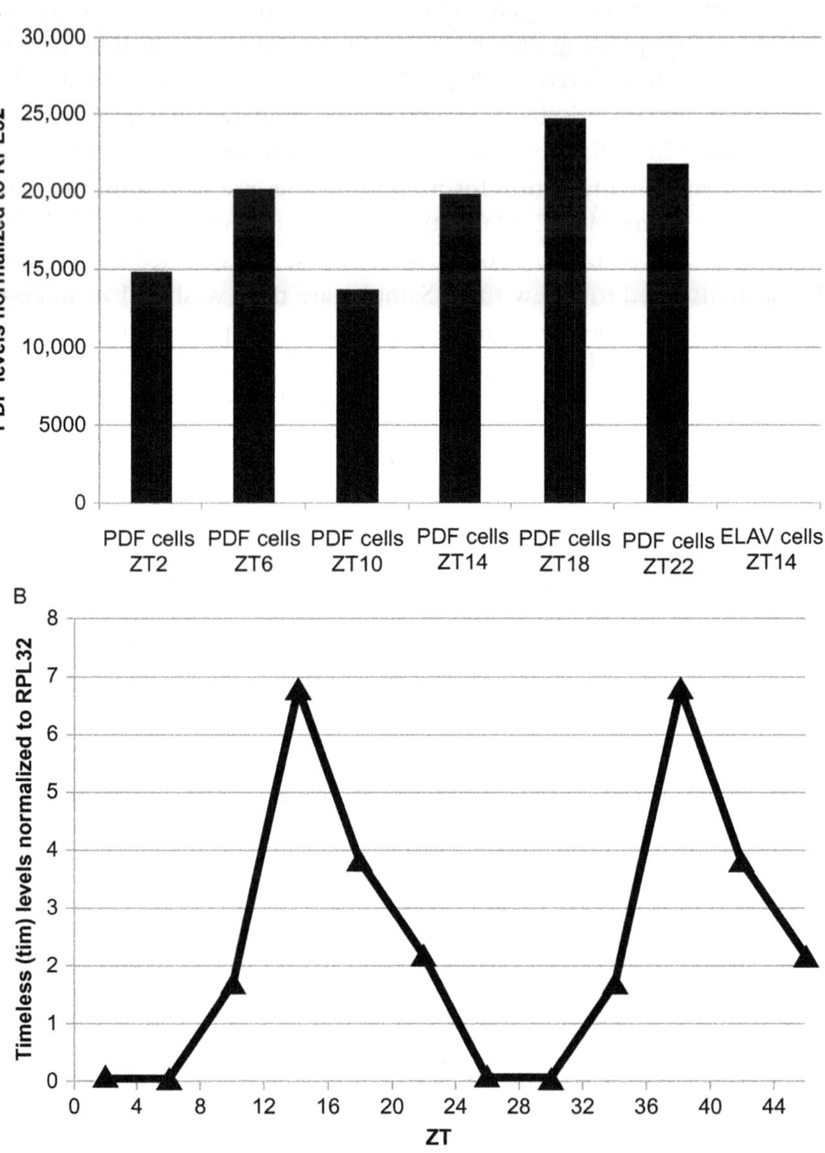

Figure 2 cDNA isolated from M-cells shows both an enrichment for PDF and cycling of tim. Approximately 100 fluorescent cells were isolated from dissociated brains of *pdf*-GAL4, UAS-MCD8-GFP flies or *elav*-GAL4, UAS-EGFP at timepoints throughout the day. mRNA was isolated from these cells and reverse transcribed using a combination of dT- and random-T7 primers. The resulting cDNA was diluted and used as a template for q-PCR using primers for *PDF*, *tim*, and *RPL32* (for normalization purposes). (A) Cells sorted from *pdf*-GAL4, UAS-MCD8-GFP are highly enriched for PDF in comparison to cells isolated from *elav*-GAL4, UAS-EGFP. (B) tim mRNA cycles in cells isolated from *pdf*-GAL4, UAS-MCD8-GFP. Tim levels peak in early evening (ZT14). Data from six timepoints are shown double-plotted.

the NanoDrop (Thermo Scientific). 100 ng of this sample is concentrated to 2 µl using a Speed-Vac and used as input for the RNA Tru-Seq library generation protocol (Illumina). To be more cost-effective, the Tru-Seq library kit was used according to manufacturer's recommendations except that one-third volume was used for every step. The size of the libraries is verified using a High Sensitivity DNA kit on the Bioanalyzer (Agilent), and library concentration is determined by using q-PCR using the PCR primers (Illumina Tru-Seq kit, forward primer: AATGATACGGCGACCACCGA, reverse primer: CAAGCAGAAGACGGCATACGA) and a library of known concentration. A 2 n*M* mixture of six different barcoded libraries is mixed together, and 10 p*M* of this sample is sequenced in a single lane of the Illumina HiSeq 2000.

The resulting data were parsed to separate the barcoded reads, and the resulting sequence files (fastq format) were aligned to the *Drosophila* genome (version dm3) via TopHat (Trapnell, Pachter, & Salzberg, 2009) using the following criteria: −m 1 -F 0 -p 6 -g 1 –microexon-search –no-closure-search –solexa-quals -I 50000. In our initial experiments, the sequencing reads from RNA amplified from sorted neurons mapped very poorly to the *Drosophila* genome, i.e., only approximately 10% of the libraries corresponded to the *Drosophila* genome due to contamination (see Section 3). By cleaning all work surfaces (PCR machines, centrifuges, pipettes, etc.) extremely carefully prior to each experiment (DNA-Off; Takara), we were able to increase the percentage of our libraries that mapped to the transcriptome (non-rRNA) to ~40–70%. Visualization files were generated by converting the bam output file (TopHat) to a bigwig file using a custom script. These files were viewed using the Integrated Genomics Viewer (Robinson et al., 2011; Thorvaldsdottir, Robinson, & Mesirov, 2013).

To validate the cell sorting and deep-sequencing approach, we visually examined the sequencing results for known neuronal group specific genes. As a simple example, we compared the expression of the circadian gene *timeless* (*tim*) and the dopaminergic cell-enriched enzyme tyrosine hydroxylase (*TH* or *pale*) in sequencing data derived from M-cells or dopaminergic cells (*pdf*-GAL4, UAS-MCD8-GFP, and *TH*-GAL4, UAS-EGFP, respectively). Figure 3 shows the IGV browser showing the number of sequencing reads found for the housekeeping gene *rpl32* as well as *TH* and *tim*. Similar levels of Rpl32 are found in both M-cells and dopaminergic cells. As expected, only dopaminergic cells express *pale* (*TH*). There are two main isoforms of *ple*; dopaminergic cells preferentially express the central nervous system-specific isoform that lacks exons 3 and 4 and not the hypodermal

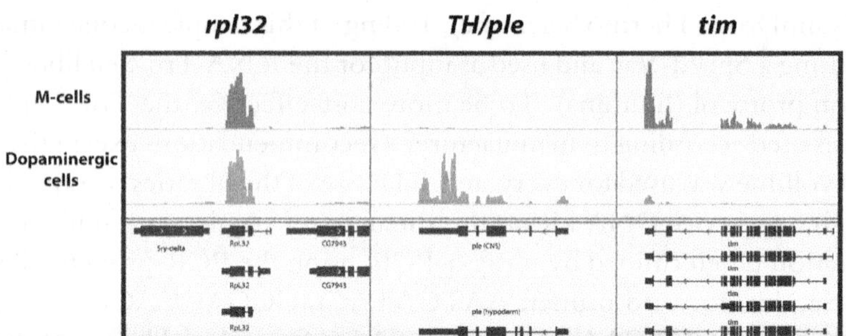

Figure 3 RNA sequencing data from neuronal groups show specificity. Sequencing data from sorted M-cells (*pdf*-GAL4, UAS-MCD8-GFP) or dopaminergic cells (*TH*-GAL4) are visualized on the IGV. Rpl32 shows similar levels in both M-cells and dopaminergic cells. In contrast, ple/TH (tyrosine hydroxylase) is only detected in dopaminergic cells. As expected, dopaminergic neurons show the central nervous system-specific isoform of *ple* that is lacking exons 3 and 4. The circadian mRNA tim is only found in M-cells and not in dopaminergic cells.

form (Fig. 3; Friggi-Grelin, Iche, & Birman, 2003). In addition, only the circadian M-cells express substantial levels of *timeless*.

It is also possible to detect cycling gene expression and differential gene expression in sequencing data from specific neuronal populations. As shown earlier (Fig. 2), cycling *tim* mRNA was observed in the cDNA made directly from M-cell mRNA. Not surprisingly, cycling *tim* mRNA is also easily detected in the sequenced libraries from *in vitro* transcription of this cDNA. Figure 4 shows *tim* gene expression in six timepoints of RNA from M-cells with peak expression at ZT14 as expected. In addition to the core clock genes, other cycling genes have been identified in M-cells. The inward-rectifying potassium channel, Ir, is one example of such a gene. *Ir* mRNA cycles in M-cells with expression peaking at ZT12 as seen with microarrays (Kula-Eversole et al., 2010; Mizrak et al., 2012). There are also genes that are expressed predominantly in either M-cells or E-cells (Fig. 5). CG18343 is an unknown gene detected in mRNA isolated from E-cells but not M-cells. The tetraspanin, *Tsp42Eo*, shows the reverse pattern; it is detected in M-cells but not E-cells.

2.3. Amplification of miRNA

The method for making miRNA libraries from isolated neurons is adapted from Hafner et al. (2012) as well as the Mello Lab Small RNA Cloning Protocol (http://www.umassmed.edu/PageFiles/43096/Mello%20lab%

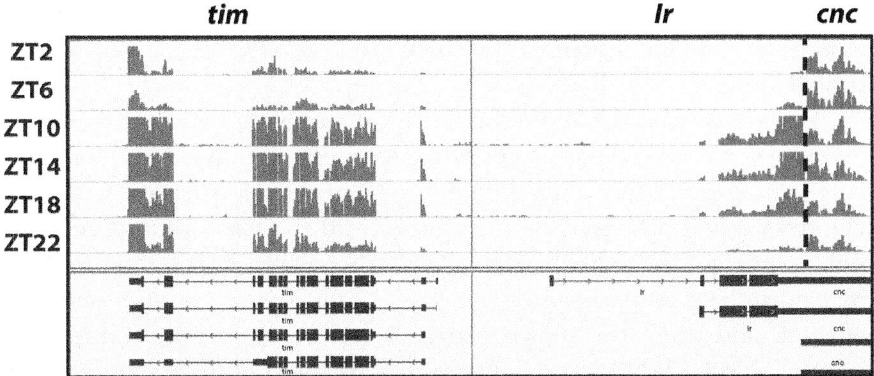

Figure 4 *Timeless (tim)* and *Ir* cycle in M-cells. RNA from M-cells isolated at six different timepoints throughout the day was isolated, amplified, and made into Illumina sequencing libraries. The resulting sequencing data were visualized using the IGV. Data are shown for *tim* (bottom strand; transcription going from right to left), *Ir* (top strand, transcription going from left to right), and *cnc* (bottom strand, transcription going from right to left). Solid blue boxes (dark gray in the print version) indicate exons and the intervening lines represent introns. The dashed line indicates the location where the *Ir* and *cnc* 3′-ends overlap. Tim levels cycle throughout the day in M-cells with the highest expression detected at ZT14. Ir levels also cycle in M-cells with peak phase at ZT14. In contrast, the adjacent gene *cnc* shows equal expression throughout the day.

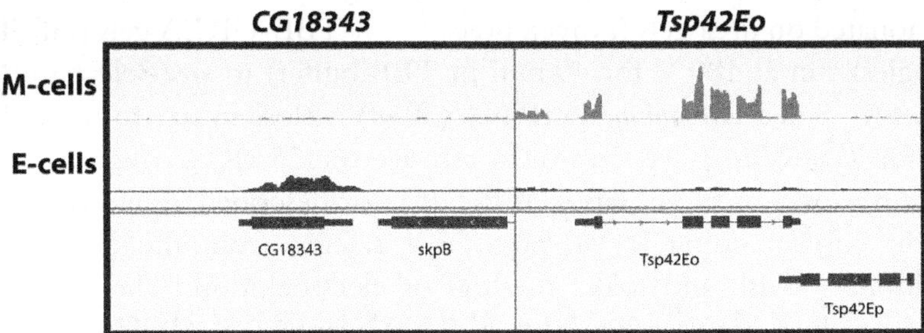

Figure 5 Identification of differentially expressed genes in E-cells and M-cells. RNA-seq libraries from isolated M-cells and E-cells were compared on the IGV. There are many genes that are differentially expressed in subsets of neurons. Two examples are shown here. CG18343 is expressed more strongly in E-cells than in M-cells. In contrast, *Tsp42Eo* is expressed more strongly in M-cells than E-cells.

20small%20RNA%20cloning%20protocol.pdf). In brief, total RNA is extracted from isolated neurons as described above, polyadenylated adaptors are ligated to the 3′-end, and the RNA is size-selected. 5′-Adaptors are then ligated, and the resulting RNAs are reverse transcribed to generate cDNA that can be amplified by PCR. A second size-selection helps to ensure that cDNA contains primarily miRNAs and not other similar-sized RNA contaminants. Finally, libraries are made and sequenced as described in detail below.

To extract total RNA, cells are thawed and lysed by pipetting up and down several times. 100 μl of TRIzol (Invitrogen) is added, and total RNA is extracted according to the manufacturer's instructions with 40 μl chloroform and precipitated overnight at −20 °C. GlycoBlue (Life Technologies) is used as coprecipitant. Total RNA is precipitated the next day by centrifuging at maximum speed at 4 °C for 30 min, washed with 75% ethanol, and dried for approximately 5 min at room temperature. The amount of total RNA obtained here may be too low to be detected with the NanoDrop (Thermo Scientific), but the quality of the total RNA should be assessed using the RNA 6000 Pico kit in combination with the Bioanalyzer (Agilent).

3′ Preadenylated adaptors are then ligated to the total RNA using T4 RNA ligase 2, truncated (NEB; rAppTGGAATTCTCGGGTGCCAAGG/ddC/; adaptor is specific to Illumina HiSeq 2000 but other adaptors could be designed for other platforms). The adaptors should be at least sixfold in excess of the total RNA to optimize ligation efficiency. 10% DMSO can be added to denature the RNA. A longer incubation time, for example, 6 h, can also increase ligation efficiency. Ligated RNA products are then fractionated on urea gels (Novex precast 15% TBE-UREA gels (Life Technologies) run at 180 V for 40 min in TBE buffer) to size-select miRNAs (Fig. 6A). Since *Drosophila* 2S rRNA (30 nt) is close in size to the miRNA fraction (18–27 nt), it is critical to remove the 2S rRNA region to avoid contamination of the libraries. If the adaptor described above is used, the region corresponding to 40–50 nt is extracted from the gel with the traditional "crush and soak" method or electroelution. The precipitated RNA products are then ligated to 5′ RNA adaptors (5′-GUUCAGAGUUCUACAGUCCGACGAUC-3′) with T4 RNA ligase (Ambion). The efficiency of this step is usually very low, and so overnight incubation at room temperature is advisable.

After 5′ ligation, RNA products are reversely transcribed with SuperScript II with primers that are complementary to the 3′ adaptors. cDNAs are then amplified with Phusion High-Fidelity DNA polymerase (NEB) with universal forward primers and indexed reverse primers for barcoding (Illumina Customer Sequence Letter, Section: Oligonucleotide sequences for Tru-Seq™ Small RNA Sample Prep Kits) for 12–15 cycles. The number of cycles in the PCR program is critical, because overamplification will

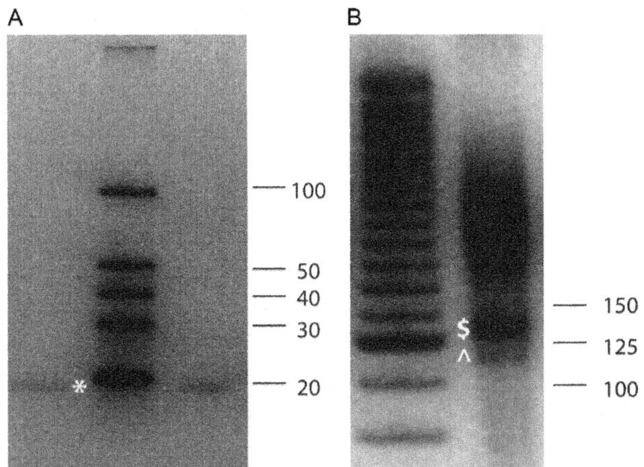

Figure 6 Preparation and amplification of miRNAs from sorted neurons. (A) Adaptor-bound total RNA is fractionated on a 15% TBE-urea gel (Novex) stained with cyber gold (Life Technologies). * indicates excess 3′ preadenylated adaptors. The 3′ ligated products are too little to be observed at this step. A region corresponding to ~40–50 bp is cut from the gel. (B) Amplified products are size-selected on a 4–12% TBE gel (Novex; Life Technologies) stained with cyber gold. ^indicates self-ligated adaptors. $ indicates amplified miRNA libraries. Some high-molecular weight smear is visible here due to overamplification.

generate large amounts of a high-molecular weight smear, which affects library quality.

The amplified products are then subjected to a second size-selection to remove unwanted products, including self-ligated adaptors, residual amplified 2S rRNA products, and the high-molecular weight smear (Fig. 6B). If the exact adaptors and primers are used as described in this protocol, the DNA from the region containing 130–140 bp products should be extracted as previously described. If no bands or only faint bands are observed in the gel in the 130–140 bp region, a wider region (125–150 bp) can be extracted and amplified a second time.

Quality and quantity of miRNA libraries are assessed with an Agilent bioanalyzer using either the DNA 1000 (Fig. 7A) or a high-sensitivity DNA Chip. miRNA libraries should then be diluted to 2 n*M* with EB buffer and mixed together equally as described for mRNA libraries. Libraries are sequenced on an Illumina HiSeq 2000 and the resulting data analyzed using custom scripts. Data from a miRNA library made from sorted EGFP-labeled M-cells are shown in the IGV (Fig. 7B, top). The miRNA library shows sequencing reads for mir-14 but not for any of the protein-encoding genes

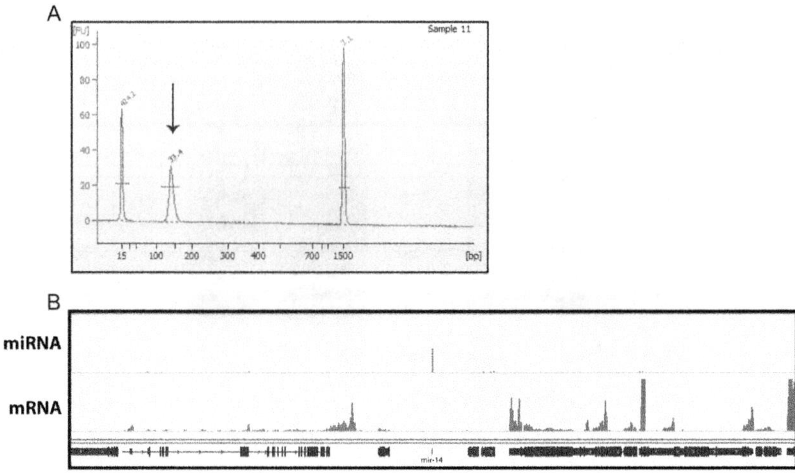

Figure 7 Visualization of miRNA sequencing libraries. (A) Bioanalyzer profile of miRNA sequencing library on a DNA 1000 chip (Agilent). The library (labeled with a black arrow) is ~150 bp in size and shows a clear, defined peak. (B) miRNA libraries viewed on the IGV. Sequencing data from a miRNA and mRNA library made from isolated M-cells are shown in the IGV centered on the miR-14 locus. miR-14 is visible in the miRNA library but there is no signal from the surrounding mRNAs. In contrast, mir-14 signal is not present in the mRNA library.

surrounding it. In contrast, a mRNA library from M-cells is shown below in green (gray in the print version) (Fig. 7B).

3. DISCUSSION

Recent studies suggest that the control and regulation of circadian behavior is due to the coordinated response of several different groups of circadian neurons acting as a network. One critical part of dissecting this network (or any network) is learning more about the role of specific cells and groups of cells. We present here a method for isolating specific groups of neurons from the *Drosophila* brain and using deep sequencing to profile their miRNA and mRNA populations present in these cells. Any neuronal population marked with a fluorescent protein can be manually sorted from dissociated *Drosophila* brains. In this chapter, we show data from both a medium (dopaminergic) and a small (M-cells) population, about 130 and 20 cells/brain, respectively. The two populations show the expected divergent gene expression patterns for two well-characterized genes and many others (not shown), i.e., only M-cells express well the circadian gene *tim* and only dopaminergic cells express *TH* or *ple* (Fig. 3). It is unclear whether the small amount of *tim* signal from dopaminergic cells reflects cell contamination or bona fide clock gene expression.

In addition to identifying cell-specific gene expression patterns, we have also identified cycling and differential gene expression using this approach. Both *tim* and *Ir* mRNAs cycle throughout the day in M-cells. There are also genes that are expressed in one subgroup of neurons but not another. Two such examples are shown in Fig. 5: CG18343 and *Tsp42Eo*. We also describe the isolation of miRNAs from M-cells. This approach allows for the analysis of neuronal miRNAs, e.g., changes in level with circadian time and/or light and firing, and allows a dissection of their role in circadian and neuronal posttranscriptional gene regulation.

The ability to manually sort neurons from dissociated *Drosophila* brains is not new. Several studies published in the past years have illustrated the usefulness of this approach in combination with microarrays (Kula-Eversole et al., 2010; Mizrak et al., 2012; Nagoshi et al., 2010). The advent of deep sequencing has made it appealing to extend these previous studies and adapt existing methods for this purpose. Since *Drosophila* microarrays contain antisense probes for the 3′-ends of transcripts, amplifying mRNA for this assay only requires material from the 3′-end of the mRNA. In addition, contamination that often occurs when amplifying very small amounts of RNA (picograms) is not an issue because most contaminants, i.e., from human sources, would not hybridize to the *Drosophila* microarray probes. Therefore, adapting existing methods to deep sequencing presents two major hurdles: (1) creating an mRNA library that contains information beyond the 3′-end of the transcript and (2) preventing contamination in both mRNA and miRNA libraries.

All current mRNA amplification methods used in conjunction with deep sequencing start with a reverse transcription step in which the RNA is made into a cDNA template. In some approaches, only the 3′-ends are amplified using a T7-dT oligo (Hashimshony et al., 2012), whereas in other strategies more full-length transcripts can be obtained using strand switching (Deng, Ramskold, Reinius, & Sandberg, 2014; Picelli et al., 2014; Ramskold et al., 2012). We have tried to obtain less 3′-biased libraries by poly-A selecting mRNA from isolated cells and amplifying using a combination of dT- and random-T7 primers. Although these changes have allowed us to obtain sequencing information along the entire length of genes, there is still a 3′-bias present in the libraries. It is not clear whether this bias comes from the amplification method and/or from partial degradation of the RNA during the cell sorting procedure followed by oligo-dT selection.

Contamination of both mRNA and miRNA libraries can be a substantial problem since libraries generated can contain non-*Drosophila* sequences that dramatically reduce the amount of usable data. (In our early experiments, as much as 90% of the sequencing reads did not map to the *Drosophila* genome). It is recommended to have a separate clean room and/or a PCR workstation to avoid contamination. However, it is possible to drastically reduce contamination by working extremely carefully and cleaning all equipment prior to each experiment using DNA-OFF (Takara).

Deep-sequencing data provide us with much more information than was gleaned from microarray data. The mRNA-seq data can be analyzed to reveal much more detail, for example, isoform-specific expression, differential splicing, and mRNA editing. As we and others have examined many of these processes in larger more heterogeneous *Drosophila* tissues (Brown et al., 2014; Khodor et al., 2011; Rodriguez, Menet, & Rosbash, 2012; St Laurent et al., 2013), it is possible that studies of neuronal populations will reveal novel and perhaps important regulatory events that were masked in whole head or whole body studies.

miRNA-seq will identify miRNAs that are present in specific neurons. Targets can be experimentally identified in many ways, for example, by depleting the miRNA of interest using miRNA sponges (Ebert, Neilson, & Sharp, 2007; Loya, Lu, Van Vactor, & Fulga, 2009) and performing mRNA-seq from the same neuronal population. These approaches could also be used in conjunction with different environmental stimuli (light, feeding, etc.) to determine how the mRNA and miRNA profiles of these neurons respond to these stimuli.

New drivers will certainly be available in the near future to express EGFP in only one or two key neurons, opening the possibility of understanding the circuit at a single-neuron level. These more specific drivers would also facilitate single-neuron sequencing. We note in this context that RNA-seq has been used successfully to profile gene expression in single cells (Hashimshony et al., 2012; Ramskold et al., 2012; Tang et al., 2009). Adapting these methods to *Drosophila* neurons would allow us to accompany behavioral experiments and define each key cell by its gene expression profile, creating a complete and accurate map of where and at what circadian time specific genes are being expressed in the circadian neuronal system.

REFERENCES

Brown, J. B., Boley, N., Eisman, R., May, G. E., Stoiber, M. H., Duff, M. O., et al. (2014). Diversity and dynamics of the Drosophila transcriptome. *Nature*, *512*, 393–399.

Deng, Q., Ramskold, D., Reinius, B., & Sandberg, R. (2014). Single-cell RNA-seq reveals dynamic, random monoallelic gene expression in mammalian cells. *Science*, *343*(6167), 193–196.

Depetris-Chauvin, A., Berni, J., Aranovich, E. J., Muraro, N. I., Beckwith, E. J., & Ceriani, M. F. (2011). Adult-specific electrical silencing of pacemaker neurons uncouples molecular clock from circadian outputs. *Current Biology*, *21*(21), 1783–1793.

Ebert, M. S., Neilson, J. R., & Sharp, P. A. (2007). MicroRNA sponges: Competitive inhibitors of small RNAs in mammalian cells. *Nature Methods*, *4*(9), 721–726.

Friggi-Grelin, F., Iche, M., & Birman, S. (2003). Tissue-specific developmental requirements of Drosophila tyrosine hydroxylase isoforms. *Genesis*, *35*(4), 260–269.

Grima, B., Chelot, E., Xia, R., & Rouyer, F. (2004). Morning and evening peaks of activity rely on different clock neurons of the Drosophila brain. *Nature*, *431*(7010), 869–873.

Guo, F., Cerullo, I., Chen, X., & Rosbash, M. (2014). PDF neuron firing phase-shifts key circadian activity neurons in Drosophila. *Elife*, *3*, e02780.

Hafner, M., Renwick, N., Farazi, T. A., Mihailovic, A., Pena, J. T., & Tuschl, T. (2012). Barcoded cDNA library preparation for small RNA profiling by next-generation sequencing. *Methods*, *58*(2), 164–170.

Hashimshony, T., Wagner, F., Sher, N., & Yanai, I. (2012). CEL-Seq: Single-cell RNA-Seq by multiplexed linear amplification. *Cell Reports*, *2*(3), 666–673.

Helfrich-Forster, C. (1995). The period clock gene is expressed in central nervous system neurons which also produce a neuropeptide that reveals the projections of circadian pacemaker cells within the brain of Drosophila melanogaster. *Proceedings of the National Academy of Sciences of the United States of America*, *92*(2), 612–616.

Hempel, C. M., Sugino, K., & Nelson, S. B. (2007). A manual method for the purification of fluorescently labeled neurons from the mammalian brain. *Nature Protocols*, *2*(11), 2924–2929.

Khodor, Y. L., Rodriguez, J., Abruzzi, K. C., Tang, C. H., Marr, M. T., 2nd., & Rosbash, M. (2011). Nascent-seq indicates widespread cotranscriptional pre-mRNA splicing in Drosophila. *Genes & Development*, *25*(23), 2502–2512.

Kula-Eversole, E., Nagoshi, E., Shang, Y., Rodriguez, J., Allada, R., & Rosbash, M. (2010). Surprising gene expression patterns within and between PDF-containing circadian neurons in Drosophila. *Proceedings of the National Academy of Sciences of the United States of America*, *107*(30), 13497–13502.

Lee, T., & Luo, L. (1999). Mosaic analysis with a repressible cell marker for studies of gene function in neuronal morphogenesis. *Neuron*, *22*(3), 451–461.

Loya, C. M., Lu, C. S., Van Vactor, D., & Fulga, T. A. (2009). Transgenic microRNA inhibition with spatiotemporal specificity in intact organisms. *Nature Methods*, *6*(12), 897–903.

Mao, Z., & Davis, R. L. (2009). Eight different types of dopaminergic neurons innervate the Drosophila mushroom body neuropil: Anatomical and physiological heterogeneity. *Frontiers in Neural Circuits*, *3*, 5.

Mizrak, D., Ruben, M., Myers, G. N., Rhrissorrakrai, K., Gunsalus, K. C., & Blau, J. (2012). Electrical activity can impose time of day on the circadian transcriptome of pacemaker neurons. *Current Biology*, *22*(20), 1871–1880.

Nagoshi, E., Sugino, K., Kula, E., Okazaki, E., Tachibana, T., Nelson, S., et al. (2010). Dissecting differential gene expression within the circadian neuronal circuit of Drosophila. *Nature Neuroscience*, *13*(1), 60–68.

Nitabach, M. N., Blau, J., & Holmes, T. C. (2002). Electrical silencing of Drosophila pacemaker neurons stops the free-running circadian clock. *Cell, 109*(4), 485–495.

Picelli, S., Faridani, O. R., Bjorklund, A. K., Winberg, G., Sagasser, S., & Sandberg, R. (2014). Full-length RNA-seq from single cells using Smart-seq2. *Nature Protocols, 9*(1), 171–181.

Ramskold, D., Luo, S., Wang, Y. C., Li, R., Deng, Q., Faridani, O. R., et al. (2012). Full-length mRNA-Seq from single-cell levels of RNA and individual circulating tumor cells. *Nature Biotechnology, 30*(8), 777–782.

Robinson, J. T., Thorvaldsdottir, H., Winckler, W., Guttman, M., Lander, E. S., Getz, G., et al. (2011). Integrative genomics viewer. *Nature Biotechnology, 29*(1), 24–26.

Rodriguez, J., Menet, J. S., & Rosbash, M. (2012). Nascent-seq indicates widespread cotranscriptional RNA editing in Drosophila. *Molecular Cell, 47*(1), 27–37.

St Laurent, G., Tackett, M. R., Nechkin, S., Shtokalo, D., Antonets, D., Savva, Y. A., et al. (2013). Genome-wide analysis of A-to-I RNA editing by single-molecule sequencing in Drosophila. *Nature Structural & Molecular Biology, 20*(11), 1333–1339.

Stoleru, D., Peng, Y., Agosto, J., & Rosbash, M. (2004). Coupled oscillators control morning and evening locomotor behaviour of Drosophila. *Nature, 431*(7010), 862–868.

Sugino, K., Hempel, C. M., Miller, M. N., Hattox, A. M., Shapiro, P., Wu, C., et al. (2006). Molecular taxonomy of major neuronal classes in the adult mouse forebrain. *Nature Neuroscience, 9*(1), 99–107.

Tang, F., Barbacioru, C., Wang, Y., Nordman, E., Lee, C., Xu, N., et al. (2009). mRNA-Seq whole-transcriptome analysis of a single cell. *Nature Methods, 6*(5), 377–382.

Thorvaldsdottir, H., Robinson, J. T., & Mesirov, J. P. (2013). Integrative Genomics Viewer (IGV): High-performance genomics data visualization and exploration. *Briefings in Bioinformatics, 14*(2), 178–192.

Trapnell, C., Pachter, L., & Salzberg, S. L. (2009). TopHat: Discovering splice junctions with RNA-Seq. *Bioinformatics, 25*(9), 1105–1111.

Yoshii, T., Rieger, D., & Helfrich-Forster, C. (2012). Two clocks in the brain: An update of the morning and evening oscillator model in Drosophila. *Progress in Brain Research, 199*, 59–82.

Zhang, L., Chung, B. Y., Lear, B. C., Kilman, V. L., Liu, Y., Mahesh, G., et al. (2010). DN1 (p) circadian neurons coordinate acute light and PDF inputs to produce robust daily behavior in Drosophila. *Current Biology, 20*(7), 591–599.

Zhang, Y., Liu, Y., Bilodeau-Wentworth, D., Hardin, P. E., & Emery, P. (2010). Light and temperature control the contribution of specific DN1 neurons to Drosophila circadian behavior. *Current Biology, 20*(7), 600–605.

CHAPTER EIGHTEEN

Analysis of Circadian Regulation of Poly(A)-Tail Length

Shihoko Kojima[1,*], Carla B. Green[†]
*Department of Biological Sciences, Virginia Tech, Blacksburg, VA, USA
[†]Department of Neuroscience, University of Texas Southwestern Medical Center, Dallas, Texas, USA
[1]Corresponding author: e-mail address: skojima@vt.edu

Contents

Abstract

The poly(A) tail is found on the 3′-end of most eukaryotic mRNAs, and its length significantly contributes to the mRNAs half-life and translational competence. Circadian regulation of poly(A)-tail length is a powerful mechanism to confer rhythmicity in gene expression posttranscriptionally and provides a means to regulate protein levels independent of rhythmic transcription in the nucleus. Therefore, analysis of circadian poly(A)-tail length regulation is important for a complete understanding of rhythmic physiology, since rhythmically expressed proteins are the ultimate mediators of rhythmic function. Nevertheless, it has previously been challenging to measure changes in poly(A)-tail length, especially at a global level, due to technical constraints. However, new methodology depending on differential fractionation of mRNAs depending on the length of their tails has recently been developed. In this chapter, we describe these methods as used for examining the circadian regulation of poly(A)-tail length and provide detailed experimental procedures to measure poly(A)-tail length both at a the single mRNA level and the global level. Although this chapter concentrates on methods

Methods in Enzymology, Volume 551
ISSN 0076-6879
http://dx.doi.org/10.1016/bs.mie.2014.10.021

we used for analyzing poly(A)-tail length in the mammalian circadian system, the methods described here can be applicable to any organisms and any biological processes.

1. INTRODUCTION

Posttranscriptional gene regulatory mechanisms allow modification of gene expression after transcripts are made from DNA, and this type of regulation gives tremendous flexibility in overall gene expression including when, where, and how much protein product is generated. Posttranscriptional processes include many different mechanisms, such as capping, splicing, 3′-end cleavage and polyadenylation, localization, translation, and ultimate turnover of the mRNA, and circadian clocks have been shown to extensively utilize posttranscriptional regulation for rhythmic regulation of gene expression, influencing many of these steps (Kojima, Shingle, & Green, 2011). The changes in poly(A)-tail length of mRNAs is one of the important posttranscriptional regulatory steps that the circadian clock uses to control rhythmic gene expression. Poly(A) tails at the 3′ends are hallmarks of most eukaryotic mRNAs and are implicated in many aspects of mRNA function, such as mRNA stability and translation efficiency. The changes in poly(A)-tail length can occur throughout the lifetime of an mRNA both in the nucleus and in the cytoplasm, and the balance between deadenylation and polyadenylation ultimately determines the poly(A)-tail length (Eckmann, Rammelt, & Wahle, 2011). Dynamic variation in poly(A)-tail length is a powerful mechanism for driving rhythmic protein expression, and therefore, developing sensitive assays that can monitor changes in poly(A)-tail length is important.

To date, accurate measurement of poly(A)-tail length has been technically challenging, especially at the genomewide level, due to the homogenous nature of poly(A) tails. To overcome this issue, we developed a novel genomewide method called "Poly(A)denylome" analysis to measure changes in poly(A)-tail lengths of individual mRNAs in an unbiased manner and to identify mRNAs that have diurnal rhythmicity in their poly(A)-tail length (Kojima, Sher-Chen, & Green, 2012). Using this technique, we discovered that approximately 2.5% of mRNAs in mouse liver have rhythmic poly(A)-tail lengths, thus providing evidence that the circadian clock globally regulates this posttranscriptional modification. Most importantly, we also demonstrated that the fluctuation in the poly(A)-tail length can ultimately drive the rhythms in the amount of protein produced (Kojima et al., 2012).

2. MEASUREMENT OF POLY(A)-TAIL LENGTH AT A GENOMEWIDE LEVEL

Since the emergence of microarrays, genomic approaches have been widely utilized to examine expression of tens of thousands of genes simultaneously. Even though the primary focus of developing these tools was to analyze differences in gene expression levels of two or more independent samples, innovative applications have subsequently enabled the measurement of other events such as DNA methylation, transcription factor binding sites, alternative splicing, and poly(A)-tail length (Heller, 2002). For example, a method we developed, called "Poly(A)denylome" analysis, can identify mRNAs that have different poly(A)-tail lengths and has successfully detected mRNAs that undergo rhythmic changes in their poly(A)-tail length around the circadian clock.

Poly(A)denylome analysis consists of three different components: RNA fractionation according to poly(A)-tail size, 3′-end label assay to validate the fractionation, and microarray analysis. First, total RNAs are selected by the presence of poly(A) stretches, and then these RNAs are divided into two different populations: one has short poly(A)-tail lengths of approximately ~70 nt or less, and the other has poly(A)-tail lengths that are longer than ~60 nt. This length threshold can be modified by changing the salt concentration in the elution buffer (see below for details). The poly(A)-tail length in each population as well as the RNA integrity need to be further verified by the 3′-end label assay. Following the validation and cleanup of these RNAs, they can be subjected to microarray analysis. Needless to say, microarrays are tools to quantify mRNA levels, not to obtain poly(A)-tail length information. Therefore, relative poly(A)-tail length information is derived from the ratio of the level of each mRNA in the long poly(A)-tail length population over the level in the short poly(A)-tail length population. We have shown that these ratios accurately reflect the dynamics of poly(A)-tail length changes and have successfully used these ratios to define whether a specific transcript has a rhythmic poly(A)-tail length, based on certain threshold criteria, such as basal expression level, a fit to a sinusoid, or amplitude of rhythms (Kojima et al., 2012).

In this chapter, we will describe detailed procedures for the RNA fractionation and 3′-end labeling assay later. However, we will not cover the detailed techniques for microarray analyses here, as there are well-documented standard protocols, although we will provide some information specific to poly(A)denylome analysis.

2.1. Poly(A)-tail size RNA fractionation

The first step is to employ poly(A)-tail size RNA fractionation, which divides RNAs into populations that have either short or long poly(A)-tail length. This is done by oligo(dT) chromatography where oligo(dT)-bound poly(A)$^+$ RNAs can be differentially recovered using elution buffers with different salt concentrations, owing to the difference in affinity of long and short tails for the oligo(dT) beads (Fig. 1), as reported previously (Meijer et al., 2007). Another similar method for poly(A)-tail length fractionation that utilizes the temperature, instead of salt concentration, has also been developed (Beilharz & Preiss, 2009); however, we found that varying the salt concentration is easier and more reproducible than changing the temperature, partly because ambient temperature can be slightly different each day and hard to strictly control. It is also important to keep in mind that all the reagents must be prepared as RNA grade (i.e., Diethylpyrocarbonate (DEPC)-treated), as the success of RNA fractionation depends on the integrity of the RNAs. This is particularly important for the RNA fractionation step, since the elution step needs to be performed at room temperature (20–25 °C) overnight. RNA can be extracted by any conventional methods, but a significant amount of starting materials (we used 80 μg) are needed to be able to visualize the bulk poly(A)-tail length at a later step (see also 3′-end label assay). As a general guideline, 100 mg of mouse liver tissue with 1 ml TRIZOL (Life Technologies) yields 300–400 μg of RNA.

(1) Prepare dilution buffer by mixing the following and preheat at 70 °C.

300 μl	20 × Saline-Sodium-Citrate (SSC) buffer
10 μl	1 *M* Tris–HCl (pH 7.4)
2 μl	0.5 *M* EDTA
25 μl	10% SDS
10 μl	β-Mercaptoethanol
653 μl	DEPC-Destilled and Deionized Water (DDW)

(2) Mix 80-μg RNA sample (in a volume of 40 μl or less) with 400-μl PolyATtract GTC extraction. buffer, 8-μl β-mercaptoethanol, 15-μl biotinylated oligo(dT) (50 pmol/μl), and 816-μl dilution buffer (prepared in Step 1).

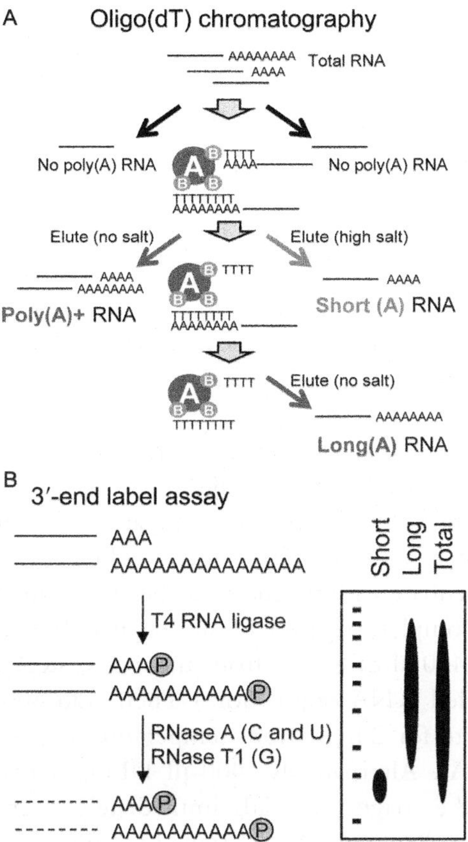

Figure 1 Method for fractionation of mRNAs with different poly(A)-tail lengths (Kojima et al., 2012). (A) Oligo(dT) chromatography was used to separate total RNA into fractions with either short or long poly(A) tails by varying salt concentrations in the elution. Total PolyA$^+$ (nonfractionated) RNAs can also be isolated as a reference. (B) Validation of fractionation by a 3′-end label assay. T4 RNA ligase catalyzes the ligation of the 5′ phosphate terminus of a nucleic acid donor to the 3′ OH terminus of a nucleic acid acceptor, thus introduces radioisotope label at the 3′-end of poly(A) tails. 3′-End-labeled mRNAs are then digested by both RNase A that recognizes cytidine and uridine residues and RNase T1 that recognizes guanidine residues, resulting in degradation of the RNA body but not poly(A) stretches. The size of these poly(A) stretches can be visualized by autoradiography after PAGE (polyacrylamide gel electrophoresis) analysis.

(3) Incubate the sample mixture at 70 °C for 10 min.

(4) Spin at 12,000 × *g* for 10 min at room temperature. Note that there might be white precipitant after the spin, which should be avoided. If there is a white-crystallized material floating instead, tap these and spin down again to remove. Collect the supernatant.

(5) Meanwhile, prewash the magnetic beads three times with 600-μl 0.5 × SSC, and keep the last 0.5 × SSC in a tube until the beads are ready for the next step.

(6) After carefully removing the 0.5 × SSC, add the supernatant from Step 4) to the magnetic beads. Try not to collect any of the white precipitants, as this could stick to magnetic beads and hinder the RNA-oligo(dT) interaction.

(7) Incubate for 15 min at room temperature with gentle mixing or rotation.

(8) Take out the supernatant and save this as an "unbound fraction" for later troubleshooting, if necessary.

(9) Wash the magnetic beads three times with 600-μl 0.5 × SSC at room temperature, with at least 5 min of shaking between each wash.

(10) After the third wash, add 200-μl 0.075 × SSC to the magnetic beads and incubate for 2 h at room temperature to elute the short-tailed RNAs. After collecting the eluent, add another 200-μl 0.075 × SSC and incubate with the magnetic beads overnight at room temperature for complete elution of short-tailed RNAs. (Therefore, the combined 400 μl of eluent from these two steps contains the short poly(A)-tailed RNA population.) Then, add 400-μl DEPC–DDW and incubate for 2 h at room temperature to collect long poly(A)-tailed RNAs. Alternatively, 400-μl DEPC–DDW can be directly added to the magnetic beads immediately after the washing step (Step 9) to collect poly(A)$^{+}$-enriched (total nonfractionated) RNAs.

(11) Recovered RNAs can be further cleaned by RNeasy MinElute Cleanup Kit, if necessary. Alternatively, these RNAs can also be precipitated with a conventional method (i.e., isopropanol precipitation) in order to further purify and concentrate the RNAs.

2.2. 3′-End labeling assay

This assay examines the poly(A)-tail length of a bulk RNA pool. Typically, we perform this assay immediately after poly(A) size fractionation to check and validate the poly(A)-tail length difference in each pool. To this end, the 3′-end of the poly(A) tail is radiolabeled followed by digestion of the body of the RNAs with both RNase A that cleaves single-stranded C and U residues and RNase T1 that cleaves G residues, leaving the poly(A) tail intact. Digested RNAs can therefore be separated on polyacrylamide gels, and the poly(A)-tail length can be visualized by autoradiography. For the labeling of the 3′-end of poly(A) tails, [5′-^{32}P] pCp molecules are ligated to the

RNAs. [5′-^{32}P] pCp is commercially available, although costly. Therefore, it is highly recommended to make [5′-^{32}P] pCp on your own by phosphorylating Cytidine Monophosphate (CMP) molecules instead, which is fairly straightforward and simple, and reduces the cost significantly. Therefore, a protocol to produce [5′-^{32}P] pCp is also provided below.

(1) To prepare [5′-^{32}P] pCp, mix the following:

2.5 μl	1 *M* Tris–HCl (pH 7.4) RNA grade
10.0 μl	50 m*M* $MgCl_2$
2.5 μl	0.1 *M* DTT
1.0 μl	7.5 m*M* CMP dissolved in DEPC-DDW (pH 9.0) and filtered (0.22 μm)
1.0 μl	75 m*M* spermine dissolved in DEPC-DDW and filtered (0.22 μm)
5.0 μl	[γ-^{32}P] ATP
2.0 μl	T4 polynucleotide kinase
26.0 μl	DEPC-DDW to make a total volume of 50 μl

(2) Incubate the mixture at 37 °C overnight and then heat-deactivate the enzymes by heating at 90 °C for 3 min. This mixture can be stored at −20 °C until use; however, handle it with caution as this contains radioisotope materials.

(3) 3′-End label RNAs by mixing the following:

5.0 μl	10 × T4 RNA ligase buffer
5.0 μl	10 m*M* ATP
5.0 μl	DMSO
5.8 μl	Glycerol (RNA grade)
10.0 μl	[5′-^{32}P] pCp (made in Steps 1 and 2 earlier)
2.0 μl	T4 RNA ligase
17.2 μl	RNA with DEPC-DDW to make a total volume of 50 μl

(4) Incubate this labeling mixture at 4 °C overnight. Note that this contains radioactive materials, so the incubation needs to be performed carefully.

(5) Digest RNA bodies by incubating the following for 30 min at 37 °C:

40.0 μl	Labeling mixture (from Section 3 and 4 earlier, save the other 10 μl for Step 7)
12.0 μl	5 *M* NaCl
2.0 μl	0.5 *M* EDTA
2.0 μl	RNase A/T1 mixture
1.0 μl	Yeast tRNA (10 mg/ml)
143.0 μl	DEPC-DDW to make a total volume of 200 μl

(6) Incubate the mixture at 37 °C for 30 min.

(7) Add 190 μl of DEPC-DDW to the 10 μl RNA leftover from 3′end labeling. These will serve as a control for the labeling and should be run on the gel with RNase-digested samples (Step 10).

(8) Both digested and undigested RNAs will be extracted by TRIZOL, precipitated by isopropanol, and dissolved in 5 μl DEPC-DDW using a standard procedure.

(9) Mix RNA samples with 2 × gel-loading buffer.

(10) Heat the samples at 95 °C for 5 min then keep on ice.

(11) Run samples on 7.5% denaturing minigels (see recipe in Section 4) until the dyes run off. Generally, it takes approximately 1 h to finish when the voltage is set at 120 V.

2.3. Microarray analysis

It is crucially important to choose the right microarray platform that does not require poly(A)$^+$ selection during the RNA/cDNA preparation process when performing poly(A)denylome analysis, as each RNA pool (short vs. long vs. nonfractionated) has different poly(A)-tail lengths and differences in poly(A)-tail length could introduce a bias during sample preparation. In addition to this technical concern, the analytical method is also critical in determining whether a given transcript exhibits the poly(A) rhythmicity. The status of relative poly(A)-tail lengths can be obtained as a ratio of gene expression level in the long poly(A) pool over the short poly(A) pool. Therefore, the circadian fluctuation of the ratio needs to be calculated to detect whether a particular transcript has a rhythmic pattern. The presence of rhythmic patterns can be determined by several algorithms such as JTK_CYCLE, ARSER, COSOPT, Fisher's *G* test, and CircWave (http://www.euclock.

org/results/item/circ-wave.html) (Hughes, Hogenesch, & Kornacker, 2010; Panda et al., 2002; Yang & Su, 2010), although each algorithm has unique characteristics, and multiple factors such as tolerance to noise (i.e., outliers), and a fit to a sinusoid can significantly affect the results. This is particularly important for poly(A)denylome analysis, because the long/short ratio, a poly(A)-tail length indicator, is far less robust in rhythmicity and much more noisy compared to circadian fluctuation of mRNA expression. Therefore, it is highly recommended to use much less stringent parameter settings compared to conventional transcriptome analyses, in order to obtain poly(A) rhythmic genes. Any false positives can be removed by independent validation of the tail length rhythms using methods described later.

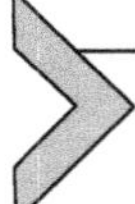

3. MEASUREMENT OF POLY(A)-TAIL LENGTH AT A SINGLE-GENE LEVEL

3.1. Poly(A) tail (PAT) assay

Compared to large-scale analyses, measuring the poly(A)-tail length at the single mRNA level has been much more well developed, and a variety of methods, generally called *P*oly(*A*) tail (PAT) assays, are currently available. A classic example, Oligo(dT)/RNase H Northern, compares the mRNAs with poly(A) tails against those devoid of poly(A) tails using oligo(dT) and RNase H, as the oligo(dT) hybridizes to the poly(A) tail of the mRNAs and RNase H then recognizes and cleaves the RNA:DNA hybrids (Salles & Strickland, 1999). A second antisense oligodeoxynucleotide that targets the sequence near the 3′end of the RNA can be also included in the reaction for better visualization of the changes in the poly(A) tail by cleaving off the majority of RNA body. This method, however, is labor-intensive and requires a large amount of RNA. In addition, it is difficult to detect small changes in poly(A)-tail length, especially when your gene of interest has a long transcript. More recent techniques are all based on RT-PCR, rather than Northern Blot, and successfully overcome these issues by providing increased sensitivity, speed, and length quantification. Several methods, such as RACE-PAT (Salles, Darrow, O'Connell, & Strickland, 1992), and ligation-mediated (LM)-PAT (Salles & Strickland, 1995), extension PAT (ePAT) (Janicke, Vancuylenberg, Boag, Traven, & Beilharz, 2012), or splint-mediated PAT (sPAT) (Minasaki, Rudel, & Eckmann, 2014) have been described, and these are slightly different in their technology in the cDNA synthesis step, each having advantages and disadvantages.

The RACE-PAT assay derives from the 3′RACE (rapid amplification of cDNA ends) protocol, where RNA is reverse transcribed with an oligo(dT) primer linked to a G/C-rich anchor sequence. Since hybridization of the oligo(dT) can occur anywhere along the entire length of the poly(A) tail, a heterogeneous pool of cDNAs primed at all possible positions along the poly(A) tail will be synthesized (Fig. 2A). In contrast, LM-PAT targets the

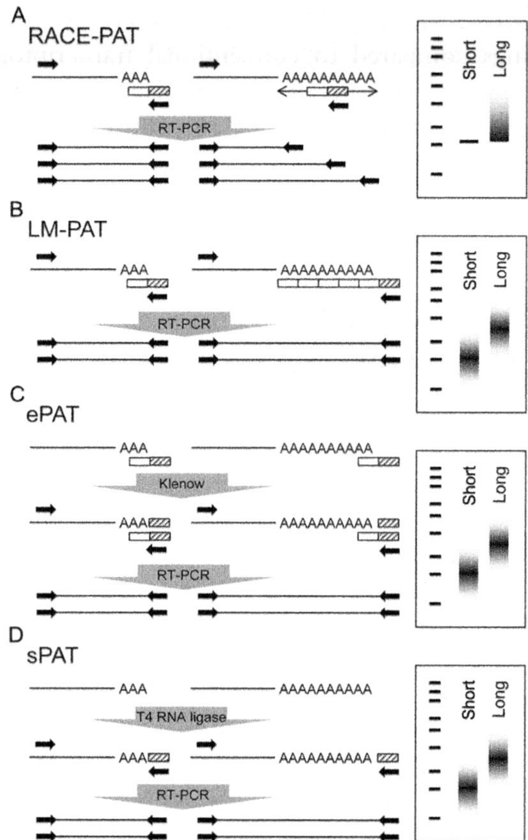

Figure 2 Schematic representations of various PAT assays and their ideal results. (A) RACE-PAT. Due to the ability of the oligo(dT) anchor to prime reverse transcription from anywhere along the poly(A) tail, the results appear as a solid smear due to annealing from the 5′-most start of the poly(A) tail to the 3′-end. While (B) LM-PAT, (C) ePAT, and (D) sPAT can easily identify discrete changes in poly(A) tail length over time, due to targeting of the oligo(dT) anchor to the 3′- end by oligo(dT) ligation. The half-white/half-diagonally hatched box represents the oligo(dT) anchor for reverse transcription, the open box (in LM-PAT) represents oligo p(dT)12–18 subunits, the hatched box (in sPAT) represents an RNA anchor, and the closed box represents the message-specific primer for PCR.

oligo(dT) anchor to the 3′-end of the poly(A) tail by saturating the poly(A) tail being annealed with phosphorylated oligo$(dT)_{12–18}$ [p$(dT)_{12–18}$], which are ligated together in the presence of T4 DNA ligase (Fig. 2B). Potential problems with these two methods include the possibility of internal priming of the oligo(dT) anchor primer to internal A-rich stretches during cDNA synthesis and/or transcripts arising from alternative polyadenylation site usage within the same gene. Theoretically, however, PCR products derived from internal poly(A) stretch priming or alternative polyadenylation should be easily distinguishable from that of the poly(A) 3′-end priming products, as these would result in bands of different size. More recent methods, ePAT and sPAT, were developed to avoid this internal priming issue. With ePAT, the oligo(dT) anchor is first annealed to poly(A) tails at a high temperature (55 °C), ensuring the anchor can only stay annealed at the end of the poly(A) tails. Klenow polymerase and dNTPs are then added to the reaction, resulting in template extension of the 3′-end of the poly(A) tract (Fig. 2C). Finally, sPAT utilizes an RNA (not DNA) anchor sequence ligated immediately after the 3′-end of the poly(A) tail using a single-strand DNA splint as a bridge (Fig. 2D), thus avoiding the internal priming while targeting the anchor to the end of poly(A) tail. Subsequent PCR amplification of this cDNA pool with a message-specific primer and the oligo(dT) anchor yields a mixture of products representing a variety of different poly(A)-tail lengths of the specific transcript of your interest.

Message-specific primers should be suitable for any types of PAT assays, although there are basic criteria that need to be met as this is the only parameter that dictates the target specificity. Ideally, primers should be designed 250–500 nt upstream of the 3′-end of the mRNA (i.e., poly(A) signal) and include a unique restriction site (within the amplified region) approximately 50–200 bp downstream from the message-specific primer to test amplification specificity. It is better to design a primer that yields a single shorter band representing the 5′- fragments and a slower-migrating heterogeneous 3′-end smear, but it is also tolerable to have longer 5′-fragments as long as they do not overlap with the poly(A)-tail smear derived from the 3′-fragment, the size of which is sometimes hard to predict. Depending on the expression level and sequence near the 3′-end of your mRNA of interest, it might take several different attempts to find primers that confer enough specificity to give rise to a single band. However, it is also possible to have two or more specific bands derived from alternative polyadenylation, which occurs in more than 50% of human genes (Tian, Hu, Zhang, & Lutz, 2005). It is also worth noting that the restriction enzyme sites might be altered due to SNPs; therefore, it is encouraged to try multiple different enzymes, if possible, especially when the message-specific primer yields a single band.

Standard PCR conditions should work for most PAT assays. The annealing temperature and cycle numbers must be empirically determined, depending on the primer sequence and your mRNA of interest. Once the PCR products are obtained, proper amplification of the mRNA of interest can be confirmed by either restriction digest analysis as described above or direct sequencing of the PCR products. As an initial approach, electrophoretic analysis can be conducted using ethidium bromide-stained 2–3% agarose gels, which provided enough resolution and sensitivity in our study. If increased resolution and sensitivity are desired, the addition of a small amount of radiolabeled dNTP is also an option in each PCR, in which case the samples should be cleaned to remove radiolabeled dNTPs.

3.2. Potential issues with PAT assays

The biggest issue common to all PAT assays is that the poly(A)-tail length measurements can be significantly affected by a difference in gene expression level between each samples, and it is extremely difficult to compare the poly(A)-tail length of transcripts from two samples when their RNA expression levels are considerably different. Considering PAT assays are qualitative, rather than quantitative, it might be preferable to adjust the amount of RNAs prior to PAT assay, so that amplification efficiency and visibility on the gels may become similar.

Another thing to keep in mind is that PCR amplification favors the generation of smaller amplicons; in other words, it is biased toward shorter poly(A) tails, and this is partly because DNA polymerases (i.e., Taq polymerase) generally work less efficiently in amplifying homogenous nucleotides, such as poly(A) stretches. This is particularly relevant for transcripts with long poly(A) tails but low in abundance, needing a higher number of amplification cycles. RACE-PAT, ePAT, and sPAT are much more sensitive to low-abundance mRNAs, compared to LM-PAT that requires the ligation process which is not highly efficient and reduces the size of the cDNA population. However, LM-PAT, ePAT, and sPAT all assure the targeting of oligo(dT) anchor to the 3′-end of poly(A) tails, regardless of the length of poly(A) tails, thus avoiding any bias toward shorter poly(A) tails.

The main technical challenge with PAT assays is achieving good PCR products. It is not uncommon to get very weak/no amplification, or high background noise, given that the message-specific primer is the only parameter that dictates the PCR specificity (as the other primer, the anchor primer, needs to be common to all target mRNAs). This problem can be sometimes be simply solved by changing PCR conditions (i.e., annealing temperature, primer concentration, a number of cycles, etc), but it often requires trying several different primers with various PCR conditions to get an optimal result. For this reason, it is highly recommended to try a test PCR using a control cDNA to validate primers and restriction enzymes, before testing the experimental samples.

3.3. LM-PAT assay

In this chapter, we provide a detailed protocol for the LM-PAT assay, as we find this technique to be most reliable and reproducible, compared to other PCR-based PAT assays. Readers interested in other methods may want to refer elsewhere (Janicke et al., 2012; Minasaki et al., 2014; Salles et al., 1992). For LM-PAT, it is highly recommended to start with poly(A)$^+$-enriched RNAs, as this not only eliminates genomic DNA contamination but also significantly increases the sensitivity and target specificity of the assay. Alternatively, it is also possible to use radioisotopes to increase the sensitivity.

(1) Prewarm a cDNA synthesis mixture of the following at 42 °C:

4 μl	5 × First strand buffer
1 μl	RNasin (20 U/μl)
2 μl	0.1 *M* DTT
1 μl	10 m*M* dNTP
1 μl	10 m*M* ATP
1 μl	T4 DNA ligase (10 U)
3 μl	DEPC-DDW

(2) Mix poly(A)$^+$ RNA (50 ng or more) and phosphorylated Oligo(dT)$_{15}$ (20 ng) in a total volume of 10 μl in a 0.2-ml tube.

(3) Incubate RNA-Oligo(dT) mixture at 65 °C for 10 min and then transfer to 42 °C immediately. Note that thermal cyclers are not recommended for this step, as this takes several seconds to shift the temperature. The sudden change in temperature is essential (i.e., use two separate water baths closely placed).

(4) Add 10 μl of the prewarmed reaction mixture in Step 1 to the tube. Mix well by pipetting and incubate at 42 °C for 30 min.

(5) At the end of the incubation, while still at 42 °C, add 1 μl of Oligo(dT)$_{12}$ anchor (5′-GCGAGCTCCGCGGCCGCGTTTTTT TTTTTT-3′) (200 ng/μl).

(6) Quickly vortex, spin, and incubate the mixture at 12 °C for 2 h, then 42 °C for 2 min. This step and hereafter can be done in a thermal cycler.

(7) While still at 42 °C, add 1 μl of SuperScript II Reverse Transcriptase (Life Technologies), and mix well.

(8) Incubate 42 °C for 1 h and inactivate enzymes by heating at 65 °C for 20 min.

(9) Perform a PCR as follows. The PAT first-stranded cDNAs can be diluted 1:50–1:200, for analysis of multiple mRNAs. Alternatively, these cDNAs can also be used undiluted, especially when the expression level of target mRNAs is low.

2.0 μl	10 × PCR buffer
0.8 μl	50 m*M* $MgCl_2$
0.4 μl	10 m*M* dNTP
0.4 μl	Oligo(dT)anchor (25 pmol/μl)
0.4 μl	Message-specific primer (25 pmol/μl)
0.2 μl	Taq polymerase
1.6 μl	PAT cDNA (original or diluted)
14.2 μl	DDW to make a total volume of 20 μl

Recommended PCR cycle as a starter:

1 ×	93 °C, 5 min
30–40 ×	93 °C, 30 s 60–65 °C, 30 s 72 °C, 1 min
1 ×	72 °C, 7 min

(10) Save 10 μl of the PCR reaction mixture at 4 °C or −20 °C until use. Add 1 μl of restriction enzyme to the remaining 10 μl PCR mixture and incubate at 37 °C for 60 min.

(11) Run the PCR reactions on agarose gels with appropriate size markers. By running digested and undigested samples together, one should be able to see the shift by enzyme digestion in amplified DNA size.

(12) Visualize the resulting DNA fragments by ethidium bromide staining.

(13) Determine poly(A)-tail lengths using densitometry software (we used Alpha Multiphotoimager II with AlphaInnotech software). Since the majority of bands will appear as smears due to the heterogeneous poly(A)-tail lengths, it is crucial to use some objective measurement of the size. We used the center of gravity of the PCR product (as determined by densitometry) to assign a poly(A)-tail size in each lane.

4. MATERIALS

4.1. Poly(A)-tail size RNA fractionation

Biotinylated Oligo(dT) (Promega Z5261)
Magnetic beads (Promega Z5481). Magnetic stands are available from Promega or Life Technologies.
RNeasy MinElute Cleanup Kit (Qiagen Cat# 74204)
PolyATtract GTC Extraction Buffer (Promega Z5531)

4.2. 3′-End labeling assay

CMP (Sigma C1133)
Spermine (Sigma S1141)
T4 polynucleotide kinase (New England Biolabs M0201)
10 m*M* ATP (Epicentre)
T4 RNA ligase (Epicentre)
RNase A/T1 mixture (Fermentas: #EN0551)
Yeast tRNA (Ambion #7118)
Gel-loading buffer (Ambion; #AM8546)
7.5% Denaturing gel (for two 8×10 cm gels, 1.0 mm): Dissolve 7.2 g urea in a mixture of 1.5 ml, $10\times$ TBE; 3.25 ml, 30% acrylamide/bis; 3.25 ml, DDW; 75 μl, 10% APS; and 15 μl TEMED.

5. CONCLUDING REMARKS

Rhythmic gene expression is controlled by a plethora of regulatory steps including both transcriptional and posttranscriptional mechanisms, and poly(A)-tail length control is one of the driving forces to achieve rhythmic protein expression. In addition to circadian biology, the relevance of regulating poly(A)-tail length has also been demonstrated in other biological processes such as cell cycle, oocyte maturation, cellular senescence, and synaptic plasticity (Charlesworth, Meijer, & de Moor, 2013). Therefore, the poly(A)denylome analysis method that we developed to originally understand the circadian poly(A)-tail length has broader applications and can be applied to other processes such as those described earlier in order to comprehensively identify which mRNAs are subject to poly(A)-tail length regulation.

It should be noted that although our protocol utilizes microarrays, similar analyses can be done using recently developed RNA-seq-based methods. For example, several recent reports have described new methods by which to comprehensively analyze the poly(A)-tail length, such as TAIL-seq (Chang, Lim, Ha, & Kim, 2014) or poly(A)-tail length (PAL)-seq (Subtelny, Eichhorn, Chen, Sive, & Bartel, 2014). Further studies using these new techniques in different tissues, cells, and organisms will be interesting for examination of the global effect of poly(A)-tail length regulation on circadian gene expression.

ACKNOWLEDGMENTS

We thank all the Green laboratory members for helpful discussions and comments on the manuscript, especially Dr. Danielle Hyman. This work was supported by the U.S. National Institutes of Health (R01GM090247, R21NS079986, and R01AG045795) to C. B. G., Tomizawa Jun-ichi, and Keiko Fund of Molecular Biology Society of Japan for Young Scientist and the Brain and Behavior Research Foundation (NARSAD) to S. K.

REFERENCES

Beilharz, T. H., & Preiss, T. (2009). Transcriptome-wide measurement of mRNA polyadenylation state. *Methods*, *48*(3), 294–300.

Chang, H., Lim, J., Ha, M., & Kim, V. N. (2014). TAIL-seq: Genome-wide determination of poly(A) tail length and 3′ end modifications. *Molecular Cell*, *53*(6), 1044–1052.

Charlesworth, A., Meijer, H. A., & de Moor, C. H. (2013). Specificity factors in cytoplasmic polyadenylation. *Wiley Interdisciplinary Reviews RNA*, *4*(4), 437–461.

Eckmann, C. R., Rammelt, C., & Wahle, E. (2011). Control of poly(A) tail length. *Wiley Interdisciplinary Reviews. RNA*, *2*(3), 348–361.

Heller, M. J. (2002). DNA microarray technology: Devices, systems, and applications. *Annual Review of Biomedical Engineering*, *4*, 129–153.

Hughes, M. E., Hogenesch, J. B., & Kornacker, K. (2010). JTK_CYCLE: An efficient nonparametric algorithm for detecting rhythmic components in genome-scale data sets. *Journal of Biological Rhythms*, *25*(5), 372–380.

Janicke, A., Vancuylenberg, J., Boag, P. R., Traven, A., & Beilharz, T. H. (2012). EPAT: A simple method to tag adenylated RNA to measure poly(A)-tail length and other 3' RACE applications. *RNA*, *18*(6), 1289–1295.

Kojima, S., Sher-Chen, E. L., & Green, C. B. (2012). Circadian control of mRNA polyadenylation dynamics regulates rhythmic protein expression. *Genes and Development*, *26*(24), 2724–2736.

Kojima, S., Shingle, D. L., & Green, C. B. (2011). Post-transcriptional control of circadian rhythms. *Journal of Cell Science*, *124*(Pt 3), 311–320.

Meijer, H. A., Bushell, M., Hill, K., Gant, T. W., Willis, A. E., Jones, P., et al. (2007). A novel method for poly(A) fractionation reveals a large population of mRNAs with a short poly(A) tail in mammalian cells. *Nucleic Acids Research*, *35*(19), e132.

Minasaki, R., Rudel, D., & Eckmann, C. R. (2014). Increased sensitivity and accuracy of a single-stranded DNA splint-mediated ligation assay (sPAT) reveals poly(A) tail length dynamics of developmentally regulated mRNAs. *RNA Biology*, *11*(2), 111–123.

Panda, S., Antoch, M. P., Miller, B. H., Su, A. I., Schook, A. B., Straume, M., et al. (2002). Coordinated transcription of key pathways in the mouse by the circadian clock. *Cell*, *109*(3), 307–320.

Salles, F. J., Darrow, A. L., O'Connell, M. L., & Strickland, S. (1992). Isolation of novel murine maternal mRNAs regulated by cytoplasmic polyadenylation. *Genes and Development*, *6*(7), 1202–1212.

Salles, F. J., & Strickland, S. (1995). Rapid and sensitive analysis of mRNA polyadenylation states by PCR. *PCR Methods and Applications*, *4*(6), 317–321.

Salles, F. J., & Strickland, S. (1999). Analysis of poly(A) tail lengths by PCR: The PAT assay. *Methods in Molecular Biology*, *118*, 441–448.

Subtelny, A. O., Eichhorn, S. W., Chen, G. R., Sive, H., & Bartel, D. P. (2014). Poly(A)-tail profiling reveals an embryonic switch in translational control. *Nature*, *508*(7494), 66–71.

Tian, B., Hu, J., Zhang, H., & Lutz, C. S. (2005). A large-scale analysis of mRNA polyadenylation of human and mouse genes. *Nucleic Acids Research*, *33*(1), 201–212.

Yang, R., & Su, Z. (2010). Analyzing circadian expression data by harmonic regression based on autoregressive spectral estimation. *Bioinformatics*, *26*(12), i168–i174.

CHAPTER NINETEEN

Sample Preparation for Phosphoproteomic Analysis of Circadian Time Series in *Arabidopsis thaliana*

Johanna Krahmer, Matthew M. Hindle, Sarah F. Martin, Thierry Le Bihan, Andrew J. Millar[1]

SynthSys and School of Biological Sciences, University of Edinburgh, Edinburgh, United Kingdom
[1]Corresponding author: e-mail address: andrew.millar@ed.ac.uk

Contents

Methods in Enzymology, Volume 551
ISSN 0076-6879
http://dx.doi.org/10.1016/bs.mie.2014.10.022

Abstract

Systems biological approaches to study the *Arabidopsis thaliana* circadian clock have mainly focused on transcriptomics while little is known about the proteome, and even less about posttranslational modifications. Evidence has emerged that posttranslational protein modifications, in particular phosphorylation, play an important role for the clock and its output. Phosphoproteomics is the method of choice for a large-scale approach to gain more knowledge about rhythmic protein phosphorylation.

Recent plant phosphoproteomics publications have identified several thousand phosphopeptides. However, the methods used in these studies are very labor-intensive and therefore not suitable to apply to a well-replicated circadian time series. To address this issue, we present and compare different strategies for sample preparation for phosphoproteomics that are compatible with large numbers of samples. Methods are compared regarding number of identifications, variability of quantitation, and functional categorization. We focus on the type of detergent used for protein extraction as well as methods for its removal. We also test a simple two-fraction separation of the protein extract.

ABBREVIATIONS

OG *n*-octyl-β-D-glucoside
PEG polyethylene glycol
SDS sodium dodecyl sulfate
TCA trichloroacetate

1. INTRODUCTION

Plant circadian transcript abundance has been extensively studied (Covington & Harmer, 2007; Edwards et al., 2006). By contrast, few studies have investigated rhythmicity of protein abundance with a systems biology approach (Robles & Mann, 2013).

Proteomic studies under diel or circadian conditions have been published for animal tissues: mouse liver (Mauvoisin et al., 2014; Reddy et al., 2006), mouse retina (Tsuji et al., 2007), and mouse SCN (Deery et al.,

2009). With the exception of Mauvoisin et al. (2014), laborious 2D gels were used by these studies, which is becoming obsolete as a technique preceding mass spectrometry (Robles & Mann, 2013). Mauvoisin et al. (2014) used a SILAC approach and detected over 5000 proteins, about 200 of which were rhythmic in diurnal conditions. Baker, Kettenbach, Loros, Gerber, and Dunlap (2009) carried out circadian interaction proteomics and phosphoproteomics specifically for the clock protein FREQUENCY in *Neurospora crassa*. A recent circadian proteomics study in *Synechococcus elongatus* (Guerreiro et al., 2014) achieved 82% coverage of the proteome (1537 proteins quantified) and detected 77 rhythmic proteins, using six-plex isobaric labeling and strong cation exchange (SCX) fractionation.

Regarding eukaryotic algae, Wagner, Fiedler, Markert, Hippler, and Mittag (2004) published a circadian proteomic time series in *Chlamydomonas reinhardtii*, and Akimoto, Wu, Kinumi, and Ohmiya (2004) studied diurnal proteomic changes in *Lingulodinium polyedrum*. Both studies used 2D-gels. In higher plants, Hwang et al. (2011) used 2D gels for a diurnal and circadian proteomic study in rice, finding approximately 3000 protein spots, 354 of which were rhythmic in diurnal conditions and 53 of these were rhythmic in constant darkness. Reiland et al. (2009) compared the chloroplast phosphoproteome at the end of day and end of night time points in a diurnal cycle. Their protocol involved SCX fractionation and two partially complementary phosphopeptide enrichment steps, yielding 16 fractions per sample.

A common finding of circadian proteomic studies is that the fraction of rhythmic proteins is lower than the fraction of rhythmic transcripts; many rhythmic transcripts do not have rhythmic proteins and vice versa (Hwang et al., 2011; Mauvoisin et al., 2014; Reiland et al., 2009; Wagner & Mittag, 2009). When RNA and protein both cycle, peak phases are often several hours apart (Mauvoisin et al., 2014). Where these have been measured, arrhythmicity of protein abundance is most often due to long protein half-lives (Mauvoisin et al., 2014).

One hypothesis created from this discrepancy between transcript and protein rhythmicity is that the total amount of a protein is not as physiologically relevant as the abundance of its posttranslational modification state at a given time of day. Phosphorylation plays a role in the plant circadian clock as well as its output (for a recent review, see Kusakina & Dodd, 2012). Phosphorylation of plant circadian clock proteins has been demonstrated by Western blotting (e.g., Farré & Kay, 2007; Sugano, Andronis, Ong, Green, & Tobin, 1999), but the pervasiveness of rhythmic phosphorylation at the *Arabidopsis* proteome scale is unknown. A circadian phosphoproteomic time series would address this question but this requires a streamlined experimental protocol.

We present an optimized sample preparation for phosphoproteomics in *Arabidopsis* that can be applied to a well-replicated time series of about 30 samples. Plant studies that yielded a large number of phosphopeptides employed techniques that are too time-consuming to be applied to 30 samples (Reiland et al. (2009): over 3000 phosphopeptides; Wang et al. (2013): over 5000; Facette, Shen, Björnsdóttir, Briggs, and Smith (2013): over 11,000).

Quantitation is obviously desirable to detect rhythmic regulation and can be facilitated by isotopic labeling. Metabolic labeling is difficult to achieve in plants due to inefficient isotope incorporation and its high cost for large numbers of samples. The latter is also the case for chemical labeling (Slade, Werth, Chao, & Hicks, 2014). Therefore, we limited our quantitation strategy to label-free quantitation.

Since only a small fraction of peptides is phosphorylated, it is difficult to identify phosphopeptides in total protein samples due to ionic suppression (Mann et al., 2002). It is therefore necessary to enrich phosphopeptides. Popular strategies are metal oxide affinity chromatography (MOAC) and immobilized metal affinity chromatography. TiO_2 beads are commonly used in a MOAC approach. Negatively charged phosphopeptides bind to positive charges on the surface of the TiO_2 particles (Dunn, Reid, & Bruening, 2010). We used commercial TiO_2 spin columns for all our analyses.

The scope of this chapter is to find a strategy for phosphoproteomic sample preparation that balances the number of phosphopeptide identifications, variability, and workload. We test different extraction buffer compositions and detergent removal methods, including an acid-labile detergent and separation of the protein extract into two fractions. All our workflows allow us to take digest aliquots before phosphoenrichment. We therefore present the analysis of global peptide digests along with the phosphoproteomic results.

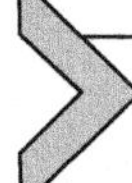

2. MATERIALS AND METHODS

2.1. Plant material

Seedlings of *Arabidopsis thaliana*, ecotype Columbia-0 were grown on plates (2.15 g/l Murashige & Skoog medium Basal Salt Mixture (Duchefa Biochemie), pH 5.8 (adjusted with KOH) and 12 g/l agar (Sigma)) for 10 days followed by 3 weeks on soil. Growth conditions were 12 h light and 12 h dark at 22 °C and a fluence rate of 120 μmol/s/m^2 (cool-white fluorescent tubes). Rosettes without roots and shoots were snap-frozen in liquid nitrogen. For sample OG trichloroacetate (TCA), rosettes were grown in the same way but harvested 1 week earlier.

2.2. Protein extraction for buffer optimization and the RapiGest™ SF experiment

Frozen plant material was ground using a dry ice cooled mortar and pestle. About 300 mg of plant material was used per replicate. 500 μl of extraction buffer of the following composition was added: for the detergent-free extraction, the extraction buffer contained 8 *M* urea, 100 m*M* Tris–HCl pH 7.5, 75 m*M* NaCl, 10 m*M* EDTA, 0.5% (v/v) polyvinylpolypyrrolidone, cross-linked (Sigma; CAS 25249-54-1), 1 × PhosStop (Roche), and 1 × Complete Protease inhibitor cocktail EDTA free (Roche).

For *n*-octyl-β-D-glucoside (OG) containing samples (refer to OG ethyl acetate, OG TCA MetOH, and OG TCA), 1% (w/v) OG was added. For sodium dodecyl sulfate (SDS) containing samples (SDS TCA, SDS ethyl acetate), 1% SDS (v/v) was added; for sample IGEPAL TCA, 1% (v/v) IGEPAL CA-630 (Sigma) was added. For the RapiGest™ SF method and its control at pH 8.5, the pH of Tris–HCl was adjusted to 8.5. 0.1% (v/v) RapiGest™ SF (Waters Corp.) was added for the RapiGest™ SF extracted samples.

All samples were homogenized on a Qiagen Tissue Lyzer with a metal ball of 3 mm diameter for 1 min at 30 Hz. After incubation on ice for 15 min, extracts were cleared at 20,000 × *g* and 4 °C for 10 min twice. Protein was quantified using the Quick Start™ Bradford Protein Assay (Bio-Rad) or the SDS compatible DC™ protein assay (Bio-Rad) for SDS-containing samples.

2.3. Fractionation with polyethylene glycol

A streamlined version of a protocol published by Aryal, Krochko, and Ross (2012) was used to deplete a protein extract fraction of RuBisCO by centrifugation. In contrast to the original protocol, only one centrifugation with polyethylene glycol (PEG), at 15%, was carried out. 10 ml of Mg/Tx-100 buffer (0.5 *M* Tris–HCl pH 8.3, 0.5% polyvinylpolypyrrolidone, 2% (v/v) Triton X-100, 20 m*M* $MgCl_2$, 2% (v/v) β-mercaptoethanol, 1 × Complete Protease Inhibitor Cocktail EDTA Free (Roche), 1 × PhosStop (Roche)) were added to 2 g of crushed plant material. Samples were incubated on ice for 1 h and centrifuged at 3220 × *g* for 20 min at 4 °C to remove cell debris. The supernatants were transferred to a fresh tube. 1.5 g PEG powder (average MW 3500, Sigma) was added to the tube. An unfractionated control sample was carried out in parallel without PEG and no second centrifugation. The samples for fractionation were incubated on ice for 30 min and centrifuged at $3220 \times g$ for 20 min. The supernatant was transferred to a fresh tube. The pellet was resuspended in 2.5 ml Mg/Tx-100 buffer.

2.4. Protein precipitation by TCA/acetone

Protein of all extracts except for the detergent-free protocols, the RapiGest™ SF samples plus their controls and the samples for ethyl acetate-based detergent removal were precipitated by a TCA and acetone-based method.

H_2O was added to the extract containing 2 mg of protein to a volume of 1250 μl. 310 μl of 100% TCA solution, 1248 μl of methanol, and 624 μl of chloroform were added. Samples were mixed and precipitated at −20 °C overnight. Centrifugation at $3220 \times g$ for 5 min at 4 °C resulted in an upper clear phase and a lower green phase. The upper phase was discarded; the green phase and interface were kept. 1 ml methanol was added; samples were mixed and centrifuged for 1 min at $3220 \times g$ at 4 °C. The supernatant was discarded and the resulting protein pellet was reconstituted in 1 ml acetone, centrifuged 1 min at $3220 \times g$ and 4 °C, repeated three times. The resulting protein pellets were finally dissolved in 8 *M* urea with the exception of peptides for method OG TCA MetOH, which were resuspended in methanol. Protein concentration was measured with a Bradford assay, except for the OG TCA MetOH.

2.5. Tryptic digest

500 μg of protein was subjected to protease digestion with trypsin as follows. For OG TCA MetOH, the volume required for 500 μg was estimated based on our experience that on average 60% of protein is lost in the TCA/acetone precipitation of OG extracted tissue. Volumes were adjusted to 250 μl with 8 *M* urea, or methanol in the case of OG TCA MetOH. 25 μl of 1 *M* ammonium bicarbonate and 25 μl of 200 m*M* DTT were added and samples were incubated to reduce disulfide bonds for 30 min at room temperature. 25 μl of 500 m*M* iodoacetamide were added to carbamidomethylate thiol groups followed by incubation at room temperature and in the dark for 1 h. 665 μl

distilled water were added and 10 μl of 1 mg/ml trypsin (Worthington). Samples were digested at room temperature overnight.

2.6. Detergent removal by ethyl acetate for sample OG ethyl acetate and SDS ethyl acetate

The digest volume was reduced to 100 μl in a Speed-vac. A method described by Yeung and Stanley (2010) was followed which uses ethyl acetate to remove detergents.

2.7. Removal of RapiGest™ SF by acidification

For hydrolysis of RapiGest™ SF, 50 μl of 20% formic acid was added to digested samples "RapiGest" and "No RapiGest pH 8.5", which lowered the pH below 2. Samples were incubated at 37 °C for 45 min and centrifuged at 13,000 × *g* at room temperature for 10 min. The supernatant was transferred to a fresh tube.

2.8. Cleanup of digests

After digest or detergent removal by ethyl acetate, peptides were desalted on a reverse phase resin using BondElute columns (25MG columns, Agilent): 1 ml methanol and two times 1 ml HPLC grade water (Thermo-Fisher) were sequentially passed through the column before loading the sample. After binding of peptides, columns were washed using 1 ml of HPLC grade water and peptides were eluted in 1 ml acetonitrile. Sample eluates were split into 490 μg (for phosphopeptide enrichment) and 10 μg (for global peptide analysis). Both aliquots were vacuum dried in a Speed-vac (RC1010, Thermo).

2.9. Phosphopeptide enrichment

Preliminary tests showed that the Titanshpere™ spin tip kit (GL Sciences Inc.) worked best in our hands (data not shown), so we consistently used this method for our experiments. Therefore, all steps from the phosphopeptide enrichment step were identical for all methods.

The following solutions were prepared: sln0 (2.5% acetonitrile, 0.5% trifluoroacetic acid (TFA)), sln1 (80% acetonitrile, 0.5% TFA), sln2 (9 x sln1 + 1 × Solution B of the enrichment kit which is 100% lactic acid), sln3 (5% (v/v) ammonium hydroxide solution, prepared just before use), sln4 (5% (v/v) pyrrolidine solution). Dried peptides were sequentially dissolved and sonicated in 50 μl of sln0 and 100 μl of sln2 and centrifuged at 20,000 × *g* for 5 min at 22 °C. The supernatant was recovered for enrichment. Spin adapters were inserted into a waste tube (2 ml Eppendorf tube without lid) and spin tips were inserted. Centrifugation of spin columns was always done at 200 × *g* at 22 °C. Spin tips were washed twice with 20 μl of sln1 for 2 min and waste tubes were exchanged. The sample was loaded and spin tips were centrifuged for 5 min. After incubation for 15 min, spin tips were spun again for 20 min to pass the entire sample through. The flow through was reloaded and the two spins with 15 min incubation were repeated. The reloading and centrifugation steps with incubation were repeated once more. Spin tips were washed with 20 μl sln1, twice with 20 μl sln2, and again twice with 20 μl sln1. For elution, the waste tube was exchanged for a clean 1.5-ml microfuge tube. Peptides were first eluted with 50 μl of sln3 for 5 min and then with 50 μl of sln4 for 8 min. Immediately, 20 μl 20% formic acid were added to lower the pH to about 3. Phosphopeptides were desalted on reverse phase resin using Bond Elut Omix tips (Agilent). The Omix tip was washed with 130 μl (all volumes were 130 μl except for elution) Buffer B (80% acetonitrile, 0.1% TFA) twice and then with buffer A (0.1% TFA in distilled water) twice. Peptides were retained on the tip by pipetting up and down 10 times, followed by two washes with buffer A and elution with 50 μl of buffer B. Samples were dried on a Speed-vac.

2.10. Mass spectrometry

Dried peptides were dissolved in 12 μl 0.05% TFA for phosphopeptide analysis, and 20 μl for global peptide analysis. Peptides were passed through Millex-LH 0.45 μm (Millipore) filters. 8 μl was run on an on-line capillary-HPLC-MSMS system consisting of a micropump (1200 binary HPLC system, Agilent, UK) coupled to a hybrid LTQ-Orbitrap XL instrument (Thermo-Fisher, UK). Reverse phase buffer used for LC-MS separation was buffer A (2.5% acetonitrile, 0.1% FA in H_2O) and buffer B (10% H_2O, 90% acetonitrile, 0.1% formic acid, 0.025% TFA). LC peptide separation was carried out on an initial 80 min long linear gradient from 0% to 35% buffer B, then a steeper gradient up to 98% buffer B over a period of 20 min then remaining constant at 98% buffer B (for 15 min) until a quick drop to 0% buffer B before the end of the run at 120 min. The tair Arabidopsis_1rep database was used for data-dependent detection.

2.11. Data analysis

For determination of the number of (phospho)peptides detected in each replicate, each raw file was converted to an mgf file and peptide searches were carried out with the following parameters: database Arabidopsis_1rep, trypsin as enzyme, allowing up to two missed cleavages, carbamidomethyl (C) as a fixed modification, Oxidation (M), Phospho (ST) and Phospho (Y), Acetyl(Protein N-term) as variable modifications, a peptide tolerance of 7 ppm, and MS/MS tolerance of 0.4 Da, peptide charges 2+, 3+, and 4+, on an ESI-trap instrument, and with decoy search. For export, an ion-cutoff of 20 was chosen. To determine the number of phosphopeptides for Figs. 1C, 2D, and 3C, the number of peptide entries containing phosphorylated residues was counted as well as the number of total peptides.

The Progenesis LC MS software was used for label-free quantitation. Raw files were imported into a label-free analysis experiment. Chromatograms were aligned by adding up to four manual vectors where peaks were misaligned, before starting automatic alignment and peak picking. Only charges 2+, 3+, and 4+ and data from 6 to 100 min of the runs were chosen for analysis. The exported file of MS/MS spectra was uploaded on the Mascot website and a search was carried out as described above for the single run searches. Protein and peptide measurements were exported and statistically analyzed. For phosphopeptide-enriched samples, peptide measurements were exported as .csv file and then saved as an .xlsx file for peptide merging (Hindle et al., in review).

The mass spectrometry proteomics data have been deposited to the ProteomeXchange Consortium (Vizcaíno et al., 2014) via the PRIDE partner repository with the dataset identifier PXD001134 and DOI 10.6019/PXD001134.

For a measure of variability of each method, the coefficient of variation (CV) was calculated from the arcsinh transformed data of all replicates for each protein from the Progenesis output file or the merged phosphopeptide file for phosphopeptides. For visualization, boxplots of these CV values are presented (Figs. 1D, 2E, and 3D–E).

2.12. Gene ontology enrichment analysis

Gene ontology (GO) term enrichment analysis was applied to both pairwise comparisons ("No Detergent" vs. "IGEPAL TCA" and "RapiGest" vs. "No RapiGest pH 8.5") in order to compare overrepresentation of functions and processes. Significance difference of the means in each pairwise

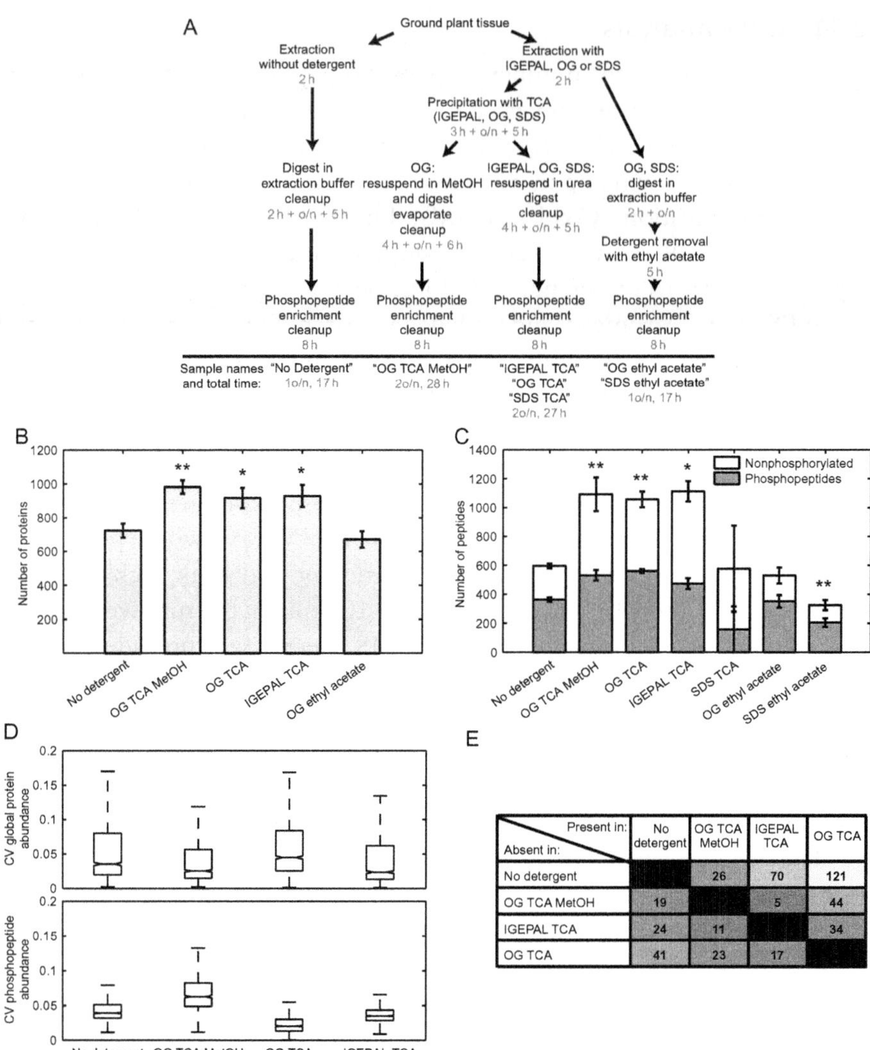

Absent in: \ Present in:	No detergent	OG TCA MetOH	IGEPAL TCA	OG TCA
No detergent		26	70	121
OG TCA MetOH	19		5	44
IGEPAL TCA	24	11		34
OG TCA	41	23	17	

Figure 1 Comparison of protein extraction methods. Protein was extracted from *Arabidopsis* rosette tissue using several detergents and detergent removal methods. After digestion with trypsin, the resulting peptides were analyzed by high-resolution mass spectrometry, with or without phosphopeptide enrichment on TiO_2 resin. Replicates were five except for sample OG TCA and SDS TCA (three replicates). For all samples, the same pool of ground tissue was used except for OG TCA, which was extracted from rosettes that were 1-week younger. (A) Workflow for all extraction methods, names of resulting samples and approximate duration of each step (red (light gray in the print version)) when processing 30 samples. (B) Average number of proteins identified by each method in the global peptide analysis. (C) Average number of phosphopeptides and nonphosphorylated peptides identified by each method in the phosphoenriched

comparison was determined by homoscedastic *t*-tests on an arcsinh transform of the normalized abundance as quantified in Progenesis. For each pairwise comparison, we tested enrichment of GO terms within protein entries based on the *p*-values for significance differences in the mean abundances. For the phosphoproteomics data, the phosphopeptide quantification with the lowest *p*-value was used if more than one phosphopeptide of the same protein were found.

The annotation of GO terms to *Arabidopsis* proteins was obtained from http://www.geneontology.org/GO.downloads.annotations.shtml (submitted: 10 February 2014). We used RStudio and the topGO library of the bioconductor package (http://bioconductor.org/biocLite.R, version: 2.16.0; Alexa & Rahnenführer, 2010). The Kolmogorov–Smirnov (KS) test was used to evaluate the significance of enrichment for each GO term. The KS statistic was applied with the weight01 algorithm to account for the topology of the GO tree and is a combination of a weighted and elim methods (Alexa, Rahnenführer, & Lengauer, 2006). All enrichment *p*-values of <0.05 were taken as significantly enriched terms.

3. RESULTS

3.1. Choice of extraction buffer and detergent removal method affects number of detected proteins and phosphopeptides

Protein extracts of mature *Arabidopsis* rosettes were prepared using four extraction buffers and two detergent removal strategies as summarized in

samples. Lower error bars: standard error of the mean (SEM) of number of phosphopeptides. Upper error bars: SEM of the number of total detected peptides. (D) Boxplot illustrating the coefficient of variation (CV) of protein abundance for the four best performing methods in the global peptide analysis (top) and the abundance of phosphopeptides in the phosphoenriched samples (bottom). The data were arcsinh transformed before calculating the CV and only entries with at least three nonzero values were used. Otherwise, zero values were ignored for determining the CV. Center of notch: median; notch: 95% confidence interval of median; box: first to third quartile (Q1–Q3); whiskers: Q1 $-$ 1.5 $\times$ interquartile range or smallest data point (bottom whiskers); Q3 $+$ 1.5 $\times$ interquartile range or largest data point (top whiskers). (E) Pairwise assessment of method specificity: each cell shows the number of phosphopeptides that were identified in all replicates of one method (top row, "present in") but in none of the replicates of the respective other method (left column, "absent in"). Cells were shaded for ease of reading: the lighter the shade, the higher the number of method-specific phosphopeptides. $^{*}p<0.05$ and $^{**}p<0.01$ in *t*-test against "No Detergent."

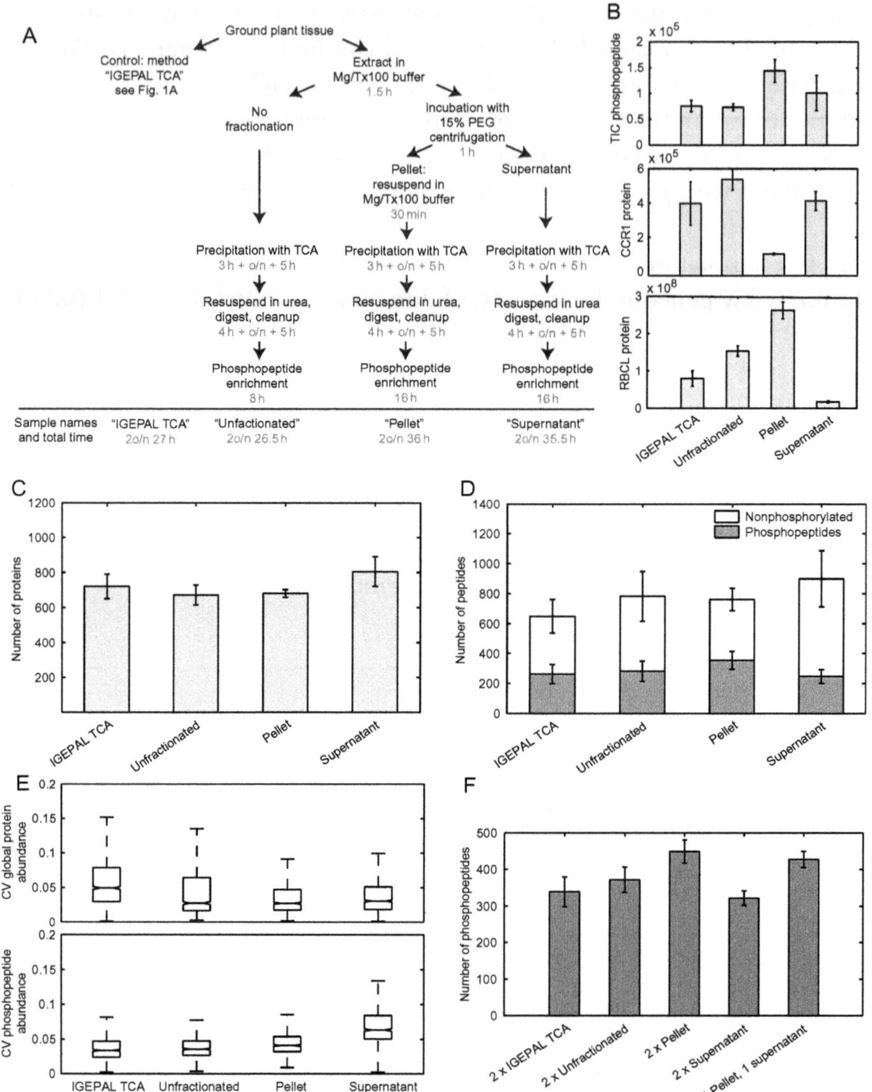

Figure 2 Protein fractionation by centrifugation with PEG. Protein was separated into two fractions ("Pellet" and "Supernatant") by centrifugation with PEG, along with a control without fractionation ("Unfractionated") and a control using the previously described method "IGEPAL TCA." Replicates were five for "Unfractionated" and "Pellet," three for the "IGEPAL TCA" control and four for "Supernatant." (A) Workflow for all methods, names of resulting samples and approximate duration of each step (red (light gray in the print version)) when processing 30 samples. (B) Examples: normalized abundance of the phosphopeptide AGSFRD**pS**PEEEGPVELPEAAR from the clock protein TIC (top) found in the analysis of phosphoenriched peptides; normalized

Fig. 1A. A urea– and Tris–HCl-based buffer (see Section 2) without detergent provided a reference that avoided any detergent removal ("No Detergent"; Fig. 1A). Proteins were digested with trypsin following a standard in-solution protocol and phosphopeptides were enriched using a titanium dioxide-based kit. The phosphoenrichment and downstream procedures were the same for all methods.

Since detergents increase the amount of extracted protein, the detergents IGEPAL, octyl glucoside (OG), or SDS were added to this buffer. The amount of extracted protein increased more than twofold for the same amount of starting material compared to "No Detergent" (data not shown); however, the detergent needs to be removed before mass spectrometry. We took two different approaches. First, we precipitated and washed proteins using a TCA/acetone-based method and resuspended in urea before trypsination, resulting in the samples "IGEPAL TCA," "OG TCA," and "SDS TCA" (Fig. 1A). In the case of OG, we also tested resuspending the protein pellet in methanol instead of urea since this may increase digestion efficiency (Russell, Park, & Russell, 2001), resulting in sample "OG TCA MetOH." The loss of protein in this step was approximately 50%. In the second approach, we digested in the presence of the detergent (OG or SDS) and removed it thereafter using an ethyl acetate-based partitioning, which produced samples "OG ethyl acetate" and "SDS ethyl acetate" (Yeung & Stanley, 2010).

In addition to the phosphoenriched peptides, an aliquot of each nonenriched protein digest was also analyzed on the mass spectrometer. Nonenriched samples from extractions using SDS were not tested to avoid problems due to remaining detergent traces.

Figure 1B shows the average number of proteins detected in the global peptide analysis. Addition of detergent and precipitation by TCA/acetone

abundance of the clock output protein CCR1 (middle) and the RuBisCO large subunit protein (bottom) from the global peptide analysis. (C) Average number of proteins identified by each sample type in the global peptide analysis; error bars: SEM. (D) Average number of phosphopeptides and nonphosphorylated peptides identified by each method in the phosphoenriched peptide analysis. Lower error bars: SEM of the number of phosphopeptides. Upper error bars: SEM of the number of total detected peptides. (E) Boxplot illustrating the CV of protein abundance in the global peptide analysis (top) and the CV of phosphopeptide abundance in the phosphoenriched sample (bottom). See legend of Fig. 1D for transformation of data and boxplot details. (F) Average number of phosphopeptides identified by any possible combination of two mass spec analyses, of either the same workflow or one pellet and one supernatant; error bars: SEM.

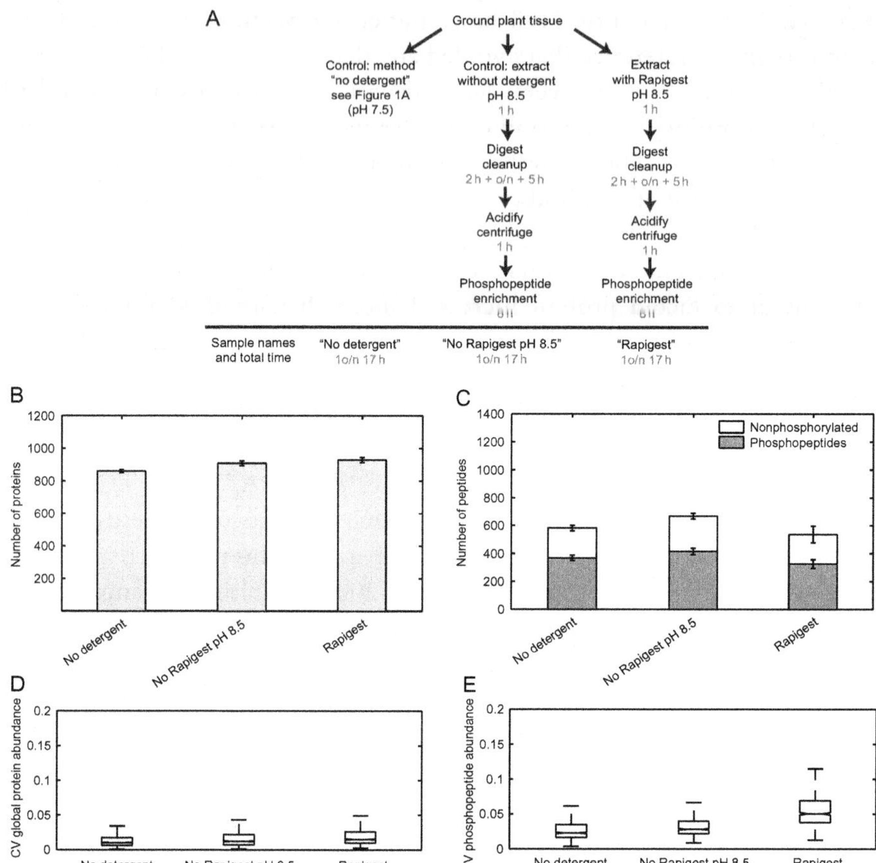

Figure 3 Protein extraction using the acid-labile detergent RapiGest™ SF. Protein was extracted with the acid-labile detergent RapiGest™ (sample "Rapigest"). (A) Workflow and names of resulting samples and approximate duration of each step (red (light gray in the print version)) for processing 30 samples. A control was treated the same but without addition of RapiGest™. Another control was performed in the same way as "No Detergent" in Fig. 1A. (B) Average number of proteins identified by each method in the global peptide analysis; error bars: SEM. (C) Average number of phosphopeptides and nonphosphorylated peptides identified by each method in the phosphoenriched peptide analysis. Lower error bars: SEM of number of phosphopeptides. Upper error bars: SEM of the number of total detected peptides. (D and E) Boxplots illustrating the CV of (D) protein abundance in the global peptide analysis and (E) the abundance of phosphopeptides in the phosphoenriched samples. See legend of Fig. 1D for transformation of data and boxplot details.

yielded significantly more protein identifications compared to "No Detergent." OG removal by ethyl acetate did not perform as well as the other methods. There was no significant difference in protein numbers between the three TCA/acetone extraction-based methods.

The situation was similar for the number of phosphopeptides (Fig. 1C): addition of detergent increased the number of identified phosphopeptides significantly if the detergent was removed by TCA/acetone precipitation. The ethyl acetate strategy did not result in as many phosphopeptides identified as the TCA/acetone-based methods. In both SDS extraction methods, the number of peptides was significantly lower than in "No Detergent." In two of the three "SDS TCA" replicates, no phosphopeptides were identified. The percentage of phosphopeptides in the total number of peptides identified was not significantly different between OG TCA MetOH, OG TCA, and IGEPAL TCA.

Since SDS TCA, SDS ethyl acetate, and OG ethyl acetate did not perform as well as the other methods, we focused on the others in further analyses.

The median CV of protein abundance in the global protein analysis was between about 0.02 and 0.05, with IGEPAL TCA being the lowest (Fig. 1D). For phosphopeptide analysis, the range of the median values was larger (between 0.02 and 0.07). OG TCA MetOH had the largest median and OG TCA the smallest.

To indicate how these methods overlap regarding the identity of detected phosphopeptides, we measured the number of phosphopeptides that were identified in all replicates of one method and in none of the other methods (Fig. 1E). For each detergent extraction, the number of specific phosphopeptides is higher when comparing to the "No Detergent" samples than when comparing to other detergent-based methods, suggesting that detergents extracted a characteristic set of additional proteins. Method "OG TCA" has more specific phosphopeptides than the other methods.

The IGEPAL TCA method and the no detergent control were compared in a GO analysis (Table 1). Fewer terms were enriched in the phosphopeptide analysis than in the global protein analysis and the IGEPAL TCA method generally enriched more terms than the no detergent control. Although terms enriched by either method are from a broad spectrum, some words consistently appeared in only one of the methods: the majority of IGEPAL TCA enriched terms for cellular component were related to photosynthesis or plastids or membranes, while No Detergent was enriched for many ribosome-related terms in the global protein analysis.

Table 1 Gene ontology analysis

Enriched in IGEPAL TCA over No Detergent		Enriched in No Detergent over IGEPAL TCA	
Global	**Phospho-enriched**	**Global**	**Phospho-enriched**
Chloroplast thylakoid membrane	Membrane	Cellular component	None
Membrane part		Ribosomal subunit	
Protein complex		Cytosolic ribosome	
Chloroplast stroma		Intracellular nonmembrane-bounded organelle	
Photosynthetic membrane		Intracellular organelle part	
Thylakoid membrane		Ribosome	
Plastid thylakoid membrane		Ribonucleoprotein complex	
Intracellular membrane-bounded organelle		Nnonmembrane-bounded organelle	
Thylakoid		Organelle part	
Chloroplast		Cytosolic part	
Thylakoid part			
Cytoplasmic part			
Chloroplast part			
Membrane			
Chloroplast thylakoid			
Plastid			
Plastid stroma			
Intracellular organelle part			
Plastid thylakoid			
Organelle subcompartment			
Membrane-bounded organelle			
Organelle part			
Plastid part			
Plastid part			

Methods "IGEPAL TCA" and "No Detergent" were compared in a GO analysis. Terms enriched by one method relative to the other are shown for the global analysis as well as the phospho-enriched peptide analysis. Only the "cellular component" GO category is shown.

Membrane-related terms were also enriched by IGEPAL TCA in the phosphopeptide analysis.

3.2. Fractionation with PEG does not increase numbers of identified peptides

Fractionation can significantly improve the number of identified peptides by reducing the complexity of the sample, thus allowing the MS instrument to identify a higher proportion of the peptides. However, it also multiplies the number of samples to be processed. This is problematic for circadian time series, which already have a large number of samples. We explored a method that produces only two fractions per sample by a centrifugation step with PEG. This pellets denser proteins, in particular the RuBisCO subunits, and thereby is expected to reduce the range of protein abundance in the supernatant fraction (Aryal et al., 2012).

As shown in Fig. 2A, we extracted with a magnesium and Triton X-100 containing buffer called "Mg/Tx-100 buffer" (Aryal et al., 2012). After incubating the protein extract with PEG, centrifugation resulted in a pellet and a supernatant. The pellet was resuspended in Mg/Tx-100 buffer, and both were TCA precipitated, digested, and phosphoenriched.

As a control, we conducted one extraction in the same way but without the addition of PEG, as an unfractionated control. To control for the fact that the extraction buffer could make a difference compared to the other methods we have used, we also extracted using the method "IGEPAL TCA" (see Fig. 1A) in parallel. All extractions used tissue from the same pool of ground tissue. The number of replicates was three for "IGEPAL TCA", five for "Unfractionated," and "Pellet," four for "Supernatant."

The fractionation reduced the amount of RBCL in the supernatant 15-fold (Fig. 2B bottom). The amount of the clock output protein CCR1 is significantly higher in the supernatant fraction than the pellet (Fig. 2B middle). The only clock protein of which a phosphopeptide was detected in the phosphoenriched sample was TIME FOR COFFEE (TIC, Fig. 2B top). We did not find a significant difference in its abundance between the two fractions.

There was no statistically significant difference in the average number of quantified proteins between any of the four methods (Fig. 2C).

There is no statistically significant difference in the average number of identified phosphopeptides between the different workflows and fractions, which shows that the Mg/Tx-100 buffer is generally suitable for protein extraction but the PEG treatment did not increase the number of identified phosphopeptides in the supernatant sample even though some abundant proteins were present at lower levels.

Therefore, it is not beneficial to use only the supernatant sample instead of an unfractionated sample. However, we wanted to find out if the number of phosphopeptides increased when both fractions are analyzed. Due to the quantification procedure in Progenesis, increasing the number of samples even if they are replicates from the same extraction method will increase the number of detected peptides. Therefore, we determined the number of phosphopeptides found in combinations of two runs of the same treatment or of one pellet and one supernatant analysis. Figure 2F shows the average number of phosphopeptides for all possible pairs. The highest average was found in pairs of just the pellet fraction. The average number for a supernatant pair was significantly lower than for a pellet pair ($p=0.012$, *t*-test). The average number for a supernatant and pellet sample pair was not significantly different from the pellet pair or the unfractionated pair, but a pellet and supernatant pair had significantly more phosphopeptides than two supernatant samples ($p=0.020$, *t*-test).

With respect to variability of the phosphopeptide analysis, the box plots in Fig. 2E (bottom) show that the supernatant sample has a higher median CV than the other methods. This is not the case for the median CV of protein abundance calculated from the global peptide analysis, where the CV of the IGEPAL TCA control has the highest median.

3.3. The acid-labile detergent RapiGest™ does not increase the number of detected phosphopeptides

As detergents extracted additional proteins and phosphopeptides but required time-consuming desalting steps (Fig. 1), we tested whether detergent removal could be avoided using an acid-labile detergent. We added RapiGest™ SF surfactant to the same buffer as used in the "No Detergent" sample in Fig. 1A, except the pH was adjusted to 8.5 to make sure RapiGest™ is not hydrolyzed prematurely.

We conducted a control without RapiGest™ but still at pH 8.5. The control was also subjected to acidification after digest (workflow see Fig. 3A). To rule out the possibility that a higher pH and acidification after digest affects performance, we also carried out an extraction in the same way as "No Detergent" in Fig. 1A in parallel and from the same tissue powder as the two other methods.

Figure 3B and C shows that neither the number of proteins in the global peptide analysis nor the number of phosphopeptides identified in the phosphoenriched samples is significantly different between any of the three treatments. The median CV is slightly higher for the RapiGest™ samples in the phosphoenriched samples (Fig. 3E) but not in the protein quantification from the global peptide analysis (Fig. 3D).

In a GO analysis of the global protein data, photosynthesis and membrane-related terms were enriched in the RapiGest™ SF data compared to the control at pH 8.5. In the latter, ribosome- and translation-related terms were enriched.

4. DISCUSSION

When developing or choosing a sample preparation method for enriching a specific class of peptides (in this case phosphopeptides), several factors have to be considered: (1) the number of peptides identified (phosphorylated peptides in this case) (2) if quantitation is needed then the variability between replicates has to be low, and (3) the capability to enrich for interesting peptides. One major challenge in proteomics is its bias toward highly abundant species often masking more interesting lower abundant proteins. A common approach to tackle this issue is to fractionate a sample into several subfractions. Although this approach is appropriate for few samples to analyze, it can be quite a tedious one for more complex experimental designs, such as different drug treatments or genotypes, or circadian time series. Extensive fractionation is not a viable option in the case of circadian phosphoproteomic time series. Methods employed by published studies that achieved high phosphoproteome coverage, such as Reiland et al. (2009) and Wang et al. (2013), are too laborious for a well-replicated circadian time series.

For a circadian (phospho)proteomics experiment, a high number of identifications as well as reliable quantification are required across several time points.

Sample preparation for quantitative phosphoproteomic experiments generally requires long protocols to be carried out accurate and carefully. In addition, there are plant-specific challenges: the presence of cell walls makes cells more resilient to protein extraction; protein content is typically lower than in animal tissues so more material is required; plant cells contain interfering compounds such as polyphenols and a range of secondary metabolites; the enzyme RuBisCO amounts to 30–60% of cellular protein, making less abundant peptides more difficult to be detected (Koroleva & Bindschedler, 2011; Slade et al., 2014).

Our aim was to compare the performance of several sample preparation methods for proteomics and phosphoproteomics, which are compatible with a well-replicated circadian time series experimental design.

4.1. Extraction with a nonionic detergent and precipitation with TCA/acetone outperforms other strategies

We demonstrate a comparison of seven different extraction methods. A detergent-free buffer is appealing since time-consuming detergent removal introduces additional steps, which is likely to increase the technical variation. In this work, we show that reliable proteomics and phosphoproteomics can be done without detergent. However, regarding the number of proteins and phosphopeptides identified, the detergent-free method yields significantly fewer phosphopeptides than methods using a nonionic detergent (IGEPAL or OG) in the extraction buffer and precipitation of protein with a TCA/acetone-based protocol.

We chose to test the strong ionic detergent SDS and the nonionic detergents IGEPAL and OG. IGEPAL (also known as NP-40) is widely used for protein extraction. OG was found to be superior to other nonionic detergents by Arachea et al. (2012). In our hands, buffers containing IGEPAL, OG and SDS extracted more than twice as much protein as the detergent-free buffer from a given amount of tissue (data not shown). However, it should be taken into account that about 50% of protein material is lost during the TCA/acetone precipitation, so roughly equal amounts of starting material are required.

In contrast to our expectations we did not observe higher variability with the additional step for detergent removal by TCA/acetone. The exception is method "OG TCA MetOH," which resulted in phosphopeptide quantification with significantly more variability than in the other methods (Fig. 1D). We had expected this since resuspension with methanol is not compatible with the standard quantitation methods, so the appropriate amount of protein for digestion was estimated according to typical losses experienced in the OG TCA method. According to Chen, Cociorva, Norris, and Yates (2008), trypsination efficiency is higher in mixed organic-aqueous systems compared to aqueous buffers. Our data do not support this since we did not find a significant difference in phosphopeptide or protein numbers between OG TCA and OG TCA MetOH. However, it is possible that factors other than digestion are limiting the number of identifications in our case, making increased digestion efficiency unnoticeable.

An additional benefit of the precipitation step is removal of nonprotein contaminants from either the plant cell extract or buffer components, which allows better control of the sample composition and compatibility with the electrospray ionization. We have experienced contamination with polymers in detergent-free samples but not in TCA/acetone-precipitated samples.

Regarding the choice of detergent, the differences in global protein and phosphopeptide numbers identified in OG TCA and IGEPAL TCA are not significant. IGEPAL is cheaper and may therefore be the preferred choice. The ionic detergent SDS did not prove useful in our hands since the number of phosphopeptides identified was significantly lower than the detergent-free procedure (Fig. 1C). Possible reasons include incomplete removal and therefore potential incompatibility with the phosphoenrichment strategy or the mass spectrometric analysis.

Presence of detergents can also increase the efficiency of tryptic digestion (Chen et al., 2008). Therefore we tested a strategy of extracting with detergent and removing it after digest using ethyl acetate (Yeung & Stanley, 2010). The particularity of this method is that the detergent is removed at the peptide level, not the protein level. When using OG as a detergent, the method resulted in approximately as many global protein and phosphopeptide identifications as the detergent-free method but significantly fewer than the OG extracted and TCA/acetone-precipitated samples. Therefore, extraction with OG and its removal by ethyl acetate does not have any advantages over the simpler detergent-free extraction.

The use of acid-labile detergents such as RapiGest™ SF is a very attractive way to avoid the time-consuming detergent removal process, while benefitting from the ability of the detergent to solubilize membrane proteins and increase digestion efficiency (Chen et al., 2008). After digest, the detergent is hydrolyzed by acidification, resulting in a soluble and an insoluble product. The latter is removed by centrifugation.

We expected an increase in the number of identifications compared to a control that was subjected to the same procedure but without the addition of RapiGest™ SF. However, we found no statistically significant differences in the number of identifications in either the global protein or phosphopeptide analyses (Fig. 3B and C) and the variability in the phosphoproteomics data of the RapiGest™ SF treated samples was significantly higher than that of the other methods. Therefore, with respect to numbers and variability, there is no benefit from investing in this expensive detergent.

Our GO analysis suggests that RapiGest™ SF at 0.1% succeeds in extracting proteins typical of detergent-containing buffers. Both IGEPAL

TCA and RapiGest™ SF analyses enriched terms related to photosynthesis and membranes, compared to their respective controls.

4.2. The nonionic detergent IGEPAL extracts more membrane- and chloroplast-related proteins

Our method specificity analysis (Fig. 1E) suggests that with methods based on a nonionic detergent and TCA/acetone precipitation, a set of phosphoproteins is identified that was not found when extracting without detergent. According to our GO analysis, more photosynthesis/chloroplast related as well as membrane-related proteins are extracted by the IGEPAL TCA method than the detergent-free control, while the detergent-free approach should be preferred if ribosome-related proteins are of particular interest. A likely explanation is that the detergent solubilizes both the cell membrane and the plastid membranes, releasing the proteins inside the organelle. Abundant cytoplasmic proteins such as ribosome components will thereby be diluted. An enrichment of transmembrane proteins by the IGEPAL TCA method is also supported by an analysis using the TMHMM Server (v. 2.0; http://www.cbs.dtu.dk/services/TMHMM/; Krogh, Larsson, von Heijne, & Sonnhammer, 2001), which predicted significantly more transmembrane proteins in the IGEPAL TCA global protein results than the no detergent results ($p = 0028$, χ^2 test of independence).

Phosphopeptides from proteins related to the circadian clock or phosphorylation that were consistently found in the IGEPAL extracted samples but not the "No Detergent" method were, for example, from MAP kinase 2, calcium-dependent kinase 19, casein kinase 1-like protein 2, BR-signaling kinase 1, and trehalose phosphatase/synthase 7.

The list of phosphopeptides that are likely to be found with the detergent-free method but not the IGEPAL TCA methods included three transcription factors. Lists of proteins and peptides found by each method will be available on the PRIDE database upon publication.

4.3. Fractionation by density using PEG is not superior to increasing replicate number

Mass spectrometric analysis of multiple sample fractions can be tedious for well-replicated circadian time series. We tested a separation into two fractions, which doubles the number of samples but only adds about one more working day to the procedure (due to doubling the number of digests to be phosphoenriched). Aryal et al. (2012) developed a fractionation procedure based on several centrifugation steps with PEG at different concentrations, which results in a pellet and a supernatant in each centrifugation step. They demonstrated that in the supernatant fraction, the RuBisCO subunits are depleted and 23% more peptides were identified than in an unfractionated sample. We streamlined this procedure (see Section 2) to one centrifugation, yielding two fractions, and precipitated with TCA/acetone instead of phenol/chloroform. We confirmed that the RuBisCO subunits are pelleted by PEG (Fig. 2B; small subunit not shown). Regarding clock-related proteins, the supernatant enriches proteins of CCR1 (Fig. 2B center) and PATHOGEN AND CIRCADIAN CONTROLLED 1 (data not shown), so if these particular proteins are of interest, the supernatant fraction is useful. There is no significant difference in the abundance of a TIC phosphopeptide between the pellet and supernatant fractions (Fig. 2B bottom).

Using the fractionation approach, we did not identify more proteins or phosphopeptides in the supernatant sample compared to the unfractionated control or the pellet. Therefore, our effort to deplete the most abundant protein(s) did not increase protein identifications significantly. Moreover, this additional step increases the variability of quantification (Fig. 2E).

Although the number of identified peptides can be quite low per sample run on LC-MS, merging the different replicates and experiment significantly increases the number of identified peptides. We therefore asked whether we can increase the number of identifications by analyzing both fractions, rather than analyzing two replicates of an unfractionated sample. Assuming that two different fractions of a same sample analyzed by LC-MS and the identification output combined will likely result in more identifications than one, we compared pairs of runs of the same method with supernatant-pellet pairs. The pellet-supernatant combination performs better than a supernatant-pair and a pellet pair performs better than a supernatant pair, but combining one of each fraction does not perform better than two unfractionated replicates.

We conclude that analysis of both protein fractions does not result in more identifications than analysis of two different replicates. Workload may be smaller for fractionation than for more replication. However, we predict that this fractionation does not justify carrying out half the number of replicates because data analysis will become more unreliable for fractions compared to the same number of replicates.

Therefore, we recommend a reasonable number of replicates rather than fractionation with PEG, unless a subset of proteins enriched by one of the fractions is of particular interest.

4.4. Alternative strategies

There are other conceivable strategies for acceptable (phospho)proteomic analysis of a circadian time series. General strategies for proteomics in plants have been reviewed by Koroleva and Bindschedler (2011), and phosphoproteomics methods by Slade et al. (2014).

A simple fractionation approach is removal of chloroplasts, which will deplete the remaining fraction of several highly abundant proteins (Hiltbrunner et al., 2001). However, this requires careful and consistent homogenization without rupturing chloroplasts prior to separation. This in turn requires specialized equipment and is time consuming, making it less attractive for a large set of samples.

Plant samples can be depleted of RuBisCO using commercial antibody-based strategies (Cellar et al., 2008). The cost of these bead-coupled antibodies is unattractive for circadian time series. The same applies to equalizer beads (Righetti, Boschetti, Lomas, & Citterio, 2006), which can be used to reduce the range of protein abundance independent of protein identity. One potential issue with flattening of the dynamic range is reducing the sensitivity of relative protein quantitation.

The FASP principle, in which washing and digestion take place on a filter (Wisniewski, Zougman, Nagaraj, & Mann, 2009), is appealing to conveniently remove detergent within the digestion process, however the success of this method seems to depend very much on the specific experiment conducted with possible loss for low abundant proteins.

In-gel digestion is very commonly used in proteomics, is efficient at removing detergent, and offers the possibility of fractionating. However, the extract volumes required here exceed gel well volumes by far. Using multiple wells and several gel slices per well will lead to longer working time.

A strategy involving long LC separation during analysis could represent a possibility to increase the number of identifications.

5. CONCLUSIONS

We conclude that extraction with a nonionic detergent such as OG or IGEPAL and precipitation with TCA/acetone without fractionation is a successful, reliable, and feasible strategy for conducting phosphoproteomic time series. Time-consuming fractionation or expensive acid-labile detergents are not beneficial. We expect that these results can be transferred to proteomic analysis of other posttranslational modifications.

ACKNOWLEDGMENTS

We thank Lisa Imrie and Katalin Kis for expert technical support.

Funding: This work was supported by the Wellcome Trust [096995/Z/11/Z] and BBSRC, and EPSRC awards [BB/D019621] and [BB/J009423].

REFERENCES

Akimoto, H., Wu, C., Kinumi, T., & Ohmiya, Y. (2004). Biological rhythmicity in expressed proteins of the marine dinoflagellate *Lingulodinium polyedrum* demonstrated by chronological proteomics. *Biochemical and Biophysical Research Communications, 315*, 306–312.

Alexa, A., & Rahnenführer, J. (2010). *topGO: Enrichment analysis for Gene Ontology. R package version 2.16.0.*

Alexa, A., Rahnenführer, J., & Lengauer, T. (2006). Improved scoring of functional groups from gene expression data by decorrelating GO graph structure. *Bioinformatics, 22*, 1600–1607.

Arachea, B. T., Sun, Z., Potente, N., Malik, R., Isailovic, D., & Viola, R. E. (2012). Detergent selection for enhanced extraction of membrane proteins. *Protein Expression and Purification, 86*, 12–20.

Aryal, U. K., Krochko, J. E., & Ross, A. R. S. (2012). Identification of phosphoproteins in Arabidopsis thaliana leaves using polyethylene glycol fractionation, immobilized metal-ion affinity chromatography, two-dimensional gel electrophoresis and mass spectrometry. *Journal of Proteome Research, 11*, 425–437.

Baker, C. L., Kettenbach, A. N., Loros, J. J., Gerber, S. A., & Dunlap, J. C. (2009). Quantitative proteomics reveals a dynamic interactome and phase-specific phosphorylation in the *Neurospora* circadian clock. *Molecular Cell, 34*, 354–363.

Cellar, N. A., Kuppannan, K., Langhorst, M. L., Ni, W., Xu, P., & Young, S. A. (2008). Cross species applicability of abundant protein depletion columns for ribulose-1,5-bisphosphate carboxylase/oxygenase. *Journal of Chromatography B, Analytical Technologies in the Biomedical and Life Sciences, 861*, 29–39.

Chen, E. I., Cociorva, D., Norris, J. L., & Yates, J. R., III (2008). Optimization of mass spectrometry compatible surfactants for shotgun proteomics. *Journal of Proteome Research, 6*, 2529–2538.

Covington, M. F., & Harmer, S. L. (2007). The circadian clock regulates auxin signaling and responses in *Arabidopsis*. *PLoS Biology, 5*, 1773–1784.

Deery, M. J., Maywood, E. S., Chesham, J. E., Sládek, M., Karp, N., Green, E. W., et al. (2009). Proteomic analysis reveals the role of synaptic vesicle cycling in sustaining the suprachiasmatic circadian clock. *Current Biology, 19*, 2031–2036.

Dunn, J. D., Reid, G. E., & Bruening, M. L. (2010). Techniques for phosphopeptide enrichment prior to analysis by mass spectrometry. *Mass Spectrometry Reviews, 29*, 29–54.

Edwards, K. D., Anderson, P. E., Hall, A., Salathia, N. S., Locke, J. C. W., Lynn, J. R., et al. (2006). FLOWERING LOCUS C mediates natural variation in the high-temperature response of the Arabidopsis circadian clock. *Plant Cell, 18*, 639–650.

Facette, M. R., Shen, Z., Björnsdóttir, F. R., Briggs, S. P., & Smith, L. G. (2013). Parallel proteomic and phosphoproteomic analyses of successive stages of maize leaf development. *Plant Cell, 25*, 2798–2812.

Farré, E. M., & Kay, S. A. (2007). PRR7 protein levels are regulated by light and the circadian clock in Arabidopsis. *The Plant Journal, 52*, 548–560.

Guerreiro, A. C. L., Benevento, M., Lehmann, R., van Breukelen, B., Post, H., Giansanti, P., et al. (2014). Daily rhythms in the cyanobacterium *Synechococcus elongatus* probed by high-resolution mass spectrometry-based proteomics reveals a small-defined set of cyclic proteins. *Molecular and Cellular Proteomics*, *13*, 2042–2055.

Hiltbrunner, A., Bauer, J., Vidi, P.-A., Infanger, S., Weibel, P., Hohwy, M., et al. (2001). Targeting of an abundant cytosolic form of the protein import receptor at Toc159 to the outer chloroplast membrane. *The Journal of Cell Biology*, *154*, 309–316.

Hindle, M. M., Le Bihan, T., Krahmer, J., Martin, S. F., Noordally, Z. B., Simpson, T. I., et al. (in review). qPMerge: Galaxy tools to reduce motif redundancy in proteomics.

Hwang, H., Cho, M.-H., Hahn, B.-S., Lim, H., Kwon, Y.-K., Hahn, T.-R., et al. (2011). Proteomic identification of rhythmic proteins in rice seedlings. *Biochimica et Biophysica Acta*, *1814*, 470–479.

Koroleva, O. A., & Bindschedler, L. V. (2011). In A. R. Ivanov, & A. V. Lazarev (Eds.), *Sample preparation in biological mass spectrometry* (pp. 381–409). Netherlands: Springer.

Krogh, A., Larsson, B., von Heijne, G., & Sonnhammer, E. L. (2001). Predicting transmembrane protein topology with a hidden Markov model: application to complete genomes. *Journal of Molecular Biology*, *305*, 567–580.

Kusakina, J., & Dodd, A. N. (2012). Phosphorylation in the plant circadian system. *Trends in Plant Science*, *17*, 575–583.

Mann, M., Ong, S., Grønborg, M., Steen, H., Jensen, O. N., & Pandey, A. (2002). Analysis of protein phosphorylation using mass spectrometry: Deciphering the phosphoproteome. *Trends in Biotechnology*, *20*, 261–268.

Mauvoisin, D., Wang, J., Jouffe, C., Martin, E., Atger, F., Waridel, P., et al. (2014). Circadian clock-dependent and -independent rhythmic proteomes implement distinct diurnal functions in mouse liver. *Proceedings of the National Academy of Sciences of the United States of America*, *111*, 167–172.

Reddy, A. B., Karp, N. A., Maywood, E. S., Sage, E. A., Deery, M., O'Neill, J. S., et al. (2006). Circadian orchestration of the hepatic proteome. *Current Biology*, *16*, 1107–1115.

Reiland, S., Messerli, G., Baerenfaller, K., Gerrits, B., Endler, A., Grossmann, J., et al. (2009). Large-scale Arabidopsis phosphoproteome profiling reveals novel chloroplast kinase substrates and phosphorylation networks. *Plant Physiology*, *150*, 889–903.

Righetti, P. G., Boschetti, E., Lomas, L., & Citterio, A. (2006). Protein Equalizer Technology: The quest for a "democratic proteome" *Proteomics*, *6*, 3980–3992.

Robles, M. S., & Mann, M. (2013). A. Kramer, & M. Merrow (Eds.), *Handbook experimental pharmacology: Vol. 217*. (pp. 389–407). Berlin Heidelberg: Springer.

Russell, W. K., Park, Z. Y., & Russell, D. H. (2001). Proteolysis in mixed organic-aqueous solvent systems: Applications for peptide mass mapping using mass spectrometry. *Analytical Chemistry*, *73*, 2682–2685.

Slade, W. O., Werth, E. G., Chao, A., & Hicks, L. M. (2014). Phosphoproteomics in photosynthetic organisms. *Electrophoresis*. http://dx.doi.org/10.1002/elps.201400154.

Sugano, S., Andronis, C., Ong, M. S., Green, R. M., & Tobin, E. M. (1999). The protein kinase CK2 is involved in regulation of circadian rhythms in Arabidopsis. *Proceedings of the National Academy of Sciences of the United States of America*, *96*, 12362–12366.

Tsuji, T., Hirota, T., Takemori, N., Komori, N., Yoshitane, H., Fukuda, M., et al. (2007). Circadian proteomics of the mouse retina. *Proteomics*, 7, 3500–3508.

Vizcaíno, J. A., Deutsch, E. W., Wang, R., Csordas, A., Reisinger, F., Ríos, D., et al. (2014). ProteomeXchange provides globally co-ordinated proteomics data submission and dissemination. *Nature Biotechnology*, *30*(3), 223–226.

Wagner, V., Fiedler, M., Markert, C., Hippler, M., & Mittag, M. (2004). Functional proteomics of circadian expressed proteins from *Chlamydomonas reinhardtii*. *FEBS Letters*, *559*, 129–135.

Wagner, V., & Mittag, M. (2009). Probing Circadian Rhythms in *Chlamydomonas reinhardtii* by Functional Proteomics. In T. Pfannschmidt (Ed.), *Plant Signal Transduction: Vol. 479.* (pp 173–188). Humana Press.

Wang, P., Xue, L., Batelli, G., Lee, S., Hou, Y.-J., Van Oosten, M. J., et al. (2013). Quantitative phosphoproteomics identifies SnRK2 protein kinase substrates and reveals the effectors of abscisic acid action. *Proceedings of the National Academy of Sciences of the United States of America, 110,* 11205–11210.

Wisniewski, J. R., Zougman, A., Nagaraj, N., & Mann, M. (2009). Universal sample preparation method for proteome analysis. *Nature Methods, 6,* 3–7.

Yeung, Y.-G., & Stanley, E. R. (2010). Rapid detergent removal from peptide samples with ethyl acetate for mass spectrometry analysis. *Current Protocols in Protein Science,* 59:16.12:16.12.1-16.12.5.

AUTHOR INDEX

Note: Page numbers followed by "*f*" indicate figures.

A

B

C

D

E

F

I

L

M

N

O

P

S

T

U

X

Y

Z

SUBJECT INDEX

Note: Page numbers followed by "*f*" indicate figures and "*t*" indicate tables.

D

E

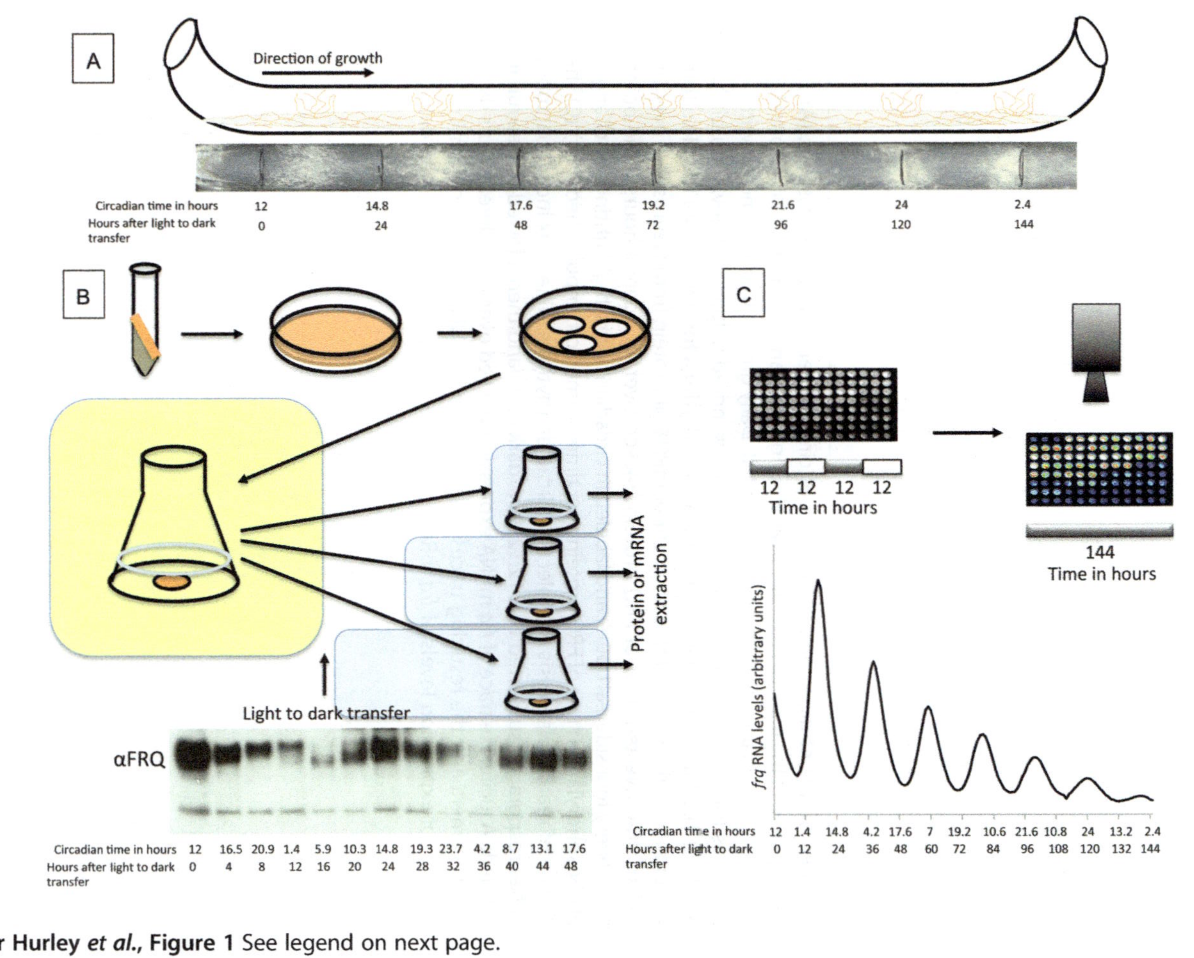

Jennifer Hurley *et al.*, Figure 1 See legend on next page.

Jennifer Hurley *et al.*, Figure 1 Methods of circadian analysis in Neurospora. (A) The basic outline of the use of a race tube in analyzing circadian rhythms in Neurospora along with the image of an actual race tube beneath it. Daily growth fronts are noted with vertical lines (sidereal time and in circadian time are noted below; J. Hurley, unpublished data). (B) The basics of analysis of the molecular rhythms for Neurospora liquid culture. The outline of the protocol to extract either mRNA or protein from Neurospora over circadian time. Western blot of FRQ protein tracked over 48 sidereal hours (time points taken every four sidereal hours) and labeled in circadian hours, highlighting the changes of phosphorylation state of FRQ protein over time (J. Emerson, unpublished data). (C) The outline of real-time analysis of molecular circadian rhythms. Ninety-six individual tubes of Neurospora are subjected to a 12:12 light/dark cycle and then allowed to free run in the dark. A luciferase trace of *frq* mRNA expression tracked over 144 sidereal hours using a CCD camera and the resulting traces are labeled in circadian hours, highlighting the changes of expression levels of *frq* mRNA over time (J. Emerson, unpublished data).

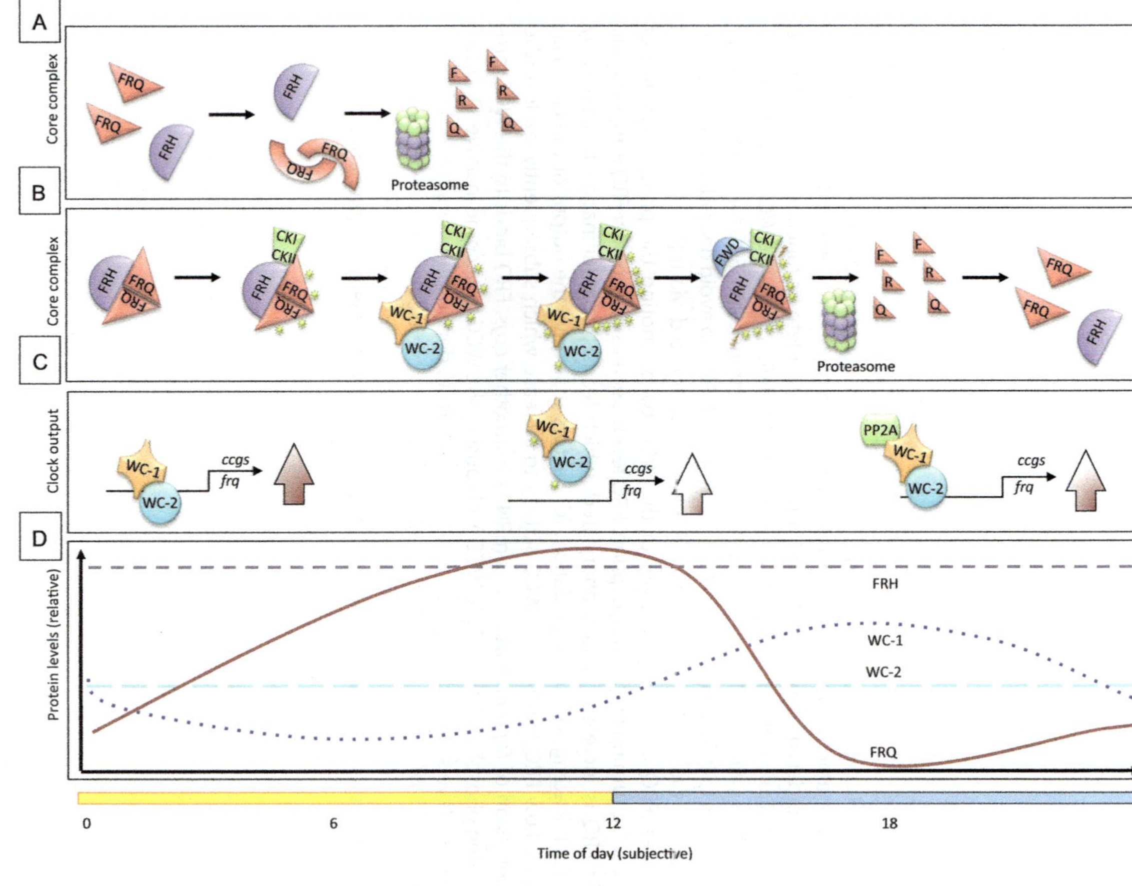

Jennifer Hurley *et al.*, Figure 2 See legend on next page.

Jennifer Hurley *et al.*, Figure 2 Neurospora circadian cycle at the molecular level. (A) If FRQ is not able to bind to its stabilizer, FRH, it is degraded by default due to the inherently disordered nature of FRQ and is unable to complete its function in the circadian clock. (B) During the late subjective night of the circadian cycle, the WCC induces expression of *frq* mRNA, leading to a rapid increase in FRQ translation. FRQ forms a homodimer and binds to its stabilizer FRH, allowing for the IDP FRQ to avoid degradation by default. As the circadian day progresses, FRQ is phosphorylated via interaction with several kinases. FRQ inhibits the activity of the WCC by promoting the phosphorylation of the WCC, turning off *frq* transcription. FRQ levels decrease as no new FRQ is made while old FRQ is increasingly phosphorylated, which leads to ubiquitination facilitated by FWD-1, leading to FRQ degradation. (C) Factors that drive the output of the circadian clock. Low FRQ levels cause WCC activity to increase which subsequently leads to the expression of *frq* mRNA as well as mRNAs from other *ccgs*. FRQ binds to the WCC, promoting phosphorylation of the WCC and causing the WCC to become inactive. Decreasing FRQ levels allow phosphatases to bind the WCC, dephosphorylating the WCC, and increasing WCC activation. (D) Protein levels of the core clock components. While FRH and WC-2 remain constant, FRQ and WC-1 oscillate in opposite phases to one another. Stars represent phosphorylation and lightning bolts represent ubiquitination.

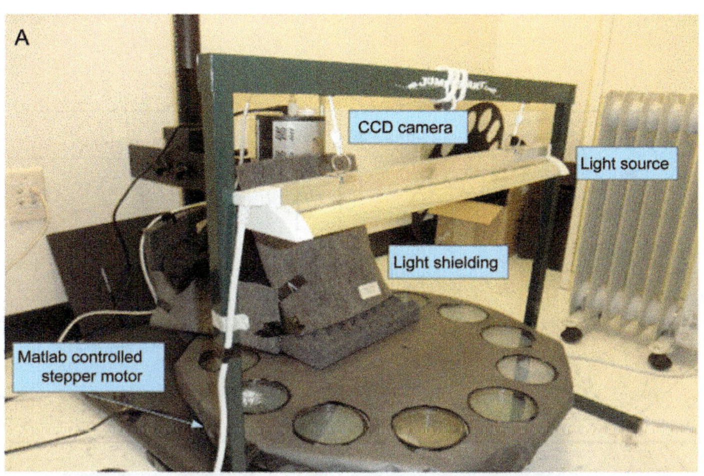

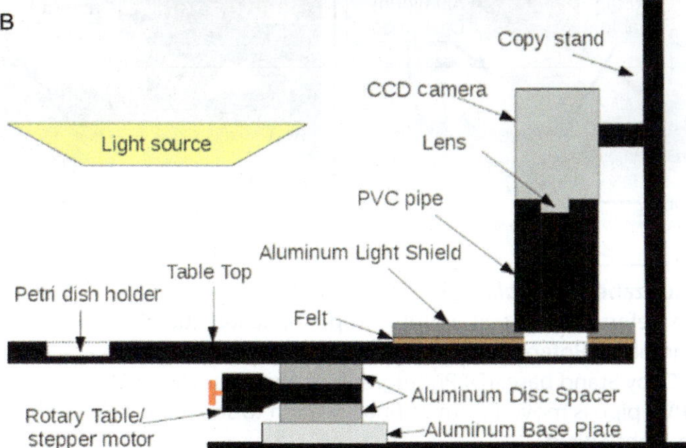

Ryan K. Shultzaberger *et al.*, Figure 1 Fully constructed turntable. (A) Photograph of assembled computer-controlled turntable. (B) Cross-sectional schematic of assembled computer-controlled turntable.

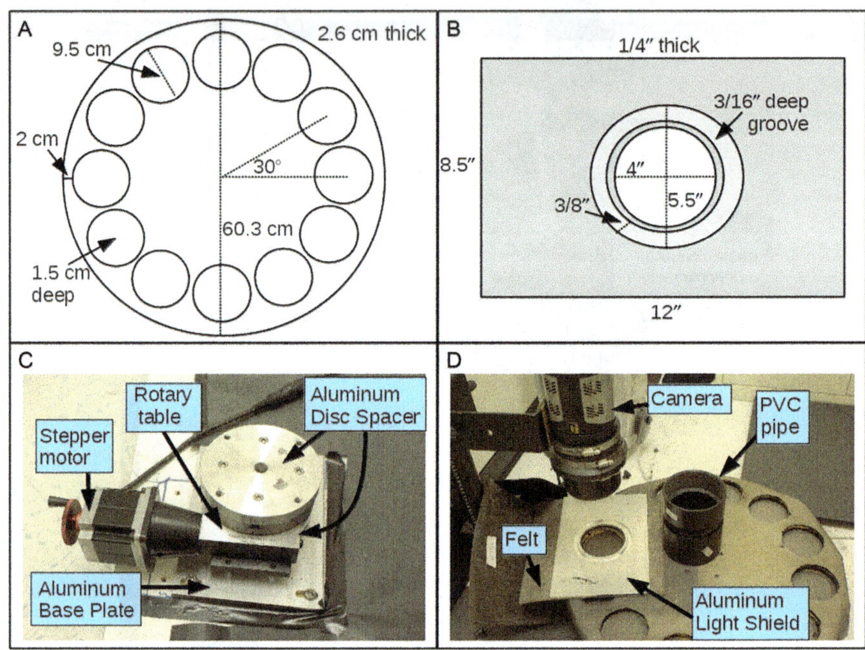

Ryan K. Shultzaberger *et al.*, Figure 2 Turntable components. (A) Schematic of turntable surface (referred to in text as Table Top). (B) Schematic of Aluminum Light Shield. (C) Photograph of Rotary Table attached to Aluminum Disc Spacers, Aluminum Base Plate, and Copy Stand base. (D) Photograph of Aluminum Light Shield on turntable surface. The PVC pipe is moved from its final location to show the hole in the center of the Aluminum Light Shield.

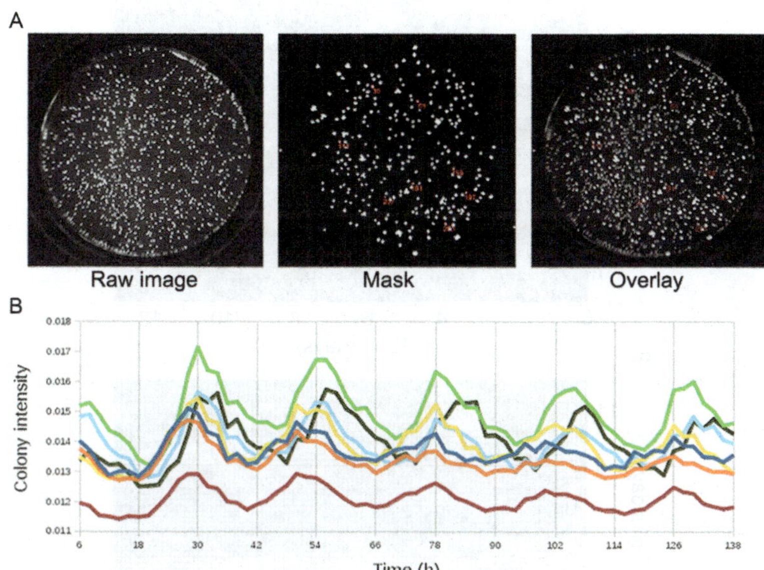

Ryan K. Shultzaberger *et al.*, Figure 3 Example data from time course experiment. (A) The image on the left is a raw image of a plate with *luxAB-luxCDE* expressing cyanobacteria. The image in the center is the mask generated by the `RCFinder.R` script. Each white spot represents an identified colony. Those spots that are numbered and circled in red were identified as rhythmic. The number is displaced down and to the right of the spot. The right image is an overlay of the first two images to show which colonies on the plate are rhythmic. (B) Bioluminescence data for five rhythmic colonies found in (A). Colony intensity is a measure of the average pixel intensity for a colony object and varies between 0 and 1 (arbitrary units).

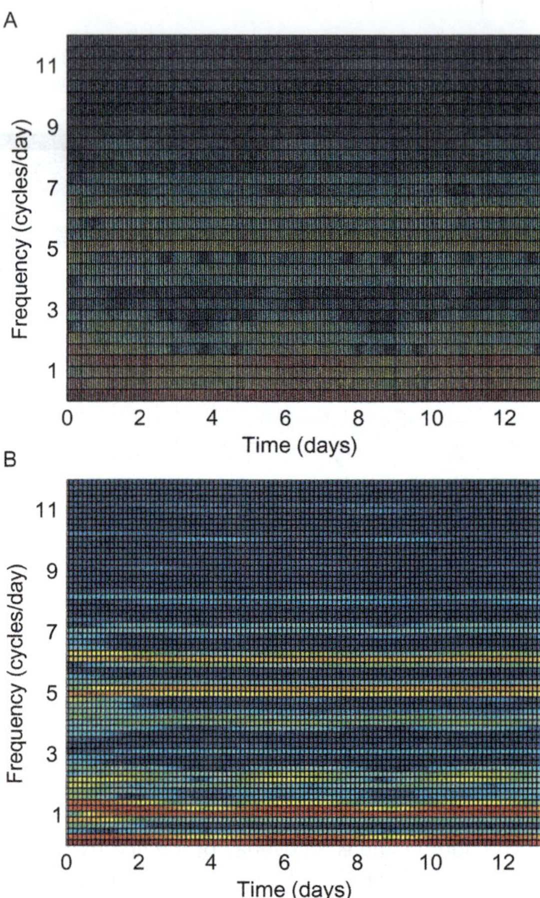

Tanya L. Leise, Figure 2 Spectrograms of simulated time series shown in Fig. 1A. Window lengths of (A) 64 h and (B) 128 h were employed to generate the spectrograms, both with a Hamming window, in MATLAB. The shorter window more clearly reveals the 5-day pattern in the circadian rhythm, while the longer window has tighter frequency resolution.

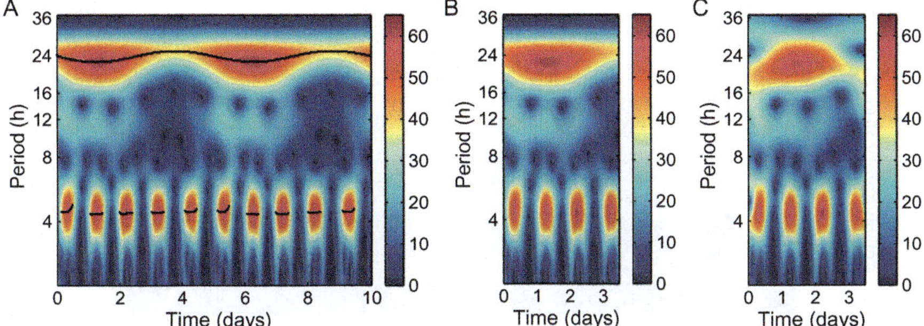

Tanya L. Leise, Figure 4 Analytic wavelet transform of the simulated time series. (A) Heat map (scalogram) of the AWT coefficient magnitudes for the simulated signal shown in Fig. 1A. The black curves mark the wavelet ridges, indicating what periods are present in the signal at each time point. The amplitude of each component is indicated by the color of the heat map along the corresponding ridge curve. (B) Accurate scalogram of first 3.5 days of simulated data. (C) Scalogram of the same segment of the time series, exhibiting significant edge effects due to poor choice of boundary conditions. Scalograms were generated in MATLAB using JLAB (Lilly, 2012).

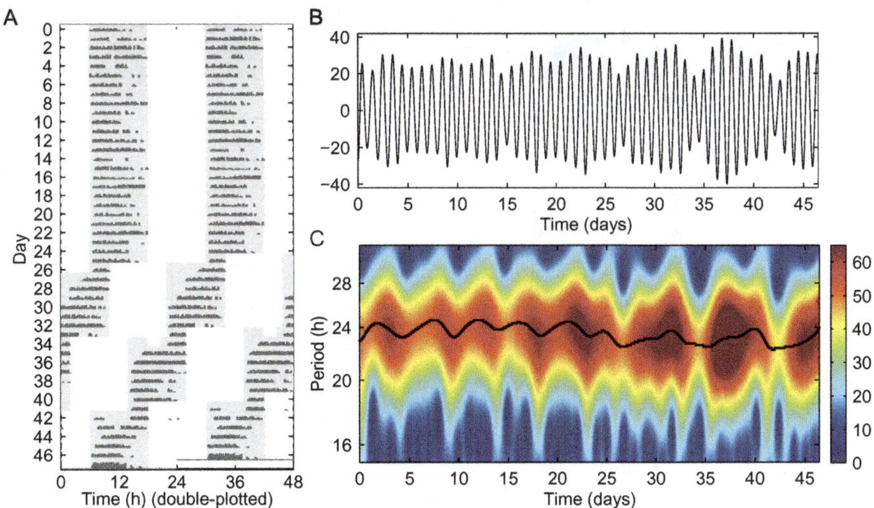

Tanya L. Leise, Figure 9 Wheel-running record of female mouse entrained to a 12:12 LD cycle, followed by repeated 8 h advances of the light–dark cycle, record from study described in Leise and Harrington (2011). (A) Actogram of wheel-running activity. (B) Circadian component of activity derived by the DWT. (C) AWT scalogram with ridge.

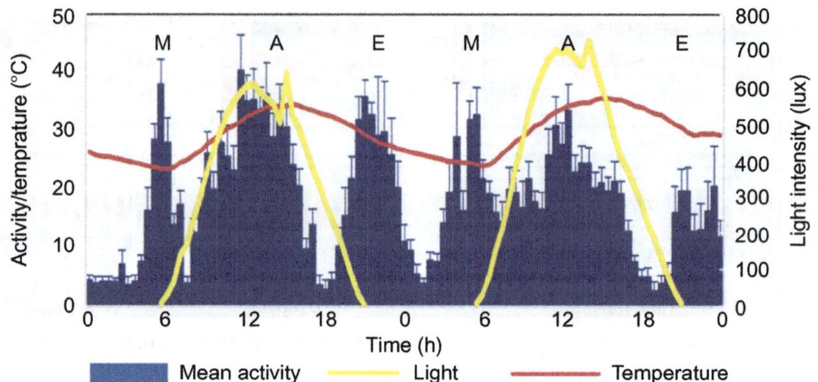

Edward W. Green *et al.*, Figure 1 TriKinetics-generated natural circadian locomotor profile collected from Treviso (Italy) between June 24 and 25, 2009. Mean and SEMs of locomotor activity from 15-min time bins are shown. The M, A, and E components are shown in addition to the light intensity (yellow, right-hand *Y*-axis) and temperature (red, left-hand *Y*-axis). The photoperiod is approximately LD16:8 and the temperature maximum is 35.6 °C, minimum 24.1 °C.

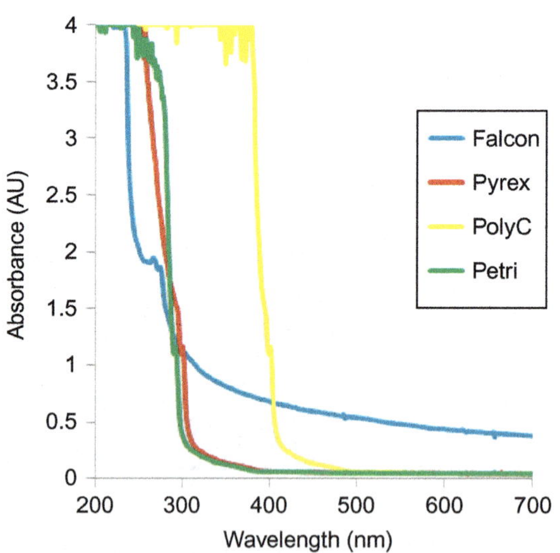

Edward W. Green *et al.*, Figure 2 Absorbance (in arbitrary units, AU) of different types of transparent materials. Note how PolyC (polycarbonate) absorbs UVA and blue wavelengths. *Data from M. Pegoraro.*

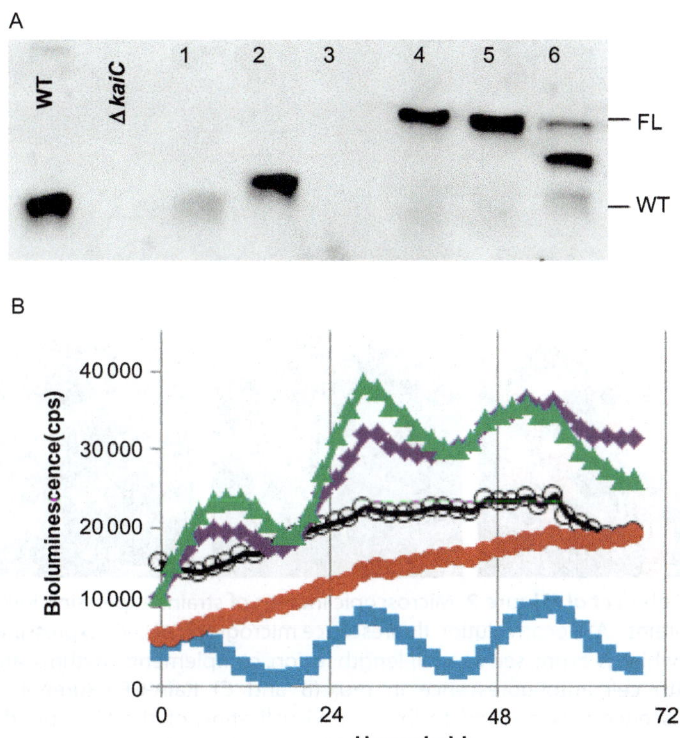

Susan E. Cohen *et al.*, Figure 1 Screening KaiC fusions for quality, quantity, and functionality. (A) Immunoblot of soluble extracts incubated with αKaiC antiserum. Six KaiC fusions (1–6) consisting of N-terminal fusions (1, 3, 4, and 5) and C-terminal fusions (2 and 6) with various linker lengths were tested. WT denotes the expected size for a wild-type untagged KaiC protein and FL denotes the predicted size for a full-length fusion protein. (B) Monitoring rhythms of gene expression from a P_{kaiBC}–*luc* reporter for strains expressing fusions 1–6 as the only copy of *kaiC*. Representative traces for WT (blue squares), Δ*kaiC* (red circles), fusions 1, 3, and 6 (which were indistinguishable, black open circles), fusion 2 (green triangles), and fusions 4 and 5 (which were indistinguishable, purple diamonds). Fusions 4 and 5 produce full-length fusion proteins and complement rhythms of gene expression. In contrast, Fusion 6 produces a full-length fusion protein in addition to truncated products near in size to untagged KaiC, none of which support rhythmicity. Fusion 2 produces near WT rhythms; however, it is not expressed as a full-length fusion protein, and a truncated product near in size to untagged KaiC is observed.

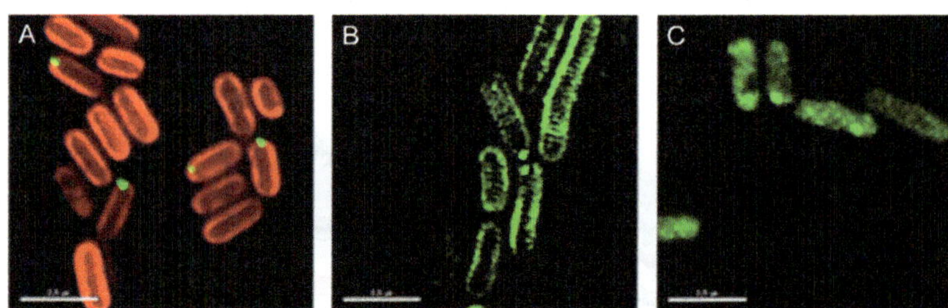

Susan E. Cohen *et al.*, Figure 2 Microscopic images of strains expressing KaiC and KaiA fusion proteins. (A) Deconvolution fluorescence micrograph of cells expressing YFP-KaiC (fusion 4) which is expressed as a full-length fusion, complements rhythms, and appears green, with cell autofluorescence in red; (B and C) KaiA-GFP (green), for which autofluorescence was omitted to improve visualization of the low-abundance KaiA fusion. (B) 3D-SIM micrograph using a FITC filter set where bleed-through from the thylakoid fluorescence is obvious and (C) narrow band-pass GFP filter set on a deconvolution fluorescence microscope to reduce bleed-through from the photosynthetic pigments. Scale bar = 2.5 μm.

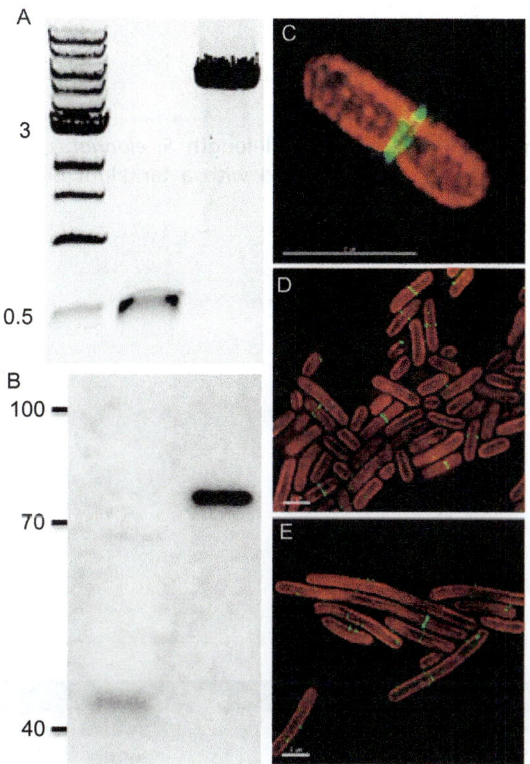

Susan E. Cohen *et al.*, Figure 3 Characterization of a YFP–FtsZ fusion protein. (A) PCR analysis of *ftsZ* locus demonstrates that YFP–FtsZ fusion expressed from the P_{trc} promoter can replace the chromosomal copy of *ftsZ*. Lane 1, 1 kb DNA ladder (NEB); Lane 2, amplification of a 444 bp region of the *ftsZ* locus (from 104 bp upstream to 340 bp into *ftsZ*); Lane 3, amplification of the same chromosomal locus where a construct expressing the spectinomycin resistance cassette-Lac*I*–P_{trc}–YFP–FtsZ has replaced the native *ftsZ*. Homogenous segregation demonstrates that the *yfp–ftsZ* allele can completely replace endogenous *ftsZ*. (B) Immunoblot of soluble extracts incubated with αFtsZ. Lane 1, 13 μg extract from WT; Lane 2, 0.5 μg extract from YFP–FtsZ-expressing cells. YFP–FtsZ is expressed as a full-length fusion protein; however, it is ~55-fold overexpressed compared to the endogenous FtsZ. (C–E) 3D-SIM micrographs of strains expressing YFP–FtsZ as the only source of FtsZ. (C) Representative individual cell in which a Z-ring has formed near mid-cell. Field of cells expressing YFP–FtsZ in a (D) otherwise WT background, normal cell shape, and Z-ring formation is uniformly observed or (E) Δ*cikA* mutant background, where cells are elongated and FtsZ appears mislocalized. Scale bars = 2 μm.

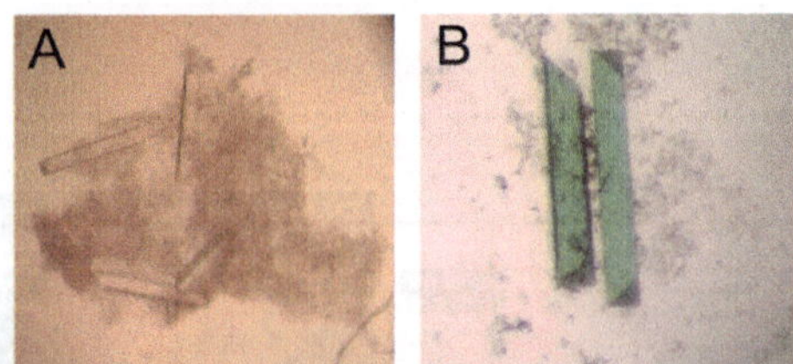

Martin Egli, Figure 7 Crystals of (A) full-length *S. elongatus* KaiC (N1-519) with a C-terminal $(His)_6$ tag, and (B) derivatized with a tantalum bromide cluster $Ta_6Br_{12}^{2+}$ compound.

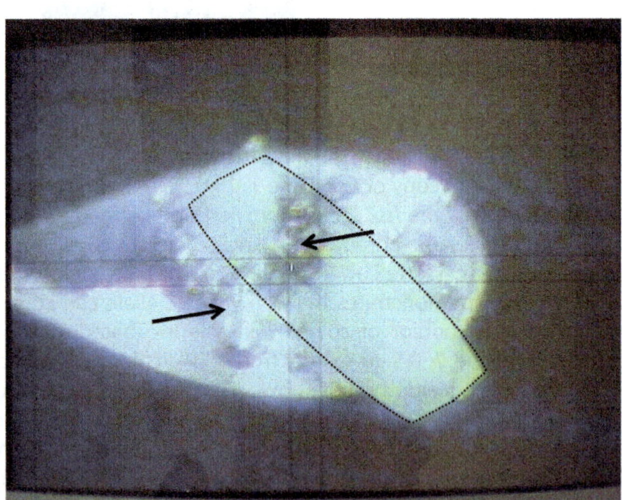

Martin Egli, Figure 8 Annealing of a KaiC crystal (outlined with a dashed line) in the cryo-loop. The liquid N_2 stream is temporarily blocked, and the image taken from a monitor at the beam line depicts bubbling mother liquor (arrows) around the crystal as it is warming up.

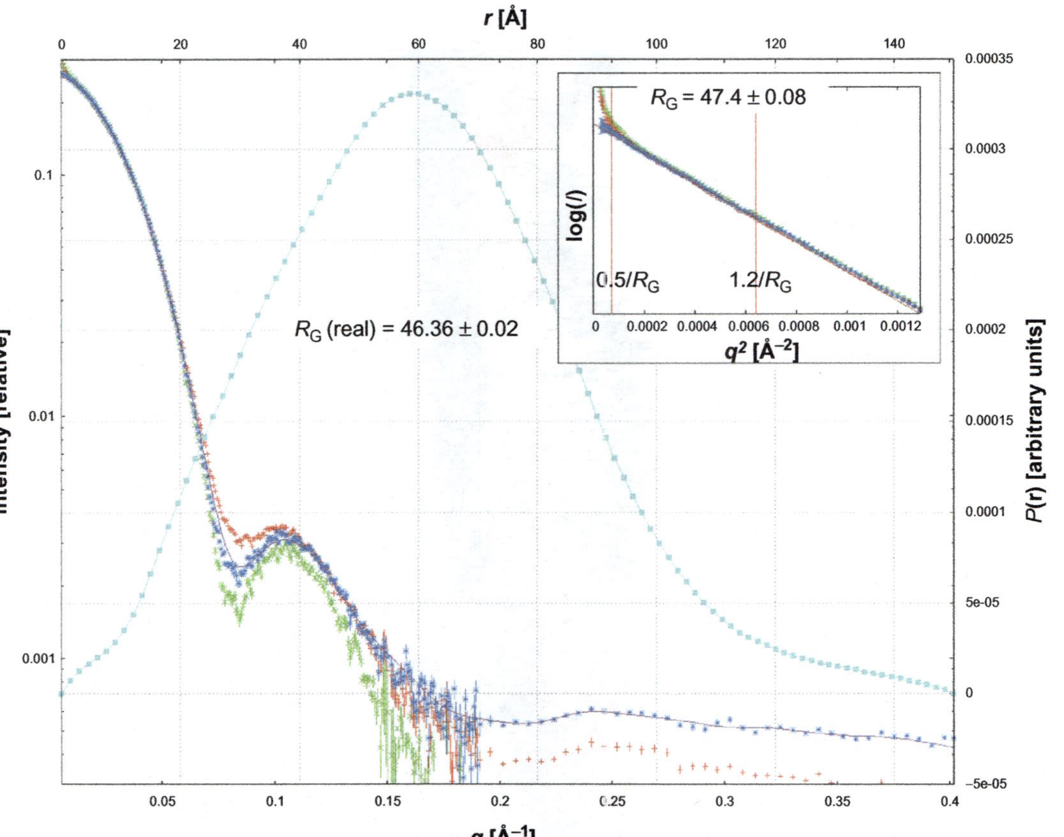

Martin Egli, Figure 10 SAXS scattering curves *I*(*q*) (red, high concentration; blue, medium concentration, and green, low concentration), pairwise function *P*(*r*) (cyan), and Guinier plots (inset) for *S. elongatus* KaiC. *From Pattanayek et al. (2011).*

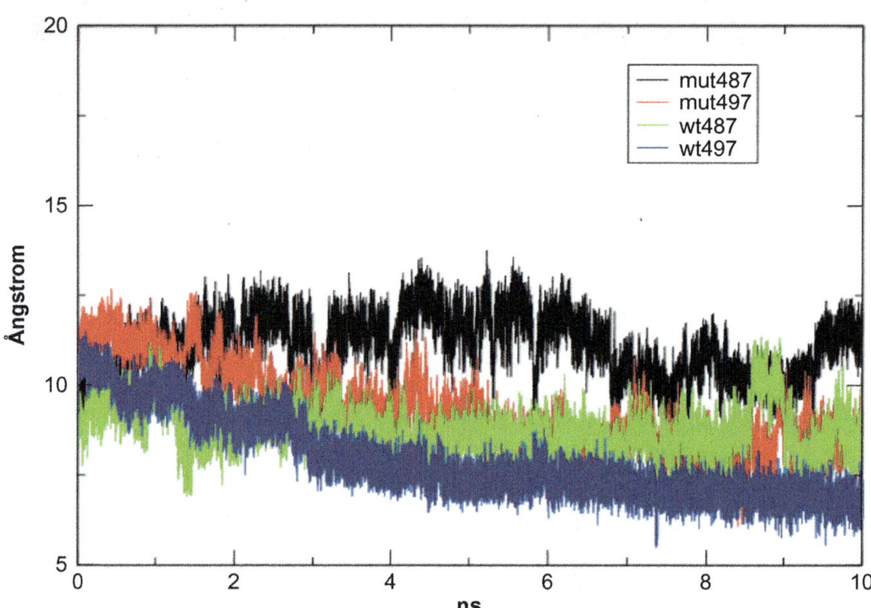

Martin Egli, Figure 13 Variations in the distance between the Cα positions of Ser-417 and Gly-421 monitored over 10 ns of MD simulations for four KaiC hexamers: KaiC 1–497, KaiC 1–487 (lacking the so-called A-loop) and the 1–497 and 1–487 KaiCs with an additional A422V mutation. The 422-loop region encompassing residues 417–421 was hypothesized to become more flexible as a result of removal of the A-loop. Indeed, the black and green traces for the 1–487 systems without the A-loop indicate increased distance fluctuations.

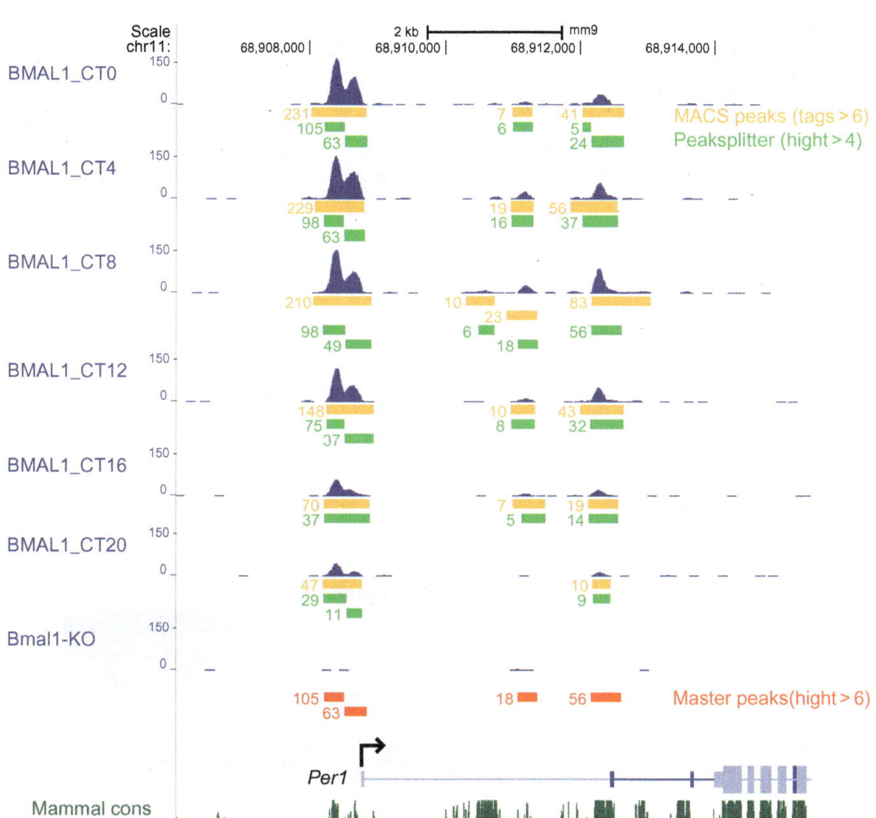

Joseph S. Takahashi *et al.*, Figure 1 UCSC genome browser view of MACS peak calls for six-timed BMAL1 ChIP-seq occupancy at the *Per1* gene. Orange bars indicate the MACS peak calls and green bars indicate the peak after using Peaksplitter. The numbers to the left of each bar refer to the peak height. Red bars at the bottom show the final consolidated peaks used to construct the master peak list. The Peaksplitter peak with the largest peak height in the region of overlap of peaks is chosen to represent this peak in the master peak list. *Data adapted from Koike et al. (2012).*

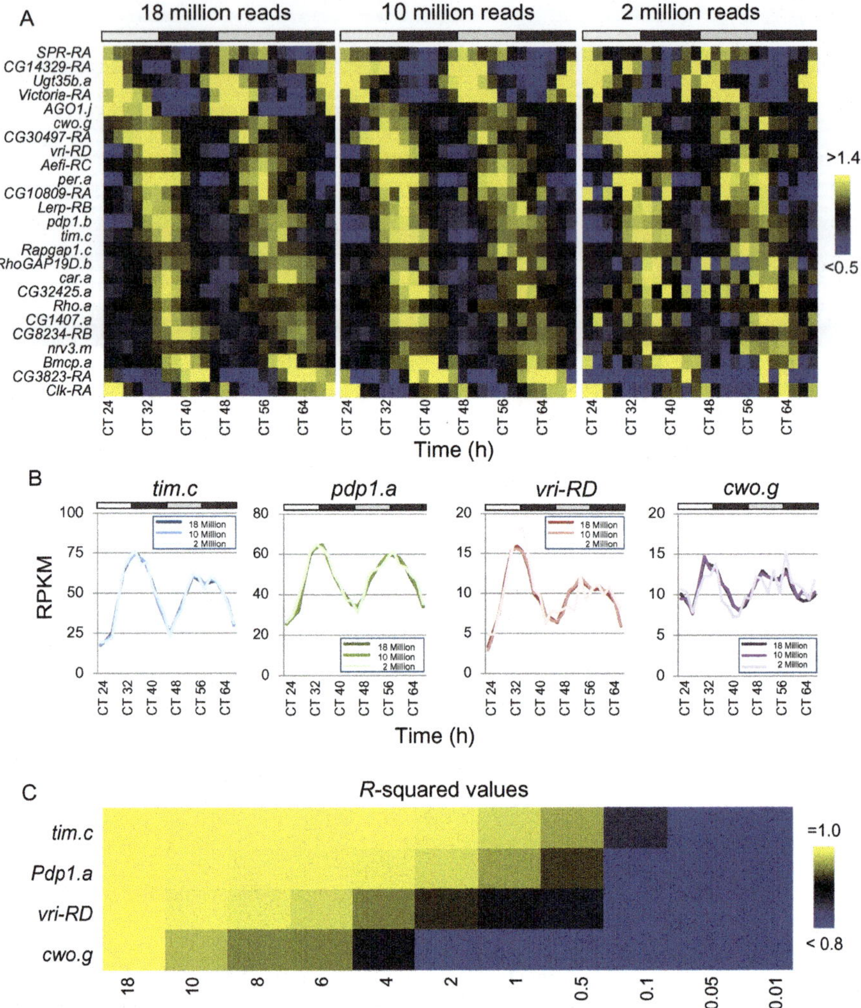

Jiajia Li *et al.*, Figure 2 Many cycling transcripts are detectable with far fewer reads per sample than seen in legacy data sets. (A) The expression pattern of 25 cycling transcripts in the fruit fly brain at different read depths are shown as heat maps. The top gray and black bars represent subjective day and night, respectively. Every transcript's expression profile has been median-normalized. Expression levels greater than 1.4-fold are shown as bright yellow; expression levels less than 0.5-fold are shown as bright blue (legend on the far right). (B) The rhythmic expression pattern of four representative transcripts, *timless.c*, *pdp1.a*, *vrille-RD*, and *cwo.g* are shown at different read depths. The top gray and black bars represent subjective day and night, respectively. The vertical axis represents RPKM values; the horizontal axis indicates time points which are marked on the bottom. (C) Pearson correlation coefficients between maximal read depth and subsampled data of expression values of the four transcripts in (B). The horizontal axis indicates different read depth. Correlations equal to 1.0 are shown as bright yellow; correlations less than 0.8 are shown as bright blue (legend on the far right).

Printed by Printforce, the Netherlands